100 years
of Planck's
Quantum

100 years of Planck's Quantum

Ian Duck
Rice University, Houston, USA

E C G Sudarshan
University of Texas at Austin, USA

World Scientific
Singapore • New Jersey • London • Hong Kong

Published by

World Scientific Publishing Co. Pte. Ltd.

P O Box 128, Farrer Road, Singapore 912805

USA office: Suite 1B, 1060 Main Street, River Edge, NJ 07661

UK office: 57 Shelton Street, Covent Garden, London WC2H 9HE

British Library Cataloguing-in-Publication Data
A catalogue record for this book is available from the British Library.

100 YEARS OF PLANCK'S QUANTUM

ISBN 981-02-4309-X ✓

Printed in Singapore by Uto-Print

DEDICATION

We dedicate this book to the memory of MAX PLANCK. The profound purity of his creation, the clarity, depth and deceptive simplicity of his thinking, and the lifelong nobility of his character and his uncompromised principles are his everlasting monument.

The AUTHORS:

Ian Morley Duck, b. 1933 Kamloops, Canada; BSc 55 Queen's, Canada; Res Asst 55-56 UBC; PhD 61 Caltech; Res Assoc. 61-63 USC; Research Associate to Professor of Physics 1963- Rice University. Theoretical research in nuclear and particle physics from ev to Gev; radiative capture of light nuclei, muon capture, bremsstrahlung in p-p scattering, Fadeev equations, $N - \Delta$ process, quark model of nucleon structure, models of color confinement, quark-gluon plasma excitation in antiproton annihilation in heavy nuclei, fine structure of the baryon octet-decuplet spectrum. Co-author with ECGS of (1) *Pauli and the Spin-Statistics Theorem* (World Scientific, 1997) and (2) *Toward an Understanding of the Spin-Statistics Theorem*, Am. J. Phys. **66**(4), 284 (1998).

Ennackel Chandy George Sudarshan, b. 1931 Kottayam, India; BSc 51, MA 52 Madras; Res Asst 52-55 Tata Inst; PhD 58 Rochester; Research Fellow 57-59 Harvard; to Associate Professor 59-64 Rochester; Professor 64-69 Syracuse; Professor of Physics, Center of Theoretical Physics, Texas 1969-; concurrent positions at Bern 63-64, Brandeis 64, Madras 70-71, Bangalore 72-. Honors: DSc from Wisconsin, Delhi, Chalmers, Madras, Burdwan, Cochin; Medals: Order of the Lotus, Bose Medal, Sarvadhikari medal and many others. Research: co-inventor with R.E. Marshak of the universal V-A weak interaction (1957); author of ten books and over 350 journal articles in the areas of elementary particle theory, quantum theory, group theory, quantum optics, and the foundations, philosophy and history of contemporary physics.

TABLE of CONTENTS

Introduction; Bohr-Heisenberg Personality Conflicts; Concluding Remarks.

Paper XI·1: Excerpt from W. Heisenberg, Zeits. f. Phys. 43, 172 (1927).

Paper XI·2: Excerpt from W. Heisenberg, 'The Development of the Interpretation of the Quantum Theory' in *Niels Bohr and the Development of Physics* (Pergamon, London, 1955) W. Pauli, Ed., Pp. 12–29.

Introduction and Discussion.

Paper XII·1: Excerpt from N. Bohr, Nature 121, 580 (1928).

Introduction and Discussion.

Paper XIII·1: Excerpt from A. Einstein, B. Podolsky, and N. Rosen, Phys. Rev. 47, 777 (1935).

Paper XIII·2: Excerpt from N. Bohr, Phys. Rev. 48, 696 (1935).

Paper XIII·3: Excerpt from N. Bohr in 'Atomic Physics and Human Knowledge' (Wiley, NY, 1958).

PART FOUR

Introduction; Bell's Inequality; Aspect's Experiment; 'Alice to Bob' Teleportation; Concluding Remarks; Biographical Notes.

Introduction; Feynman Path Integrals; Influence Functional and Decoherence; Concluding Remarks.

Paper XV·1: Excerpt from R.P. Feynman, Rev. Mod. Phys. 20, 267 (1948).

Paper XV·2: Excerpt from R.P. Feynman and F.L. Vernon, Annals of Physics 24, 118 (1963).

Paper XV·3: Excerpt from P.A.M. Dirac, Phys. Zeits. Sowjetunion 3, 64 (1933).

Decoherent Histories – A) Definitions, B) FPI Decoherence Functional, C) Recovering Conventional Quantum Mechanics;

WERNER HEISENBERG

RICHARD FEYNMAN

Introduction

When in the year 5000 people look back three thousand years to our era, we all hope that they will find some epochal events as myth-making for them as the Trojan War for us. Many events of such lasting significance are to be found among the achievements of our twentieth century scientific revolution in physics. The very first of these revolutionary events – certainly in time and arguably in significance – is Planck's invention of the energy quantum in 1900.

It is customary to undervalue Planck's achievement as perhaps a lucky guess, a shot-in-the-dark by someone incapable of understanding its real significance, simply an ill-understood parametrization of data, a singular achievement but one that would have been found sooner or not significantly later by someone else – probably Einstein. We take the opposite view. Planck's discovery was indeed manifested in the first instant as an empirical fit to data, but he was so profoundly prepared and deeply involved that he immediately understood an essential motivation for the derivation of the result. That profound motivation was the *necessity* of *discrete countability* of the states of the black-body radiation in order that it conform to the Second Law of Thermodynamics and have a *definable* entropy.

It is our view that Planck opened the door to an utterly new, totally unanticipated, wonderfully strange and mysterious but absolutely necessary ultimate reality of the world in the mechanics of the quantum – that is to say in quantum mechanics. Furthermore, that reality still remains all those things today even after one hundred years of the most intense scrutiny by some of the most knowledgeable, brilliant, imaginative and creative people the world has ever produced, functioning at an apex of intellectual intensity unsurpassed in the history of humankind. Where will it lead? And how will it end? We cannot know where it will lead and we cannot believe that it will end. For profound reasons which we leave for you to expand upon, we in fact *must* believe that it will *not* end.

We are devoted to the belief that the *original* accounts of great ideas are themselves not only great sources of further ideas but also unrivaled sources of inspiration and understanding [1]. Scientists at every stage of their careers are the richer for knowing intimately and at first hand the classic works of the immortals

1

of their subject. For this reason we have here taken a completely historical approach to the still lively and still controversial subject of the meaning of quantum mechanics. In Parts I-III we explore in detail the headwaters of this great river in the first mention by Planck of the quantum h, then we scrutinize in detail the increasingly rapid succession of discoveries by Einstein, Compton, Bohr, de Broglie, and $\cdots\cdots\cdots$

To speak of classics and original sources is not to suggest that we all do the moral equivalent of reading Homer's Iliad in the ancient Greek. Our focus is ever on the physics, not the poetry. For most people the language barrier is a complete deterrent not only because few people (neither of us) are facile in German, French, Italian etc, but also because the language (even when it is English) is itself frequently difficult. A notorious case in point is the writing of Niels Bohr, even in English, even in later life, which is maddeningly obtuse. The reasons for this almost universal communication problem amongst the founders of quantum mechanics are surely many. The work was new and abstract and as difficult to explain as to understand. In addition, the old greats were extremely careful to anticipate arguments where doubt existed, with the result that their writing was full of qualifications and caveats when one aches for a straightforward definitive statement which – of course – was not possible. And in their explorations, they frequently wandered down secondary paths where to follow would surely exhaust us and defeat our primary purpose. The compromise we have reached is to regard every source as a document to be translated into contemporary English, to be abridged and compressed to suit our purposes, and to be highlighted with explanatory notes where we thought they would be useful to point out the moment of creation, or at least the first mention of a new idea. Let us reassure everyone who might be offended by these elisions in the original texts, that they are made with a profound respect approaching reverence for the original authors. Our sole purpose is to make their classic work more accessible and widely used.

On Understanding Quantum Mechanics

The fascination of the quantum realm is without limit. Multitudes of questions constantly and repeatedly arise which test our understanding and even redefine the concept of understanding the new in terms of the familiar. The discovery of the quantum realm was forced on Planck by the failure of classical thermodynamics to produce so much as a *finite* result for the universal spectrum of radiation leaking from a hot oven. The myth is that Planck got lucky and fudged a connection between Wien's infra-red and Maxwell-Boltzmann's ultra-violet spectra which fit the data. That much is true, but Planck's immediate understanding was profound, and based on the fundamental definition of the entropy $S = k \ln w$ which requires definition of the probability w for an equilibrium configuration of the radiation with its sources. Planck decided that probabilities are only definable for countable alternatives. The probability w of a configuration is the total number of 'complexions' included in it. To make these complexions countable, Planck was forced to introduce his energy quantum $E = hf$ (see ChI).

Einstein further explained that Planck's quantum implied the physical result that light occur in irreducible energy quanta (ChII). Then in giant steps Bohr created his picture of the quantum atom as a micro-solar system, and introduced the concept of 'quantum states' with quantized energy and quantized angular momentum, and of 'transitions' between two such energy states with the emission of a light quantum (ChIII). The understanding of these transitions – indeed their understandability – remains a question. Technically, of course, we can calculate the result to a fair-thee-well, and we are instructed on what to believe, and even what questions we may ask [3]. Whether or not this is the end of the story is still being debated [4].

The explosion of progress triggered by Heisenberg's invention of quantum algebra (ChVI) – immediately recognized by Born and Jordan (ChVII) as matrix multiplication and, almost simultaneously and almost independently, further developed by Dirac (ChVIII) and by Schrödinger (ChIX) – changed forever the way to think about the fundamental structure of the microscopic world (ChX-XII).

People who look in casually on the sport of quantum mechanics are excited primarily by the buzz words, catch phrases, and paradoxes; the ones which plague us most include 'Is the moon there when no one looks?', 'Schrödinger's cat is both alive and dead – until you look!', and many more. These are problems of interpretation which occur at the interface between the microscopic quantum realm and the macroscopic classical world of our gross perceptions. They obviously *are* bothersome questions because they and ones like them bothered Einstein (ChXIII), and because they received rather peremptory responses from authority figures such as Bohr and Heisenberg. Although such incomplete understanding never slowed or even really affected the break-neck pace of research into the then accessible applications of quantum mechanics, such questions have always been somewhat embarrassing even to pragmatic physicists, and – of course – have been the primary focus of the philosophical among us. Only recently have questions of the dynamics of the interface between the quantum and the classical worlds become a very interesting and even important subject of *experimental* as well as theoretical research (ChXIV).

The Quantum View

Quantum mechanics as introduced in the very first instance by Born and Jordan (ChVIII) requires a very formal – to some a dauntingly abstract but rigid – view of the world. In the quantum extension of classical Hamiltonian mechanics, the classical number-valued canonical variables q, p are replaced by matrix operators in an abstract Hilbert space of 'state'-vectors characteristic of the particular quantum system being considered. The space is spanned by a set – sometimes non-denumerably infinite – of mutually perpendicular unit vectors chosen as reference axes. Any vector in the space is $1 \leftrightarrow 1$ with some quantum state of the system. For the simplest case of a spin-1/2 particle, there are two axes which can be chosen as spin-up and spin-down along some arbitrarily chosen z-direction.

A still simple example is the 1-dimensional simple-harmonic oscillator. Now a choice of axes could be the energy eigenstates $|n\rangle$, or the position eigenstates $|x\rangle$ or momentum eigenstates $|p\rangle$, each with its own orthogonality and completeness relations, and each expressible in terms of others by the expansion coefficients, e.g., the familiar oscillator coordinate-space wavefunction $\psi_n(x) = \langle x|n\rangle$. Now we can imagine a pure state of the oscillator. It might be a pure energy eigenstate $|n\rangle$, in which case a measurement of the energy would always produce the result E_n – the energy eigenstate is stationary, and only changes its phase. A measurement of the position would produce the result x with the probability $|\psi_n(x)|^2$. Suppose we prepare the oscillator (somehow) in a position eigenstate $|x_0\rangle$. A measurement of the energy will produce the result E_m with probability $|\psi_m(x_0)|^2$. A later measurement of the position however produces a hodge-podge.

It is interesting to imagine the introduction of a small non-linearity into the restoring force of the oscillator, described by an interaction Hamiltonian H_{int}. Suppose the oscillator initially to be in an energy eigenstate of the simple-harmonic oscillator (hereafter called an SHO-state). We emphasize that it is NO LONGER in an ENERGY eigenstate as defined by the FULL Hamiltonian. A later measurement (how?) of the SHO-state of the non-linear oscillator will no longer produce with certainty the original initial SHO-state. The state-vector remains normalized, but it no longer points in the same direction with respect to the SHO-axes. The state vector 'rotates' in the infinite-dimensional Hilbert space of SHO-states – it 'evolves' by a unitary transformation generated by the interaction Hamiltonian H_{int}. The probability of finding the non-linear oscillator in a different SHO-state will now be $\sim |\langle n_f|H_{int}|n_i\rangle|^2$.

It is noteworthy that the non-linear oscillator continues to evolve as a pure quantum state; the state vector is always of unit length but not simply along one of the initial SHO-axes.

The problem of what happens to the oscillator during (because of) any particular measurement has been separated from the purely quantum mechanical evolution of the oscillator in a way that has caused much consternation. This dichotomy is finally being addressed in recently devised 'quantum non-demolition' experiments (ChXIV).

An older problem is how the above description applies to a classical oscillator. Schrödinger (ChIX) thought to describe the classical oscillator as a 'coherent state' – still a pure state, but a carefully constructed special superposition of many quantum states with precarious phase and amplitude relations. This was immediately recognized by Heisenberg as insufficient (ChXI): the classical oscillator must be described by very rapid dephasing and decoherence caused typically by dissipative interactions with a 'classical' environment. (ChXV, XVI).

The concept of the quantum state reveals other aspects of nature besides quantization of energy and angular momentum. Dirac pointed out that the basic symmetries of nature are present already in the classical Action and the classical equations of motion. But it is only in the quantum *states* that the symmetries of nature are directly *realized*. Even the elementary symmetries of translation, rotation, and reflection invariance – present in all fundamental classical equations – are rarely if ever observed in classical states.

Apologia

Quantum mechanics – like physics in general – is seemingly infinite in all directions. In our own experience we have progressed (?) from reading 'everything' to reading 'nothing' while continuing to read the same amount. In such a world it is impossible to do justice to a whole subject. We have followed one path – primarily non-relativistic and non-field theoretic, hopefully quantum mechanics *per se* rather than specific applications – through a vast jungle, guided more by our limitations than by our ambitions. In doing so, our primary concern has been to traverse the jungle of the first **100 Years of Planck's Quantum.** Much that is both beautiful and important has been omitted and for that we apologize.

In our final chapter, we have tried to anticipate the future course of our subject over the next hundred years. We find ourselves quite incapable of sensible predictions even over the next ten years, and have included our foolishness merely for the sake of amusement.

Footnotes and References:

1) Our inspiration for this book and our earlier one [2] is Julian Schwinger's invaluable collection *Quantum Electrodynamics* (Dover, New York, 1958).

2) I. Duck and E.C.G. Sudarshan, *Pauli and the Spin-Statistics Theorem* (World Scientific, Singapore, 1997).

3) Most recently by R. Omnès, *Understanding Quantum Mechanics* (Princeton, Princeton NJ, 1999); and R.B. Griffiths and R. Omnès, Physics Today **52**-8, 26 (1999).

4) For recent contributions to this debate see the exchange of letters in Physics Today **52**-2, 11 (1999) in response to the views of S. Goldstein, Physics Today **51**-3, 42 (1998); **51**-4, 38 (1998).

PART ONE. Chapter I

PLANCK:

die Wahrscheinlichkeit zu finden, dass die N Resonatoren ingesamt Schwingungsenergie U_N besitzen, U_N nicht als eine unbeschränkt teilbare, sondern als eine ganzen Zahl von endlichen gleichen Teilen aufzufassen ···.

(Planck: To find the probability that N oscillators have a total energy U_N, it is necessary to suppose that U_N is not continuously divisible, but is an integral multiple of finite equal parts. ··· We call such a part an energy-quantum ϵ. (Ann. d. Phys. **4**, 553 (1901); see our **Paper I·4**.))

§I-1. Introduction.

Planck's discovery of the Quantum of Action 'h' in November 1900 [1] is an archetype of scientific revolutions, but Planck was more a Marx than a Lenin, and certainly no Stalin. He himself never pursued the meaning or significance of his parametrization. When he introduced the new fundamental constant 'h' – known forevermore as 'Planck's constant' – to fit the wavelength and temperature dependence of the universal energy spectrum of the radiant heat emitted by hot 'black' bodies, his success was greeted by a resounding silence. Even the few participants who had been involved in the intense controversy over this central subject for the preceding five years were – after a brief flurry of bickering from Wien [2] – silenced by the precision of Planck's result. The subject vanished almost completely from the pages of the *Annalen der Physik*. For four years no further mention of the word 'quantum' can be found in the pages of this premier journal where Planck had first introduced it, and no speculations on its implications, until Einstein's revelations of March 1905 [3].

In fact, for most it was something even less than business as usual after January

7

1901. Wien – after his initial objections – seems to have ignored the subject and shifted his interests to other areas. Planck himself spent a year tidying up his preliminary work on the application of the Second Law to radiation using his new result for the energy distribution [4], but he meticulously avoided use of the word 'quantum' in doing so. With amazing prescience falling just short of an Olympian vision, in a treatise on the nature of white light, Planck [5] did take the Fourier analysis of incoherent classical electromagnetic fields to the very edge of a semiclassical description of the photon field. He introduced the expression 'partial wave' in a sense different from what we use today, effectively to describe the field of an individual uncorrelated photon; he used the concept of 'ray bundles', again effectively to describe coherent photons; he described light rays from the point of view of emission and of detection; he emphasized the analogy "$\cdots$ the partial waves can be compared with the motion of molecules $\cdots$ the principal difference being the molecules of the gas are individual, whose mass and other properties have no variation, whereas the partial waves have not only their amplitude and phase, but also their period"; and most of all "the energies (square of the amplitudes) $\cdots$ of the individual partial waves into which the light wave can be decomposed, can be interpreted in the sense of probability theory"; and finally "$\cdots$ the nature of the light of the spectral lines is the most difficult and complicated problem which now faces Optics and Electrodynamics." And never a mention of the word 'quantum'.

In the midst of this uncertainty, confusion, competition, and conflict, Planck had the purest insight to introduce the quantum of energy, but not the insight to develop the idea any further. We now explore in depth Planck's great achievement recorded in his three classic papers in *Annalen der Physik* of 1900 and 1901.

§I-2. Blackbody Radiation.

The Planck Spectrum of Blackbody Radiation is the primary cornerstone of modern physics and its importance cannot be overstated. Its manifestations range from the phonons of solid state physics, to the photons of the cosmic background radiation, from the Bose-Einstein condensation of supercooled atoms to the elusive phase transition in the quark-gluon plasma, with many fascinating and important variants in between. It was the motivating force in the development of quantum statistics which has become the subject matter of every undergraduate text on modern physics. By the introduction of the quantum of energy hf it was directly responsible for Einstein's elucidation of the photon, for the Bohr atom, for de Broglie's postulate of the wave nature of matter, and for the eventual development of quantum mechanics in all its glory. Our purpose here is to take our streamlined and sophisticated understanding of Planck's result and go back from this modern vantage point to understand as best we can, in as deep and intimate a way as possible, the steps that led Planck to the quantum.

It is clearly incorrect – but nonetheless tempting – to undervalue Planck's accomplishment as somehow a lucky trick that worked, something that fell to him by chance as one of many fiddling about to parametrize the data. And even with the hindsight of 43 years, reflecting on his discovery at age 85 in wartime Berlin [6], Planck could barely explain his insight more deeply than at the first instant. It was done, he tells us, as a necessary first stratagem to define an entropy for the radiation field in terms of the probability of thermodynamic states, as dictated by Boltzmann. Probabilities, he decided, can be calculated only for discrete and countable alternatives.

Planck's Blackbody Energy Spectrum is the energy density $\mathcal{U}$ of the radiant heat in the 'Hohlraum' or cavity of a furnace, as a function of the frequency f or the wavelength $\lambda = c/f$ of the radiation and of the equilibrium temperature T of the furnace walls (assumed emitting and absorbing at all wavelengths), the heating element, and the radiation, all monitored through a peep-hole and measured by bolometers. The result we all know (in terms of $\hbar = h/2\pi$, the photon energy E and momentum p with $E = cp = \hbar\omega = hf$) is:

$$\mathcal{U} = \frac{2}{(2\pi)^3} \int 4\pi p^2 dp \frac{hf}{\exp{(hf/kT)} - 1} \equiv \int_0^\infty df u_f(T) \text{ or } \int_0^\infty d\lambda u_\lambda(T), \qquad (1)$$

where the spectral densities are defined as

$$u_f(T) = \frac{8\pi h}{c^3} \frac{f^3}{e^{hf/kT} - 1} \quad \text{and} \quad u_\lambda(T) = \frac{8\pi h}{c^2 \lambda^5} \frac{1}{e^{hc/\lambda kT} - 1}, \qquad (2)$$

which involve Planck's constant h, Boltzmann's constant k, and the speed of light *c.* The integral gives Stefan's Law

$$\mathcal{U} = \sigma T^4 \quad \text{with} \quad \sigma = \frac{8\pi^5 k^4}{15 h^3 c^3}. \tag{3}$$

The peak in the energy distribution $u_f(T)$ occurs at f_m with

$$\frac{h f_m}{kT} = 2.8214,$$

and that in $u_\lambda(T)$ at λ_m with

$$\frac{hc}{\lambda_m kT} = 4.9651.$$

At frequencies large compared to f_m, the exponential term in the spectral density dominates and Planck's distribution reduces to Wien's speculated result:

$$u_f(T)_{f \gg f_m} \rightarrow \frac{8\pi h f^3}{c} e^{-hf/kT}.$$

Einstein was to explain this [3] as the result of high energy photons undergoing only quasi-elastic scattering at low temperatures, and retaining their existence almost unchanged, like a classical gas of material molecules.

In the opposite limit, Rayleigh's result from classical electrodynamics and equipartition is obtained:

$$u_f(T)_{f \ll f_m} \rightarrow \frac{8\pi f^2}{c} kT.$$

Here the thermal energies kT are so much larger than the quantum energies hf that the quanta are produced almost without energy constraints and the classical wave theory is adequate.

Today we have the powerful machinery of quantum statistical mechanics [7] to derive the Planck distribution – from which, in fact, it originated. The Grand Partition Function, or equivalently the density matrix, defined as the sum of Boltzmann factors over all possible states κ is simply

$$\mathcal{Q} = \sum_\kappa e^{-\beta E_\kappa},$$

with $\beta = 1/kT$. We apply this to a cell of momentum $\vec{p}$ and polarization $\hat{\epsilon}$ and all possible number $n(\vec{p}, \hat{\epsilon})$ of photons, as an open system in thermodynamic equilibrium with a reservoir at temperature T. We are instructed to count each state $n(\vec{p}, \hat{\epsilon}) = 1, 2, \cdots$ once, with weight unity. Then, in what has become a ritual,

$$\mathcal{Q} = \sum_{\vec{p}} \sum_{\hat{\epsilon}} \mathcal{Q}(\vec{p}, \hat{\epsilon}),$$

the grand partition function for an individual cell is

$$\mathcal{Q}(\vec{p}, \hat{\epsilon}) = e^{-\beta E_p} + e^{-2\beta E_p} + e^{-3\beta E_p} + \cdots = \frac{1}{e^{\beta E_p} - 1}.$$

From the standard formalism, the average occupation number of a cell of momentum $\vec{p}$ (including both polarization states $\hat{\epsilon} = \pm 1$) is

$$< n_{\vec{p}} > = -\frac{1}{\beta} \frac{\partial}{\partial E_{\vec{p}}} \log \mathcal{Q} = \frac{2}{e^{\beta E_p} - 1}; \tag{4}$$

and the internal energy density is

$$\mathcal{U} = -\frac{\partial}{\partial \beta} \log \mathcal{Q} = \sum_{\vec{p}} E_p < n_{\vec{p}} >, \tag{5}$$

which reduces to the above integral over Planck's spectral density $u_f(T)$ with $E_p = \hbar \omega_p = h f_p$.

So we understand the Planck distribution from today's vantage point. We now turn to a detailed scrutiny of Planck's original discovery.

§I-3. Planck's Discovery of the Blackbody Formula.

For 20 years Planck had struggled to understand the full implications of the Second Law of Thermodynamics [6]. For four years he had tried to understand the equilibrium between radiant heat – tentatively recognized as electromagnetic radiation – and its sources – idealized as an array of simple harmonic oscillators radiating at all frequencies. As a devoted student of thermodynamics, Planck set himself the task to understand the equilibrium state of radiant heat and its sources as a problem akin to chemical equilibrium. The radical departure was the idea to apply thermodynamics to empty space! Spurred on by Wien's exponential fit to the observed spectrum [8], Planck [9] was able to *'define'* an entropy for an individual oscillator of frequency f and energy U, which he extended to an entropy for the radiation field of the same frequency and of intensity $\mathcal{R}$. He then defined an equilibrium temperature T the same for all frequencies by requiring the entropy to be a minimum, and from his postulated form of the entropy to derive the Wien distribution.

The successes of the Wien spectrum were impressive. Already with it, Planck was able to deduce values for what would become the fundamental constants k and h, which even without the impending correction to the spectrum could have led him to definitive values for Avogadro's number, the atomic mass unit, and the electron charge.

Wien criticized Planck's assumption of thermodynamic equilibrium as requiring a reversibility of fundamental radiative processes which was ruled out by examples of coherent processes which he conjured up. This objection is answered finally in Planck's 1943 *Reflections* by invoking a sort of molecular disorder for the radiation field as well as for the oscillators. But even after all these years, Planck never acknowledged the contribution of Gibbs who resolved all such paradoxes by invoking the Gibbs' ensemble [10] – a distribution in phase space, not a single point – as the fundamental physical state to which thermodynamic arguments apply.

Wien's exponential spectral density was soon put to a severe test by experiments at long wavelengths in the classical Rayleigh regime, and found to fail completely. The first somewhat tentative reports of discrepancies came from Lummer and Pringsheim [11] and from Paschen [12], but these were controversial. Then – as Planck recalled the exact chronology in his 1943 *Reflections* – Rubens and Kurlbaum, at the Deutschen Physikalischen Gesellschaft meeting of 19 October 1900, reported measurements of almost incredible precision [13] which showed spectral densities at long wavelengths increasing linearly with the temperature, in complete disagreement with Wien's spectrum. Planck had been told about the result a few days before Kurlbaum's lecture and had some time to prepare his response.

Planck's first step was to set $U = C \cdot T$ and from

$$\frac{\partial S}{\partial U} \equiv \frac{1}{T}, \tag{6}$$

then $S = C \cdot \log U$ and

$$\frac{\partial^2 S}{\partial U^2} = -\frac{C}{U^2} \tag{7}$$

valid for large U. From the entropy previously constructed to give Wien's distribution

$$S_W = -\frac{U}{af} \log \frac{U}{ebf}, \tag{8}$$

on the other hand, we get

$$\frac{\partial^2 S_W}{\partial U^2} = -\frac{1}{afU} \tag{9}$$

valid for small U. Planck combined these in a simple heuristic way to give

$$\frac{\partial^2 S}{\partial U^2} = -\frac{1}{afU + U^2/C}, \tag{10}$$

valid in both limits. By integration

$$\frac{\partial S}{\partial U} \equiv \frac{1}{T} = \frac{1}{af} \log \left(1 + \frac{aCf}{U}\right), \tag{11}$$

which we can solve for [14]

$$\mathcal{R}_f = \frac{f^2}{c^2} U = \frac{hf^3}{c^2} \frac{1}{e^{hf/kT} - 1}, \tag{12}$$

after comparison with Planck's earlier derivation [9] of the Wien distribution.

This was the basis of the distribution put forward by Planck in the 'lively' discussion following Kurlbaum's lecture. Rubens immediately verified that this form was in complete agreement with their results, as he reported to Planck the next morning [6].

Planck's preliminary work of 19 October 1900 was completed and reported first on 14 December 1900, and submitted for publication in Annalen der Physik on 7 January 1901. His path to the derivation is recalled very briefly and modestly in his 1943 *Reflections*. The formal derivation of the Planck spectrum which actually appears in his published result makes no mention of the heuristic result he says in his *Reflections* to have been cobbled together in response to Rubens and Kurlbaum's lecture. On the contrary, the formal derivation makes full use of the necessity of the countability of the states of the radiation in order to define the entropy. In his 1901 paper, Planck states "··· the theory (Note added: of 1900, see **Papers I·1,2,3**) requires revision ··· the whole question is to find S as a function of U ··· I had written down S directly ··· without foundation ··· it cannot be true"; and " ··· another condition must be introduced to make the entropy calculable ··· necessary to have a more fundamental conception of the idea of entropy."

From his *Reflections*: "Boltzmann ··· had the deepest understanding of entropy ··· a measure of the probability of that state ···." Planck had ignored this directive because he believed any calculation of probability involved assumptions and he "firmly believed ··· the Second Law was free of assumptions." But then, " that a similar fundamental hypothesis about the radiation is just as necessary and plays exactly the same role there as that of molecular disorder in the theory of gases, occurred to me at once with perfect clarity."

Again in 1901: "··· set the entropy proportional to the logarithm of the probability W ···. In fact we have ··· no criterion to speak of such a probability ··· the prescription is suggested by its simplicity ··· it is necessary to suppose that U_N is ··· a discrete, even an integral, multiple of finite equal parts ···. We call such a part an energy-quantum ···."

And in his *Reflections*: "Since such a probability-like quantity can only be found by counting, then it was necessary for the energy U_N to be expressible as a sum of discrete identical energy elements ϵ."

In his 1901 discovery paper, Planck made almost no comment about the deeper significance of the energy-quantum. One remark does appear in a note immediately following on 9 January 1901 [15]. Here Planck applies his new theory to derive values of unprecedented precision for the electron charge, the hydrogen atomic weight, and Avogadro's number among others. With this supporting evidence he does declare that the theory must be correct in general, and with absolute validity. In a longer paper almost a year later [4], Planck showed – following his first paper of the same title – that the new energy distribution was stationary, but made no further interpretation or even any specific mention of the quantum. And again in 1902 [5] in a long discussion of the nature of white light, although he emphasized the outstanding problem of the interpretation of the nature of spectral lines, he made no mention of the energy-quantum.

In 1943: "··· there arose the all-important problem, to assign this remarkable constant a physical meaning. ··· But the nature of the energy-quantum remained unclear." And finally: " For many years I continued to do further research, trying somehow to fit the action quantum into the system of classical physics. But it seems to me that this is not possible."

§I-4. Planck's Discovery as Prolog.

In the remainder of this book we collect and discuss principal landmark contributions to the discoveries of quantum theory which sprang directly – if at first slowly and reluctantly – from Planck's invention of the energy quantum. We have chosen to follow a fundamental but rather narrow path leading from Planck to Einstein, Bohr, and de Broglie; then to the formulations of quantum mechanics due to Heisenberg, Born and Jordan, and to Schrödinger, and to Dirac. Then we enter into the interpretation of quantum mechanics, again following original contributions, here of Born, Heisenberg, and Bohr. Then we vault forward to contemporary contributions from Bell and Aspect and others, originating in the Einstein-Podolsky-Rosen paradox.

We complete our account of the first one hundred years of Planck's h-quantum with an introduction to recent advances in the theory of measurement and decoherence. We conclude with brief speculations on the next one hundred years and what remains to be achieved within the limited but fundamental context of non-relativistic quantum mechanics.

We now commend to you the original papers in which Max Planck first developed his ideas and then reported his discovery of the Blackbody Spectrum.

Footnotes and References:
1) M. Planck, Ann. d. Phys. **4**, 553 (1901); **4**, 564 (1901) (here as **Papers I·4,5**).

2) W. Wien, Ann. d. Phys. **4**, 422 (1901) (here as **Paper I·3**).

3) A. Einstein, Ann. d. Phys. **17**, 17 (1905) (here as **Paper II·1**).

4) M. Planck, Ann. d. Phys. **6**, 818 (1901).

5) M. Planck, Ann. d. Phys. **7**, 390 (1902).

6) M. Planck, Die Naturwissenschaften **14/15**, 153 (1943) (here as **Paper I·6**).

7) K. Huang, *Statistical Mechanics*, John Wiley & Sons, New York (1987), p.280.

8) W. Wien, Wied. Ann. **58**, 69 (1900).

9) M. Planck, Ann. d. Phys. **1**, 69 (1900); **1**, 719 (1900) (here as **Papers I·1,2**).

10) e.g., R.C. Tolman, *The Principles of Statistical Mechanics* (Oxford, London 1942), Pp.70, 325. This is an endlessly quarrelsome subject.

11) O. Lummer and E. Pringsheim, Verhandl. d. Deutsch. Physikal. Gesellsch. **1**, p.31 (1900); **2**, p.163 (1900); **3**, p.37 (1900); Ann. d. Phys. **4**, 225 (1901).

12) F. Paschen, Ann. d. Phys. **4**, 277 (1901); **6**, 646 (1901).

13) H. Rubens and F. Kurlbaum, Sitzungsber. d. k. Akad. d. Wissensch. zu Berlin **41**, 928 (1900); Ann. d. Phys. **4**, 649 (1901) (from which many references can be traced).

14) The relation $\mathcal{R}_f = f^2 U/c^2$ can be understood on dimensional grounds. Planck expends great effort to establish this connection between the equilibrium energy of the charged dipole oscillator U and the energy flux $\mathcal{R}_f$ of the equilibrium field by equating the emission rate $\sim fU$ and the absorption rate $\sim c^2 \mathcal{R}_f/f$. We forgo this lengthy discussion in Paper I·1.

15) M. Planck, Ann. d. Phys. **4**, 564 (1901) (here as **Paper I·5**).

Biographical Note on Planck.

Max Karl Ernst Ludwig Planck (1858-1947) – From the London Times 6OCT47 we learn about Planck's long, distinguished and much documented life. Planck was the son of a Professor of Constitutional Law at Kiel and Göttingen, distinguished as joint author of the Prussian Civil Code. Planck studied physics and mathematics, but especially thermodynamics, under Helmholtz, Kirchoff and Weierstrauss and in 1879 obtained his doctorate *summa cum laude* from Munich for his studies on the Second Law of Thermodynamics. In 1884 he became Professor in Kiel, in 1889 Professor Extraordinarius in Berlin, and in 1892 succeeded Kirchoff there in the chair of Experimental Physics. He became Permanent Secretary to the Prussian Academy of Science (1912), Rector of the University of Berlin (1913-14), Nobel Laureate (1918), President (1920) of the Kaiser Wilhelm Society (now the Max Planck Society). He climbed the Jungfrau at age 72. Planck's Berlin villa was destroyed by bombs in WWII, so what his circumstances were when he wrote his

1943 *Reflections* is an interesting question. His son Erwin was executed by the Gestapo for his part in the July 1944 attempted assassination of Hitler.

Yourgrau and Mandelstam [†] describe the extreme irony of Planck's personal views and his scientific discovery of the *h*-quantum, which he himself was first to recognize as the 'elementary quantum of action'. Planck was in a quandary "Can it be that the astonishing simplicity of ··· [the Principle of Least Action] ··· rests ··· on chance? It is ··· difficult to believe this. On the contrary ··· Leibniz's Principle of Least Action ··· [leads] ··· to a deeper understanding of the quantum of action." They characterize Planck's arguments as "··· metaphysical ··· put forward with zeal and persistence." Yourgrau and Mandelstam recommend instead Born's attitude "··· I do not like this metaphysics ··· that there is a definite goal to be reached and often claims to have reached it. Metaphysical systematization means formalization and petrification ··· there *are* metaphysical problems ··· they are 'beyond physics' indeed and demand an act of faith. We have to accept this fact to be honest."

These comments are almost fifty years old, and it is interesting to compare them to what we believe today in light of our efforts to forecast the next 100 Years of Planck's *h* Quantum.

Yourgrau and Mandelstam express equal regard for the action and for the equations of motion, on the grounds that Sturm-Liouville differential equations of motion are equivalent to a variation principle. This was at a time (1955~1963) when S-matrix theory was in, and quantum field theory out of fashion. The revolution of the Standard Model restored full faith in field theory based on an Action Principle constructed to express profound invariance properties, symmetries both full and broken, and renormalizability. The only conceivable access to the construction of such theories, and then to calculation with them is via the Action Principle. The equations of motion are available but not of primary interest. As Born feared, it seems that there *is* a definite goal, there *are* frequent claims to have reached it, and the hope of systematization, formalization, and we suppose the remote specter of ultimate petrification. Then this remote, ultimate, intricate 'reality' and all its elements will require an existential – possibly only metaphysical – understanding.

† – W. Yourgrau and S. Mandelstam, *Variational Principles in Dynamics and Quantum Theory* (Pitman, London, 1960), Pp.146, 160.

Paper I·1: Excerpt from Annalen der Physik **1**, 69 (1900).

On Irreversible Radiation Processes

von **Max Planck**

(Nach den Sitzungsbr. d. k. Akad. d. Wissensch. zu Berlin vom 4. Februar 1897, 8. Juli 1897, 16. December 1897, 7. Juli 1898, 18. Mai 1899 and from a lecture given at the 71st Natural-Philosophy Meeting in Munich, edited for the Annalen der Physik by the author.)

The following is an exposition of the principal results of my investigations into the meaning of the Second Law of Thermodynamics as applied to radiant heat considered from the standpoint of electromagnetic theory.

That even radiant heat must respect the requirements of this fundamental Law – for example, the balance of radiation between bodies of different temperature is always in the direction to equalize their temperatures – is generally undisputed. The electromagnetic nature of radiant heat makes it a most urgent problem to understand and even to prove the Second Law of Thermodynamics as applied to heat radiated purely electromagnetically. The prerequisite to this is that one should understand the emission and absorption of radiant heat as an electromagnetic process; that one consider the emission of radiant heat as the emission of electromagnetic waves by elementary oscillators which have some connection with the actual atoms of the radiating matter; and further, that one consider the absorption not as the result of an electrical resistance or some kind of friction, but as a resonance phenomenon, in which the oscillators not only emit waves, but also are stimulated into vibrations by incident waves. ······

Chapter One: Emission and Absorption of Electromagnetic Radiation by an Oscillator.

······

* * * * * * * * * * * * * * * * * * **

Note added in translation: We cannot follow every strand back to its beginning but must accept some results as being "obvious" although they are by no means trivial or uninteresting. Planck's oscillator was an electric dipole with a natural frequency f; its energy U was electric and magnetic energy related to the square of its amplitude. On average, the oscillator energy decreases due to 'spontaneous' radiation at a rate proportional to fU and increases due to absorption of the incident flux $\mathcal{R}_f$ of radiant energy at the resonant frequency. On dimensional grounds the rate of absorption must be proportional to $c^2 \mathcal{R}_f / f$. Planck establishes the equality at equilibrium, $\mathcal{R}_f = f^2 U / c^2$.

＊＊＊＊＊＊＊＊＊＊＊＊＊＊＊＊＊＊＊＊

Chapter Two: Conservation of Energy and Increase of Entropy.

. . . . . .

§17. Definition of the Electromagnetic Entropy.

By analogy to the total energy of the system U, we define a new quantity, the *total electromagnetic entropy* of the system:

$$S_t = \sum S + \int s d\tau.$$

The sum is over all the oscillators, the integration over all volume elements $d\tau$ occupied by the fields. S is the entropy of an individual oscillator and s the entropy density at a point in the fields.

The entropy S of an oscillator of frequency f and energy U we *define* as follows:

$$S = -\frac{U}{af} \log \frac{U}{ebf}, \tag{41}$$

where a and b are two universal positive constants $\cdots$; and $e = 2.71 \cdots$ is included for later convenience. $\cdots\cdots$.

＊＊＊＊＊＊＊＊＊＊＊＊＊＊＊＊＊＊＊＊

Planck's choice for the entropy of an oscillator mimics the Boltzmann H function as a function of U, chosen as the only available characteristic of the oscillator. The factors af and bf could be arbitrary positive functions of frequency in this state of the theory, but are chosen linear anticipating Wien's fit to the data. Planck's choice for the field entropy – which eventually has to be modified by the quantum hypothesis – is understood when we recognize that the field energy per unit volume for an unpolarized, isotropic distribution is

$$U_F \equiv \int \frac{d^3 f}{c^3} \mathcal{U}_f = \int \frac{2 \cdot 4\pi}{c^3} f^2 df \frac{c^2}{f^2} \mathcal{R}_f.$$

The factor 2 comes from the sum over the polarizations and the factor $4\pi f^2 df/c^3$ from the integration over the wave-number $\vec{k} = f\vec{c}$. The equality of Eqn49 $\mathcal{R}_f = f^2 U/c^2$ which originated from the energy equilibrium of the oscillator-field system, is now a result of the entropy maximum with the above definitions of the entropy.

✳ ✳ ✳ ✳ ✳ ✳ ✳ ✳ ✳ ✳ ✳ ✳ ✳ ✳ ✳ ✳ ✳ ✳ ✳✳

We $\cdots$ define the entropy of a monochromatic linearly polarized beam of intensity $\mathcal{R}$ by:

$$\mathcal{L} = -\frac{\mathcal{R}}{af} \log \frac{c^2 \mathcal{R}}{eb f^3}. \tag{43}$$

$\cdots\cdots$ In the special case that the radiation through the point is unpolarized and independent of direction, the entropy density is $\cdots$

$$s = \frac{8\pi}{c} \cdot \int_0^\infty df \cdot \mathcal{L}. \qquad \cdots\cdots \tag{45}$$

§19. Requirements on the Stationary State.

That state of the system which corresponds to the maximum total entropy, is referred to as the stationary state; from the Law of Increase of Entropy, it is no longer possible for it to undergo further changes, so long as there are no interactions from outside the system. Another necessary condition for the stationary state is that the total entropy itself no longer changes with the time, so that all the inequalities of the preceding paragraphs become equalities. This condition is fulfilled, as can easily be shown, when at every point and for every direction of propagation and polarization

$$\mathcal{R} = \frac{f^2}{c^2} U. \tag{49}$$

The total radiation of each frequency is assumed unpolarized and isotropic.

But the conditions necessary for the absolute maximum of the total entropy go still further. For each infinitesimally small virtual change of state of the system, the change in the total entropy S_t must vanish. If we imagine a virtual change in which an infinitesimal amount of energy of an oscillator with frequency f_1 is transferred to another oscillator of frequency f_2, while everything else remains unchanged, then we must have

$$\delta S_t = \delta S_1 + \delta S_2 = 0 \quad \text{and} \quad \delta U_1 + \delta U_2 = 0.$$

These equations with Eqn41 give:

$$-\frac{1}{af_1}\log\frac{U_1}{bf_1} = -\frac{1}{af_2}\log\frac{U_2}{bf_2};$$

If we abbreviate

$$-\frac{1}{af_1}\log\frac{U_1}{bf_1} \equiv \frac{1}{T}, \tag{50}$$

then it follows, since f_1 and f_2 are completely independent, that the value of T in the stationary state must be the same for all oscillators in the system. Since the value of U determines that of $\mathcal{R}$ by Eqn49, then the stationary state of the whole system in all its parts depends only on the single parameter T.

We can now express the value of all quantities in the stationary state in terms of the parameter T. First, from Eqn50, the energy of an oscillator of frequency f:

$$U = bfe^{-af/T},$$

and from Eqn49, the intensity of a linearly polarized beam of frequency f:

$$\mathcal{R} = \frac{bf^3}{c^2}e^{-af/T}, \tag{51}$$

and from Eqn25 for the total intensity of radiation in a given direction:

$$K = 2\int_0^\infty \mathcal{R}df = \frac{12bT^4}{c^2a^4} \tag{52}$$

and from Eqn28 for the energy density of the field:

$$u = \frac{4\pi K}{c} = \frac{48\pi bT^4}{c^3a^4}. \tag{53}$$

This energy density consists of the sum of the energy densities $\mathcal{U}(f)$ of the individual frequencies in the following way:

$$u = \int_0^\infty \mathcal{U}(f)df, \text{ with } \mathcal{U}(f) = \frac{8\pi\mathcal{R}}{c} = \frac{8\pi b f^3}{c^3}e^{-af/T}.$$

From Eqns41 and 50, the entropy of an oscillator of frequency f is:

$$S = b\left(\frac{f}{T} + \frac{1}{a}\right)e^{-af/T},$$

and from Eqns43 and 50, that of the radiation of frequency f in a given direction is:

$$\mathcal{L} = \frac{bf^2}{c^2}\left(\frac{f}{T} + \frac{1}{a}\right)e^{-af/T}.$$

The total intensity of the radiation entropy in a given direction is, from Eqn43:

$$L = 2\int_0^\infty \mathcal{L}df = \frac{16bT^3}{c^2a^4},$$

and finally the total density of the field entropy from Eqn45:

$$s = \frac{4\pi L}{c} = \frac{64\pi bT^3}{c^3a^4}, \tag{54}$$

which follows from the entropy density $\mathcal{S}(f)$ of a single frequency:

$$s = \int_0^\infty \mathcal{S}(f)df, \text{ with } with\mathcal{S} = \frac{8\pi\mathcal{L}}{c} = \frac{8\pi b f^2}{c^3}\left(\frac{f}{T} + \frac{1}{a}\right)e^{-af/T}.$$

That the total entropy S_t actually has its maximum value can easily be shown from the first and second variations.

Chapter Three: Thermodynamic Conclusions.

§20. Thermodynamic Entropy of The Radiation.

If one accepts the electrodynamic nature of light and radiant heat, one gets the stationary radiation states of the above chapter. These have a fundamental thermodynamic significance. According to a law derived by Kirchoff and used in this context by Wien [1], the heat radiation, which is surrounded by equal temperature matter completely containing it, is independent of the nature of the matter, and completely determined by a single parameter: the temperature. The radiation is the same as if the surrounding matter was completely "black".

The same law holds when the walls of the vacuum are perfectly reflecting and when the matter is distributed in any manner in the vacuum. One does have to assume that at least some finite part of the matter's radiation is emitted into each spectral interval. If this condition is not met, then an essentially unstable radiation state can occur in which a single color is completely missing.

According to the known laws the amount and the particular nature of the emitting and absorbing matter is completely irrelevant to the equilibrium state of the radiation, so one is forced to the conclusion *that the · · · · · · stationary radiation state of the vacuum fulfills all the conditions of the radiation of black bodies, completely without regard to the question whether or not the assumed electromagnetic oscillators are the actual sources of heat radiation in any particular matter.*

This conclusion leads to a somewhat different point of view: The Second Law of Thermodynamics requires that not only the ambient heat, but also the radiant heat must correspond to a definite entropy; when a body loses heat by radiation its entropy decreases, and – according to the Principle of Increasing Entropy – there must occur somewhere a compensating entropy increase, which in this case can only be in the resulting heat radiation. If thermal and electromagnetic radiation are identical, then nothing is left for the thermal radiation entropy, which must be determined solely by the state of the radiation, except to be identified with the electromagnetic entropy. When we do this, we arrive at the further conclusion that the absolute maximum of the entropy corresponding to the stationary radiation state or equivalently to the equilibrium state of the heat radiation, gives the blackbody radiation.

From the identification of thermodynamic and electromagnetic entropy there follows a number of connections between thermal and electrical quantities, whose significance is pointed out in the following paragraphs.

§**21.** Electromagnetic Definition of Temperature.

From the entropy of a system in thermodynamic equilibrium its temperature is also defined, because the absolute temperature is the ratio of an infinitesimal amount of heat added to the system to the resulting change in the entropy, provided the system is kept in thermodynamic equilibrium during the change of state. If we assume a unit volume of the vacuum filled with the stationary radiation and hold the volume constant and the radiation at equilibrium, then the infinitesimal change in the energy u of the system is equal to the infinitesimal amount of heat added dQ, and from Eqn53:

$$dQ = du = \frac{192\pi b T^3}{c^3 a^4} dT = T ds,$$

where the resulting change of the entropy s of the system is ds, and the absolute temperature is T. Defined in this way, the absolute temperature of the stationary radiation state in the vacuum is identical to the purely electromagnetically defined parameter T previously introduced, which determines all properties of these states in the above way. From Eqn51 the inverse temperature of a linearly polarized monochromatic beam of frequency f and intensity $\mathcal{R}$ is:

$$\frac{1}{T} = \frac{1}{af} \log \frac{bf^3}{c^2 \mathcal{R}}.$$

When the conditions for the stationary state are not met, but when other radiation behavior occurs in the vacuum, then one can no longer speak of the temperature at a particular place, or of the temperature of the radiation in a given direction, but one must ascribe a separate temperature to each individual linearly polarized monochromatic beam, which is determined by its intensity and its frequency according to the above equation [2]. The beam retains its temperature during propagation with its intensity unchanged – except, for example, if it passes through a focal point – until it is either divided or absorbed.

On the other hand each oscillator has a given temperature completely determined by Eqn50. The stationary radiation state can therefore be characterized as all oscillators and all monochromatic rays having the same temperature.

§**22.** Dependence of the Total Radiation on the Temperature.

The total intensity of the radiant energy in any given direction is given by the expression for K in Eqn52, which is proportional to the fourth power of the temperature and expresses the well-known Stefan-Boltzmann Law, whose validity was established by Boltzmann [3] from thermodynamics and recently in the investigations of Lummer and Pringsheim [4], at least within the temperature interval from $T = 290K$ to $T = 1560K$, has received a remarkable experimental confirmation.

§23. Distribution of Energy in the Normal Spectrum.

The Law by which the total stationary radiation intensity K is distributed among the radiation intensities $\mathcal{R}$ of individual frequencies f, is given by Eqn51. This Law is usually expressed in terms of the wavelength λ rather than the frequency f, so from Eqn52:

$$K = 2 \int_0^\infty \mathcal{R} df \equiv \int_0^\infty E_\lambda d\lambda.$$

With $\lambda = c/f$, we have

$$E_\lambda = 2\mathcal{R}|\frac{df}{d\lambda}| = \frac{2c\mathcal{R}}{\lambda^2}, \text{ or from Eqn51: } E_\lambda = \frac{2c^2 b}{\lambda^5} \cdot e^{-ac/\lambda T}.$$

This is just Wien's Energy Distribution Law [5], whose at least approximate validity has recently been been demonstrated in the experiments of Paschen [6], Paschen and Wanner [7], and Lummer and Pringsheim [8].

Wien derived his Law on the basis of certain assumptions about the number of radiating oscillators per unit volume and about the details of their motion; in the theory developed here these quantities play no role, but the Law appears as a necessary result of the definition developed in §17 of the electromagnetic entropy of the radiation. The question of the validity of the Law depends entirely upon the validity of that definition. *I have tried repeatedly to derive Eqn41 for the electromagnetic entropy of an oscillator – from which Eqn43 for the entropy of the radiation is determined – by modifying or generalizing it so that it has the same validity as other theoretically well established electromagnetic and thermodynamic Laws, but I have not succeeded.* (Note: Italics added.)

Thus for example one could define the entropy of an oscillator, instead of by Eqn41, in the following way:

$$S = -\frac{U}{\psi(f)} \cdot \log \frac{U}{\phi(f)},$$

where $\psi(f)$ and $\phi(f)$ are undetermined positive functions of the frequency f. Then for the electromagnetic processes discussed in §18 the Principle of Increase of Entropy is definitely fulfilled, but one gets instead of Eqn50 a different expression for the inverse temperature of the oscillators:

$$\frac{1}{T} = -\frac{1}{\psi(f)} \cdot \log \frac{eU}{\phi(f)}, \text{ and then: } U = \frac{\phi(f)}{e} \cdot e^{-\psi(f)/T}$$

and from Eqn49 an Energy Distribution Law instead of Eqn51:

$$\mathcal{R} = \frac{f^2 \phi(f)}{c^2 e} \cdot e^{-\psi(f)/T}. \tag{56}$$

This is exactly the same form as found by Wien on the basis of his special considerations. On the other hand, if one starts from some other form of the Energy Distribution Law and calculates the entropy from it, then one always gets a contradiction with the Principle of Increase of Entropy.

We must conclude that the definition of radiation entropy given in §17, and with it also Wien's Energy Distribution Law, is a necessary consequence of the application of the Principle of Increase of Entropy to the electromagnetic theory of radiation, and that the limit of the validity of these laws – if indeed such a limit exists – can only be that of the Second Law of Thermodynamics. Naturally, the further experimental test of these Laws is of the greatest fundamental interest.

. . . . . .

(Eingegangen 7. November 1899.)

Footnotes and References:
1) W. Wien, Wied. Ann. **52**, 133 (1894).
2) The necessity of such a generalization of the idea of temperature was noted first by E. Wiedemann, Wied. Ann. **34**, 448 (1888); see also W. Wien, loc.cit., p.132.
3) L. Boltzmann, Wied. Ann. **22**, 291 (1884).
4) O. Lummer and E. Pringsheim, Wied. Ann. **63**, 395 (1897).
5) W. Wien, Wied. Ann. **58**, 662 (1896).
6) F. Paschen, Wied. Ann. **60**, 662 (1897); Sitzungsber. d. k. Akad. d. Wissensch. zu Berlin, p.405, p.893 (1899).
7) F. Paschen and H. Wanner, Sitsungsber. d. k. Wissensch. zu Berlin, p.5 (1899).
8) O. Lummer and E. Pringsheim, Verhandl. d. Deutsch. Physikal. Gesellsch. **1**, p.28,

p.215 (1899). Also see H. Beckmann, Inaug.-Diss., Tübingen (1898); and H. Rubens, Wied. Ann. **69**, 582 (1899). These works discuss variants of Wien's Law.

Paper I·2: Excerpt from Annalen der Physik **1**, 719 (1900).

Entropy and Temperature of Radiant Heat

von **Max Planck**

§1. Introduction and Summary.

In a recently published paper [1], I have constructed an expression for the entropy of radiant heat, which complies with all the requirements on the properties of this quantity arising on the one hand from thermodynamics, and on the other from electromagnetic theory. $\cdots\cdots$ the result, and thus also Wien's Law, is a necessary consequence of the application of the Principle of the Increase of Entropy to the theory of electromagnetic radiation. $\cdots\cdots$

§6. Complete Derivation of the Entropy Function.

$\cdots$ an oscillator in its stationary radiation field $\cdots$ the energy of all the oscillators is the sum of their individual energies, $U_n = nU$, and the entropy is the sum of their individual entropies, $S_n = nS$. We set

$$\frac{d^2S}{dU^2} = -\mathcal{F}(U), \tag{12}$$

where $\mathcal{F}$ is a positive function of U. If one sets

$$U_n = n \cdot U \text{ everywhere, then } \mathcal{F}(nU) = \frac{1}{n}\mathcal{F}(U)$$

and the solution of this functional equation is [2]:

$$\mathcal{F}(U) = \frac{\text{const.}}{U} \text{ or from (12): } \frac{d^2S}{dU^2} = -\frac{\alpha}{U},$$

where α can only be a positive function of the frequency f. From this expression it follows by a double integration:

$$S = -\alpha U \log(\beta U), \tag{14}$$

where β is a second positive function depending only on the frequency f. A further positive integration constant is suppressed since it can have no physical significance. Finally $\cdots$ (Note: from Eqn49 of [1]) it follows that:

$$\mathcal{L} = -\frac{f^2}{c^2}\alpha U \log(\beta U) = -\alpha \mathcal{R} \log\left(\frac{\beta c^2}{f^2}\mathcal{R}\right).$$ (15)

If one further assumes the temperature for the oscillators T to be:

$$\frac{1}{T} = \frac{dS}{dU}, \text{ and for the radiation } \mathcal{R} \text{ to be: } \frac{1}{T} = \frac{d\mathcal{L}}{d\mathcal{R}},$$

then:

$$\frac{1}{T} = -\alpha \log(\beta eU) \text{ and } = -\alpha \log\left(\frac{\beta e c^2}{f^2}\mathcal{R}\right),$$ (16)

or:

$$\mathcal{R} = \frac{f^2}{\beta e c^2}e^{-1/\alpha T}$$

and this equation with the substitution:

$$\frac{1}{\alpha} = \psi(f), \qquad \frac{1}{\beta} = \phi(f)$$

gives Eqn56 of my original paper. According to the thermodynamic results of Wien [2], both $\psi(f)$ and $\phi(f)$ are proportional to f, so we can set:

$$\frac{1}{\alpha} = af, \qquad \frac{1}{\beta} = ebf,$$

where a and b are universal positive constants which, together with Eqns14 and 15, complete the definition of the oscillator entropy S and the radiation entropy $\mathcal{L}$. From Eqn16 then:

$$\frac{1}{T} = \frac{1}{af} \log\left(\frac{bf^3}{c^2\mathcal{R}}\right).$$ (17)

From the measurements of Kurlbaum and of Paschen, the numerical values of a and b are [3]:

$$a = 0.4818 \times 10^{-10}[\text{sec} \times \text{deg}],$$
$$b = 6.885 \times 10^{-27}[\text{erg} \times \text{sec}].$$

Berlin, März 1900.

(Eingegangen 22. März 1900.)

Footnotes and References:
1) M. Planck, Ann. d. Phys. **1**, 69 (1900).
2) W. Wien, Wied. Ann. **58**, 662 (1896).
3) (Note added: 1997 values are

$$a \equiv \frac{h}{k} = .47994 \times 10^{-10}[\text{sec} \times \text{deg}] \quad \text{and} \quad b \equiv h = 6.6261 \times 10^{-27}[\text{erg} \times \text{sec}].)$$

Paper I·3: Excerpt from Annalen der Physik 4, 422 (1901).

On the Theory of Radiation –
Comments on Planck's Criticism

von **W. Wien**

Planck has criticized my earlier work on blackbody radiation which supports reversibility of the radiation for free propagation but asserts non-reversibility of any coherent emission processes, whereas Planck holds that such processes should also be reversible. In discussion it has been agreed that this difference of opinion is eliminated when a practical definition of an irreversible process is given. The difference of opinion arose essentially because we did not mean the same thing by an irreversible process.

If we put a radiating surface element at the center of a reflecting container, all rays will be reflected back and we can consider the reflected radiation itself as re-emitted in a larger volume and reflected from a sphere with a larger radius. This process is certainly reversible if the reflecting element is sufficiently small that all rays can be considered as radial. On the other hand, when the size of the element is not negligible, then rays strike the spherical surface at different angles and differently reflected rays will have different path lengths back and forth to the second sphere. As a result the rays will not return in phase to the first sphere. If we consider two radiation packets which traverse different length paths in their first passage, then these will only arrive in phase when the ratio of their wavelengths is

a rational number. For an extended radiation source the original state will never exactly recur. Even this process can be considered as reversible, if in a finite time the original state can recur approximately. If the time is unlimited, then there is no limit for the approximation. When one realizes that conservative processes which take place in even the greatest disorder cannot bring about irreversibility, then one is free to expand in the above sense what must be considered reversible.
········· [2]···

I approve gladly that such a sharp criticism should be directed against so fundamental a problem but a renewed investigation into the question, how in this case the reversibility occurs, seems urgently desirable. I myself have not yet found a satisfactory answer.

(Eingegangen 29. December 1900.)

Note added in Proof: Jahnke, Lummer and Pringsheim [3] report a deviation from my theory [4] which essentially repeats their earlier claims. I consider it unnecessary to go into this once more. Their earlier remarks, which were prompted by my conjecture that one might expect a scaling of the radiation law, rested on a misunderstanding. I have expressly noted that deviations from the basic requirement of essential molecular disorder must occur with increasing temperature at smaller wavelengths. I have repeatedly stated that for a given wavelength, any increase in the temperature must lead to other molecular effects which for lower temperature only entered for longer wavelengths [5].

Footnotes and References:
1) M. Planck, Ann. d. Phys. **3**, p.764 (1900).
2) (Note added in translation: Here Wien embarks on a discussion of uncompensated depolarization of scattered rays, leading in his opinion to irreversibility. Planck had extended an unstated hypothesis of molecular disorder to the radiation field, cutting the Gordian knot of all such coherence effects.)
3) E. Jahnke, O. Lummer and E. Pringsheim, Ann. d. Phys. **4**, 225 (1900).
4) W. Wien, Ann. d. Phys. **3**, 530 (1900).
5) (Note added in translation: Here Wien reproduced the data of Jahnke, Lummer and Pringsheim which showed a small deviation at high temperature, in the direction of Planck's modification of Wien's Law.)

Paper I·4: Excerpt from Annalen der Physik **4**, 553 (1901).

On the Energy Distribution
in the Blackbody Spectrum

von **Max Planck**

(Communicated also in the Deutschen Physikalischen Gesellschaft, Sitzung vom 19. October und vom 14. December 1900, Section **2**, p.202 and p.237, 1900.)

Introduction.

New spectral measurements of Lummer and Pringsheim [1] and even more remarkably those of Rubens and Kurlbaum [2], which confirm an earlier result of Beckmann [3], show that the law first derived by Wien from molecular kinetics and later by me from the theory of electromagnetic radiation, has no general validity.

In each case, the theory requires revision. In the following, I make an attempt from the basis of my previous work. It turns out, in the course of a derivation which leads to Wien's Energy Distribution Law, that one discovers a term for which a new prescription is rewarding; then it becomes a question to identify this term in the derivation and to make a suitable interpretation of it.

The physical basis of my radiation theory which includes the hypothesis of "spontaneous radiation", has strong critics with opposite views, as I have described in my recent article [4]. Nonetheless, the calculation to my knowledge contains no errors. It completely determines the Law of Energy Distribution in the Blackbody Spectrum, when one calculates the entropy S of oscillators radiating at a fixed frequency f, as a function of their vibration energy U. From the relation

$$\frac{dS}{dU} = \frac{1}{T},$$

one gets the dependence of the energy U on the temperature T. Since the energy U is simply related to the radiation density, it too is given in terms of the temperature. The Blackbody Energy Distribution is that one for which the radiation density of all different frequencies has the given temperature.

So the whole question is to find S as a function of U, and the following work describes our solution to this problem. In my first treatment of this matter, I had written down S directly by definition, without further foundation, as a simple function of U, and was content to prove that this form of the entropy satisfied all the requirements of thermodynamics. I thought then that it was unique and that Wien's Law, which necessarily follows from it, had general validity. But a later more careful investigation [5] showed there could be another derivation, which also made possible the restriction required to make S well behaved. I then saw this requirement to have been based directly on the plausible hypothesis that – for an infinitesimal irreversible change in a system of N identical oscillators almost in thermodynamic equilibrium in identical stationary radiation fields – the increase in their total entropy $S_N = NS$ only depends on the total energy $U_N = NU$ and on their changes, but not on the energy U of the individual oscillators. This hypothesis led in turn to Wien's Energy Distribution Law. But now, from recent experiments, it cannot be true. One is forced to the conclusion that even this hypothesis cannot be generally valid and must be removed from the theory [6].

Therefore another condition must be introduced to make the entropy calculable: a more fundamental conception of the idea of entropy is necessary, which does lead to a new simple expression for the entropy and thereby to a new radiation formula which appears to agree with all known facts.

I. Calculation of the Entropy of an Oscillator as a Function of its Energy.

§1. Entropy denotes disorder, and this disorder – according to the electromagnetic radiation theory of an oscillator of given frequency, in a stationary radiation field – depends on the irregularity with which changes of its amplitude and its phase occur over time-intervals large compared to the period of the oscillations, but small compared to the time of a measurement. If the amplitude and phase were absolutely constant, then the oscillations would be perfectly uniform, their entropy would be zero, and their energy would be perfectly freely transformable to work. The constant energy U of an individual oscillator is only an interrupted time-average, or equivalently a simultaneous mean value of the energies of a large number N of identical oscillators, all in the same stationary radiation field and far enough from each other to have no direct mutual influence. In this sense, we speak of the mean energy U of individual oscillators. The total energy of the system of N oscillators is

$$U_N = NU \tag{1}$$

corresponding to a fixed total entropy

$$S_N = NS, \tag{2}$$

where S is the mean entropy of an individual oscillator. The entropy S_N represents the disorder with which the total energy U_N is distributed among the N individual oscillators.

§2. We set the entropy S_N of the system, to within an arbitrary additive constant, proportional to the logarithm of the probability W with which the N oscillators can possess the total energy U_N, that is:

$$S_N = k \log W + \text{ const.} \tag{3}$$

This prescription in my opinion is based on the definition of the probability W. In fact we have – on the assumption that electromagnetic theory is the basis of the radiation – no criterion to speak of such a probability in any definite sense. The suitability of the prescription introduced next is suggested first by its simplicity, and second by its similarity to a corresponding prescription from the kinetic theory of gases [7].

§3. *To find the probability that the N oscillators have the total energy U_N, it is necessary to suppose that U_N is not a continuous, endlessly divisible quantity, but a discrete integral multiple of finite equal parts combining to the whole. We call such a part an energy-quantum ϵ, and set*

$$U_N = P \cdot \epsilon, \tag{4}$$

where P is an integer, expected to be large, whereas the value of ϵ is as yet undecided.

It is clear that the partition of P energy-quanta among N oscillators is possible in a finite exactly determined number of ways. Each individual partition we call a "complexion", following similar ideas of Boltzmann. If one labels the oscillators by the integers $1, 2, \cdots, N$ in order; and the number of energy-quanta possessed by each as $\{p_1, p_2, \cdots, p_N\}$ with $p_1 + p_2 + \cdots + p_N = P$; then each different choice of p's is a separate complexion. The number $\mathcal{Z}$ of all possible complexions is equal to the number of all possible ways to distribute P identical objects (quanta) in N different cells. It should be remarked that two complexions are considered different when the same numbers p occur but in different order; so for $N = 2, P = 10$, the complexion $\{3, 7\}$ is different from $\{7, 3\}$.

From Combinatorial Theory the number $\mathcal{Z}$ of all possible complexions for given N, P is

$$\mathcal{Z} = \frac{(N + P - 1)!}{(N - 1)!P!}.$$

Using Stirling's Approximation – $N! = N^N$ – we get

$$\mathcal{Z} = \frac{(N + P)^{N+P}}{N^N \cdot P^P}.$$

§4. The hypothesis which we use as the basis for further calculations, runs as follows: The probability W that the N oscillators have total energy U_N is proportional to the number $\mathcal{Z}$ of all possible complexions for the partition of the energy U_N among the N oscillators; in other words, any particular complexion is just as probable as any other particular complexion. Whether this hypothesis actually corresponds to Nature, can be proved only by experiment. $\cdots\cdots$

§5. From Eqn3, the entropy of a system of N oscillators with energy $U_N = NU = P\epsilon$, with a suitable choice of additive constant, is:

$$S_N = k \log \mathcal{Z} = k \left\{ (N+P) \log (N+P) - N \log N - P \log P \right\}, \qquad (5)$$

and using Eqns4 and 1:

$$S_N \equiv NS = kN \left\{ \left(1 + \frac{U}{\epsilon} \right) \log \left(1 + \frac{U}{\epsilon} \right) - \frac{U}{\epsilon} \log \frac{U}{\epsilon} \right\}, \qquad (6)$$

expressing the entropy S of an individual oscillator in terms of its mean energy U.

II. Introduction of Wien's Displacement Law.

§6. Next to Kirchoff's Law of the equality of the emission- and absorption-strength of a radiator, Wien's Displacement Law [9] – based on a special application of the Stefan-Boltzmann Law for the temperature dependence of the total radiant energy – is the most valuable relation in the firmly established theory of radiant heat. In the form given by M. Thiesen [10], it is

$$E_\lambda \cdot d\lambda = T^5 \psi(\lambda T) \cdot d\lambda,$$

where λ is the wavelength, $E_\lambda d\lambda$ the spatial density of "black"-radiation in the interval λ to $\lambda + d\lambda$, T the temperature, and $\psi(x)$ a universal function of the single argument x.

§7. Now we investigate what Wien's Displacement Law has to say about the dependence of the entropy S of our oscillators on their energy U and their frequency f, in the general case that the oscillator is in a random thermal medium. For this purpose, we first generalize Thiesen's form of the law to radiation propagating with the speed of light c and introduce the frequency variable f instead of the wavelength λ. Introducing the spatial density of energy in the frequency interval f to $f + df$ as $\mathcal{U} df$, and the substitutions $E \cdot d\lambda \to \mathcal{U} \cdot df$, $\lambda \to c/f$, and $d\lambda \to c df / f^2$, Thiesen's form gives

$$\mathcal{U} = T^5 \cdot \frac{c}{f^2} \cdot \psi \left(\frac{cT}{f} \right).$$

Now from the Kirchoff-Clausius Law, the rate of emission of energy by a black surface in a thermal medium, at temperature T and frequency f, is inversely proportional to c^2; therefore the energy density $\mathcal{U}$ is inversely proportional to c^3, and we get

$$\mathcal{U} = \frac{T^5}{f^2 c^3} \xi_1 \left(\frac{T}{f}\right),$$

where the constants in the function ξ_1 are independent of c.

* * * * * * * * * * * * * * * * * * **

Thiesen's result has the thermodynamic foundation of the Stefan-Boltzmann Law $E = \sigma T^4$ combined with the empirical result of Wien's Displacement Law that

$$E_\lambda = \frac{dE}{d(\lambda \cdot T)} = \sigma T^4 \times \psi(\lambda \cdot T),$$

where $\psi(\lambda \cdot T)$ is a universal function of the single argument $x = \lambda \cdot T$. Here Planck uses these classical results to deduce that the energy-quantum must be a linear function of the frequency $\epsilon = hf$.

* * * * * * * * * * * * * * * * * * **

Instead of this we could also define a new function $\xi_2(x) \equiv x^5 \xi_1(x)$ and write

$$\mathcal{U} = \frac{f^3}{c^3} \xi_2 \left(\frac{T}{f}\right), \qquad (7)$$

from which we conclude that the product $\mathcal{U}\lambda^3$ is a universal function of T/f for all media in thermodynamic equilibrium.

§8. Now to go from the spatial density of radiation $\mathcal{U}$, to the energy U of an oscillator with the resonant frequency f in the radiation field, we use Eqn34 of my earlier paper on irreversible radiative processes [11] to obtain the expression:

$$\mathcal{R} = \frac{f^2}{c^2} U$$

where $\mathcal{R}$ is the intensity of monochromatic, linearly polarized radiation. Together with the known equation $\mathcal{U} = 8\pi\mathcal{R}/c$, this gives

$$\mathcal{U} = \frac{8\pi f^2}{c^3} U. \qquad (8)$$

From here and from Eqn7 follows:

$$U = \frac{f}{8\pi}\xi_2\left(\frac{T}{f}\right),$$

where c no longer occurs atall. Instead of this we can also write

$$T = f\xi_3\left(\frac{U}{f}\right)$$

in terms of still another universal function ξ_3.

§**9.** Finally we introduce the entropy per oscillator, for which we set

$$\frac{1}{T} = \frac{dS}{dU}, \tag{9}$$

which gives

$$\frac{dS}{dU} = \frac{1}{f}\xi_4\left(\frac{U}{f}\right) \text{ and integrating } S = \xi\left(\frac{U}{f}\right), \tag{10}$$

that is, the entropy of the oscillators in a medium in thermodynamic equilibrium depends only on the single variable U/f, and on universal constants. This, to me, is the simplest way to express Wien's Displacement Law.

§**10.** If we apply Wien's Displacement Law in this form to Eqn5 for the entropy, then we recognize that the energy-quantum ϵ must be proportional to the frequency, that is $\epsilon = hf$, and consequently:

$$S = k\left\{\left(1 + \frac{U}{hf}\right)\log\left(1 + \frac{U}{hf}\right) - \frac{U}{hf}\log\frac{U}{hf}\right\}.$$

Here h and k are universal constants.

* * * * * * * * * * * * * * * * * **

Note that this is the first use of the symbol 'h', and the first recognition of the role of this universal constant in the definition of the quantum of energy, which concept is introduced in this paper for the first time.

* * * * * * * * * * * * * * * * * **

By substitution in Eqn9 one obtains

$$\frac{1}{T} = \frac{k}{hf}\log\left(1 + \frac{hf}{U}\right), \text{ or } U = \frac{hf}{e^{hf/kT} - 1} \tag{11}$$

and from Eqn8 the sought after energy distribution law is

$$U = \frac{8\pi h f^3}{c^3} \cdot \frac{1}{e^{hf/kT} - 1} \tag{12}$$

or also, if one uses the wavelength λ instead of the frequency f:

$$E = \frac{8\pi ch}{\lambda^5} \frac{1}{e^{ch/k\lambda T} - 1}. \tag{13}$$

The expression for the intensity and for the entropy of the radiation propagating in the thermal medium, as well as the law of increase of the total entropy for the nonstationary radiation processes, I hope to discuss in another place.

III. Numerical Values.

§**11.** The value of the two universal constants h and k can be calculated exactly with the help of measurements which have already been made. F. Kurlbaum [12] has found that the total energy $\mathcal{E}_t$ which is radiated from 1 cm^2 in 1 sec by a blackbody at the temperature t^0C satisfies:

$$\mathcal{E}_{100} - \mathcal{E}_0 = 0.0731 \frac{\text{Watts}}{\text{cm}^2} = 7.31 \times 10^5 \frac{\text{ergs}}{\text{cm}^2\text{sec}}.$$

This gives the spatial density of the total radiant energy at the absolute temperature 1^0K:

$$\frac{4 \cdot 7.31 \times 10^5}{3 \times 10^{10}(373^4 - 273^4)} = 7.061 \times 10^{-15} \frac{\text{ergs}}{\text{cm}^2\text{sec}}.$$

On the other hand, from Eqn12 the spatial density of the total radiant energy for $T = 1$:

$$\begin{aligned}
u &= \int_0^\infty U df = \frac{8\pi h}{c^3} \int_0^\infty \frac{f^3 df}{e^{hf/k} - 1} \\
&= \frac{8\pi h}{c^3} \int_0^\infty f^3 df \left(e^{-hf/k} + e^{-2hf/k} + e^{-3hf/k} + \cdots \right).
\end{aligned}$$

and by term-wise integration

$$u = \frac{8\pi h}{c^3} \cdot \left(\frac{k}{h} \right)^4 \left(1 + \frac{1}{2^4} + \frac{1}{3^4} + \cdots \right) = \frac{48\pi k^4}{c^3 h^3} \cdot 1.0823.$$

If one sets this equal to 7.061×10^{-15}, then with $c = 3 \times 10^{10}$,

$$\frac{k^4}{h^3} = 1.1682 \times 10^{15}. \tag{14}$$

§**12.** O. Lummer and E. Pringsheim [13] have measured the product $\lambda_m T$, where λ_m is the wavelength for the maximum of E at the temperature T, to be

$$\lambda_m T = 0.2940 \text{cm} \cdot \text{deg K}.$$

By differentiating Eqn12,

$$\frac{dE}{d\lambda} = 0 \Rightarrow (1 - ch/5k\lambda_m T) \cdot e^{ch/k\lambda_m T} = 1$$

and from this transcendental equation

$$\lambda_m T = \frac{ch}{4.9651 \cdot k},$$

and

$$\frac{h}{k} = \frac{4.9651 \cdot 0.294}{3 \times 10^{10}} = 4.866 \times 10^{-11}.$$

From this and Eqn14, the fundamental constants have the values

$$h = 6.55 \times 10^{-27} \text{erg} \cdot \text{sec, and } k = 1.346 \times 10^{-16} \frac{\text{erg}}{\text{deg K}}. \tag{15}$$

These are the same numbers which I gave in my earlier article [14].

(Eingegangen 7. Januar 1901.)

Footnotes and References:

1) O. Lummer u. E. Pringsheim, Verhandl. der Deutsch. Physikal. Gesellsch. **2**, 163 (1900).

2) H. Rubens und F. Kurlbaum, Sitzungsber. d. k. Akad. d. Wissensch. zu Berlin vom 25 October, 1900, p.929.

3) H. Beckmann, Inaug.-Dissertation, Tűbingen 1898. See also H. Rubens, Wied. Ann. **69**, 582 (1899).

4) M. Planck, Ann. d. Phys. **1**, 719 (1900). Here as **Paper I·2.**

5) M. Planck, l.c. p.730 ff.

6) One should note here the criticism that this prescription has received: by W. Wien (Report of the Paris Conference **2**, 40 (1900)), and by O. Lummer (l.c. **2**, 92 (1900)).

7) L. Boltzmann, Sitzungsber. d. k. Akad. d. Wissensch. zu Wien (II) **76**, 428 (1877).

8) Joh. v. Kries, *Die Principien der Wahrscheinlichkeitsrechnung*, p.36, Freiburg, 1886.

9) W. Wien, Sitzungsber. d. k. Akad. Wissensch. zu Berlin vom 9 Februar. 1893, p.55.

10) M. Thiesen, Verhandl. d. Deutsch. Phys. Gesellsch. **2**, 66 (1900).

11) M. Planck, Ann. d. Phys. **1**, 69 (1900). Here as **Paper I·1**.

12) F. Kurlbaum, Wied. Ann. **65**, 759 (1898).

13) O. Lummer u. E. Pringsheim, Verhandl. der Deutsch. Physikal. Gesellsch. **2**, 176 (1900).

14) [Note added in translation: This statement is not quite accurate. The value of h/k has been increased 1% by evaluating it from the maximum of the blackbody spectrum rather than from the exponential slope of the Wien distribution; the value used for the total intensity of the blackbody spectrum between $273K$ and $373K$ has been reduced 1% but enters to the fourth power; the Planck distribution results in an 8% increase in the spectral integral at $1K$; for a net reduction of $8 - 4 + 1 = 5\%$ in h from the previous value 6.886×10^{-27} (4% high) to 6.55×10^{-27} (1% low). At the same time, Boltzmann's constant has gone from 3.5% high to 2.5% low.]

Paper I·5: Excerpt from Annalen der Physik **4**, 564 (1901).

On the Elementary Quanta
of Matter and Electricity

von **Max Planck**

(Aus den Verhandlungen der Deutschen Physikalischen Gesellschaft, **2**, p.244. 1900, mitgeteilt vom Verfasser.)

In his fundamental paper "On the Relation Between the Second Law of Thermodynamics and the Probability Basis of the Law of Thermal Equilibrium", Boltzmann [1] found the equilibrium entropy of a monatomic gas to be the logarithm of the probability of the state. He proves the relation:

$$\int \frac{dQ}{K} = \frac{2}{3} \log \mathcal{P}.$$

Here dQ is the external heat added to the system, K is the mean kinetic energy of the atoms, and $\mathcal{P}$ is the number of "complexions" which measures the probability of the stationary velocity distribution of the atoms.

Now if M is the mass of a gram-mole of atoms, m that of an individual atom, and $< v^2 >$ the mean-square atomic velocity:

$$K = \frac{M}{m} \cdot \frac{1}{2}m < v^2 >, \text{ and further: } < v^2 > = \frac{3RT}{M},$$

where R is the gas constant (8.31×10^7 for $O = 16$), and T the temperature; so the entropy of the gas is

$$\int \frac{dQ}{T} = \frac{m}{M} R \log \mathcal{P}.$$

On the other hand, in my recently developed Electromagnetic Theory of Radiant Heat, for the entropy of a large number of linear oscillators vibrating independently in a stationary radiation field, I found the expression [2] $k \log \mathcal{R}$, where $\mathcal{R}$ is the number of possible complexions and k is 1.346×10^{-16}[erg $\times$ deg].

This connection between the entropy and the probability only has a physical meaning if it holds in general, not just for the atomic motion and the oscillator vibrations separately, but for both occurring together. If the oscillators in the gas are also radiating, then the entropy of the whole system must be proportional to the logarithm of the number of all possible complexions of particle velocity and radiation combined. Moreover, since according to the electromagnetic theory of radiation, the velocities of the atoms are completely independent of the distribution of the radiant energy, then the total number of complexions is simply equal to the product of the separate numbers for the motion and for the radiation, so that the total entropy is:

$$k' \log (\mathcal{P}\mathcal{R}) = k' \log \mathcal{P} + k' \log \mathcal{R}.$$

The first term is the entropy of the particle motion, the second that of the radiation. By comparison with the original expressions we get:

$$k' = \frac{m}{M} R = k, \text{ or } \frac{m}{M} = \frac{k}{R} = 1.62 \times 10^{-24},$$

that is, a single molecule is the 1.62×10^{-24} part of a gram-mole, or: one hydrogen atom weighs 1.64×10^{-24} gram, since $M_H = 1.01$g, or a gram-mole of each substance contains $M/m = 6.175 \times 10^{23}$ molecules. Meyer [3] calculates this number to be 6.10×10^{23}, in essential agreement. Loschmidt's Number $\mathcal{N}$, the number of gas molecules in one cc. at $0°C$ and one atmosphere pressure is:

$$\mathcal{N} = \frac{1013200}{R \cdot 273 \cdot m/M} = 2.76 \times 10^{19}.$$

Drude [4] finds $\mathcal{N} = 2.1 \times 10^{19}$.

The Boltzmann-Drude Consant α, the mean kinetic energy of an atom at the absolute temperature of $1°K$, is

$$\alpha = \frac{3}{2}\frac{m}{M}R = \frac{3}{2}k = 2.02 \times 10^{-16}.$$

Drude [4] finds $\alpha = 2.65 \times 10^{-16}$.

The elementary quantum of electricity e, i.e. the electric charge of a positive singly charged ion or electron is, with ϵ the known charge of a gram-mole of singly charged ions:

$$e = \epsilon\frac{m}{M} = 4.69 \times 10^{-10} \text{esu}.$$

Richarz [5] finds 1.29×10^{-10}, and Thomson [6] recently found 6.5×10^{-10}.

All these relations require that the theory should be correct in general, not just approximately but with absolute validity. The precision of all the calculated constants is essentially the same as that of Boltzmann's constant k, and exceeds by far all previous determinations of these quantities. Its proof by direct methods is a necessary and important task for further research.

(Eingegangen 9. Januar 1901.)

Footnotes and References:

1) L. Boltzmann, Sitzungsber. d. k. Akad. d. Wissensch. zu Wien (II) **76**, Pp.373, 428 (1877).
2) M. Planck, see the preceding paper, Eqns5,15. Here as **Paper I·4**.
3) O.E. Meyer, Die kinetische Theorie der Gase, 2. Aufl. p.337 (1899).
4) P. Drude, Ann. d. Phys. **1**, p.578 (1900).
5) F. Richarz, Wied. Ann. **52**, p.397 (1898).
6) J.J. Thomson, Phil. Mag. (5) **46**, p.528 (1898).

Paper I·6: Excerpt from Die Naturwissenschaften, **14/15**, 153 (1943).

Reflections on the Discovery of the Quantum of Action

von MAX PLANCK, Berlin.

A new epoch of physical science began with the discovery of the fundamental quantum of action. I feel the need and the obligation to record the way I arrived at the calculation of this fundamental constant, at least as it seems compressed in my memory.

I

To this end I must go back to my student years. What interested me above everything else in Physics was the great general laws which have significance for all natural processes, independent of the properties of the matter taking part in the particular process, and independent of any assumptions which one might make about its structure. This seemed to me to be particularly true of the great Laws of Thermodynamics. Whereas the First Law – the law of conservation of energy – has a very simple and easily understood meaning, and there is no reason to subject it to a particular scrutiny, the full understanding of the Second Law requires careful study. I learned this law in my last year of studies (1878, at age 20) in the lectures of R. CLAUSIUS ······ The various attacks which lead to the CLAUSIUS proof, lead also to a deep understanding of the complete significance of his hypotheses.

In the struggle to make this point completely clear to myself, I came to a formulation of the hypothesis which seemed to me more simple and more compelling. It stated: "The process of heat conduction is 'in no way' 'completely' reversible." With this statement the same result is derivable as from the statement of CLAUSIUS, but without requiring a particular mechanism. One must only interpret appropriately the words 'auf keinerlei Weise' and 'vollständig'. It must be understood, that in the attempts to make the process reversible, absolutely any auxiliary agent can be used – mechanical, thermal, electrical, chemical – but always with the restriction that at the end of the complete process the auxiliary agent itself must be returned to the same state that one assumed for its original use. Such a process, which can 'in no way' be made 'completely' reversible, I called "irreversible."

  But the error which one makes by a too narrow interpretation of the CLAU-
SIUS statement of the Second Law, and which I have sought to correct all my
life, is, it seems, still not eradicated. Even at the present time, instead of the
above definition of irreversibility, I find the following: "A process is irreversible if
it cannot run in the opposite direction." That is not correct. ······ In the deeper
sense of irreversibility, the Second Law holds not only for radiant heat, but also
for all conceivable natural processes. ······

II

  WIEN's Law for the normal energy distribution, derived already in 1896, was
in essential agreement with all existing measurements (Mai 1899). Everything
seemed to be in satisfactory order.

  But soon, first by LUMMER and PRINGSHEIM, and later by PASCHEN, a
deviation from WIEN's distribution law was noticed, which had been found when
extending it to longer wavelengths and increasing the precision of the existing
measurements. It was clear that the general validity of Eqn4 must be considered
in serious doubt. ······

  At that time a large number of prominent physicists, both on the experimental as well as on the theoretical side, turned their attention to the problem of
the energy distribution in the normal spectrum. But all looked only at the result
for the radiation intensity $\mathcal{R}_f$ as a function of temperature, whereas I questioned
the deeper consistency of the dependence of the entropy S on the energy U. Because the significance of the entropy had not yet found its correct mathematical
expression – which concerned no one until the use of my method – I could pursue
my calculations in peace and quiet without the need to fear any disturbance or
competition.

  Let us now take a deeper look into the properties of the entropy. First I calculated in complete generality, without making use of Eqn2 (Note added: equations
referred to but omitted from the abridged text are in the footnotes.), the total entropy
change when an oscillator in a stationary radiation field *absorbs* the energy dU

from the field, whose energy at the same time *increases* by a small amount ΔU. The entropy change is

$$\delta S = \frac{3}{5} \cdot \frac{d^2 S}{dU^2} \cdot \Delta U \cdot dU.$$

For any change allowed in nature, the corresponding energy changes dU and ΔU always have the opposite sign, and from the Second Law $\delta S > 0$, then it follows necessarily that

$$\frac{d^2 S}{dU^2} < 0.$$

In fact, Eqn2 for the entropy, which leads to WIEN's distribution, yields:

$$\frac{d^2 S}{dU^2} = -\frac{1}{\alpha f U}. \tag{5}$$

The remarkable simplicity of this result led me $\cdots\cdots$ again to WIEN's distribution. $\cdots$ Thus my efforts to improve Eqn2 arrived at a dead end, and I had to abandon my idea.

Then began an event which would be a decisive turning point. In the Deutschen Physikalischen Gesellschaft meeting of 19 October 1900, KURLBAUM presented his results obtained with RUBENS on energy measurements for very long wavelengths, which among other things showed that with increasing temperature, the radiation intensity of black bodies was always approximately proportional to the temperature T, in complete contradiction to WIEN's distribution Eqn4, for which the radiation intensity must always remain finite. Since this result had already been known to me for a few days before the meeting from talking with the authors, I had time to think about its implications for my way of calculating the entropy of the corresponding oscillators.

If for high temperature T the radiation intensity $\mathcal{R}_f$ is proportional to the temperature, then from Eqn1 the energy of the oscillators is also proportional to it,

$$U = C \cdot T, \text{ and from Eqn3 by integration: } S = C \cdot \log U.$$

Consequently

$$\frac{d^2 S}{dU^2} = -\frac{C}{U^2}. \tag{6}$$

This result should hold for large values of U, in contrast to Eqn5 which is valid for small values of U. If one seeks now a generalized relation which contains both Eqns5 and 6 as limiting cases, then the simplest suggestion would be:

$$\frac{d^2 S}{dU^2} = -\frac{1}{\alpha f U + U^2/C}$$

and by integration:

$$\frac{dS}{dU} = \frac{1}{T} = \frac{1}{\alpha f} \cdot \log\left(1 + \frac{\alpha' f}{U}\right), \tag{7}$$

where $\alpha C = \alpha'$ are constants.

When one replaces U by $\mathcal{R}_f$ using Eqn1, this is the formula for the energy distribution – expressed in wavelengths – which I proposed and tried to justify in the course of the lively discussion following KURLBAUM's lecture.

Next morning RUBENS sought me out and told me that in the Proceedings of the meeting he would include a comparison of my formula and his actual data which he had found to be in complete satisfactory agreement. Also LUMMER and PRINGSHEIM, who were the first to claim deviations, soon withdrew their objections since, as PRINGSHEIM told me, it turned out the deviations found by them were the result of a computational error. Through later measurements the formula of Eqn7 was verified repeatedly as so accurate, that it stimulated refinements of the experimental methods.

III

There remains the question of how definitively accurate the spectral energy distribution law of the Blackbody Radiation really is. However now we return to the theoretically most important problem: to give a proper foundation for this law and – what is an even greater problem – to give a theoretical derivation of the expression for the entropy of an oscillator as in Eqn7. It can be written in the following form:

$$S = \frac{\alpha'}{\alpha}\left\{\left(\frac{U}{\alpha' f} + 1\right)\log\left(\frac{U}{\alpha' f} + 1\right) - \frac{U}{\alpha' f}\log\frac{U}{\alpha' f}\right\}. \tag{8}$$

In order to give this expression a physical interpretation, it was necessary to give a completely new meaning to the concept of entropy, as applied in the domain of electrodynamics.

Among all the physicists of that time, LUDWIG BOLTZMANN was the one who had the deepest understanding of entropy. He explained that *the entropy of a physical system in a particular state is a measure of the probability of that state* and he summed up the content of the Second Law in the statement: the system during any change occurring in nature goes over into a more probable state. In fact, it was his great achievement in the kinetic theory of gases, to define a state function H which has the property that for each change of state occurring in nature, it decreases its value; it can therefore be recognized as the negative of the so-called entropy. It is possible to justify, from the proof of this famous theorem, the validity of the fundamental hypothesis that the state of the gas is "molecular disorder." [Note added: This problem is still not resolved to everyone's satisfaction. We appeal to Gibbs ensembles, see Ref.10 in our introductory remarks.]

I myself had not considered before then the connection between entropy and probability; it had no interest for me, since every probability calculation always involved assumptions, and I firmly believed that the validity of the Second Law was free of assumptions. That my proof of irreversibility which considered radiation processes only under the postulate of "natural radiation" could succeed, and that a similar fundamental hypothesis about the radiation is just as necessary and plays exactly the same role there as that of molecular disorder in the theory of gases, occurred to me at once with perfect clarity.

Since there now seemed to me no other way out, I tried BOLTZMANN's method and with complete generality for a given state of a given system set:

$$S = k \cdot \log W, \tag{9}$$

where W is the probability of the particular state.

Since these relations should have truly general validity, and since the entropy is an additive quantity then the probability must be a multiplicative quantity and the constant k must be a universal number, depending only on the units. It is usually referred to as BOLTZMANN's constant but it is fair to remark that

BOLTZMANN neither introduced it, nor to my knowledge ever even thought to ask about its numerical value. At that time it had to be multiplied by the number of effective molecules – a problem he left entirely to his colleague LOSCHMIDT. He himself always kept in sight the possibility of its calculation, but it represented only one objective of the kinetic theory of gases, and it was enough for him to stop at the molar level.

In order to apply Eqn9 to the present situation, I imagined a system consisting of a very large number N of completely similar oscillators, and set out to calculate the probability that the system should have a given energy U_N. Since a probability-like quantity can only be found by counting, then it was necessary for the energy U_N to be expressible as a sum of discrete identical energy elements ϵ, whose number likewise could be designated by a very large number P.

Therefore

$$U_N = N \cdot U = P \cdot \epsilon, \tag{10}$$

where U is the average energy of an oscillator.

A measure of the required probability W is the number of different ways in which P energy quanta can be partitioned among N oscillators: that is

$$W = \frac{(P+N)!}{P!N!}. \tag{11}$$

According to Eqn9 the entropy of the oscillator system is:

$$S_N = N \cdot S = k \cdot \log \frac{(P+N)!}{P!N!}$$

and with STIRLING's approximation:

$$S = k \cdot \left\{ \left(\frac{P}{N} + 1 \right) \log \left(\frac{P}{N} + 1 \right) - \frac{P}{N} \log \frac{P}{N} \right\}. \tag{12}$$

The similarity between Eqns8 and 12 is obvious. It only remains to make the identifications:

$$k = \frac{\alpha'}{\alpha} \quad \text{and} \quad \frac{P}{N} = \frac{U}{\alpha' f}.$$

Then it follows from Eqn10 that the magnitude of the energy element is $\epsilon = \alpha' f$. The constant α' – which I denoted as h and called the elementary quantum of action or the action quantum, in contrast to the energy quantum hf. From the measured values of the constants α and α' the radiation law of Eqn7 gives the value of k and h:

$$k = 1.346 \times 10^{-16} \text{ ergs/degK}, \quad \text{and} \quad h = 6.55 \times 10^{-27} \text{ erg sec.}$$

What was of next concern was the experimental proof of this theory, which at that time was only possible in a very limited way. Before this the only constant thought of was k, for whose numerical value only the order of magnitude was known in some degree. $\cdots\cdots$

Now however there arose the theoretically all-important problem, to assign this remarkable constant a physical meaning. Its introduction constituted a break with the classical theory, which to many was too radical, as I had initially anticipated. So also was the way in which the entropy as a measure of the probability in the sense of BOLTZMANN was definitively determined even for the radiation. $\cdots\cdots$ But the nature of the energy quantum remained unclear.

For many years I continued to do further research, trying somehow to fit the action quantum into the system of classical physics. But it seems to me that this is not possible. In any case, as is well known, there has been the development of the Quantum physics by the younger forces, of whom I mention here only the names EINSTEIN, BOHR, BORN, JORDAN, HEISENBERG, DE BROGLIE, SCHRÖDINGER, DIRAC, $\cdots\cdots$

Footnotes and References:
1) Eqn1: $U = \frac{c^2}{f^2} \cdot \mathcal{R}_f$.
2) Eqn2: $S = -\frac{U}{\alpha f} \cdot \log \frac{U}{\alpha f}$.
3) Eqn3: $\frac{dS}{dU} = \frac{1}{T}$.
4) Eqn4: $\mathcal{R}_f = \frac{bf^3}{c^2} \cdot e^{-\alpha f/T}$.
$\cdots\cdots$

Chapter II

GOD said: Let there be light.

EINSTEIN said: There is light.

and COMPTON saw the light.

EINSTEIN: "⋯ von einem Punkte ausgehenden Lichtstrahles die energie nicht Kontinuierlich auf grösser und grösser weidende Raume verteilt, sondern es besteht dieselbe aus einer endlichen Zahl von in Raumspunkten lokalisierten Energiequanten, welche sich bewegen, ohne sich zu teilen und nur als Ganze absorbiert und erzeugt werden können." (Translation in italics in paragraph 2 of **Paper II·1**.)

COMPTON: "This remarkable *agreement between experiment and theory* indicates clearly that scattering is a quantum phenomenon ⋯ also that a radiation quantum carries ⋯ momentum as well as energy ⋯." (Paragraph 1 of **Paper II·4**.)

§II-1. Introduction.

Einstein at last in 1905 revived Planck's idea of the energy quantum which had lain dormant and virtually unmentioned since it was first presented more than four years earlier in December 1900.

Einstein recognized that all the experimental evidence supporting the wave theory of light and Maxwell's electromagnetic theory, involved only the time-average phenomena of propagation, interference, and diffraction; but that a whole class of phenomena pointed to the opposite conclusion, that light or electromagnetic energy did not spread out infinitely, as it would in a wave theory. Experiments – including the Planck blackbody spectrum and photo-electron emission and absorption – which involve the radiation and transformation of light, are better understood, Einstein maintained, by supposing that the light energy is encapsulated in Planck's indivisible *quanta* which behave in many respects like material particles. Einstein advocated the hypothesis that *"the energy of a light ray ⋯ consists of ⋯ localized energy quanta ⋯ moves without dividing ⋯ absorbed and emitted as a whole."*

He explored and defended this assumption in a series of brief essays [1] making up his 1905 Annalen der Physik paper "A Heuristic Interpretation of the Radiation and Transformation of Light" (included here as **Paper II·1**).

Einstein showed first of all, that the classical Rayleigh-Jeans-Maxwell description of the blackbody spectrum (valid, as we know from Planck, for low frequency, long wavelength radiation with $hf << kT$) was doomed to fail because the equilibrium spectral energy density $\rho_f \sim f^2 kT$ integrated to an infinite radiant energy. So the Rayleigh-Jeans spectrum could not be correct in any case.

Einstein then examined the opposite extreme, the Wien limit $E \geq kT$, where the energy of the oscillators (and the radiation) obey an exponential Boltzmann distribution. He was there able to show that the entropy of the radiation had the same volume dependence as that of a low density molecular gas *provided* the energy was $E = n \times hf$, with n playing the role of molecular number. He concluded that "$\cdots$ *monochromatic radiation acts like a gas of discrete quanta of energy hf*".

And in absorption, he points out – as we now believe at his instruction - that light is absorbed in gulps of one quantum and that the quantum cannot be dispersed infinitely as we would suppose in a wave theory. " $\cdots$ *there is no lower limit for the intensity of the incident light below which it would be incapable of causing emission.*" Einstein concludes with his photoelectric formula – cited in his 1921 Nobel Prize award – in which he assumes that each electron emitted has completely converted light energy hf into kinetic energy which is measured by a maximum cutoff voltage $eV \leq hf$.

In spite of the stimulus of Einstein's 1905 paper on the dynamic nature of the light quantum, progress was slow. A flavor of the attitudes and difficulties prevailing at the time are contained in **Paper II·2** [2], a post-lecture discussion in 1909 pitting Planck against Einstein, with Stark in the middle. They all gave hints of what was to come: Planck of the still to be discovered quantum description of the quantum harmonic oscillator; Stark trying to mollify Planck's opposition to free light quanta, but giving – between the lines – a preview of Copenhagen practicality still to come; and Einstein's ideas of coherence and wave-particle duality are almost articulated here.

In **Paper II·3**, Einstein makes explicit the role of spontaneous emission and detailed balance in his elegant and compelling derivation of the Planck distribution [3].

Compton finally, between 1921 and 1925, in his Nobel Prize winning work [4], did his great historic experiment of scattering X-rays by electrons, which defined precisely and beyond any reasonable doubt the role of the quantum nature of light in the fundamental emission and absorption processes responsible for the scattering. Einstein had already in 1917 (here in **Paper II·3A**) described in qualitative detail the kinematics of energy and momentum conservation in the fundamental elementary processes of absorption and emission of directed light quanta during molecular transitions. It seems that these remarks were either ignored or not fully understood in relation to the classical description of emission in terms of spherical waves. In either case, Compton made no mention of Einstein's prior description of the kinematical principles that were the basis of his great accomplishment.

Compton's achievement cannot be over-stated as an example of brilliant theoretical insight combined with ground breaking experimental innovation. This was a time when many people, even – or rather, especially – including Niels Bohr, argued whether the absorption and emission of quanta conserved energy or even involved momentum, and also debated what connection the angular dependence of Maxwellian radiation of waves had with the direction of emission of quanta. But Compton was able to see clearly that the scattering of X-rays by electrons should be described as the complete absorption of the incoming light quantum by a single free electron, which then emits the complete outgoing quantum.

It all seems so trivial to us now – that the massless photon is incident along the beam direction with energy hf and momentum directed along the beam hf/c and wavelength $\lambda = c/f$; where it is absorbed by an electron of mass m at rest; which then radiates a photon in a direction θ with energy hf_θ and momentum directed along θ of magnitude $hf_\theta/c = h/\lambda_\theta$; the electron recoils with a speed $v = \beta c$ and a direction ϕ determined by energy and momentum conservation.

It is obvious that the emitted photon is shifted to lower energy and longer wavelength, because of the energy lost to the recoiling electron; clearly the scattered photons are predominantly in the forward direction, but Doppler shifted by their recoiling sources.

Don't worry about more than one electron acting at a time. Don't worry about the initial velocity distribution of the electrons; don't worry about the binding or rescattering of the recoiling electron; don't worry about all the possible directions of the outgoing photon, just one at a time. Don't worry about the details of the absorption and emission process. Don't try to calculate model dependent things.

Obviously it would be nicer to do the kinematics completely relativistically from the start, even though the initial electron is at rest; fortunately it is easier to do so anyway.

Who among us would have had the intellect and the courage to break with every single misperception that paralyzed everyone else. And once it was done it was all so obvious.

All this just to get Compton's master equation for the wavelength of the scattered quantum:

$$\lambda_\theta - \lambda_0 = \frac{2h}{mc} \sin^2 \frac{1}{2}\theta.$$

Now the real work begins, because Compton does not – as most before him did and as hundreds after him would do – simply send off a report of his theoretical result. Instead he <u>does</u> the experiment, and in order to do it he invents elementary particle physics. Not from nothing of course. Rutherford and Geiger and many others had done something similar which we prefer to characterize as nuclear physics, but the step taken here – Compton scattering of photons from electrons with measurement of energy-angle correlations, and eventually with recoil correlations – is distinctly different and more appropriately recognized as the realm of elementary particles.

The first experiments [4] (included here as **Paper II·4**) measure the increase in the wavelength of the Mo Kα X-ray scattered at 90° from a graphite target to be $.730A - .708A = .022A$ in close agreement with the result $.0484 \sin^2 45^o = .024A$ predicted from the kinematics of quantum scattering from a free electron. Further experiments [5] (**Paper II·5**) using tungsten K X-rays of wavelength $\sim .2A$ and γ-rays of wavelength $\sim .02A$ produced similar results. The experimental uncertainties of these numbers are not mentioned, but some idea can be inferred from the original data reproduced in Compton's figures.

Finally, Compton was able to detect the recoil electron with the scattered photon (**Paper II·6**) both in a two-counter coincidence experiment and in cloud chamber photographs [6]. The coincidence experiment began with the experimenter listening for coincident clicks on earphones, but progressed to data recorded on a paper tape. These experiments confirm the electron recoil angle ϕ occurring when the light quantum of wavelength λ is scattered through angle θ, to be that required by the quantum kinematics:

$$\tan \phi = \left(1 + \frac{h}{mc\lambda}\right) \cot \frac{1}{2}\theta.$$

The magnitude of the Compton revelation can best be judged by comparing the work of Bohr, Kramers and Slater [7] (presented in **Paper IV·1**) where – in their struggle to define the quantum *dynamics* of radiation processes – they are still hindered by confusion over the quantum *kinematics*.

Footnotes and References:

1) A. Einstein, Ann. d. Phys. **17**, 132 (1905) (included here as **Paper II·1**).

2) A. Einstein, Phys. Zeits. **10**, 817 (1909) (here as **Paper II·2**).

3) A. Einstein, Deutsche Phys. Ges. Verhand. **18**, 318 (1916) (here as **Paper II·3**); Physikalische Zeitschrift **18**, 121 (1917) (here as **Paper II·3A**).

4) Arthur H. Compton, Phys. Rev. **21**, 483 (1923) (here as **Paper II·4**).

5) Arthur H. Compton, Phys. Rev. **22**, 409 (1923) (here as **Paper II·5**).

6) Arthur H. Compton, Proc. N.A.S. **11**, 303 (1925) (here as **Paper II·6**).

7) N. Bohr, H.A. Kramers and J.C. Slater, Zeits. f. Phys. **24**, 69 (1924) (here as **Paper IV·1**.)

Biographical Notes.

Albert Einstein (1879-1955) – Einstein needs no introduction to anyone, but there are some interesting points in the present very limited context. Einstein's very first paper – "Implications of Capillarity", Ann. d. Phys. **4**, 513 (1901) – published when he was 21, appeared in the same issue of Annalen der Physik as Planck's 1901 announcement of his discovery of the quantum. Einstein was quite active in the intervening years and his 1905 paper reviving Planck's quantum (included here as **Paper II·1**) was his sixth publication. In his 1921 Nobel Prize award citation, the 1905 work was the one singled out for special mention and even then there was no specific mention of the quantum but only the rather general remarks " for services to theoretical physics and especially for his discovery of the law of the photoelectric effect." The year 1905 was THE year of Einstein's illustrious career in which he published [a] the photoelectric paper (18 March), his thesis on molecular dimensions (30 April), on Brownian motion (11 May), special relativity (30 June), the velocity dependence of the mass (27 Sept), and again on Brownian motion (19 Dec). And of course there was much more to come.

a) *'The Collected Papers of Albert Einstein'* Vol.2, John Stachel, Editor (Princeton University Press 1989) Pp.ix-xii.

Arthur H. Compton (1892-1962)[a] – educated at the College of Wooster, with a PhD from Princeton (1916); subsequent appointments at Minnesota, Cambridge, Oxford, Washington University, and the University of Chicago. He was the Wayman Crowe Professor of Physics and Chairman of the Physics Department at Washington University (1921-23); Professor of Physics (1923-), Chairman of the Physics Department and Dean of Natural Sciences (1940-) and Director of the Manhattan District's Plutonium Research Project (1942-) at the University of Chicago (1923-45); Chancellor (1945-53), Distinguished Professor of Natural Philosophy (1954-61), and Professor-at-large (1961-62) at Washington University. His wartime research experiences are described in his book *Atomic Quest, a Personal Narrative (1956).*

Compton won the 1927 Nobel Prize in Physics (shared with C.T.R. Wilson) for his experiments on the elastic scattering of photons by electrons. In earlier work, he was the first to find the diffraction patterns of X-rays at grazing incidence on ruled gratings. This permitted successively: 1) standardization of X-ray wavelengths; 2) determination of lattice constants using Bragg diffraction; and 3) a determination of Avogadro's number and the electron charge.

Compton anticipated by five years the spectroscopic discovery by Goudsmit and Uhlenbeck of the electron spin. Compton concluded that the electron must be "spinning like a tiny gyroscope" and have "an angular momentum $\hbar$" $\cdots$ "offering an explanation for the large value of e/m observed" $\cdots$ "about one-third the magnetic moment of the iron atom" in order to generate the magnetic properties of matter [b].

Compton also anticipated the relation between momentum and wavelength

$p = h/\lambda$ usually credited solely to de Broglie [c].

He was President of the American Physical Society, was awarded some twenty medals and prizes, and twenty-five honorary degrees.

a) PHYSICS TODAY, May 62, p.88; also D.L. Livesay, *Atomic and Nuclear Physics* (Blaisdell, Waltham, 1966), Pp.34,95,107.

b) Arthur H. Compton, Journal of the Franklin Institute **192**-2, 145 (1921); this work is not mentioned in the usually acknowledged discovery work of G.E. Uhlenbeck and S. Goudsmit, Die Naturwissenschaften **47**, 953 (1925).

c) A.H. Compton, Phys. Rev. **23**, 118 (1924). According to Martins and Brown [d], the result was presented at the Chicago meeting of the American Physical Society on 1 December 1923 almost one year *before* the presentation of de Broglie's thesis (included here as **Paper IV·1**, L. de Broglie, Phil. Mag. **47**, 446 (1924)). Only the abstract of Compton's report was published, and it is reproduced below (following [d]):

* * * * * * * * * * * * * * * * * **

"*A quantum theory of uniform rectilinear motion.* Arthur H. Compton, University of Chicago. – For uniform rectilinear motion, the quantum postulate $\int p\,dq = nh$ states that the momentum of the system is $p = nh/q_1$, where $q_1 = \int dq$ is the displacement required to bring the system back to its initial condition. The fact that a thing in uniform rectilinear motion repeats its initial condition at regular space intervals makes it in the general sense a train of waves, for which $\lambda = q_1$. Using Bohr's correspondence principle, each value of n is identified with the order of a harmonic component of the wave. For the nth harmonic, $\lambda_n = q_1/n$. Thus in general, for sine waves $p = h/\lambda$. But the momentum of a wave train of energy E and velocity v is $p = E/v$. Thus $E = hv/\lambda = hf$. The application of these equations to electromagnetic radiation is confirmed by the change of wavelength of X-rays when scattered by the photo-electric effect. Thus also on the quantum theory, radiation consists of trains of waves. Considering a moving diffraction grating of grating space D as a train of waves, its momentum is similarly nD/h, which is the basic hypothesis of Duane's quantum theory of diffraction."

* * * * * * * * * * * * * * * * * **

d) H. R. Brown and Roberto de A. Martins, Am. J. Phys. **52**(12), 1130 (1984). These authors follow de Broglie waves through their many transmogrifications, due to de Broglie himself and to his many successors.

Paper II·1: Excerpt from Annalen der Physik **17**, 132 (1905).

A Heuristic Interpretation of the Radiation and Transformation of Light

von **A. Einstein.**

The theoretical notions which any physicist has about a gas or a massive body, and about Maxwell's Theory of the electromagnetic processes in so-called empty space, differ in a fundamental formal way. We describe the state of a body by the perfectly definite position and velocity of a very large but finite number of atoms and electrons, whereas for the specification of the electromagnetic state in a volume we make use of continuous spatial functions, so a finite number of quantities is not sufficient for the complete specification of the electromagnetic state. From Maxwell's Theory for purely electromagnetic phenomena, and therefore also for light, the energy is understood as a continuous function of space, whereas the energy of a massive body – according to our present understanding – is represented by a sum over the atoms and electrons. The energy of a massive body cannot divide into many arbitrarily small parts. In contrast, for the light from a point-like source sending out light rays according to Maxwell's Theory (or in general according to any wave theory), the energy of the light can spread out continuously over an ever increasing volume.

The wave theory of light based on continuous space functions has been thoroughly tested for purely optical phenomena and cannot be replaced by any other theory. It should still be kept in mind that the optical observations depend on time average values but not on instantaneous values, and in spite of the successful confirmation of the theory by experiments involving diffraction, reflection, refraction, dispersion, etc., it is conceivable that the present wave theory of light might lead to contradictions with the facts when applied to radiation and transformation of light. It seems to me in fact, that the observations of "blackbody" radiation, photoluminescence, the production of cathode rays by ultraviolet light, and other phenomena involving radiation and transformation of light appear to be better understood with the assumption that the energy of the light is distributed discontinuously in space. According to the assumption made here, *the energy of a light ray outgoing from a point is not continuously spread over greater and greater volumes, but it consists of a finite number of energy quanta localized at space points, each of which moves without dividing and can only be absorbed and emitted as a whole.* (Note: Italics added.)

* * * * * * * * * * * * * * * * * * **

Planck's energy 'element' is reborn here as Einstein's dynamical light quantum – yet to be named the photon. The era of quantum dynamics is set in motion in Einstein's critical insights epitomized here.

* * * * * * * * * * * * * * * * * * **

In the following I will describe the train of thought and the motivating facts which have led me to this conclusion, in the hope that the underlying point of view might be found useful by others.

§1. On a Difficulty in the Theory of Blackbody Radiation.

We first take the point of view of Maxwell's Theory and electron theory and consider the following case. A space enclosed by perfectly reflecting walls contains a number of gas molecules and electrons, which are free to move and which exert conservative forces when they come very close to one another, i.e. as close as molecules can according to the kinetic theory of gases [1]. A number of "oscillator"-electrons are imagined to be linked to far distant points of the space by elastic forces directed to these points and proportional to the distance. In addition, these oscillator-electrons (hereafter referred to as oscillators) are presumed to have a conservative interaction with the free molecules and electrons, when they come very close to each other. The oscillators emit and absorb electromagnetic radiation of a definite frequency.

According to the present view of the nature of light, the radiation must be distributed in space in a way determined by Maxwell's equations for the case of dynamical equilibrium, which is identical to the blackbody radiation – at least when oscillators of all frequencies are assumed to exist.

We disregard for the moment the emission and absorption of radiation by the oscillators and ask how the interaction (the collisions) of the molecules and electrons corresponds to dynamical equilibrium. The kinetic theory of gases states that the average kinetic energy of an oscillator must be equal to the average kinetic energy of the translational motion of the gas molecules. If we resolve the motion of the oscillators into three perpendicular oscillations, then we find for the mean value $\bar{E}$ of the energy of each such linear oscillation

$$\bar{E} = \frac{R}{N}T,$$

where R is the gas constant, N the number of "effective molecules" in a gram-mole, and T is the absolute temperature. The energy $\bar{E}$ is the same as the time-average of the kinetic and potential energy of the oscillators and 2/3 times the average kinetic energy of a free monatomic gas molecule. If for some reason – in our case, radiation exchange – the energy of an oscillator should be greater or smaller than the time average $\bar{E}$, then the collisions with the free electrons and molecules would lead to a zero average difference between the energy emitted to the gas or conversely, absorbed from the gas. It is only possible to have dynamical equilibrium when each oscillator has the mean energy $\bar{E}$.

We get a similar situation for the interaction of the oscillators and the radiation in the surrounding space. For this case, Planck has derived [2] the condition for dynamical equilibrium from the assumption that the radiation can be considered as a random process [3]. He found [Note added: Eqn7 of Ref(4) with Planck's $U \to \bar{E}_f$ and $\mathcal{U} \to \rho_f$.]:

$$\bar{E}_f = \frac{c^3}{8\pi f^2}\rho_f,$$

where $\bar{E}_f$ is the average energy of an oscillator of frequency f (per vibration component), c is the speed of light, f the frequency, and $\rho_f df$ the energy per unit volume of the radiation with frequency between f and $f + df$.

If the radiant energy at the frequency f is constant on average, neither decreasing nor increasing, then we have

$$\frac{R}{N}T = \bar{E} = \bar{E}_f = \frac{c^3}{8\pi f^2}\rho_f, \text{ or } \rho_f = \frac{R}{N}\frac{8\pi f^2}{c^3}T.$$

This condition for dynamical equilibrium not only disagrees with experiment but also states that in our description, a specific energy division between space and matter cannot be made. Moreover, the greater the frequency of the oscillators, the greater the radiant energy in the space, and we get in the limit

$$\int_0^\infty \rho_f df = \frac{R}{N}\frac{8\pi}{c^3}T\int_0^\infty f^2 df = \infty.$$

* * * * * * * * * * * * * * * * * * **

Einstein concludes that the Rayleigh-Jeans-Maxwell classical limit on the radiation is incompatible with thermodynamic equilibrium and must be incorrect even without any further knowledge of the blackbody spectrum. This is a point not fully emphasized by Planck.

* * * * * * * * * * * * * * * * * * **

§2. On Planck's Definition of the Elementary Quantum.

In the following we wish to show that the definition of the elementary quantum given by Planck, is in a sense independent of his derivation of the blackbody spectrum.

* * * * * * * * * * * * * * * * * * **

This point was still the subject of debate between Planck and Einstein in discussions following a 1909 lecture by Einstein (see **Paper II·2** included next).

* * * * * * * * * * * * * * * * * * **

The apparently correct Planck formula for ρ_f [4] is

$$\rho_f = \frac{\alpha f^3}{e^{\beta f/T} - 1}, \quad \text{where } \alpha = 6.10 \times 10^{-57}, \quad \text{and} \quad \beta = 4.866 \times 10^{-11}.$$

For large values of T/f, that is for long wavelengths and large radiation density (high temperature), this formula goes into:

$$\rho_f = \frac{\alpha}{\beta}f^2 T,$$

which agrees with that derived from Maxwell's equations and electron theory in §1. By comparison of the coefficients of the two formulas one obtains:

$$\frac{R}{N}\frac{8\pi}{c^3} = \frac{\alpha}{\beta} \text{ or } N = \frac{\beta}{\alpha}\frac{8\pi R}{c^3} = 6.17 \times 10^{23};$$

That is, an atom of hydrogen weighs $1/N$ gram $= 1.62 \times 10^{-24}$ grams. This is exactly the value found by Planck, and agrees well with the results of other methods.

We arrive then at the conclusion: the greater the energy density (temperature) and the wavelength of the radiation, the more useful the pre-existing theoretical foundation; for radiation of small wavelength and small density (low temperature) however, that theory completely fails.

* * * * * * * * * * * * * * * * * * **

In summary, Einstein concludes that the Rayleigh-Jeans-Maxwell classical limit holds for

$$hf \equiv \frac{hc}{\lambda} << kT;$$

but for the opposite limit we must abandon the classical theory.

* * * * * * * * * * * * * * * * * * **

§3. On the Entropy of Radiation.

The following discussion is taken from a famous paper by Wien and here there is room only for the conclusions.

Consider radiation in a volume V. We assume that the measurable properties of the radiation are completely specified when the radiation density $\rho(f)$ is given for all frequencies [5]. The radiations of different frequencies are assumed to be separable from one another without doing work and without addition of heat, so the entropy can be expressed in the form

$$S = V \int_0^\infty \phi(\rho, f)df,$$

where ϕ is a function of the variables ρ and f. ϕ can be reduced to a function of only one variable by making use of the fact that the entropy is not changed by adiabatic compression of the radiation between reflecting walls. However, we will not dwell on this point, but instead investigate what can be deduced about the function from the blackbody radiation law.

For blackbody radiation, ρ is a function of f such that the entropy for a given energy is a maximum, that is:

$$\delta \int_0^\infty \phi(\rho, f) df = 0, \text{ when } \delta \int_0^\infty \rho df = 0.$$

From this it follows that for any choice of $\delta\rho$ as a function of f

$$\int_0^\infty \left(\frac{\partial \phi}{\partial \rho} - \lambda\right) \delta\rho df = 0,$$

where λ is independent of f. For blackbody radiation then, $\partial\phi/\partial\rho$ is also independent of f.

For a temperature increase dT, the entropy of a unit volume of blackbody radiation increases by

$$dS = \int_{f=0}^{f=\infty} \frac{\partial \phi}{\partial \rho} d\rho df,$$

or, since $\partial\phi/\partial\rho$ is independent of f:

$$dS = \frac{d\phi}{d\rho} dE.$$

Since dE is the heat introduced and the process is reversible, then:

$$dS = \frac{1}{T} dE; \text{ so we get } \frac{d\phi}{d\rho} = \frac{1}{T}.$$

This is the law of blackbody radiation. By integration, we find that ϕ vanishes for $\rho = 0$.

§4. Limiting Relation for the Entropy of Monochromatic Radiation at Low Energy Density.

From the subsequent observations on blackbody radiation, it appears that the original relation proposed by Wien for the blackbody spectrum

$$\rho = \alpha f^3 e^{-\beta f/T}$$

is not exactly correct. At the same time however, for large values of f/T, it is in very good agreement with experiment. We will use this formula as the basis for our arguments, keeping in mind that our results only hold within such limits.

From this formula we get

$$\frac{1}{T} = -\frac{1}{\beta f} \log\left(\frac{\rho}{\alpha f^3}\right),$$

and using the earlier results we find:

$$\phi(\rho, f) = -\frac{\rho}{\beta f}\left\{\log\left(\frac{\rho}{\alpha f^3}\right) - 1\right\}.$$

Consider now a volume V of radiation with frequency between f and $f + df$ and energy $E = V\rho df$. The entropy of this radiation is:

$$S = V\phi(\rho, f)df = -\frac{E}{\beta f}\left\{\log\left(\frac{E}{V\alpha f^3 df}\right) - 1\right\}.$$

If we consider only the dependence of the entropy on the volume occupied by the radiation, and call the entropy S_0 when the volume is V_0, then we get:

$$S - S_0 = \frac{E}{\beta f} \log\left(\frac{V}{V_0}\right).$$

This equation shows that the entropy of monochromatic radiation of sufficiently small density depends on the volume in the same way as the entropy of an ideal gas or of a dilute solution. This result will be interpreted in the following on the basis of the Boltzmann Principle, that the entropy of a system is a function of the probability of its state.

* * * * * * * * * * * * * * * * * * * **

Einstein's next step will be to compare this entropy of the electromagnetic field – in the Wien regime where

$$hf \gg kT \quad \text{and} \quad \rho_f \sim e^{-hf/kT}$$

– to that for a classical molecular gas.

* * * * * * * * * * * * * * * * * * * **

§5. Molecular Theoretic Investigation of the Dependence
of the Entropy of Gases and of Dilute Solutions
on Their Volumes.

If one speaks of the probability of the state of a system, and understands entropy increase as a transition to a more probable state, then the entropy S_1 of a system must be a function of the probability W_1 of its instantaneous states. Let there be two systems S_1 and S_2, so that one sets

$$S_1 = \phi_1(W_1) \quad \text{and} \quad S_2 = \phi_2(W_2).$$

If one considers the two systems as a single system of entropy S and probability W, then

$$S = S_1 + S_2 = \phi(W) \text{ and } W = W_1 \cdot W_2.$$

The last equation says that the states are independent of one another. It follows that

$$\phi(W) = \phi(W_1 \cdot W_2) = \phi_1(W_1) + \phi_2(W_2)$$

and finally

$$
\begin{aligned}
\phi_1(W_1) &= C \log(W_1) + \text{const.,} \\
\phi_2(W_2) &= C \log(W_2) + \text{const.,} \\
\phi(W) &= C \log(W) + \text{const.}
\end{aligned}
$$

The quantity C is therefore a universal constant; its value follows from the kinetic theory of gases as R/N, where the constants R and N are defined above. If S_0 is the entropy of a particular state and W is the relative probability of a state of entropy S, then we have in general

$$S - S_0 = \frac{R}{N} \log(W).$$

Next consider the following special case: In a volume V_0 there are n moving point particles (for example, molecules). In addition to these there can be arbitrarily many particles of some other kinds. No assumptions need be made about the laws governing the motion of these particles, except that there should be no preferred sub-volumes and no preferred directions. Furthermore the number of particles should be so small that any interactions with one another can be ignored.

The system – which could be, for example, an ideal gas or a dilute solution – corresponds to an entropy S_0. We consider a sub-volume V of the whole V_0 and suppose all the particles are in the volume V, without the system being otherwise changed. This state corresponds to another value of the entropy (S), and we will now calculate the entropy difference with the help of Boltzmann's Principle.

We ask: How great is the probability of the latter state relative to the original? Or: How great is the probability, that at a randomly chosen moment all n particles moving randomly and independently of one another in a volume V_0 should be found in the sub-volume V? For this probability we get the obvious value:

$$W = \left(\frac{V}{V_0}\right)^n;$$

and by application of the Boltzmann Principle:

$$S - S_0 = \frac{R}{N} n \log\left(\frac{V}{V_0}\right).$$

Remarkably, the derivation of this equation – from which the Boyle/Gay-Lussac Law and the similar Law of Osmotic Pressure can be derived thermodynamically [6] – required no assumptions about the laws by which the molecules move.

* * * * * * * * * * * * * * * * * **
Einstein is now ready to identify the Boltzmann entropy of the molecular gas with the Wien entropy of the electromagnetic field which requires the Planck energy quantum $E = n \times hf$.
* * * * * * * * * * * * * * * * **

§6. Interpretation of the Dependence of the Entropy of Monochromatic Radiation on the Volume.

In §4 for the entropy of monochromatic radiation, we found the expression:

$$S - S_0 = \frac{E}{\beta f} \log\left(\frac{V}{V_0}\right).$$

If one writes this equation in the form:

$$S - S_0 = k \log \left[\left(\frac{V}{V_0} \right)^{E/k\beta f} \right]$$

and compares it with the general result from Boltzmann's Principle

$$S - S_0 = k \log W,$$

with $k = R/N$, then one arrives at the following conclusion: If monochromatic radiation of frequency f and energy E is confined (by reflecting walls) in the volume V_0, then the probability that at an arbitrary moment the whole of the radiation energy will be found in the sub-volume V of the volume V_0 is:

$$W = \left(\frac{V}{V_0} \right)^{E/k\beta f}.$$

From this we conclude that: Monochromatic radiation of low density (in the validity range of Wien's Radiation Law) behaves in a heat theoretic sense, as if it consisted of independent energy quanta of energy $k\beta f$.

Next we compare the average energy per quantum in blackbody radiation with the average kinetic energy of the center of mass motion of a molecule at the same temperature, which is $\frac{3}{2}(R/N)T$. The average energy per quantum assuming Wien's Law is:

$$\frac{\int_0^\infty \alpha f^3 e^{-\beta f/T} df}{\int_0^\infty \frac{N}{R\beta f} \alpha f^3 e^{-\beta f/T} df} = 3\frac{R}{N}T.$$

So monochromatic radiation (of sufficiently low density), as far as the volume dependence of the entropy is concerned, acts like a gas of discrete quanta of energy $k\beta f$. This suggests that we investigate whether even the laws of radiation and transformation of light can be interpreted as if the light consists of such energy quanta. We pursue this question in what follows.

* * * * * * * * * * * * * * * * * **

Now at last we come to Einstein's great generalization from thermo-dynamics to quantum electrodynamics.

* * * * * * * * * * * * * * * * * **

§7. On Stokes' Law.

Monochromatic light is changed by photoluminescence into light of another frequency and according to the above result, both the incident and the emitted light consist of quanta of energy $k\beta f$, where f is the frequency involved. The transformation process is then to be understood in the following way. Each incident quantum of frequency f_1 is absorbed and there is – at least for sufficiently low density of the incident quanta – for each one, the emission of a light quantum of frequency f_2; in certain cases, after the absorption of the incident light quantum, there are also light quanta of frequencies f_3, f_4 etc. produced, as well as energy of other kinds (e.g., heat). Any agency which takes place as an intermediate process is immaterial to the final result. If the photoluminescent substance is not a source of energy, then by energy conservation the energy of an emitted quantum cannot be greater than that of the incident quantum; so the relation $f_2 \leq f_1$ must hold, which is Stokes' Law.

In particular, it should be emphasized that for low intensity of the incident light, the intensity of the light emitted must – according to our assumptions – be proportional, since each emitted quantum originates from an elementary process of the above kind, independent of the interaction of other emitted quanta. More specifically, *there is no lower limit for the intensity of the incident light below which it would be incapable of causing emission.* (Note: Italics added.)

* * * * * * * * * * * * * * * * * * **
This is a trivial point for us now, but a key observation then.
* * * * * * * * * * * * * * * * * * **

Violations of Stokes' Law are conceivable according to the above explanation 1) when the number of energy quanta simultaneously being converted per unit volume is so great that an energy quantum of the emitted light can get its energy from several incident quanta; and 2) if the incident (or emitted) light is outside the range of validity of Wien's Law, or if the incident light is irradiating a substance of sufficiently high temperature, that Wien's Law might no longer hold.

This last possibility deserves special attention. According to our explanation it is not out of the question that a "non-Wien" spectrum might result in an even greater reduction in the energy distribution than a blackbody spectrum, even in the validity range of Wien's Law.

§8. On the Production of Cathode Rays by the Irradiation of a Solid.

The usual idea that the energy of light is continuously spread over space, has particular difficulty to explain the photoelectric phenomenon, as described in the pioneering work of Lenard [7].

According to the concept that light consists of quanta of energy $k\beta f$, the production of cathode rays by light can be understood in the following way. Energy quanta penetrate into the surface layer of a solid, where their energy changes at least partly into kinetic energy of electrons. The simplest possibility is that the light quantum gives its whole energy to a single electron; we will assume that this happens. However it should not be ruled out, that an electron only partially absorbs the energy of the light quantum. If an electron inside the solid absorbs energy, after it reaches the surface it has lost part of its kinetic energy. It will be assumed that each electron does work P (characteristic of the solid) while leaving the surface of the solid. Electrons leave the solid with the greatest normal velocity if they are ejected at the surface and normal to it. The kinetic energy of such electrons is

$$\frac{R}{N}\beta f - P.$$

If the solid is raised to the positive potential Π and enclosed by a conductor at zero potential and if Π is just able to prevent any electric current from the solid, then we must have

$$e\Pi = \frac{R}{N}\beta f - P,$$

where e is the magnitude of the electron charge, or

$$E\Pi = R\beta f - P',$$

where $E = Ne$ is the charge of a gram-mole of singly charged ions and P' is the potential of this amount of negative charge with respect to the solid [8].

If one sets $E = 9.6 \times 10^3$, then $10^{-8}\Pi$ is the stopping potential in volts for the solid in vacuum.

In order to see if the derived result agrees in order of magnitude with experiments, we set $P' = 0$, $f = 1.03 \times 10^{15}$ (corresponding to the limit of the sun's spectrum in the ultraviolet), and $\beta = 4.866 \times 10^{-11}$. We get $10^7 \Pi = 4.3$ Volts, which agrees with the order of magnitude of Lenard's results [9].

If the derived formula is correct, then Π must be a linear function of the frequency of the irradiating light, whose slope is independent of the nature of the investigated solid.

> **
> The above equation is the humble, uncomplicated, empirical formula chosen from amongst all Einstein's profound results for explicit citation in his 1921 Nobel Prize awarded "··· especially for his discovery of the law of the photoelectric effect."
> * * * * * * * * * * * * * * * * * * **

Our theory, so far as I can see, is not in disagreement with the properties of the photoelectric effect observed by Lenard. When each quantum of irradiating light gives all its energy independently to one electron, then the velocity distribution of the electrons, i.e. the quality of the emitted cathode rays, should be independent of the intensity of the light; on the other hand, the number of electrons leaving the solid should be proportional to the intensity of the incident light [10].

The limits to the validity of the above law are similar to those of Stokes' Law.

In the preceding it is assumed that at least part of the energy of the irradiating light quantum is given to an individual electron. Without this obvious assumption one gets the following inequality:

$$E\Pi + P' \leq R\beta f.$$

For cathode luminescence, which is the inverse of the process considered above, one gets by an analogous argument

$$E\Pi + P' \geq R\beta f.$$

For the substances investigated by Lenard, $E\Pi$ is always much greater than $R\beta f$, and the potential which the cathode rays must fall through to produce even visible light was in one case 100, and in another case 1000 Volts [11]. It must be assumed that the kinetic energy of each electron is given up to the production of many light quanta.

§9. On the Ionization of a Gas by Ultraviolet Light.

We assume in the ionization of a gas by ultraviolet light, that for each molecule ionized one quantum is absorbed. From this it follows that the ionization energy (that is the work theoretically necessary for ionization) of a molecule cannot be greater than the energy of a reactive light quantum. With J the ionization energy per gram-mole, we must have

$$R\beta f \geq J.$$

According to Lenard's measurements the greatest reactive wavelength for air is around 1.9×10^{-5}cm, so

$$R\beta f = 6.4 \times 10^{12}\text{erg} \geq J.$$

An upper limit for the ionization energy is obtained from the ionization potential in rarefied gases. According to Stark [12] the smallest measurable ionization potential (with a platinum cathode) for air is around 10 Volts. This gives an upper limit for J of 9.6×10^{12}, which is close to that found above. This gives still another result, whose experimental proof seems to me to be of even greater importance. If each absorbed light quantum ionizes one molecule, then there must exist, between the total absorbed light energy L and the number j of gram-moles ionized by it, the relation:

$$j = \frac{L}{R\beta f}.$$

This relation, if our explanation corresponds to reality, must hold for every gas, for which (at the particular frequency) all absorption is accompanied by ionization.

(Bern, den 17. März 1905.)

Footnotes and References:
1) This assumption is just as important as the postulate that the mean kinetic energy

of gas molecules and electrons at temperature equilibrium should be the same. With the latter, Drude has understood theoretically the known behavior of thermal and electrical conduction in metals.

2) M. Planck, Ann. d. Phys. **1**, 69 (1900). Here as **Paper I·1**.

3) This hypothesis can be formulated as follows. The z-component of the electrical force F at a given point, at time t between 0 and T (where T is a time very large compared to all natural periods of concern) is expanded in a Fourier series

$$F = \sum_{f=1}^{f=\infty} A_f \sin\left(2\pi f \frac{t}{T} + \alpha_f\right),$$

where $A_f \geq 0$ and $0 \leq \alpha_f \leq 2\pi$. If one further imagines making a similar expansion for a different space-point, or a different time-origin, then one obtains different values for the quantities A_f and α_f. There exists then, for all the different combinations of A_f's and α_f's, a (statistical) probability distribution of the form

$$dW = \phi(A_1 A_2 \cdots \alpha_1 \alpha_2 \cdots) dA_1 dA_2 \cdots d\alpha_1 d\alpha_2 \cdots.$$

The radiation can be considered disordered when

$$\phi(A_1 A_2 \cdots \alpha_1 \alpha_2 \cdots) = \Phi_1(A_1)\Phi_2(A_2) \cdots \phi_1(\alpha_1)\phi_2(\alpha_2) \cdots$$

that is, when the probabilities of a particular value of one of the quantities A or α is independent of the values of the other A's and α's. With the further stipulation that the emission- and absorption-processes of a particular resonant group depend only on the individual pairs of quantities A_f and α_f, the radiation can be seen as "completely random".

4) M. Planck, Ann. d. Phys. **4**, 553 (1901). Here as **Paper I·4**.

5) This assumption is rather arbitrary. One will naturally keep this simplest assumption as long as experiments do not compel us to abandon it.

6) If E is the energy of the system, then we get:

$$-d(E - TS) = PdV = TdS = \frac{R}{N}n\frac{dV}{V} \quad \text{where} \quad PV = \frac{R}{N}nT.$$

7) P. Lenard, Ann. d. Phys. **8**, 169,170 (1902).

8) If one assumes that the individual electron must be ejected from a neutral molecule at the expense of a fixed energy from the light quantum, one does not have to change the above result; but then P' must be interpreted as the sum of two terms.

9) P. Lenard, Ann. d. Phys. **8**, 165,184 (1902).

10) P. Lenard, Ann. d. Phys. **8**, 150,166 (1902).

11) P. Lenard, Ann. d. Phys. **12**, 469 (1903).

12) J. Stark, *Electricity in Gases*, p.57. Leipzig 1902.

Paper II·2: Excerpt from Physikalische Zeits. **10**, 817 (1909).

On the Evolution of our Ideas about the Nature and the Constitution of Radiation

von A. Einstein.

(Vorgetragen in der Sitzung der physikal. Abteilung der 81. Versammlung Deutscher Naturforscher und Ärtz zu Salzburg am 21.September 1909.)

(We skip the long manuscript and go direct to the discussion following Einstein's lecture, which involved pointed remarks by Planck, Stark and Rubens followed by a prescient response by Einstein.) · · · · · ·

(Eingegangen 14.Oktober 1909.)

Diskussion.

Planck: I would like to join in the thanks of the whole audience which has just listened to Einstein's lecture with the greatest interest. If any further comments are in order, perhaps they can be considered now. I would naturally like to add that I have a different opinion than the speaker. For the most part, what the speaker has said is not subject to any dispute. I would emphasize the necessity of the introduction of the fixed quanta. Furthermore, we can get no further with the whole radiation theory unless we introduce the energy in the definite sense of quanta, which are to be thought of as effective atoms. The next question is: Where should one find these quanta? According to the latest instructions of Herr Einstein it would be necessary for the free rays in vacuum, and even the light waves themselves, to behave as if they were atomistically constructed, and therefore for us to give up Maxwell's equations. That appears to me to be a step which is not absolutely necessary. I will not go into it in detail but only make the following reply.

In his remarks, Herr Einstein related the motion of matter to the vibrations of the free radiation in the pure vacuum. This conclusion seems to me to be supportable only when one knows completely the interactions between the radiation in the vacuum and the motion of the matter; when that is not the case, when the connection is missing, then it is simply necessary to ignore the effect on the motion of the intensity of the incident radiation. It seems to me that this interaction between the free electrical energy in the vacuum and the motion of the atoms of the matter is known to be very small. It arises essentially from the emission and absorption of the light. Certainly the radiation pressure is small, at least it is according to the usual dispersion theory, which includes the reflection during emission and absorption. Now the actual emission and absorption are the obscure part, about which we know very little. About the absorption we know a bit, but what about emission? One understands it as arising from the acceleration of the electrons, but this point is the weakest of the whole theory. One represents the electron as a given volume and charge distribution $\cdots$ that however contradicts the $\cdots$ atomistic assumptions of electricity $\cdots$

At this point one could, I imagine, introduce the use of quantum theory. We can only speak of the laws for large times, but for small times and large accelerations one runs into a void which requires new hypotheses. Perhaps one could assume that an oscillator does not possess a continuously variable energy but that its energy is simple multiples of each elementary quantum. I imagine that with this prescription one might construct a satisfactory radiation theory. Now the question always is: What does it mean? $\cdots$ In mechanics and in the usual electrodynamics we have no discrete action elements and therefore we cannot even describe such a mechanical or electrodynamic model.

* * * * * * * * * * * * * * * * * * *
Planck's vision is not so different from the matrix solution of the quantum harmonic oscillator still 17 years in the future.
* * * * * * * * * * * * * * * * * * *

Mechanically then everything seems impossible and one must deal with that. All our attempts to represent light mechanically are completely stalled. Moreover one must represent the electrical current mechanically. In that case we have had the comparison with a water current, but even that must be abandoned; and when one has fixed that, one must do the same for the oscillator. Obviously this theory must still be worked out in much greater detail as I have already outlined; perhaps there is someone else as lucky as I.

In any case I believe that one must next investigate the whole complexity of the quantum theory in order to understand the problem of the interaction between matter and radiation; the example in pure vacuum can still be explained as before with Maxwell's equations.

H. Ziegler: If one represents the *ur*-atom of matter as a tiny invisible sphere which always has the speed of light, then all the interactions of matter and radiation can be represented and also the connection sought by Herr Planck between the material and the non-material would be achieved.

Stark: Herr Planck has emphasized that we have no prior reason to accept Einstein's result that radiation in space which is free of matter should be considered as localized. Originally I was also of the same opinion as Herr Planck: that one could only limit it in that way beforehand, in that the quantum prescription referred back to a definite interaction mode of the oscillator. But I see now that it does require electromagnetic radiation in the absence of matter to be localized in space. The particular example which led me to this realization is that of the electromagnetic radiation emitted from an X-ray tube into the surrounding space. Even at large distances, up to 10 meters, it can still be localized for interaction with a single electron. I conclude therefore, that this fact is an answer to the question under consideration: that indeed the electromagnetic radiation energy should be understood as localized, and even further, that this conclusion does not depend on the presence or absence of matter.

* * * * * * * * * * * * * * * * * * **

Stark's remark is marvelously close to the eventual Copenhagen philosophy of quantum mechanics, which would hold that it is meaningless and ultimately not permitted to speak of unobservable attributes of physical quantities. In this instance, we cannot conjure up a separate existence for electromagnetic fields in some unseeable state of vacuum. How good a vacuum? Is one electron every 10 meters a vacuum? What you see is what there is, and that's all there is. We are left to argue what *seeing* is.

* * * * * * * * * * * * * * * * * **

Rubens: From the extended interpretation of Herr Einstein there would be an observable consequence, which could be verified experimentally. As we all know, both α-rays and β-rays produce a scintillation effect on fluorescent materials. According to the view just presented this must also be the case for γ-rays and X-rays.

Planck: For X-rays, it is a separate case as I have pointed out recently. Stark has mentioned something in favor of the quantum theory, I will mention something against it: there are the interferences from the huge phase differences of hundreds of thousands of wavelengths. When a quantum interferes with itself, it must have an extension of hundreds of thousands of wavelengths. That, however, is certainly a difficulty.

Stark: Interference phenomena certainly can be held against the quantum hypothesis. But when one considers it with a greater sympathy for the quantum hypothesis, then one can find an explanation even for interference, which I have described as encouraging. What is involved, which must be re-emphasized, in the experiment about which Herr Planck has just spoken, is very strong radiation, so that very many quanta of the same frequency are concentrated in the light bundle; that must be carefully allowed for in the treatment of any interference. With very weak radiation the interference phenomenon would be completely different.

* * * * * * * * * * * * * * * * * **

Here, and in Einstein's following remark, we are so close to a full early expression of the wave-particle duality.

* * * * * * * * * * * * * * * * * **

Einstein: The interference phenomena could well be not so difficult to observe as one might imagine, for the following reason: one should not assume that the rays consist of quanta which are not in interaction; that would make it impossible to explain the interference phenomena.

****

Here "not in interaction" presumably means "incoherent" or "independent"; so the dense radiation would consist of a bundle of incoherent rays, each of which corresponds to one quantum (or, if more, to coherent quanta) which interferes only with itself like the classical rays.

****

I personally think of a quantum as a singularity surrounded by an extended vector field. A vector field can be constructed from a large number of quanta, each of which differs a little from the others, as we assume for the rays. I would imagine that the scattering of radiation at an interface takes place by the effect of the surface separating the quanta, perhaps according to the phase of the resulting fields with which the quanta reach the surface. The equations for the resulting field would differ but little between the two theories. We would then not have to change much in the theory of the interference phenomena, as it was just described.

****

Presumably the "little" by which the rays differ is principally a random phase.

****

I imagine something similar for the change to the molecularization of the carrier of the electrostatic field. The field produced by electrically charged particles is not very essentially different from the previous conception, and it is not out of the question that it will appear somewhat similar in the eventual radiation theory. I do not see a major difficulty in the interference phenomena.

****

There is a strong hint here that Einstein intended what is now termed "quantization in the radiation gauge" with an instantaneous *classical* coulomb field, as indicated by the Planck distribution with just two polarization states.

* * * * * * * * * * * * * * * * * * *

Paper II·3: Excerpt from Deutsche Physikal. Gesells. Verhandlungen **18**, 318 (1916).

Emission and Absorption of Radiation According to Quantum Theory†

von **A. Einstein.**

(Eingangen am 17.Juli 1916.)

When Planck brought quantum theory to life 16 years ago, and deduced his radiation formula, he succeeded in the following way. He calculated the average energy $\bar{E}$ of the oscillators as a function of temperature according to the new quantum theoretic principle which he introduced. He then derived the radiation density ρ as a function of frequency f and temperature T by borrowing from electromagnetic theory the relation between radiation density ρ and oscillator energy $\bar{E}$ as:

$$\bar{E} = \frac{c^3 \rho}{8\pi f^2}. \tag{1}$$

His derivation was of unparalleled boldness, and found brilliant confirmation. Not only was the radiation formula itself and the value of the elementary quantum derived from it correct, but so was the quantum theoretic value for $\bar{E}$ as calculated from later investigations on the specific heat. It depended however on Eqn1 which is known only in purely electromagnetic ways. It still remains unsatisfactory that the electromagnetic-mechanical derivation which led to (1) is not compatible with the fundamental idea of quantum theory. It is not surprising then, that Planck himself and all other theoreticians who deal with this matter should continue to be troubled that the theory must be altered, before its assumptions can be trusted to be free of contradictions.

Ever since the Bohr theory of spectra produced its great successes, it has seemed beyond doubt that the basic ideas of quantum theory must be respected. It seems therefore that the universality of the theory must be accepted, and that the electromagnetic-mechanical analysis which led Planck to Eqn1 must be replaced by quantum theoretic considerations for the interaction of matter and radiation. Toward this goal I present the following argument, which is distinguished by its simplicity and generality.

§1. The Planck Oscillator in the Radiation Field.

The behavior of monochromatic oscillators in a radiation field, according to classical theory, can be easily understood by following the theory of Brownian motion first developed using such analysis. Let E be the instantaneous energy of the oscillator; we ask what the energy is after an elapsed time τ, which is large compared to the period of the oscillator, but also small enough that the change of E during τ can be regarded as infinitesimal. Then changes of two kinds can be distinguished. First is the change

$$\Delta_1 E = -AE\tau,$$

which is caused by emission of radiation; second is a change $\Delta_2 E$ which comes about from the work done by the electric field on the oscillator. This second change increases with increasing radiation density and has a "random" value and a "random" sign. An electromagnetic-statistical analysis leads to the resulting mean-value

$$\overline{\Delta_2 E} = B\rho\tau.$$

The constants A and B are calculable in a known way. We call $\Delta_1 E$ the energy change by emission and $\Delta_2 E$ the energy change by absorption. Since the multitude of oscillators should have a mean value of E which is independent of the time, then we must have

$$\overline{E + \Delta_1 E + \Delta_2 E} = \bar{E} \text{ or } \bar{E} = \frac{B}{A}\rho.$$

If one calculates A and B in the known way from electromagnetic theory and mechanics, then one recovers the result of Eqn1.

We now construct a corresponding derivation but on a quantum-theoretic basis and without special assumptions about the particles – called "molecules" in what follows – which are in interaction with the radiation.

§2. Quantum Theory and Radiation.

Consider a gas of identical molecules in statistical equilibrium with heat radiation at temperature T. Each molecule is supposed to have only a discrete series of states $Z_1, Z_2, \cdots$ with energies $\epsilon_1, \epsilon_2, \cdots$. Then it follows in a known way (analogous to statistical mechanics, or directly from Boltzmann's Principle, or ultimately from thermodynamic considerations) that the probability W_n of the state Z_n – in particular the relative number of molecules in the state Z_n – is

$$W_n = P_n e^{-\epsilon_n/kT}. \tag{2}$$

Here k is Boltzmann's constant, and P_n is the statistical "weight" of the state Z_n – a constant independent of the temperature and characteristic of the particular quantum state of the molecule.

We now assume that a molecule can go from state Z_n to state Z_m by the absorption of radiation of a fixed frequency f_{mn}, and from Z_m to Z_n by emission of radiation of the same frequency. The radiation energy involved is $\epsilon_m - \epsilon_n \geq 0$. This will be possible in general for each combination of the two indices m and n. In order for each of these elementary reactions to be in temperature equilibrium, statistical equilibrium must exist. We turn then to the question of a unique elementary process which may be considered for just a particular index pair (n, m).

For temperature equilibrium, the same number of molecules per unit time go from state Z_n to state Z_m by absorption of radiation as from state Z_m to state Z_n by emission. For these transitions we make a simple hypothesis to which we have been led by the limiting case of the classical theory, which has been briefly sketched above.

We distinguish two different processes.

a) Emission: This occurs in a transition from the state Z_m to the state Z_n with the emission of energy $\epsilon_m - \epsilon_n$. This transition occurs without any external cause. One can think of it as hardly different from a kind of radioactive decay. The number of such transitions per unit time will be set equal to

$$A_m^n N_m,$$

where A_m^n is a factor characteristic of the two states Z_m and Z_n, and N_m is the number of molecules in state Z_m.

b) Absorption: Absorption is caused by the radiation in which the molecules find themselves; it will be proportional to the radiation density ρ at the effective frequency f_{mn}. In any case, the oscillators can just as well undergo an energy decrease as an energy increase; in our case a transition $Z_n \to Z_m$ as well as transition $Z_m \to Z_n$ can occur. The number of transitions $Z_n \to Z_m$ per unit time is expressed as

$$B_n^m N_n \rho,$$

and the number of transitions $Z_m \to Z_n$ as

$$B_m^n N_m \rho,$$

where B_m^n and B_n^m are factors which depend on the states Z_n and Z_m.

The condition for statistical equilibrium of the reactions $Z_n \leftrightarrow Z_m$ is therefore the equation

$$A_m^n N_m + B_m^n N_m \rho = B_n^m N_n \rho. \tag{3}$$

From Eqn2,

$$\frac{N_n}{N_m} = \frac{P_n}{P_m} e^{(\epsilon_m - \epsilon_n)/kT}. \tag{4}$$

From (3) and (4)

$$A_m^n P_m = \rho \left(B_n^m P_n e^{(\epsilon_m - \epsilon_n)/kT} - B_m^n P_m \right). \tag{5}$$

ρ is the radiation density at the frequency f_{mn} which is emitted in the transition $Z_n \to Z_m$ and absorbed in the transition $Z_m \to Z_n$. Our equation gives the relation between T and ρ at this frequency. *If we assume that ρ should increase indefinitely with T* (Note: Italics added.), then we must have

$$B_n^m P_n = B_m^n P_m. \tag{6}$$

If we abbreviate

$$\frac{A_m^n}{B_m^n} = \alpha_{mn}, \tag{7}$$

then

$$\rho = \frac{\alpha_{mn}}{e^{(\epsilon_m - \epsilon_n)/kT} - 1}. \tag{8}$$

This is the Planck relation between ρ and T with undetermined factors defined. The factors A_m^n and B_m^n can be calculated directly, when we are in possession of an electrodynamics and mechanics modified in the sense of the quantum hypothesis.

* * * * * * * * * * * * * * * * * * **

The detailed balance result of Eqn6, that the intrinsic rates for stimulated emission and for absorption must be the same, is contained implicitly in Kirchoff's Theory of Emissive and Absorptive Power of Matter, which is the foundation of Planck's derivation of Eqn1 (above) in his paper On Irreversible Radiation Processes (**Paper I·1**). Einstein achieves the result in an amazingly ingenious, elegant and fruitful way because of the presence of the spontaneous emission contribution in his condition for statistical equilibrium, Eqn3 above. Elementary quantum mechanics achieves the same result using the equality of the emission and absorption matrix elements

$$|\langle m|\vec{r}|n\rangle|^2 = |\langle n|\vec{r}|m\rangle|^2.$$

* * * * * * * * * * * * * * * * * * **

That α_{mn} and $\epsilon_m - \epsilon_n$ cannot depend on the particular nature of the molecules, but only on the relevant frequency f, follows from the fact that ρ must be a universal function of T and f. Further, it follows from Wien's Displacement Law that α_{mn} must be proportional to f^3, and $\epsilon_m - \epsilon_n$ to f. Accordingly one has

$$\epsilon_m - \epsilon_n = hf, \tag{9}$$

where h is a constant.

I should naturally like to emphasize that the three hypotheses concerning the emission and absorption which lead to the Planck radiation formula, are in no way designed to produce the result. Moreover, the simplicity of the hypotheses, the generality with which the analysis can be freely carried out, as well as the natural connection in the limiting case to the linear Planck oscillators (in the sense of classical electrodynamics and mechanics), all make it seem to me as very probable that this should be the foundation for future theoretical discussion. It also speaks in favor of the theory, that the statistical law assumed for the radiation is not different from Rutherford's law of radioactive decay, and that (9) in combination with (8) leads to a result which is identical with the second fundamental hypothesis of Bohr's theory of spectra.

§3. Remark on the Photochemical Equivalence Law. ⋯

† – translated from A. Einstein, Deutsche Physikalische Gesellschaft. Verhandlungen **18**, 318 (1916) (as available in *The Collected Papers of* **Albert Einstein**, *Volume 6, The Berlin Years: Writings, 1914-1917* (Princeton University Press, 1996) A.J. Knox, Martin J. Klein, and Robert Schulmann, Editors. Pp.364-369).

Paper II·3A: Excerpt from Phys. Zeits. **18**, 121 (1917).

Zur Quantentheorie der Strahlung††
von **A. Einstein**.

* * * * * * * * * * * * * * * * * * **

This paper is a later version of **Paper II·3** (received 3 March 1917 and published 15 March 1917), but containing the discussion of momentum conservation in individual absorption and emission processes, which is all we quote here. In spite of Einstein's completely unequivocal statements, the subject of energy and momentum conservation – especially in photon emission – would remain in dispute for almost seven more years (see the confusion in Bohr, Kramers and Slater (**Paper III·3**)0. Einstein's remarks are fundamental to Compton's kinematics, but Compton makes no reference to them.

* * * * * * * * * * * * * * * * * * * **

§2. Hypotheses on Energy Exchange During Irradiation.

$\cdots\cdots$ The question is, what momentum is transferred to the molecule during the change of state. Let us begin with the process of absorption. Let a beam of radiation from a particular direction act on a Planck resonator, and transfer the appropriate energy. This energy transfer corresponds also – according to the momentum law – to a momentum transfer from the beam to the resonator. The resonator therefore must experience an effective force in the direction of the beam. If the transition energy is negative, then the effective force on the resonator must be in the opposite direction. On the quantum hypothesis, this obviously means the following. If during the irradiation by the beam, the absorption process $n \to m$ occurs, then the molecule will absorb momentum $(\epsilon_m - \epsilon_n)/c$ in the direction of the beam. For the emission process $m \to n$, the momentum change of the molecule has the same magnitude but the opposite direction. For the case that the molecule is simultaneously exposed to many radiations, we assume that the whole energy $(\epsilon_m - \epsilon_n)$ results from one elementary process corresponding to one of these radiations, so that even in this case the momentum $(\epsilon_m - \epsilon_n)/c$ will be transferred to the molecule.

In the case of Planck resonators, no momentum is transferred on the whole to the resonators during energy loss by emission of radiation, because in the classical theory the outgoing radiation takes place in a spherical wave. We have just remarked, however, that we can only arrive at a contradiction free quantum theory of radiation if we assume that even the process of emission must be a directed process. Only then will each elementary process of emission $m \to n$ correspond to the transfer to the molecule of a momentum of the magnitude $(\epsilon_m - \epsilon_n)/c$. If the radiation appears to be isotropic, then we must assume all directions of outgoing radiation to be equally probable. $\cdots\cdots$

†† – translated from A. Einstein, Physikalische Zeitschrift **18**, 121 (1917) (as available in *The Collected Papers of* **Albert Einstein**, *Volume 6, The Berlin Years: Writings, 1914-1917* (Princeton University Press, 1996) A.J.Knox, Martin J. Klein, and Robert Schulmann, Editors. Pp.386-387).

Paper II·4: Excerpt from Physical Review **21**, 483 (1923).

A QUANTUM THEORY OF THE SCATTERING OF X-RAYS BY LIGHT ELEMENTS

BY ARTHUR H. COMPTON

ABSTRACT

A quantum theory of the scattering of X-rays and γ-rays by light elements. – The hypothesis is suggested that when an X-ray quantum is scattered it spends all of its energy and momentum upon some particular electron. This electron in turn scatters the ray in some definite direction. The change in momentum of the X-ray quantum due to the change in its direction of propagation results in a recoil of the scattering electron. The energy in the scattered quantum is thus less than the energy in the primary quantum by the kinetic energy of recoil of the scattering electron. The corresponding *increase in the wavelength of the scattered beam* is

$$\lambda_\theta - \lambda_0 = (2h/mc)\sin^2 \frac{1}{2}\theta = 0.0484\sin^2 \frac{1}{2}\theta,$$

where h is Planck's constant, m is the mass of the scattering electron, c is the velocity of light, and θ is the angle between the incident and the scattered ray. Hence the increase is independent of the wavelength. *The distribution of the scattered radiation* is found $\cdots$ to be concentrated in the forward direction $\cdots\cdots$. Unpublished experimental results are given which show that for graphite and the Mo-K radiation the scattered radiation is longer than the primary, the observed difference $(\lambda_{\pi/2} - \lambda_0 = .022A)$ being close to the computed value $.024A$. In the case of scattered γ-rays, the wavelength has been found to vary with θ in agreement with the theory, increasing from $.022A$ (primary) to $.068A$ ($\theta = 135°$). Also the velocity of secondary β-rays excited in light elements by γ-rays agrees with the suggestion that they are recoil electrons. As for the predicted variation of absorption with λ $\cdots\cdots$. This remarkable *agreement between experiment and theory indicates clearly that scattering is a quantum phenomenon* and can be explained without introducing any new hypothesis as to the size of the electron or any new constants; also that a radiation quantum carries with it momentum as well as energy. $\cdots\cdots$

* * * * * * * * * * * * * * * * * **

Compton's amazing clarity of vision cut through all the extraneous intellectual baggage which encumbered everyone else at that time. His analysis of the relativistic kinematics of the scattering of an X-ray by an electron is a truly beautiful accomplishment which might seem simple to us now but which boggled the minds of the greats of the time. It must be kept in mind that even Niels Bohr was at this time preoccupied with the notion that energy and momentum were not conserved in individual quantum processes but only on the average over many such processes; and that others were similarly worried about fitting the quantum concept into the classical angular distribution of the radiation. Compton cut through this confusion in a way that has not been improved upon.

That the self-same person had the breadth of talent then actually to **do** the experiment and almost literally to invent the field of elementary particle physics will forever remain awe inspiring.

* * * * * * * * * * * * * * * * * **

J.J. Thomson's classical theory of the scattering of X-rays, though supported by the early experiments of Barkla and others, has been found incapable of explaining many of the more recent experiments. This theory, based upon the usual electrodynamics, leads to the result that the energy scattered by an electron in an X-ray beam of unit intensity is the same whatever the wavelength of the incident rays. Moreover, when the X-rays traverse a thin layer of matter, the intensity of the scattered radiation on the two sides of the layer should be the same. Experiments on the scattering of X-rays by light elements have shown that these predictions are correct when X-rays of moderate hardness are employed; but when very hard X-rays or γ-rays are employed, the scattered energy is found to be decidedly less than Thomson's theoretical value, and to be strongly concentrated on the forward side of the scattering plate.

Several years ago I suggested that this reduced scattering of the very short wavelength X-rays might be the result of interference between the rays scattered by different parts of the electron, if the electron's diameter is comparable with the wavelength of the radiation. By assuming the proper radius for the electron, this hypothesis supplied a quantitative explanation of the scattering for any particular wavelength. But recent experiments have shown that the size of the electron which must be assumed increases with the wavelength of the X-rays employed [1], and the concept of an electron whose size varies with the wavelength of the incident rays is difficult to defend.

* * * * * * * * * * * * * * * * * * **

Compton begins by withdrawing his own earlier misconceptions.

* * * * * * * * * * * * * * * * * * **

Recently an even more serious difficulty with the classical theory of X-ray scattering has appeared. It has long been known that secondary γ-rays are softer than the primary rays which excite them, and recent experiments have shown that this is also true of X-rays. By a spectroscopic examination of the secondary X-rays from graphite, I have indeed been able to show that only a small part, if any, of the secondary X-radiation is of the same wavelength as the primary [2]. While the energy of the secondary X-radiation is so nearly equal to that calculated from Thomson's classical theory that it is difficult to attribute it to anything other than true scattering [3], these results show that if there is any scattering comparable in magnitude with that predicted by Thomson, it is of a greater wavelength than the primary X-rays.

Such a change in wavelength is directly counter to Thomson's theory of scattering, for this demands that the scattering electrons, radiating as they do because of their vibrations forced by a primary X-ray, should give rise to radiation of exactly the same frequency as that of the radiation falling upon them. Nor does any modification of the theory such as the hypothesis of a large electron suggest a way out of the difficulty. This failure makes it appear improbable that a satisfactory explanation of the scattering of X-rays can be reached on the basis of classical electrodynamics.

THE QUANTUM HYPOTHESIS of SCATTERING

According to the classical theory, each X-ray affects every electron in the matter traversed, and the scattering observed is that due to the combined effects of all the electrons. From the point of view of the quantum theory, we may suppose that any particular quantum of X-rays is not scattered by all electrons in the radiator, but spends all of its energy upon some particular electron. This electron will in turn scatter the ray in some definite direction, at an angle with the incident beam. This bending of the path of the quantum of radiation results in a change in its momentum. As a consequence, the electron will recoil with a momentum equal to the change in momentum of the X-ray. The energy in the scattered ray will be equal to that in the incident ray minus the kinetic energy of the recoil of the scattering electron; and since the scattered ray must be a complete quantum, the frequency will be reduced in the same ratio as the energy. Thus on the quantum theory we should expect the wavelength of the scattered X-rays to be greater than that of the incident rays.

The effect of the momentum of the X-ray quantum is to set the scattering electron in motion at an angle of less than $90°$ with the primary beam. But it is well known that the energy radiated by a moving body is greater in the direction of its motion. We should therefore expect, as is experimentally observed, that the intensity of the scattered radiation should be greater in the general direction of the primary X-rays than in the reverse direction.

* * * * * * * * * * * * * * * * * * **

In these two great paragraphs, Compton defines the absolutely pristine quantum scattering process which bears his name. In doing so, he avoids every pitfall: he ignores multiple scattering of the X-rays; he has each incident quantum completely absorbed and then completely re-emitted as an outgoing quantum by only one individual electron acting freely; he ignores the effect of electron binding and initial velocity; he does not confuse the issue by worrying about the classical spreading of the outgoing X-rays; he recognizes that a full relativistic treatment of the kinematics is preferable to treating the electron as nonrelativistic; he recognizes the forward hemisphere recoil of the electron as the mechanism of the observed X-ray energy loss and of the forward peaking of the outgoing X-ray.

Compton's results anticipated by five years similar experimental results based on inelastic scattering of optical quanta in solids by Landsberg and Mandelstamm [*Naturw.* **16**, 557,772 (1928)], and in liquids by Raman and Krishnan [*Nature* **121**, 501 (1928)].

* * * * * * * * * * * * * * * * * * **

The change in wavelength due to scattering. – Imagine that an X-ray quantum of frequency f_0 is scattered by an electron of mass m. The momentum of the incident ray will be hf_0/c, and that of the scattered ray is hf_θ/c at an angle θ with the initial momentum. The principle of the conservation of momentum accordingly demands that the momentum of recoil of the scattering electron should equal the vector difference between the momenta of these two rays, as in Fig.1B. The momentum of the electron, $m\beta c/\sqrt{1-\beta^2}$ where βc is the velocity of recoil of the electron, is thus given by the relation

$$\left(\frac{m\beta c}{\sqrt{1-\beta^2}}\right)^2 = \left(\frac{hf_0}{c}\right)^2 + \left(\frac{hf_\theta}{c}\right)^2 + 2\frac{hf_0}{c}\frac{hf_\theta}{c}\cos\theta. \tag{1}$$

But the energy hf_θ in the scattered quantum is equal to that of the incident quantum hf_0 less the kinetic energy of recoil of the scattering electron, *i.e.*,

$$hf_\theta = hf_0 - mc^2\left(\frac{1}{\sqrt{1-\beta^2}} - 1\right). \tag{2}$$

We thus have two independent equations containing the two unknown quantities β and f_θ. On solving the equations we find

$$f_\theta = f_0 / \left(1 + 2\alpha\sin^2\frac{1}{2}\theta\right), \tag{3}$$

where

$$\alpha = hf_0/mc^2 = h/mc\lambda_0. \tag{4}$$

Or in terms of wavelength instead of frequency,

$$\lambda_\theta = \lambda_0 + (2h/mc)\sin^2\frac{1}{2}\theta. \tag{5}$$

$\cdots$ Eqn5 indicates an increase in wavelength due to the scattering process which varies from a few percent in the case of ordinary X-rays to more than 200% in the case of γ-rays scattered backward. At the same time the velocity of the recoil of the scattering electron varies from zero when the ray is scattered directly forward to about 80% of the speed of light when a γ-ray is scattered at a large angle.$\cdots$

* * * * * * * * * * * * * * * * * * **

We omit a lengthy discussion showing that a Doppler shift of the classical radiation from the reference frame of the recoiling electron to that of the laboratory produces the same result as Eqn5 above; and also discussions of models of X-ray scattering, in order to concentrate on the fundamental kinematic relations, their experimental verification and their quantum significance. The Doppler shift calculation was incorrectly viewed by Bohr as a justification of the correspondence principle description of the scattering process. We now recognize it as the consequence of applying a succession of Lorentz transformations to the separate absorption and emission processes – each of which treats the kinematics in a fully quantum and relativistic way - which inevitably reproduces Compton's complete result.

* * * * * * * * * * * * * * * * * * **

EXPERIMENTAL TEST

Let us now investigate the agreement of these various formulas with the experiments on the change of wavelength due to scattering $\cdots$

Wavelength of the scattered rays. – If in Eqn5 we substitute the accepted values of h, m, and c, we obtain

$$\lambda_\theta = \lambda_0 + 0.0484 \sin^2 \frac{1}{2}\theta, \tag{31}$$

if λ is expressed in Angström units. It is perhaps surprising that the increase should be the same for all wavelengths. Yet, as a result of an extensive experimental study of the change in wavelengths on scattering, I have concluded [4] that "over the range of primary rays from 0.7 to 0.025A, the wavelengths of the secondary X-rays at 90° with the incident beam is roughly 0.03A greater than that of the primary beam which excites it." Thus the experiments support the theory in showing a wavelength increase which seems independent of the incident wavelength, and which also is of the proper order of magnitude.

A quantitative test of the accuracy of Eqn31 is possible using the K X-rays from molybdenum scattered by graphite. In Fig.4 is shown a spectrum of the X-rays scattered by graphite at right angles with the primary beam, when the graphite is traversed by X-rays from a molybdenum target [5]. The solid line represents the spectrum of these scattered rays, and is to be compared with the broken line, which represents the spectrum of the primary rays, using the same slits and crystal, and the same potential on the tube. The primary spectrum is, of course, plotted on a much smaller scale than the secondary. The zero point of the spectrum of both the primary and secondary X-rays was determined by finding the position of the first order lines on both sides of the zero point.

It will be seen that the wavelength of the scattered rays is unquestionably greater than that part of the primary rays which excite them. Thus the Kα line from molybdenum has a wavelength 0.708A. The wavelength of this line in the scattered beam is found in these experiments, however, to be 0.730A. That is,

$$\lambda_\theta - \lambda_0 = 0.022A \qquad \text{(experiment)}.$$

But according to the present theory (Eqn5),

$$\lambda_\theta - \lambda_0 = 0.0484 \sin^2 45° = 0.024A \qquad \text{(theory)},$$

which is a very satisfactory agreement.

The variation in wavelength of the scattered beam with the angle is illustrated in the case of γ-rays. I have measured [6] the mass absorption coefficient in lead of the rays scattered at different angles when various substances are traversed by the hard γ-rays from RaC. The mean results for iron, aluminum and paraffin are given in column 2 of Table I. This variation in absorption coefficient corresponds to a difference in wavelength at the different angles. Using the value given by Hull and Rice for the mass absorption coefficient in lead for wavelength 0.122A and 3.0A, remembering [7] that the characteristic fluorescent absorption τ/ρ is proportional to λ^3, and estimating the part of the absorption due to scattering by the method described below, I find for the wavelengths corresponding to these absorption coefficients the values given in the fourth column of Table I. That this extrapolation is very nearly correct is indicated by the fact that it gives for the primary beam a wavelength $0.022A$. This is in good accord with my value $0.025A$, calculated from the scattering of γ-rays by lead at small angles [8], and with Ellis's measurements from his β-ray spectra, showing lines of wavelength $.045$, $.025$, $.021$ and $.020A$, with the line $.020A$ the strongest [9]. Taking $\lambda_0 = 0.022A$, the wavelengths at the other angles may be calculated from Eqn31. The results, given in the last column of Table I are in satisfactory accord with the measured values. There is thus good reason for believing that Eqn5 represents accurately the wavelength of the X-rays and γ-rays scattered by light elements.

Velocity of recoil of the scattering electrons. – The electrons which recoil in the process of the scattering of ordinary X-rays have not been observed. This is probably because their number and velocity is usually small $\cdots\cdots$ I have pointed out elsewhere [10] that there is good reason for believing that most of the secondary β-rays excited in light elements by the action of γ-rays are such recoil electrons. $\cdots\cdots$

DISCUSSION

This remarkable agreement between our formulas and the experiments can leave little doubt that the scattering of X-rays is a quantum phenomenon. The hypothesis of a large electron to explain these effects is accordingly superfluous, for all the experiments on X-ray scattering to which this hypothesis has been applied are now seen to be explicable from the point of view of the quantum theory without introducing any new hypothesis or constants. In addition, the present theory accounts satisfactorily for the change in wavelength due to scattering, which was left unaccounted for on the hypothesis of the large electron. From the standpoint of the scattering of X-rays and γ-rays, therefore, there is no longer any support for the hypothesis of an electron whose diameter is comparable with the wavelength of hard X-rays.

The present theory depends essentially upon the assumption that each electron which is effective in the scattering scatters a complete quantum. It involves also the hypothesis that the quanta of radiation are received from definite directions and are scattered in definite directions. The experimental support of the theory indicates very convincingly that a radiation quantum carries with it directed momentum as well as energy.

* * * * * * * * * * * * * * * * * * **

In this small paragraph, Compton summarizes the conclusions compelled by his scattering experiment. "··· the quanta ··· are received from definite directions and are scattered in definite directions. Experimental support ··· that a ··· quantum carries ··· momentum as well as energy."

The contrast between Compton's direct and positive statements and the agonizing ruminations of Bohr's endless and convoluted philosophizing on the subject (see excerpts of Bohr's work in **Paper IV·1**) is stark in the extreme. Compton's no-nonsense experimental confrontation of the problem is like a lighthouse beacon in a sea of fog.

* * * * * * * * * * * * * * * * * * **

Emphasis has been laid upon the fact that in its present form the quantum theory of scattering applies only to light elements. The reason for this restriction is that we have tacitly assumed that there are no forces of constraint acting upon the scattering electrons. This assumption is probably legitimate in the case of very light elements, but cannot be true for the heavy elements. For if the kinetic energy of recoil of an electron is less than the energy required to remove the electron from the atom, there is no chance for the electron to recoil in the manner we have supposed. The conditions of scattering in such a case remain to be investigated.

The manner in which interference occurs, as for example in the cases of excess scattering and X-ray reflection, is not yet clear. Perhaps if an electron is bound in the atom too firmly to recoil, the incident quantum of radiation may spread itself over a large number of electrons, distributing its energy and momentum among them, thus making interference impossible. In any case, the problem of scattering is so closely allied with those of reflection and interference that a study of the problem may very possibly shed some light upon the difficult question of the relation between interference and the quantum theory.

Many of the ideas involved in this paper have been developed in discussion with Professor G.E.M. Jauncey of this department.

WASHINGTON UNIVERSITY, SAINT LOUIS

DECEMBER 13, 1922

Footnotes and References:

1) A.H. Compton, Bull. Nat. Research Council **20**, p.10 (Oct., 1922).

2) In previous papers (Phil. Mag. **41**, 749 (1921); Phys. Rev. **18**, 96 (1921)) I have defended the view that the softening of the secondary X-radiation was due to a considerable admixture of a form of fluorescent radiation. Gray (Phil. Mag. **26**, 611 (1913); Frank. Inst. Journ., Nov., 1920, p.643) and Florance (Phil. Mag. **27**, 225 (1914)) have considered that the evidence favored true scattering, and that the softening is in some way an accompaniment of the scattering process. The considerations brought forward in the present paper indicate that the latter view is the correct one.

3) A.H. Compton, loc. cit., p.16.

4) A.H. Compton, loc. cit., p.17.

5) I hope to publish soon a description of the experiments on which this figure is based.

6) A.H. Compton, Phil. Mag. **41**, 760 (1921).

7) Cf. L. de Broglie, Jour. de Phys. et Rad. **3**, 33 (1922); A.H. Compton, loc. cit., p.43.

8) A.H. Compton, Phil. Mag. **41**, 777 (1921).

9) C.D. Ellis, Proc. Roy. Soc. A, **101**, 6 (1922).

10) A.H. Compton, loc. cit., p.27.

Paper II·5: Excerpt from Physical Review **22**, 409 (1923).

THE SPECTRUM OF SCATTERED X-RAYS

BY ARTHUR H. COMPTON

ABSTRACT

The spectrum of molybdenum K X-rays scattered by graphite at $45°$, $90°$ and $135°$ has been compared with the spectrum of the primary beam. A primary spectrum line when scattered is broken up into two lines, an "unmodified" line whose wave-length remains unchanged, and a "modified" line whose wavelength is greater than that of the primary line. Within a probable error of about $0.001A$, the difference in the wavelengths $(\lambda_\theta\text{-}\lambda_0)$ increases with the angle θ between the primary and the secondary rays according to the quantum relation

$$(\lambda_\theta - \lambda_0) = \gamma(1 - \cos\theta),$$

where $\gamma = h/mc = 0.0242A$. This wavelength change is confirmed also by absorption measurements. The modified ray does not seem to be as homogeneous as the unmodified ray; it is less intense at small angles and more intense at large angles than is the unmodified ray.

An X-ray tube of small diameter and with water-cooled target is described, which is suitable for giving intense X-rays.

I have recently proposed a theory of the scattering of X-rays [1], based on the postulate that each quantum of X-rays is scattered by an individual electron [2,3]. The recoil of this scattering electron, due to the change in momentum of the X-ray quantum when its direction is altered, reduces the energy and hence also the frequency of the quantum of radiation. The corresponding increase in the wavelength of the X-rays due to the scattering was shown to be

$$\lambda_\theta - \lambda_0 = \gamma(1 - \cos\theta) \tag{1}$$

where λ_θ is the wavelength of the ray scattered at an angle θ with the primary ray whose wavelength is λ_0, and

$$\gamma = h/mc = 0.0242A$$

where h is Planck's constant, m is the mass of the electron and c the velocity of light. It is the purpose of this paper to present more precise data than has previously been given regarding this change in wavelength when X-rays are scattered.

Apparatus and method. For the quantitative measurement of the change in wavelength it was clearly desirable to employ a spectroscopic method. In view of the comparatively low intensity of scattered X-rays, the apparatus had to be designed in such a way as to secure the maximum intensity in the beam whose wavelength was measured. The arrangement of the apparatus is shown diagrammatically in Fig.1. Rays proceeded from the molybdenum target T of an X-ray tube to the graphite scattering block R, which was placed in line with the slits 1 and 2. Lead diaphragms, suitably disposed, prevented stray radiation from leaving the lead box that surrounded the X-ray tube. Since the slit 1 and the diaphragms were mounted upon an insulating support, it was possible to place the X-ray tube close to the slit without danger of puncture. The X-rays, after passing through the slits, were measured by a Bragg spectrometer in the usual manner.

The X-ray tube was of special design. A water-cooled target was mounted in a narrow glass tube, as shown in Fig.2, so as to shorten as much as possible the distance between the target T and the radiator R. This distance in the experiments was about 2cm. When 1.5 kw was dissipated in the tube, the intensity of the rays reaching the radiator was thus 125 times as great as it would have been if a standard Coolidge tube with a molybdenum target had been employed. The electrodes for this tube were very kindly supplied by the General Electric Company.

In the final experiments the distance between the slits was about 18cm, their length about 2cm, and their width about 0.01cm. Using a crystal of calcite, this made possible a rather high resolving power even in the first order spectrum.

Spectra of scattered molybdenum rays. Results of the measurements, using slits of two different widths, are shown in Figs.3 and 4. Curves A represent the spectrum of the Kα line, and curves B, C and D are the spectra of this line after being scattered at angles 45°, 90° and 135° respectively with the primary beam. While in Fig.4 the experimental are a little erratic, it may be noted that in this case the intensity of the X-rays is only about 1/25,000 as great as if the spectrum of the primary beam were under examination, so small variations produce a relatively large effect.

It is clear from these curves that when a homogeneous X-ray beam is scattered by graphite it is separated into two distinct parts, one of the same wavelength as the primary beam, and the other of increased wavelength. Let us call these the *unmodified* and the *modified* rays respectively. In each curve the line P is drawn through the peak of the curve representing the primary line, and the line T is drawn at the angle at which the scattered line should appear according to Eqn1. In Fig.4, in which the settings were made with greater care, within an experimental error of less than 1 minute of arc, or about $0.001A$, the peak of the unmodified ray falls upon the line p and the peak of the modified ray falls upon the line T. The wavelength of the modified ray thus increases with the scattering angle as predicted by the quantum theory, while the wavelength of the unmodified ray is in accord with the classical theory.

There is a distinct difference between the widths of the unmodified and the modified lines. A part of the width of the modified line is due to the fact that the graphite radiator R subtends a rather large angle as viewed from the target T, so that the angles at which the rays are scattered to the spectrometer crystal vary over an appreciable range. As nearly as I can estimate, the width at the middle of the unmodified line due to this cause is that indicated in Fig.4 by the two short lines above the letter T. It does not appear, however, that this geometrical consideration is a sufficient explanation for the whole increased width of the modified line, at least for the rays scattered at 135°. It seems more probable that the modified line is heterogeneous, even in a ray scattered at a definite angle.

The unmodified ray is usually more prominent in a beam scattered at a small angle with the primary beam, and the modified ray more prominent when scattered at a large angle. A part of the unmodified ray is doubtless due to regular reflection from the minute crystals of which the graphite is composed. If this were the only source of the unmodified ray, however, we should expect its intensity to diminish more rapidly at large angles than is actually observed. The conditions which determine the distribution of energy between these two rays are those which determine whether an X-ray will be scattered according to the simple quantum law or in some other manner. I have studied this distribution experimentally by another method, and will discuss it in another paper [4]; but the reasons underlying this distribution are puzzling.

Experiments with shorter wavelengths. These experiments have been performed using a single wavelength, $\lambda = 0.711A$. In this case we find for the modified ray a change in wavelength which increases with the angle of scattering exactly in the manner described by Eqn1. While these experiments seem conclusive, the evidence would of course be more complete if similar experiments had been performed for other wavelengths. Preliminary experiments similar to those here described have been performed using the K radiation from tungsten, of wavelength about 0.2A. This work has shown a change in wavelength of the same order as that using the molybdenum Kα line. Furthermore, as described in earlier papers [5], absorption measurements have confirmed these results as to order of magnitude over a very wide range of wavelengths. This satisfactory agreement between the experiments and the theory gives confidence in the quantum formula Eqn1 for the change in wavelength due to scattering. There is, indeed, no indication of any discrepancy whatever, for the range of wavelength investigated, when this formula is applied to the wavelength of the modified ray.

WASHINGTON UNIVERSITY, SAINT LOUIS

MAY 9, 1923.

Footnotes and References:

1) A report on this work was presented before the American Physical Society, Apr. 21, 1923 (Phys. Rev. **21**, 715 (1923)).

2) A.H. Compton, Bull. Nat. Res. Coun., No.20, p.18 (October 1922); Phys. Rev. **21**, 207 (abstract) (Feb. 1923); Phys. Rev. **21**, 483 (May, 1923).

3) Cf. also P. Debye, Zeits. f. Phys. **24**, 161 (April 15, 1923).

4) A.H. Compton, Phil. Mag. (in printer's hands).

5) Cf. e.g., A.H. Compton, Phys. Rev. **21**, pp.494-6 (1923).

Paper II·6: Excerpt from Proc. N.A.S. **11**, 303 (1925).

ON THE MECHANISM OF X-RAY SCATTERING

BY ARTHUR H. COMPTON
DEPARTMENT OF PHYSICS, UNIVERSITY OF CHICAGO
Read before the National Academy April 28, 1925

It is now well known that whereas the usual form of the wave theory is inadequate to account for the alteration of the wave-length of X-rays when scattered, this may be explained simply on the hypothesis that the X-rays consist of quanta, each of which interacts with an individual electron. Strong support for this hypothesis is found in the discovery that there appear electrons each with about the momentum it should have acquired if it had deflected an X-ray quantum. That these electrons are associated with scattered X-rays is evident from the fact that there exists on the average about one such electron for each quantum of scattered X-rays. Nevertheless the great weight of the evidence from many sources for the wave theory of radiation makes the acceptance of this radiation quantum hypothesis difficult.

It is improbable that the idea of spreading waves of radiant energy can be successfully reconciled with these studies of X-ray scattering without abandoning the principles of the conservation of energy and momentum. However, by making this bold step, Bohr, Kramers and Slater have apparently been able to incorporate them within the general scheme of wave theory.

While discussing this matter in November 1923, Professor Swann called the attention of Professor Bohr and myself to the fact that a crucial test between the two theories could be made if it were possible to detect simultaneously a scattered quantum and the recoiling electron associated with it. For on any spreading wave theory, including that of Bohr, Kramers and Slater, there should be no correlation between the direction of ejection of the recoil electron and that of the effect of the scattered quantum. On the quantum view, however, if the recoiling electron is ejected at an angle θ with the incident X-ray beam, the scattered quantum should appear at an angle ϕ such that

$$\tan \phi = -\frac{1}{1+\alpha} \cot \frac{1}{2}\theta,$$

where $\alpha \equiv h/mc\lambda$.

Two methods of making this test suggest themselves. One is to use two point counters of the type developed by Geiger, Kovarik and others, one to receive the recoil electron and the other to catch the scattered quantum. The quantum hypothesis would predict simultaneous impulses in both chambers when set at the proper angles. This method has been employed by Mr. R.D. Bennett in our laboratory, and has recently been suggested also by Bothe and Geiger [1] as a means of testing Bohr's theory.

The second method which Mr. Simon and I have been using, is to take cloud expansion photographs showing simultaneously a recoil electron and a secondary β-ray track produced by the scattered X-ray quantum. To increase the probability of emission of a β-ray by a scattered quantum, diaphragms of thin lead foil are placed inside the chamber. Thus about 1 in 50 of the recoil electrons is found to have a secondary β particle associated with it.

Both methods have yielded provisional results. Mr. Bennett has connected each of a pair of headphones with one of the counting chambers through a 3 stage amplifier, and has listened for simultaneous impulses in the two phones. He uses a thin strip of mica to scatter the X-rays, and places the counting chambers inside an evacuated vessel to avoid the effects due to the curving of recoil electrons in air. His results for three different settings of the electron counter are shown in Fig.1. The fact that the coincidences are more frequent when the quantum counter is near the theoretical angle is in support of the quantum theory. Mr. Bennett hopes to be able to publish soon results obtained by a recording method, to avoid the uncertainties inherent in auditory observations of this type.

* * * * * * * * * * * * * * * * * * **

Compton here pioneers coincidence experiments in a two-counter experiment and in a cloud-chamber experiment to prove that each X-ray scattered at an angle θ is associated with a simultaneous recoil electron at a given angle $\phi(\theta)$. One hardly knows whether to laugh or cry at the plight of Mr. Bennett listening for coincidences. Nonetheless, these experiments closed the door finally on the defenders of the non-quantum theory of radiation.

* * * * * * * * * * * * * * * * * * **

Mr. Simon and I have taken about 750 stereoscopic cloud expansion photographs, using apparatus such as that shown diagrammatically in Fig.2. Our chief difficulty has been to eliminate stray X-rays, which give rise to meaningless tracks often indistinguishable from those due to scattered quanta. The results of the last 350 plates, in which these rays were reduced to a relatively low intensity, are shown in Fig.3. Here I have plotted the deviation Δ of the observed tracks from the theoretical angle. On the spreading wave theory, the values of Δ should be nearly uniformly distributed between 0 and 180 degrees. On the quantum theory they should be concentrated near 0, as is obviously the case. The occurrence of tracks at other angles is explicable as due in part to stray X-rays and in part to the method of plotting the results.

A more detailed account of the work will be published when further experiments, which are now in progress, have been completed. The results already obtained, however, permit us to state with very little uncertainty that the direction in which a quantum of scattered X-rays can produce an effect is determined at the moment it is scattered, and can be predicted from the direction of motion of the recoil electrons. In other words, *scattered X-rays proceed in directed quanta.*

It is possible to clothe this statement in the language of the wave-theory if we keep in mind that a wave with a single quantum of energy can produce an effect in only one direction [2].

Footnotes and References:
1) W. Bothe and H. Geiger, Zeits. f. Phys. **26**, 44 (1924).
2) Since this paper was read before the Academy, I have received a letter from Dr. Bothe informing me that H. Geiger and he have also observed the coincidences demanded by the quantum theory but contrary to the theory of Bohr, Kramers and Slater. That work will soon be described in *Die Naturwissenschaften.*

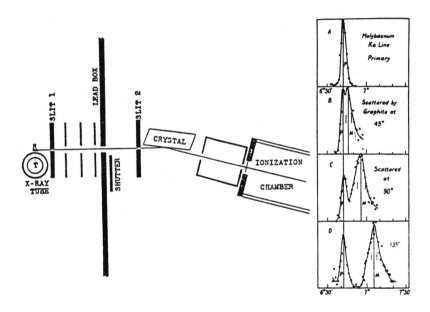

Schematic diagram of Compton's Apparatus. Spectra of the molybdenum K_α line after being scattered at different angles by the electrons of carbon. [from A.H. Compton, *X-Rays and Electrons* (Van Norstrand, NY, 1926) Pp.262,263.]

Chapter III

BOHR: "If we give 'n' different integer values, we get a series of configurations ⋯ We assume that these are states of the system in which there is no radiation."

Phil. Mag. **26**, *1 (1913), p.5.*

§ III-1. Introduction.

In April 1913, Niels Bohr took his place at center stage of the Quantum Physics drama with the invention of the Bohr atom and its stationary quantum states, whose transition radiations reproduced in exquisite detail and for the first time the Balmer lines of the spectrum of the hydrogen atom (here as **Paper III·1**).

Bohr modified Planck's radiation theory of atomic oscillators – vibrating at constant frequencies under elastic forces independent of their energy content – to accommodate Rutherford's atom with electrons bound by Coulomb inverse square forces. Dimensional analysis of quantum Coulomb orbits differs critically from that of the classical Thomson atom where there is no characteristic length defined for electron orbits in the purely Coulomb classical dynamics of the Rutherford atom. However, the introduction of Planck's quantum of action $\hbar$ changes the dimensional arguments completely and allows the definition of a new characteristic length – the Bohr radius $R_B = (\hbar c/e^2)(\hbar c/m_e c^2) = .528A$ – and a new characteristic energy – $e^2/2R_B = 13$ev $\equiv 1$ Rydberg – appropriate for atomic magnitudes.

Bohr modified and extended Planck's postulates to define the quantum dynamics of the Rutherford atom:
1) The dynamical equilibrium of the atomic system in "stationary" states is governed by the ordinary laws of mechanics, which however do not hold for the "transitions" of the atomic system between the various stationary states. 2) Energy is

radiated (or absorbed) not in a continuous way as in ordinary electrodynamics, but only during the transitions between different stationary states. 3) The radiation emitted during the transition between two stationary states has a frequency $f = \omega/2\pi$ determined by the total energy emitted E through the relation $E = \hbar\omega$, where $\hbar$ is Planck's constant. 4) The angular momentum of the electron around the nucleus is an integral multiple of $\hbar$. These postulates led directly from Planck and Rutherford to an atomic theory in perfect accord with the spectral results of Balmer and Rydberg.

Eleven years and a Nobel Prize later, from Olympian heights, with his great shaggy head, rumbling voice, tortuous thinking, armed primarily with his correspondence principle (which never quite earned capitals) and his great conscience and total integrity, Bohr summarized the confused state of the quantum theory of radiation in his paper with Kramers and Slater (here as **Paper III·2**).

Bohr's maddening – but ultimately infinitely fruitful – ruminations about the correspondence principle and the effort to infer new physics from the old [1], are included here lest anyone think that physics was easier in a simpler day, or that Einsteins and Comptons might be found on every street corner. Confusion, failed postulates and tortured conclusions were the order of that day too, even in the accomplished hands of Niels Bohr. Or maybe especially in the hands of Niels Bohr because of his stubborn refusal to abandon any part of classical physics until absolutely compelled to do so. In Bohr's defense, his ideas and his way of expressing them have left their impact to this day – for better or for worse. The Copenhagen interpretation of quantum mechanics eventually hammered out at his decree, bridges the gulf between the literally undescribable quantum world with its discontinuous transitions, via the correspondence principle, to the intuitively comprehensible classical world. More and more, however, we find ourselves rationalizing the classical in terms of our now more highly developed quantum intuitions.

What is fascinating but also frustrating about Bohr is the way he seems to drift in and out of a dreamlike state: in it he is able, with great labor, to spell out immortal insights; out of it, he is – well – out of it! The man seemed unable to make a snap judgement or to ignore any fact even on an interim basis while in pursuit of greater objectives. He ends this paper with a long paragraph on the unresolved nature of the apparent continuity of radiation pressure, followed by another on the nature of emission of radio waves from an antenna.

In his dreamstate, Bohr sets forth the defining terms by which we understand the quantum mechanics that was yet to come: "··· stationary states ··· otherwise stable ··· classical electrodynamics ··· not valid ··· a transition between two stationary states ··· by ··· emission ··· of a pulse of frequency $hf = E_1 - E_2$ ··· it seems necessary ··· to be content with a probabilistic interpretation ··· the correspondence principle ··· provides ··· the estimation of transition probabilities. ··· a problem with the time interval in which the transition actually radiates ···."

Out of it – chaos reigns. " ··· spontaneous transitions are induced by the virtual radiation field self-coupled with its own virtual oscillators ··· induced transitions ··· are due to virtual radiation originating on other atoms ··· different from the classical picture ··· we give up any causal connection ··· and also give up the ··· principles of energy and momentum conservation ··· there is definite disagreement with the conservation of energy and momentum ··· conservation of both energy and momentum ··· statistical laws ··· radiation is an essentially continuous phenomenon ··· we see the scattering of the radiation by the electron as a continuous process ··· discontinuous changes ··· ascribed to radiation pressure." And so on and on.

Who could have guessed that Bohr – twelve years later, at age 40, scooped on the mathematical creation of the quantum theory – stood on the threshold of a great second chapter of his career, but now as the philosopher sage of Quantum Mechanics.

Footnotes and References:
1) N. Bohr, Phil. Mag. **26**, 1-25 (1913) (here as **Paper III·1**). This is Part 1 "On the Constitution of Atoms and Molecules" of the famed *Trilogy* which was completed in the same volume of *Philosophical Magazine* with Part 2 "Systems Containing Only a Single Nucleus" (pp. 476-502), and Part 3 "Systems Containing Several Nuclei". These are conveniently reproduced (also in an extremely abridged form) in the memorial volume:

NIELS BOHR A Centenary Volume (Harvard University Press, Cambridge, Mass. and London, England, 1985), Edited by A.P. French and P.J. Kennedy, pp.80-90. The struggle and human drama underlying Bohr's great achievement is highlighted in the biography: *NIELS BOHR, The Man, His Science and the World They Changed* by Ruth E. Moore (Alfred E. Knopf, New York, 1966; The MIT Press, Cambridge, Mass. and London, England, 1985). Each of these books is a treasure trove of photographs, historical detail, and insights into the personal life of the great man.

Paper III·1: Excerpt from Philosophical Magazine **26**, 1 (1913).

On the Constitution of Atoms and Molecules

By N. Bohr, *Dr. Phil. Copenhagen*[*]

Introduction.

To explain experiments on scattering of α-rays, Rutherford [1] has given a theory of the structure of atoms. In his theory, the neutral atom is a positively charged nucleus surrounded by negatively charged electrons kept together by attractive forces from the nucleus. The nucleus is assumed to be the essential part of the mass of the atom, and to have dimensions extremely small compared to the dimensions of the whole atom. The number of electrons is approximately half the atomic weight. Rutherford has shown that the existence of the nucleus is necessary to explain large angle scattering of the α-rays [2].

To explain the properties of matter with this model, we meet with serious difficulties from the apparent instability of the system of electrons: difficulties purposely avoided in the Thomson model [3]. ······ The principal difference between the models of Thomson and Rutherford is that the Thomson model allows certain electron configurations which have a stable equilibrium; apparently such stable configurations do not exist for the Rutherford model. The difference between the models is the existence of a characteristic length – the radius of the positive sphere – defining the linear dimension of the Thomson atom. Such a length does not appear in the Rutherford atom, *i.e.*, a length cannot be defined from the charges and masses of the electrons and nucleus.

Considerations of this kind have undergone essential alterations in recent years in the development of the theory of radiation and its direct affirmation by experiments on such phenomena as specific heats, photoelectric effect, X-rays, *etc.* The result is a general acknowledgement of the inadequacy of classical electrodynamics in describing systems of atomic size. ······ it seems necessary to introduce a quantity foreign to classical electrodynamics, *i.e.* Planck's constant $\hbar$, the elementary quantum of action. [Note added: To clarify Bohr's results, we use modern notation, including $\hbar$ in place of h.] By the introduction of this quantity the question of the stable configuration of the electrons in the atom is essentially changed. With it and the mass and charge of the particles, a length of suitable magnitude can be defined.

* * * * * * * * * * * * * * * * * * **

With only the electron charge e and mass m of classical nonrelativistic theory, no characteristic length or time can be defined. With the speed of light and Einstein's mass-energy relation, the classical radius of the electron is defined:

$$R_e = \frac{e^2}{mc^2} = 2.8 \times 10^{-13} \text{cm}.$$

With Planck's quantum of action $\hbar$, two more lengths – the Compton wavelength of the electron

$$R_c = \frac{\hbar c}{e^2} R_e = 137 \times R_e = 3.8 \times 10^{-11} \text{cm},$$

and the Bohr radius

$$R_b = \left(\frac{\hbar c}{e^2}\right)^2 R_e = 137^2 \times R_e = .53 \times 10^{-8} \text{cm}$$

become available. Only the Bohr radius has a magnitude appropriate for atomic dimensions.

* * * * * * * * * * * * * * * * * * **

This paper applies the above ideas to Rutherford's model as a basis for a theory of atoms. In the present paper the mechanism of the binding of electrons by a positive nucleus is discussed in relation to Planck's theory. It is possible to account in a simple way for the line spectrum of hydrogen.

I wish here to express my thanks to Prof. Rutherford for his kind and encouraging interest in this work.

PART I. – BINDING OF ELECTRONS BY POSITIVE NUCLEI.

§1. *General Considerations.*

The inability of classical electrodynamics to account for the properties of atoms from Rutherford's model is clear if we consider a simple system of a positively charged nucleus of small dimensions and an electron in a closed orbit around it. Let us assume that the mass m of the electron is small compared to that of the nucleus, and that its velocity is small compared to that of light. If we ignore energy radiation, the electron will describe stationary elliptical orbits with frequency ω, major-axis $2R$, and binding energy E. [Note added: To avoid circumlocution and messy formulas, all frequencies will be angular frequencies $\omega = 2\pi f$ with f the circular frequency.] With electron charge $-e$ and nucleus charge eZ, we get

$$\omega = \frac{2\sqrt{2}}{e^2 Z} \frac{E^{3/2}}{\sqrt{m}}, \quad 2R = \frac{e^2 Z}{E}. \tag{1}$$

It can easily be shown that the mean value of the kinetic energy of the electron is equal to E. If the value of E is not given, there will be no values of ω and R defined for the system.

Now take the effect of the energy radiation into account, calculated in the ordinary way from the acceleration of the electron. The electron orbit will no longer be stationary. E will increase continuously, the electron will approach the nucleus describing orbits of smaller and smaller dimensions, and with greater and greater frequency; the electron gains kinetic energy as the whole system radiates energy. The process will go on until the dimension of the orbit is the same magnitude as the dimension of the electron or the nucleus. The energy radiated during the process will be enormously larger than that radiated by ordinary atomic processes.

It is obvious that the behavior of such a system is very different from that of an atom occurring in nature. $\cdots\cdots$ The essential point in Planck's theory of radiation [4] is that the radiation from an atomic system does not take place in the continuous way assumed in ordinary electrodynamics, but takes place in distinctly separate emissions, the amount of energy radiated from an atomic oscillator of frequency ω being equal to $n\hbar\omega$, where n is an integer, and $\hbar$ a universal constant [5].

Returning to the simple case above, let us assume that the electron starts at rest at a great distance from the nucleus. Assume further that the electron finally settles down into a stationary circular orbit around the nucleus.

Now let us assume that during the binding of the electron uniform radiation is emitted at a frequency ω, equal to half the frequency of revolution of the electron in its final orbit; then from Planck's theory, we might expect that the amount of energy emitted during the process is $n\hbar\omega$, where n is some integer. $\cdots\cdots$ The question of the rigorous validity of these assumptions, and also of the application made of Planck's theory, will be discussed further in §3.

Putting

$$E = \frac{1}{2}n\hbar\omega, \tag{2}$$

[Note added: The factor 1/2 is clearly after-the-fact fine tuning.] we get from Eqn1

$$E = \frac{mc^2}{2n^2}\left(\frac{e^2 Z}{\hbar c}\right)^2, \quad \text{and } 2R = 2n^2\left(\frac{\hbar}{mc}\right)\left(\frac{\hbar c}{e^2 Z}\right). \tag{3}$$

If we give n different integer values, we get a series of values for E, R and ω corresponding to a series of configurations of the system. *We are led to assume that these configurations are states of the system in which there is no radiation of energy; states which will be stationary as long as the system is not disturbed from outside.* [Note: Emphasis added.] E is greatest for $n = 1$, which will correspond to the most stable state of the system, *i.e.* will correspond to the ionization energy of the most stable state. Putting $n = 1$ and $Z = 1$, and introducing the experimental values

$$e = 4.4 \cdot 10^{-10}, \quad \frac{e}{m} = 5.31 \cdot 10^{17}, \quad h = 6.5 \cdot 10^{-27},$$

we get

$$2R = 1.1 \cdot 10^{-8} \text{cm}, \quad \omega = .98 \cdot 10^{15}/\text{sec.}, \quad \frac{E}{e} = 13 \text{ volts.}$$

[Note added: Hereafter we abbreviate $\alpha \equiv e^2/\hbar c \simeq 1/137$.]

These are the same order of magnitude as the dimension, frequency and ionization-potential of atomic hydrogen. $\cdots\cdots$ first pointed out by Haas [6]. $\cdots\cdots$ Systems of this kind are discussed by Nicholson [7]. $\cdots\cdots$ The atoms are supposed to consist simply of a ring of a few electrons surrounding a positive nucleus of negligibly small dimension. $\cdots$ Nicholson has obtained a relation to Planck's theory showing that the ratios of the wavelength of different sets of lines in the coronal spectrum can be understood with great accuracy by assuming that the ratio between the energy of the system and the frequency of rotation of the ring is equal to an integral multiple of Planck's constant. $\cdots$ Serious objections $\cdots$ [Note added: We will not pursue Nicholson's work.] $\cdots$ the difficulties disappear if we consider the problems from the point of view taken in this paper. The principal assumptions are:

1) that the stationary states can be discussed by ordinary mechanics, but the passage [Note added: Hereafter 'transition'.] between different stationary states cannot be treated on this basis.

2) That the transition is accompanied by the emission of *homogeneous* radiation, with the relation between frequency and energy given by Planck's theory.

$\cdots$ ordinary mechanics cannot have an absolute validity, but will only hold in calculations of certain mean values of the electrons. $\cdots\cdots$ The second assumption is in obvious contrast to the ordinary ideas of electrodynamics, but appears to be necessary in order to account for experimental facts. $\cdots\cdots$ We shall first show how we can account for the line-spectrum of hydrogen.

§2. *Emission of Line-Spectra.*

Spectrum of Hydrogen.– General evidence indicates that an atom of hydrogen consists simply of a single electron rotating around a positive nucleus of charge e [8]. $\cdots\cdots$ If we put $Z = 1$ in Eqn3, we get the total amount of energy radiated during the formation of one of the bound states,

$$E_n = \frac{mc^2}{2n^2}\alpha^2.$$

The amount of energy emitted ΔE in the transition from a state n_1 to the state n_2 is

$$\Delta E(n_1 \to n_2) = E(n_2) - E(n_1) = \frac{mc^2}{2}\alpha^2 \left(\frac{1}{n_2^2} - \frac{1}{n_1^2}\right).$$

If we now suppose the radiation to be homogeneous of frequency ω, we get

$$\omega = \frac{\Delta E}{\hbar} = \frac{mc^2}{2\hbar}\alpha^2 \left(\frac{1}{n_2^2} - \frac{1}{n_1^2}\right). \tag{4}$$

This expression accounts for the lines in the spectrum of hydrogen. For $n_2 = 2$ and varying n_1, we get the ordinary Balmer series. With $n_2 = 3$, we get the series in the infra-red observed by Paschen and previously suspected by Ritz. If we put $n_2 = 1$ and $= 4, 5 \cdots$ we get series in the ultra-violet and the extreme infra-red, respectively, which have not been observed but may be expected.

$\cdots\cdots$ The agreement between theoretical and observed values is inside the uncertainty due to experimental errors in the constants entering the theoretical values. [Note added: Around 6% in this Part 1, but revised in Part 2 to $\sim 1\%$, principally on the basis of Millikan's (1912) 1.7% increase in the electron charge.] $\cdots\cdots$

§3. *General Considerations continued.*

We now return to the discussion of the special assumptions used in deducing Eqn3 for the stationary states. We have assumed that the different stationary states correspond to an emission of a different number of energy quanta. $\cdots\cdots$ this assumption may be regarded as improbable; as soon as one quantum is sent out the frequency is altered. We can actually change this assumption and still retain Eqn2 and the formal analogy with Planck's theory.

First of all, it has not been necessary to assume that radiation has been emitted corresponding to more than a single quantum. $\cdots\cdots$ Let us assume that the ratio of the total energy emitted and the angular frequency of revolution of the electron is given by $E(n) = f(n) \cdot \hbar\omega_{rev}$, instead of (2). In this case, we get

$$E(n) = \frac{mc^2}{8f^2(n)}(\alpha Z)^2, \quad \hbar\omega_{rev} = \frac{mc^2}{8f^3(n)}(\alpha Z)^2.$$

Instead of Eqn4 we get for the radiation

$$\hbar\omega_{rad} = E(n_2) - E(n_1) = \frac{mc^2}{8}(\alpha Z)^2 \left(\frac{1}{f^2(n_2)} - \frac{1}{f^2(n_1)} \right).$$

In order to get the Balmer series we must put $f(n) = x \cdot n$. To determine the parameter x, consider the radiation in the transition $n_1 = N \to n_2 = N - 1$ for $N \gg 1$. The frequency of the radiation emitted

$$\hbar\omega_{rad}(N \to N - 1) = \frac{mc^2}{8x^2}(\alpha Z)^2 \cdot \frac{2N - 1}{N^2(N - 1)^2} \simeq \frac{mc^2}{8x^2}(\alpha Z)^2 \cdot \frac{2}{N^3},$$

must coincide for large N with the orbital frequency of the electron

$$\hbar\omega_{rev}(N) = \frac{mc^2}{8x^3}(\alpha Z)^2 \cdot \frac{1}{N^3} \simeq \hbar\omega_{rev}(N - 1),$$

which requires that $x = \frac{1}{2}$, leading again to Eqn2 and 3 for stationary states.

* * * * * * * * * * * * * * * * * * **

This is the first note in a recurrent theme in Bohr's work – that of the Correspondence Principle – which requires the quantum theory to merge smoothly with the classical theory in the limit of large quantum numbers and small energy differences.

* * * * * * * * * * * * * * * * * * **

If we consider the transition $N \to N - n$ where $N \gg n \sim 1$, we get $f = n/2$. .... may be interpreted as an electron rotating around a nucleus in an elliptical orbit emitting radiation of frequencies $n\omega_{rev}$ with ω_{rev} the angular frequency of revolution.

We are led to assume that the interpretation of Eqn2 is not that the different stationary states correspond to the emission of different numbers of quanta, but that the frequency of the energy quanta emitted during the transition from the state of zero energy to one of the different stationary states, is equal to different integer multiples of $\omega_{rev}/2$, where ω_{rev} is the frequency of revolution of the electron in the state considered. From this assumption we get exactly the same expression for the hydrogen spectrum. We may regard our preliminary considerations as a simple form of the theory.

······ we arrive at exactly the same expression if we assume

1) the radiation is emitted in quanta of energy $\hbar\omega$, and

2) the frequency of the radiation emitted during transition between successive stationary states coincides with the frequency of revolution of the electron, in the region of slow vibrations.

······ we are justified in expecting – if the whole argument is a sound one – an absolute agreement between the values calculated and observed, and not only an approximate agreement. Eqn4 may therefore be of value for the experimental determination of the constants e, m and $\hbar$.

While there obviously can be no question of a mechanical foundation of the calculations given in this paper, it is possible to give a very simple interpretation of the result. The angular momentum L of the electron around the nucleus, satisfies for a circular orbit

$$L = \frac{2T}{\omega_{rev}},$$

where ω_{rev} is the angular frequency of the revolution and T is the kinetic energy. For a circular orbit we have $T = E$, and from (2) we get $L(n) = nL_0$ where $L_0 = \hbar$.

If we assume that the orbit of the electron in the stationary states is circular, the result can be expressed by the simple condition: The angular momentum of the electron around the nucleus in a stationary state of the system is an integer multiple of a universal value, independent of the charge on the nucleus. The possible importance of the angular momentum in the discussion of atomic systems in relation to Planck's theory is emphasized by Nicholson [9]. ·········

* Communicated by Prof. E. Rutherford, F.R.S. April 5, 1913.

Footnotes and References:

1) E. Rutherford, Phil. Mag, **21**, 669 (1911).

2) Geiger and Marsden, Phil. Mag. April 1913.

3) J.J. Thomson, Phil. Mag. **7**, 237 (1904).

4) M. Planck, Ann. d. Phys. **31**, 758 (1910); **38**, 642 (1912).

5) A. Einstein, Ann. d. Phys. **17**, 132 (1905) (included here as **Paper II·1**); **20**, 199 (1906); **22**, 180 (1907).

6) A.E. Haas, Jahrb. d. Rad. u. El. **7**, 261 (1910). [See also: A. Haas, *Introduction to Theoretical Physics,* Vol.II (Constable, London 1925) p.36.]

7) J.W.Nicholson, Month. Not. Roy. Astr. Soc. **72**, 49,139,177,693,729 (1912).

8) N. Bohr, Phil Mag. **25**, 24 (1913). See also J.J. Thomson, Phil. Mag. **24**, 672 (1912).

9) J.W. Nicholson, *loc. cit.* p.679.

Paper III·2: Excerpt from Zeits. f. Phys. **24**, 69 (1924).

Über die Quantentheorie der Strahlung

von **N. Bohr, H.A. Kramers** und **J.C. Slater**

in Kopenhagen. ·

(Eingegangen am 2. Februar 1924.)

We describe the mechanism of optical spectra with the quantum theoretic explanation but keeping the classical laws of propagation of radiation. The continuous emission of radiation is combined with the discontinuous atomic processes by probability laws introduced by Einstein. Virtual oscillators are used to describe the discontinuous processes by the correspondence principle, but interpreting this principle in a way somewhat different from usual.

Introduction. Understanding the interaction between radiation and matter requires that we explain two different and apparently contradictory features: the interference phenomenon on which all optical instruments depend, which is a continuous effect characteristic of the wave theory of light as in electromagnetic theory; and the exchange of energy and momentum between radiation and matter, which is an essentially discontinuous event. *The theory of light quanta appears to contradict the wave theory of light.* [Note added: Our italics.] At the present state of knowledge it seems impossible to explain the nature of atomic processes. The result is that one must give up for the present trying to describe in detail as a discontinuous process the quantum mechanism for transitions between stationary states. Nevertheless, it does seem possible – using the correspondence principle – to develop a logical picture of the discontinuous atomic process based in a somewhat unconventional way on the continuous radiation theory.

* * * * * * * * * * * * * * * * * **

The change – to this day – has been one of interpretation only. What changes continuously is the decreasing probability-amplitude for the system to consist of the excited atom, and the increasing probability-amplitude for the system to consist of the de-excited atom accompanied by an outgoing photon. These continuously variable probability-amplitudes are sometimes sensibly described by a correspondence principle calculation of the radiating system. What is discontinuous is the experimental answer to the question: Which state is the system in? What to this day – after seventy-five years of the most intense scrutiny – remains quantum mechanically undescribable is the (non-)state of the system when it has emitted some fraction of the photon, or said another way, there is still no more detailed description of the 'actual' emission process. This flat statement will of course precipitate wails of dispute from the progenitors of would-be variants of quantum mechanics, all of which are *ad hoc* special pleading involving various 'Rube Goldberg' devices attached to the original so-called canonical quantum theory. See later comments in this regard by Heisenberg (1955).

* * * * * * * * * * * * * * * * * **

The assumption that the atom undergoes transition processes by the action of a virtual radiation field interacting with distant atoms was described by Slater [1]. In this way, a better agreement could be obtained between the physical picture of the electrodynamic theory of light and the theory of light quanta, in which the emission and absorption processes would occur by interacting atoms coupled together in pairs. Kramers then noted that the assumption of close coupling for this process was unnecessary, and concluded that the transition processes in separated atoms are better described as independent of one another. In this paper we discuss the significance which these assumptions might have in the future development of quantum theory; they supplement a recent work by Bohr [2].

* * * * * * * * * * * * * * * * * * **

Slater's idea of admitting only the completed process of emission and absorption lives on in the concept of virtual, and in fact even energy non-conserving but unobserved, intermediate states. For large separations of emitter and absorber, the 'on-energy-shell' part of an amplitude may dominate; but in any case we are concerned only with overall energy conserving processes as observable processes. Even this formal lack of energy conservation can be eliminated in formulations which are 'on-energy-shell' at every step, and involve the unobservable 'off-mass-shell' particles only in intermediate states.

* * * * * * * * * * * * * * * * * * * **

§1. **The Principles of Quantum Theory.** Electromagnetic theory describes not only the propagation of light through free space, but also the interactions of radiation and matter. These include emission, absorption, refraction, scattering and dispersion phenomena; all based on the assumption that the atom contains electrically charged particles which execute harmonic vibrations in stable equilibrium, and which exchange energy and momentum with the radiation field in accordance with the classical electrodynamic laws. The actual observed phenomena exhibit a number of features which contradict the results of classical electrodynamics. One such contradiction occurs – beyond any doubt – in the case of the laws of heat radiation. From the classical theory for emission and absorption of radiation by a harmonic oscillator, Planck found that the observations of heat radiation could be understood only by a new assumption which goes beyond the requirement that only the states of the oscillating particles need be included in the equilibrium distribution. Independent of the radiation mechanism, this result gets immediate support, as Einstein has shown, from experiments on the specific heat of solids. At the same time, Einstein also introduced the well known 'light quantum' theory, in which radiation propagates not as a continuous wave train as in the classical theory, but much more as a discrete unit in which the energy hf is contained in a small space, where h is Planck's constant and f the usual frequency.

* * * * * * * * * * * * * * * * * * **

In this labored and obtuse way, Bohr is slowly accepting Einstein's dynamic light quantum; and Slater's suggestion (a predecessor of Schwinger's 'source theory' – still fifty years in the future) in which the light quantum is a transitory unobserved state between real, realized and observable states of material particles; all refined to Kramers' idea of what became 'virtual' quanta.

* * * * * * * * * * * * * * * * * * **

Even though the great intuitive value of this hypothesis is now clearly obvious in the confirmation of Einstein's predictions on the photoelectric effect, *the theory of light quanta cannot yet be accepted as a satisfactory solution to the problem of light propagation.* [Note added: Our italics.] This is especially true because the frequency f characterizing the quantum of radiation is only defined by experiments involving interference, which is an obvious manifestation of the wave nature of light.

The fundamental obstacle to the quantum theory is the difficulty of combining this idea with accepted results for the structure of the atom. The problem is to explain the emission and absorption spectra of the elements. It must be possible to understand the behavior of so-called 'stationary states' – which are otherwise stable – for which the ideas of classical electrodynamics are not valid. *This stability has the consequence that any change in the state of the atom must always consist of a complete transition process from one stationary state to another.* [Note added: Our italics.] This possibility is dependent on the further assumption, that a transition between two stationary states is accompanied by emission of radiation, which consists of a pulse of harmonic waves whose frequency is

$$hf = E_1 - E_2, \tag{1}$$

where E_1 and E_2 are the energies of the initial and final states of the atom. The reverse transition occurs by absorption. Acceptance of this explanation of the spectra is due to the fact that it has proven possible in all cases to calculate energy values for the stationary states of isolated atoms which describe the observed spectra (G.d.Q., Kap.I, §1). But the concepts of electrodynamics do not allow us to describe the details of the transition mechanism. [Note added: – beyond the continuous growth of the transition amplitude, and the definition of a transition *rate*. See Dirac, **Paper VIII·1**.]

---
* * * * * * * * * * * * * * * * * * **

All of these problems persist to this day and define the problem of quantum measurement. Only if Bohr wants details of *completed* transition processes, are the answers available.

---
* * * * * * * * * * * * * * * * * * **

As to what the transition process means, it seems necessary to be content with a probabilistic interpretation. Such an explanation was used by Einstein [3] in an especially simple derivation of Planck's radiation formula. His derivation was based on the assumption that an atom in a particular high energy stationary state has a given probability per unit time to 'spontaneously' go over into a lower energy state and that the atom, under the influence of external radiation of the appropriate frequency, has a given probability for an 'induced' transition to another stationary state of more or less energy. Imposing the requirement of thermal equilibrium between radiation and matter, Einstein concluded that the exchange of energy hf during a transition is always accompanied by the exchange of momentum hf/c, exactly as would be the case if the transition were due to the emission or absorption of a small particle moving with the speed of light c and the energy hf. The direction of the momentum of the induced emission is the same as the direction of propagation of the irradiating wave, but for the spontaneous transition the direction of the momentum is distributed according to the laws of probability. This result, which must be seen as an argument for the physical reality of light quanta, has an important application in the explanation of the remarkable change of wavelength of radiation scattered by free electrons, in the experiments by A.H. Compton on the scattering of X-rays [4], $\cdots\cdots$ [5,6].

---
* * * * * * * * * * * * * * * * * * **

The results of Pauli [5] and Einstein [6] require a probabilistic interpretation of classical radiation theory to accommodate Compton's unequivocal quantum results.

---
* * * * * * * * * * * * * * * * * * **

The fundamental distinction between the quantum picture of the atomic process and the classical picture based on electrodynamics, must be sought as a natural generalization of the classical picture. This requires that in the limit when we consider a large number of atoms, and when we have stationary states where the difference between neighboring states is relatively small, the classical theory must agree with observations. For the emission and absorption of spectral lines this connection between the two theories led to the 'correspondence principle' which requires a correspondence for each transition between stationary states, with a given harmonic vibration in the electric moment of the atom (G.d.Q., Kap.II, §2). The correspondence principle allows the estimation of transition probabilities and relates the intensity and polarization of the spectral lines to the motion of the electrons in the atom.

The correspondence principle compares the reactions of an atom with a radiation field, and those of corresponding 'virtual' harmonic oscillators of the appropriate frequencies interacting according to classical electrodynamics (G.d.Q., Kap.III, §3). Compton used just such considerations to explain the change in wavelength of X-rays scattered by free electrons as the Doppler effect from a moving source.

* * * * * * * * * * * * * * * * * * **

Bohr evades the obvious point of Compton's result: that *quantum* kinematics provides a description of the scattering process which is clearly preferable to any *ad hoc* attempts like the Doppler shift to cobble together the result.

* * * * * * * * * * * * * * * * * * **

The correspondence principle gives an estimate of the transition probability and the average time an atom spends in a given stationary state; but it does have a problem with the time interval in which the transition actually takes place, which gives rise to great difficulties. This difficulty combines with other paradoxes of the quantum theory to reinforce doubt [7] about whether the interaction between radiation and matter can be described by a causal space-time field, as has been used up to now (G.d.Q., Kap.III, §1). Without completely abandoning the formal character of the theory, it does seem that new developments in the conception of radiation phenomena do allow a description which differs only a little from the traditional ones.

§2. **Radiation and Transition Processes.** We will assume that an atom in a stationary state interacts with other atoms through a virtual radiation field, and that harmonic oscillators – equivalent to the various possible transitions among all stationary states – can be described by classical radiation theory. We also assume that transitions of the atom are governed by this mechanism and by the probability laws of Einstein's theory for the transitions induced by external fields. From our viewpoint, the spontaneous transitions are induced by the virtual radiation field self-coupled with the motion of its own virtual oscillators. The induced transitions of Einstein's theory are due to virtual radiation originating from other atoms.

These assumptions make no difference in the correspondence principle connection with the atomic regime for the frequency, intensity and polarization of the spectral lines. They do lead to a new picture of the transition processes, different from the classical picture which ultimately failed. In this picture, a transition by a given atom from its original state will depend on the presence of other atoms with which it interacts by virtual radiation fields, but not on the occurrence of transition processes in these atoms. [Note added: This remark is a remnant of Slater's idea and is inessential, as Kramers had already pointed out.]

In the limiting case where the stationary states differ little from one another, our picture leads to a connection between the virtual radiation and the motion of the atomic particles, which gradually goes over into that of the classical theory. Neither the motion nor the radiation field in this limiting case changes significantly during the transitions between stationary states. Whatever causes the transitions – which are the essential feature of quantum theory – we give up any causal connection with transitions in distant atoms, and also give up *the direct application of the principles of energy and momentum conservation so characteristic for classical theories.* [Note added: Our italics.]

The applicability of these principles to the interaction between individual atomic systems is limited – according to our understanding of such interactions – because the atoms are so close to one another that the forces due to the radiation field are small compared to the static forces between the atoms. 'Collisions' of this kind are a typical example of the stability of the stationary state. The experimental results are in agreement with the idea that the colliding atoms are in stationary states for these processes as well (G.d.Q., Kap.I, §4) [8]. For interaction between atoms at larger distances from one another, if we assume the independence of the individual transition processes, then there is definite disagreement with the conservation of energy and momentum. We assume that an induced transition has no direct effect on a transition in a distant atom for which the transition energy is the same. If an atom contributed to the induction of a transition in a distant atom by virtual radiation fields, which is correlated with a possible transition among other stationary states, then the atom might very well undergo another of these transitions. The present work offers no immediate proof of this assumption; it should be emphasized that the assumed independence of the transitions might just be a way to obtain a simple description of the interaction between radiating atoms in which the laws of probability are dominant. This independence reduces conservation of both energy and momentum to statistical laws $\cdots\cdots$

* * * * * * * * * * * * * * * * * * **

Chaos reigns.

* * * * * * * * * * * * * * * * * * **

We therefore seek the basis of the statistical conservation of energy and momentum not in a failure of classical electrodynamic laws of wave propagation in free space, but $\cdots\cdots$ [Note added: There follows a very obscure discussion of the assumed failure of energy and momentum conservation, which we will ignore.]

§3. **Possibility of Interference of Spectral Lines.** Before we turn to the problem of the interaction between atoms and a virtual radiation field, we briefly discuss the properties of the field emitted by a single atom, and the possibility of interference of the light radiated by one and the same source. The properties of these fields have been shown to be independent of the details of the transition process, provided that its duration is small compared to a period of the radiation or the motion of the particles in the atom. These processes are supposed to occupy only the briefest of time intervals, within which the atom can interact with other atoms by means of the corresponding virtual fields.

An upper limit for the possibility of interference is the mean time the atom remains in the initial state. ·······. The possibility of interference is limited by the mean lifetimes of the stationary states. The sharpness of a spectral line depends not only on the length of the wave train of the transition, but also on any uncertainty in the frequency resulting from the Planck connection between the frequency of the spectral lines and the energy of the stationary states. It is interesting to note that the limit on the sharpness of the spectral lines corresponds to a limit in the accuracy for the definition of the energy of the stationary states. In fact the postulate of the stability of the stationary states places a limit with which the motion of these states can be described by classical electrodynamics, an *a priori* limit which makes it evident that the action of the virtual radiation fields results not in a continuous change in the motion of the atoms, but in the inducing of transitions during which the energy and momentum of the atoms undergoes a finite change (G.d.Q., Kap.II, §4).

* * * * * * * * * * * * * * * * * **

Here Bohr anticipates almost exactly the expression of Heisenberg's uncertainty principle. Bohr's philosophy of quantum mechanics – predictions must reflect the actual state of the observers knowledge rather than some idealized but unattained or even unattainable state of knowledge – is set forth here in its primitive state.

* * * * * * * * * * * * * * * * * **

In the limiting case where the stationary states differ only a little, the above limit of the coherence of the individual waves corresponds to the accuracy with which the radiation frequency is known from Eqn1, when the essential and unavoidable uncertainty in the definition of the two states is included. In the general case, where the motions in the two states can be very different from one another, the above limit on the interferability of the wave trains is closely connected with the definition of the motion in each stationary state which constitutes an initial state of the transition. Even here, we should note that the observable sharpness of the spectral lines themselves can be determined from Eqn1, but one must combine the lack of knowledge in the definition of the final state with that of the initial state as one would combine independent errors.

Just this effect of the lack of knowledge of the two stationary states on the sharpness of the spectral line makes possible the existence of a reciprocity between the shape of a line when seen in emission and when seen in absorption, since it must satisfy Kirchoff's law for thermal equilibrium. The apparent deviations from this law, particularly for the number and character of the lines, which are often evident in the striking difference between the emission and absorption spectrum of an element, can be directly explained in quantum theory by the difference in the statistical weights of the atoms in their stationary states under different external circumstances.

Closely related with the sharpness of spectral lines, is the question of the spectrum from atoms when the external forces change significantly in a mean lifetime of the stationary state, as in certain of Stark's experiments on the influence of electrical fields on spectral lines. [Note added: The time scale for adiabatic environmental changes must be large compared to atomic *periods* and to the duration of any individual quantum transition, but can still be small compared to atomic *lifetimes*, a point not grasped at this stage.] In these experiments, the radiating atoms move with great speeds and in time intervals only a small fraction of a lifetime can experience completely different field intensities. Nevertheless it is clearly found that the radiation from atoms moving at each point is influenced in the same way by the electric field at that point as would be the radiation from atoms at rest experiencing the constant effect of the field at that point. Contrary claims by various authors would give rise to difficulties because there is always a Stark effect in agreement with the picture used in this treatment. If the motion is uniform, as the atom passes through a field the field changes continuously. Nonetheless, the outgoing radiation will be the same as if the atom had moved the whole way in a field of the same intensity, provided that the radiation from other parts of the path is stopped from reaching the observer. One sees that a far reaching reciprocity between emission and absorption is guaranteed, and thanks to our picture a symmetry connects the transition process and the radiation in the two cases.

§4. **Quantum Theory of Spectra and Optical Phenomena.** In quantum theory, optical phenomena are ultimately transition processes, whereas their observation involves continuous wave trains, which are characteristic of the classical electrodynamic theory of the propagation of light through material media. In this theory, reflection, refraction and dispersion or rescattering of light are described as the result of electromagnetic forces of the radiation field driving oscillations of the electrically charged particles in the individual atoms. [Note added: This sequence is the essence of Bohr's understanding of the ultimate measurement process which always 'ends' the quantum propagation in a chaotic 'classical' shower event.] The stability of the stationary states seems at first sight to involve a fundamental difficulty with this point. However, the opposite conclusion would follow from the correspondence principle which requires the effect of an atom on a radiation field to be a reradiated field resulting from the driven virtual harmonic oscillators which are related to the various possible transitions. However, even here, there is a subtle point to understand the emission and absorption spectra in terms of the virtual oscillators. We will show in a later work that the present conception of a quantitative dispersion theory can be carried out. Here we content ourselves to emphasize again that the continuous character of these optical phenomena does seem to make impossible any representation directly involving transition processes in the propagating medium.

An instructive example of this analysis occurs in the absorption spectrum. One assumes that the absorption in a monatomic vapor of light whose frequency coincides with lines in the emission spectrum, is caused by transitions induced by each wave train in the incident radiation. That these lines do appear is the result of the absorption of the incident radiation which involves the same kind of outgoing secondary spherical wave as the irradiated atom; the induced transitions play a role by which the statistical conservation of energy is achieved. The presence of coherent secondary radiation is responsible for the anomalous dispersion of the associated absorption lines and is especially important in the selective reflection at the boundary of a metallic vapor under high pressure [9]. Induced transitions between stationary states are also observed in fluorescent radiation, where a few atoms have been irradiated to a state of higher energy. Similarly, fluorescent radiation can be induced by mixing different gases. The atoms in higher stationary states radiate during collisions, which causes an appreciable increase in the probability of atoms to return to their normal state. Similarly, the phenomena of absorption, dispersion and reflection which occur during mixing can be explained as the collision broadening of spectral lines. One also sees absorption phenomena for which an understanding cannot be found using this picture, such as absorption independent of the intensity of the radiation source as it is for reflection and refraction, where transitions do not occur in such a way (vgl. G.d.Q., Kap.III, §3).

Another interesting example is the scattering of light by free electrons. As shown by Compton in the reflection of X-rays by crystals, these X-rays undergo a frequency shift which is different in different directions, and which is in agreement with the properties expected from the classical theory of radiation from a moving source. Compton derived a formal expression for this frequency shift on the basis of the quantum theory of light by assuming that an electron absorbs a quantum of the incident light, and at the same time emits another light quantum in another direction. By this process the electron gets a given velocity in a given direction which is determined – depending on the frequency of the later emitted light – by the law of conservation of energy and momentum, where each light quantum has energy hf and momentum hf/c. As a result of this picture, we see the scattering of the radiation by the electron as a continuous process in which each electron participates by the emission of coherent secondary waves; the incident virtual radiation acts on each electron which undergoes driven oscillations, similar to the scattering one would expect from the classical theory for the velocity that the light source possesses. In this case, the virtual oscillator itself moves with a velocity which is different for each of the irradiated electrons, and corresponds directly to the situation in which – quite surprisingly – the classical ideas survive. In regard to the fundamental deviations from the classical space-time description, where the inherent idea of virtual oscillators is hardly certain in the present state of the theory, a formal interpretation obviously is required. Such an interpretation is even more necessary when one takes account of the contrast with the familiar phenomena for whose description only the wave concept of radiation plays an essential role.

In Compton's theory we assume that the irradiated electron has a certain probability to undergo – at some moment – a certain finite change of its momentum. By this effect, which in quantum theory takes the place of a continuous change of momentum which would accompany a classical scattering of the described kind, the statistical conservation of momentum is guaranteed, analogous to the above described statistical conservation of energy in the case of the absorption spectrum.

* * * * * * * * * * * * * * * * * * **

Compton's result is purely a consequence of relativistic kinematics for the electron and photon and says nothing about the dynamical mechanism of the radiative processes involved. From the Klein-Nishina formula we see that the quantum mechanical amplitude for Compton scattering has a pre-emission contribution which *anticipates* the absorption of the incident photon, so any classical interpretation is subtle at best.

Bohr – in his inimitable fashion – is stubbornly refusing to give up his correspondence principle in a situation where it cannot apply, and is prolonging the confusion which Compton had resolved experimentally (in Bohr's reference [4], and here as **Paper II·4**), and which should have been resolved by Einstein's original remarks on momentum conservation (in 1917, here in Bohr's own reference [3] and also in **Paper II·3A**).

* * * * * * * * * * * * * * * * * * **

In fact, the probability law derived by Pauli for the exchange of momentum during the interaction of free electrons and radiation has essentially the same form as Einstein's law for the transitions between well defined stationary states of an atomic system. ···

A problem which is similar to the scattering of light by free electrons is the scattering by an atom of light whose frequency is high enough to induce transitions in which an electron is completely ejected from the atom. In order to obtain the statistical conservation of momentum, we must assume – as already pointed out by Pauli and again by Smekal [10] – that the transition can occur for which the momentum of the irradiated atom undergoes a finite change without – as for the transition processes usually considered in spectral theory – the motion of the particles in the atom changing. One should note that in our picture of transition processes, the inner motion of the atom does change. ··· It is of primary importance that the transition of the momentum is a discontinuous process, whereas the radiation itself is an essentially continuous phenomenon, in which all irradiated atoms take part, independent of the intensity of the incident radiation. The discontinuous changes in the momentum of the atom are the result of the observable effect on the atom, which is usually ascribed to radiation pressure. This idea is often used to explain the thermal equilibrium between a virtual radiation field and a reflecting surface, as was deduced by Einstein [11], in which he saw support for the quantum theory of light. At the same time, it can hardly be emphasized enough that it was also consistent with the apparent continuity of the actual observations on the radiation pressure. In particular, when we consider a solid body, a change hf/c in its total momentum is completely unnoticeable and so are the unmeasurable changes in the momentum of a body in thermal equilibrium with its surroundings. For actual experiments, the frequency of such events is so great that we must question whether we can ignore the time duration of the individual transitions or, in other words, whether the limit is exceeded in which the principles of quantum theory are valid (vgl. G.d.Q., Kap.II, §5).

The last example we consider, is one in which our quantum concept of optical phenomena merges in a natural way with the usual continuous description of macroscopic phenomena provided by Maxwell's theory. The advantage of our view of the principles of quantum theory, compared to the usual one, can be clearly illustrated in the emission of electromagnetic waves by a radio antenna. A logical description of the phenomenon as outgoing radiation from transition processes – imagined separate from one another – between stationary states of the antenna is *not possible*. From the minuteness of the energy transitions, and the magnitude of the energy radiated per unit time by the antenna, one sees that the duration of individual transition processes can be only an extraordinarily small fraction of a vibration period of the antenna, and that consequently it is not legitimate to describe the result of each transition process as the emission of a wave train of this length. From our present understanding, we describe the electrical oscillations in the antenna as the result of a (virtual) radiation field which – by the probability laws – induces further changes in the motion of the electrons. These changes are practically continuous, since the individual energy steps hf are completely negligible compared to the antenna energy. The characterization of the radiation field as 'virtual', made necessary by the present state of knowledge of atomic processes, automatically loses its significance in this case, where the field has observable interactions with matter and has all the properties of a classical electromagnetic field.

Kopenhagen, Universitets Institut for teoretisk Fysik.

Footnotes and References:

1) J.C. Slater, Nature **113**, 307 (1924).

2) N. Bohr, Zeits. f. Phys. **13**, 117 (1923), cited in the following as G.d.Q. (Die Grundpostulate der Quantentheorie.)

3) A. Einstein, Phys. Zeits. **18**, 121 (1917) (included here as **Paper II·3A**).

4) A.H. Compton, Phys. Rev. **21**, 207 (1923) (see here **Paper II·4,5**); also P. Debye, Zeits. f. Phys. **24**, 161 (1923).

5) W. Pauli, Zeits. f. Phys. **18**, 272 (1923).

6) P. Ehrenfest and A. Einstein, Zeits. f. Phys. **19**, 301 (1924).

7) O.W. Richardson, *The Electron Theory of Matter* (Cambridge 1916).

8) These conclusions hold only if one disregards radiation from collisions. The energy of this radiation is usually very weak, but its very occurrence is of fundamental significance, as has been pointed out by Franck in connection with the explanation of Ramsauer's important results (Ann. d. Phys. **64**, 513; **66**, 546 (1922)). In these experiments involving the collisions between atoms and slow electrons, the electron can fly freely through the atomic volume without being stopped. In some of these cases, a large change in the motion of the electron does occur, which should involve almost as much radiation as a transition process, at least according to the correspondence principle (see F. Hund, Zeits. f. Phys. **13**, 241 (1923)). According to the conclusions reached in this work, it seemed natural that the origin of the radiation was directly in the motion of the electrons, and not primarily in transition processes. However it must be noted that we have here a case which involves significant amounts of classical deceleration radiation, which is very different from the stationary motion and the transition processes for which the existing theory has been developed.

9) R.W. Wood, Phil. Mag. **23**, 689 (1915).

10) A. Smekal, Naturwiss. **11**, 875 (1923).

11) A. Einstein, Phys. Zeits. **10**, 817 (1909) (included here as **Paper II·2**).

Chapter IV

de BROGLIE: "Le principe de Fermat appliqué à l'onde de phase est identique au principe de Maupertuis appliqué au mobile; les trajectoires dynamiquement possibles du mobile sont identique aux rayons possible de l'onde."

"Fermat's Principle applied to the phase of the wave is identical to Maupertuis' Principle applied to the motion; the trajectories dynamically possible for the motion are identical to the possible rays of the wave."

Ann. de Phys., 10e série, t. III (Janvier-Février 1925), p.56.

§ IV-1. Introduction.

de Broglie's discovery of the dual wave-particle nature of massive material particles – generalizing and completing Planck's introduction of the light quantum – was unveiled in his 1924 doctoral thesis [2,3] with marvelous depth and sophistication. One must be astounded at the profound consequences which de Broglie deduced from the elemental observation that – contained within the Einstein-Lorentz transformation of space/time and momentum/energy – there is an invariant phase associated with a moving particle, suggesting a wave associated with the motion. But then de Broglie took the courageous creative leap to give primacy to the wave motion and to search in reverse for particle-like attributes in the wave propagation.

It would have been easy for de Broglie to identify – as we now do – a plane wave with a particle in force-free motion [4], but he took a much more powerful and general qualitative approach to the question of particle trajectories from the point of view of the associated wave. de Broglie's far reaching answer, based on Fermat's Principle of ray optics, supposed that the rays of a wave field of given frequency or, equivalently, energy, in a medium of variable refraction must coincide with the possible trajectories of the associated particle. Somewhat unfortunately, de Broglie did not deal in specifics. If he had played with simple waves in simple situations, he was poised to anticipate Schrödinger and to invert the actual order of the development of quantum mechanics.

de Broglie's thesis is a treasure trove of results made possible by his Nobel Prize winning introduction of wave-particle duality. He anticipates the Bose derivation of Planck's distribution and the Bose-Einstein quantum gas for material particles in great detail, albeit without mentioning the condensation phenomenon. He was able to derive the Bohr-Sommerfeld quantization condition $\oint pdq = nh$. He gave a first explanation of interference and coherence properties of quanta. He gave an elegant covariant derivation of Compton scattering from moving electrons.

de Broglie wandered from his pioneering role and devoted his life to advocacy of his own minority view of wave-particle dualism which finds its modern realization in Bohm's guiding waves accompanying point like material particles [5]. We will return to these minority views in a 1955 review by Heisenberg, who considers them to be a useless excursion to develop an *intuitive scenario* for quantum mechanics with the express design to agree in every observable result with the more abstract interpretation of the mainstream views (created in large part, of course, by Heisenberg himself).

Footnotes and References:

1) N. Bohr, H.A. Kramers and J.C. Slater, Zeits. f. Phys. **24**, 69 (1924) (here as **Paper III·2**).

2) L. de Broglie, Phil. Mag. **24**, 446 (1924) (here as **Paper IV·1**).

3) L. de Broglie, Annales de Physique, 10^e serie, t. **III**, 22 (1925).

4) In this regard, we refer to footnote [c] following Compton's Biographical Note in Chapter 2; and also Brown and Martins in [5] immediately below.

5) For an accessible critique, see: H.R. Brown and R.de A. Martins, Am. J. Phys. **52**(12), 1130 (1984).

Biographical Note on de Broglie:

Louis de Broglie (1892-1987) – won the 1929 Nobel Prize for Physics for his extension of the notion of wave-particle duality to material particles, paving the way especially for Schrödinger's invention of his wave mechanics, an intuitively accessible formulation of quantum mechanics parallel and equivalent to Heisenberg's slightly earlier matrix mechanics. From the London Times (20Mar87) and the NYTimes (21Mar87), we learn that de Broglie was first educated in history, but under the influence of his brother Maurice, seventeen years his senior and already a prominent experimental physicist, he attended the 1911 Brussels Conference on Planck's quantum theory and was inspired to study to physics. He graduated in 1914 in time to serve as a radio-signals officer in the French army of WWI. After graduate studies in the Faculte des Sciences at the Sorbonne, he submitted his 1924 thesis *Recherches sur la theorie des quanta* which is reproduced at length in the accompanying **Paper IV·1**. He became Professor of Theoretical Physics at the Institut Henri Poincare (1932-62), member (1933) and permanent secretary (1942-75) of the Academy of Sciences, was elected to the Academy Francaise (1945) where he was inducted by his brother, and where his father and grandfather had been members.

Prince Louis Victor Pierre Raymond de Broglie [a], born at Dieppe, of a conservative military noble family originally from the Piedmont in Italy in the 17th century. The family has produced marshals, governors, ministers of state and his grandfather a Prime Minister of France in the 1870's. With a courtesy title Prince most of his life, Louis de Broglie succeeded his brother as 7th duc de Broglie in 1960.

The family was established by Francois-Marie Broglia (1611-1656) [b], count of Revel in the Piedmont, who rose in French service to be governor of the Bastille and a marshal, and was able to purchase the marquisat de Senonches (for a million livres, on an annual pension of 12000 livres) before being killed in the siege of Valencia on 2 July 1656. His older son Victor-Maurice became the count de Broglie, and Victor's third son Francois-Maurice, also a marshal, became count and then the 1st duc de Broglie.

Among Louis de Broglie's namesakes one finds [c] his great-great-grandfather Victor-Francois, 2nd duc de Broglie (1718-1804) [d], marshal, son and grandson of marshals, in action at 15, cavalry commander at 16, general at 24, marshal at 40. Minister of War for Louis XVI in 1789, he was forced to admit that his troops were unable to guarantee the king's safety and were unreliable to oppose the revolution.

At the head of an army of Prussians and emigres, he invaded Champagne in 1792. Finally, on the verge of a reconciliation with Napoleon, he died in exile at Munster. His son Charles-Louis-Victor (1756-94) served in the American war of independence, and later as a colonel on the staff of Metz in the Bourbon army. As deputy to the constituent assembly representing the nobility, he defended the cause of the people and frequently voted with the left. He was president of the assembly in 1891, and then returned to active service in the army of the Rhine under the command of Luckner. He refused to support the death sentence of the king in 1792, retiring instead. He was soon arrested, imprisoned, released for a short time, rearrested, tried by the revolutionary tribunal, condemned, and immediately executed. His brother, the physicists' great-grandfather, Victor-Amedee-Marie (1785-1870), 3rd duc de Broglie, fought against the revolution with the 'Army of the Princes' eventually as colonel of the White Rose regiment, in campaigns against France in years IV and V of the revolution, became Chavalier de Saint-Louis in year VII, and then Gentleman of Honor to the Duke of Angueleme. Returning to France, he refused to serve Napoleon, suffered electoral defeat after the restoration and retired to his chateau at Ranes. Victor's son Albert (1821-1901), 4th duc de Broglie [e] and grandfather of the physicists Maurice and Louis, demonstrated the rightist politics of his father and the literary brilliance of his grandmother, Madame de Staël [f]! He was a prolific and intellectual writer on a wide range of subjects – from Leibnitz to the later Roman Empire. Elected to the National Assembly in 1871, he helped overthrow the conservative republican government, served as prime minister (1873-74), supported the monarchist cause of Philippe d'Orleans, but was finally defeated in a second term in 1877, and with him the monarchist delusions ended. His political defeat stifled the political career of his son Louis Alphonse Victor (1846-1906), father of the physicists and 5th duc de Broglie, who had served in his father's cabinet but retired and only reappeared as a deputy in 1893.

These were the aristocratic origins of the man who gave us de Broglie waves.

The name de Broglie cannot be correctly pronounced by native English speakers, but an acceptable approximation evidently is "duh-BROY-ee". The debate seems to revolve around a subtle gargling sound centered on the "-gl-", which is possible only for native French speakers; and even amongst them there is no precise agreement. The name evolved from the Italian Broglio and has been spelled Broglia, Broglie, and even Broille which reflects the way it is often pronounced.

Footnotes and References:

a) *Who's Who in France 1983-1984* (Jacque Lafitte S.A. 75008 Paris, France, 1982), p.220.

b) *Dictionnaire du Grand Siecle* (Fayard Press, Paris, 1990) Francois Bluche, ed., p.241.

c) *Dictionnaire Historique et Bibliographique de la Revolution et de Empire 1789-1815* (Paris, Kraus Reprint, Nedeln/Lichtenstein), p.284.

d) *Historical Dictionary of the French Revolution 1789-1799* (Greenwood Press, Westport CT, 1985) S.F. Scott and B. Rothaus, eds., p.127.

e) *Historical Dictionary of the 3rd French Republic 1870-1940* (Greenwood Press, Westport CT, 1986) P.H. Hutton, ed., p.138.

f) *Les Broglie* (Paris, Fasquelle, 1950), p.160.

Paper IV·1: Excerpt from Phil. Mag. **47**, 446 (1924).

XXXV. A Tentative Theory of Light Quanta

by LOUIS DE BROGLIE*

I. *The Light Quantum.*

The experimental evidence accumulated in recent years seems to be quite conclusive in favor of the actual reality of light quanta. The photoelectric effect, $\cdots$ the recent results of A.H. Compton $\cdots$ Bohr's theory $\cdots$ that atoms can only emit radiant energy of frequency f by finite amounts of energy hf $\cdots$

I shall assume the real existence of light quanta, and try to see how to reconcile it with the strong experimental evidence on which the wave theory was based.$\cdots\cdots$

II. *The Black Radiation as a Gas of Light Quanta.*

$\cdots\cdots$ This is obviously Wien's limiting form of the radiation law. Two years ago [1] I was able to show that it was possible, by using the hypothesis made by Planck that the unit phase-space volume was $dxdydzdp_xdp_ydp_z/h^3$, to find for the radiant energy density the value

$$u_f df = \frac{8\pi h}{c^3} f^3 e^{-hf/kT} df.$$

This was an encouraging result, but not quite complete. The assumption of the unit of phase-space volume seemed to have a somewhat arbitrary and mysterious character. Moreover, Wien's law is only a limiting form of the actual radiation law, and I was obliged to suppose some kind of quanta aggregation for explaining the other terms of the series.

It seems that these difficulties are now removed, but we shall first of all explain many other ideas; we shall later on return to the "black radiation" gas.

III. *An Important Theorem on the Motion of Particles.*

Let us consider a moving particle of rest mass m_0, moving with respect to a given observer with velocity $v = \beta c$ ($\beta < 1$), and containing internal energy $m_0 c^2$. The quantum relation suggests that we ascribe to this internal energy a wave of frequency $f_0 = m_0 c^2 / h$. For the fixed observer, the whole energy is $m_0 c^2 / \sqrt{1 - \beta^2}$ and the corresponding frequency is

$$f = \frac{1}{h} \frac{m_0 c^2}{\sqrt{1 - \beta^2}} \quad \left(\equiv \frac{f_0}{\sqrt{1 - \beta^2}} \right).$$

But if the fixed observer is looking at the internal period, he will see its frequency lowered and equal to $f_1 = f_0 \sqrt{1 - \beta^2}$, so the wave seems to him to vary as $\sin 2\pi f_1 t$. The frequency f_1 is widely different from the frequency f; but they are related by an important theorem which gives us the physical interpretation of f.

Suppose that at time $t = 0$ the moving particle coincides in space with a wave of the frequency f given above and which spreads with a phase velocity

$$V_\phi = \frac{c}{\beta} = \frac{c^2}{v}.$$

According to Einstein's ideas, however, this wave cannot carry energy.

Our theorem is the following: – *"If, at the beginning, the internal wave of the moving particle is in phase with the spreading wave, this phase agreement will always persist."* In fact, at time t, the moving particle is a distance $x = vt$ from the origin and its internal wave is $\sin 2\pi f_1 x/v$; at the same place the expanding wave is

$$\sin 2\pi f \left(t - \frac{\beta x}{c} \right) = \sin 2\pi f x \left(\frac{1}{v} - \frac{\beta}{c} \right).$$

The two sines will be equal and the agreement of phase will always occur for f_1 and f as defined above:

$$f_1 = f \left(1 - \beta^2 \right) \quad \left(\equiv f_0 \sqrt{1 - \beta^2} \right).$$

This important result is implicitly contained in the Lorentz time transformation. If τ is the local time of the moving particle, the internal wave is $\sin 2\pi f_0 \tau$. By the Lorentz transformation, the fixed observer describes the same wave by

$$\sin 2\pi f_0 \frac{1}{\sqrt{1 - \beta^2}} \left(t - \frac{\beta x}{c} \right),$$

which can be interpreted as a wave of frequency $f = f_0/\sqrt{1 - \beta^2}$ spreading along the x axis with phase velocity $V_\phi = c/\beta$.

We then recognize that any moving particle can be associated uniquely with a propagating wave.

* * * * * * * * * * * * * * * * * * **

In more familiar and contemporary terms (with $\hbar$ and c set to one), we associate with a particle of mass m moving with velocity v (energy $E = m/\sqrt{1 - v^2}$, and momentum $p = \sqrt{E^2 - m^2} = Ev$) the Lorentz invariant de Broglie phase $\Phi = (Et - px)$. From this we recognize the phase velocity $V_\phi = E/p$ and the group velocity $V_g = dE/dp = p/E = v$. de Broglie's three equivalent expressions for the phase are

$$2\pi f_1 \frac{x}{v} = E \left(t - \frac{px}{E} \right) \equiv 2\pi f \frac{x}{v} \left(1 - v^2 \right),$$

and in the particle rest frame, with $(Et - px) = m\tau$,

$$2\pi f_1 t = m\tau \equiv 2\pi f_0 \tau.$$

But moving clocks run slow, so $\tau = t\sqrt{1 - v^2}$, and we get de Broglie's result

$$f_1 = f_0\sqrt{1 - v^2}.$$

* * * * * * * * * * * * * * * * * * **

This idea can also be expressed in another way. A group of waves whose frequencies are very nearly equal has a group velocity V_g, which is the velocity of energy propagation. This group velocity is related to the phase velocity V_ϕ by

$$\frac{1}{V_g} = \frac{d}{df}\left(\frac{f}{V_\phi}\right).$$

If

$$f = \frac{1}{h}\frac{m_0 c^2}{\sqrt{1 - \beta^2}} \quad \text{and} \quad V_\phi = \frac{c}{\beta},$$

we find $V_g = \beta c$ – that is, "*The velocity of the moving particle $v = \beta c$ is the energy velocity V_g of a group of waves having frequencies*

$$f = \frac{1}{h}\frac{m_0 c^2}{\sqrt{1 - \beta^2}} \quad \text{and phase velocities} \quad V_\phi = \frac{c}{\beta},$$

corresponding to very slightly different values of β."

* * * * * * * * * * * * * * * * * * **

Here de Broglie introduces the concept of wave-particle duality for material particles.

* * * * * * * * * * * * * * * * * * **

IV. *Dynamics and Geometric Optics.*

To extend these ideas to the case of variable velocity is a difficult but suggestive problem. If a particle moves on a curved path, we say that there is a field of force; at each point the potential energy can be calculated, and the particle when crossing this point has a velocity determined by the constant value of its total energy. It seems natural to suppose that the phase wave at any point must have a velocity and a frequency fixed by the value *which β would have if the particle were there*. During its propagation the phase wave has a constant frequency f and a continuously variable phase velocity V_ϕ. $\cdots\cdots$ it seems that we already know the final result:" The rays of the phase wave are identical with the paths which are dynamically possible." In fact, the paths of the rays can be computed as in a medium of variable dispersion by Fermat's Principle:

$$\delta \int d\Phi = \delta \int \frac{ds}{\lambda} = \delta \int \frac{fds}{V_\phi} = \delta \int \frac{m_0\beta c}{h\sqrt{1-\beta^2}}ds = 0;$$

in agreement with the Maupertuis form of the Principle of Least Action which gives the dynamical path of the particle by the equation

$$\begin{aligned}\delta \int fdt - f_0 d\tau &= \delta \int \frac{m_0 c^2}{h}\left(\frac{1}{\sqrt{1-\beta^2}} - \sqrt{1-\beta^2}\right)dt \\ &= \delta \int \frac{m_0\beta^2 c^2}{h\sqrt{1-\beta^2}}dt = \delta \int \frac{m_0\beta c}{h\sqrt{1-\beta^2}}ds \equiv \delta \int \frac{ds}{\lambda} = 0,\end{aligned}$$

confirming the above theorem on the phase coincidence.

This theory suggests an explanation of Bohr's quantization condition. At time $t = 0$ the electron is at point A of its trajectory. The phase wave starting at this instant will propagate around the orbit and meet the electron again at A, where it must again be in phase with the electron. That is to say: "The motion can only be stable if the phase wave is compatible with the length of the path." The requirement is:

$$\int \frac{ds}{\lambda} = \int_0^T \frac{m_0\beta^2 c^2}{h\sqrt{1-\beta^2}}dt = n,$$

with n a positive integer and T the period of the motion. $\cdots\cdots$

V. *The Propagation of Light Quanta*
and the Coherence Problem.

$\cdots$ The light quantum is in some manner a part of the wave, but for explaining interferences and other phenomena of wave optics it is necessary to see how several light quanta can be parts of the *same* wave. This is the coherence problem. $\cdots\cdots$

* * * * * * * * * * * * * * * * * * **

de Broglie here seems not quite ready to admit the dual character of quanta – particle and wave – and interference of a photon only with itself; but then immediately (in the next section, in italics) makes the leap.

* * * * * * * * * * * * * * * * * * **

VI. *Diffraction and the Inertia Principle.*

Here the theory of light quanta meets a great difficulty, known since the time of Newton. The light rays passing near an edge are no longer straight but penetrate into the geometric shadow. Newton ascribed this deflection to the action of some force exerted by the edge on the light corpuscle. It seems to me that this phenomenon deserves a more general explanation. Since an intimate connection exists between the motion of particles and the propagation of waves, and since the rays of the phase wave are the possible paths of the energy quanta, we are inclined to give up the inertia principle and to say: "A moving particle must always follow the same ray of its phase wave." In the continuous spreading of the wave, the surfaces of equal phase will change continuously and the particle will always follow the common perpendicular to two infinitely near surfaces.

When Fermat's Principle is valid for computing the ray path, then the Principle of Least Action is also valid for computing the particle path. These ideas are a synthesis of wave optics and particle dynamics.

$\cdots\cdots$ the ray – which assumes an important physical significance – is defined by the *continuous* spreading of a small part of the phase wave: it cannot be defined $\cdots$ by the $\cdots$ "energy or Poynting's vector." $\cdots$

VII. *A New Explanation of Interference Fringes.*

Consider how we detect light at a point in space $\cdots$ It seems that all these means can, in fact, be reduced to photoelectric actions $\cdots$

Next consider Young's Interference Experiment. Light quanta pass through the holes and diffract along the rays of the neighboring parts of their phase waves. In the space behind the wall, their capacity for photoelectric action will vary from point to point depending on the interference state of the two phase waves from the two holes.   *We will see interference fringes however few the number of diffracted quanta and however small the incident light intensity.* (Note: Italics added.) The light quanta do cross all the dark and bright fringes; only their ability to act on matter is constantly changing. This explanation, which removes the objections against light quanta and against energy propagation through dark fringes, may be generalized for all interference and diffraction phenomena.

---
* * * * * * * * * * * * * * * * * * **

de Broglie presents the quantum interpretation of interference phe-nomena.

* * * * * * * * * * * * * * * * * **

VIII. *The Quanta and the Kinetic Theory of Gases.*

To calculate the absolute entropy, Planck and Nernst were forced to introduce the quantum idea into the kinetic theory of gases. Planck assumed a phase-space volume element equal to

$$\frac{1}{h^3} dx\, dy\, dz\, dp_x\, dp_y\, dp_z \quad \text{or} \quad \frac{4\pi}{h^3} m_0^{3/2} \sqrt{2w}\, dw\, dx\, dy\, dz.$$

We shall now justify this assumption.

Each atom of velocity $v = \beta c$ is equivalent to waves of phase velocity $V_\phi = c/\beta$, frequency $f = m_0 c^2/h\sqrt{1 - \beta^2}$, and group velocity $V_g = \beta c$. The state of the gas can only be stable for standing waves. Following Jeans, we find the number of waves per unit volume with frequencies in the interval f to $f + df$ to be:

$$n_f df = \frac{4\pi}{V_g V_\phi^2} f^2 df = \frac{4\pi}{c^3} \beta f^2 df.$$

If w is the kinetic energy of an atom and f the corresponding frequency, then:

$$hf = \frac{m_0 c^2}{\sqrt{1 - \beta^2}} = w + m_0 c^2 = m_0 c^2 (1 + \alpha),$$

with $\alpha = w/m_o c^2$. Then $n_f df$ is given by:

$$n_f df = \frac{4\pi}{h^3} m_0^2 c (1 + \alpha)\sqrt{\alpha(2 + \alpha)}\, dw$$

Each phase wave can carry with it one, two, or more atoms, so that, according to the canonical law, the number of atoms whose energy is hf, will be proportional to:

$$n_f df \, dx \, dy \, dz \left(\sum_{n=1}^{\infty} e^{-nhf/kT} \right).$$

Consider first a classical gas whose atoms have large mass and small velocities, so we can ignore all terms in the sum except the first, and can set $1 + \alpha = 1$. The number of atoms whose kinetic energy is w will be proportional to

$$\frac{4\pi}{h^3} m_0^{3/2} \sqrt{2w} \, dw \, dx \, dy \, dz \, e^{-w/kT},$$

which justifies Planck's method and leads to Maxwell's distribution.

In the case of light quanta, α is large and we must keep the whole series, and also double the result to include both polarization states. We find the radiant energy density proportional to:

$$\frac{8\pi h}{c^2} \frac{f^3}{e^{hf/kT} - 1} df.$$

A method developed in the *Journal de Physique*, of November 1922, shows that the proportionality factor is unity, so that we obtain the actual radiation law.

* * * * * * * * * * * * * * * * * * * *

de Broglie anticipated Bose's derivation of the Planck distribution (Zeits. f. Phys. **26**, 178 (1924)) by some two years, as partially acknowledged by Einstein (S.B.d. Preuss. Akad. Wiss. Ber. **22**, 261 (1924)). Einstein, however, referred only to de Broglie's 1924 Thesis.

* * * * * * * * * * * * * * * * * * * *

IX. *Open Questions.*

The ideas stated in this paper, if correct, will require a wide modification of electromagnetic theory. The so called "electric and magnetic energies" must be only an average value, all the real energy of the fields being of a fine-grained quantum structure. The construction of a new theory seems to be a very difficult task, but we have one guide: according to the correspondence principle, the defining vectors of the old theory should give the probability of reactions in the fine-grained theory.

· · · · · · There seems to be a great analogy between the scattering of radiation and the scattering of particles · · · · · · explaining optical dispersion will be more difficult · · · What occurs when an atom passes from a stable state to another, and how does it eject a single quantum? How can we introduce the · · · quantum · · · into · · · elastic waves and into Debye's theory of specific heats?

Finally, we must remark that the quantum remains a postulate defining the constant h whose actual significance is not at all cleared up; but it seems that the quantum enigma is now reduced to this unique point.

* * * * * * * * * * * * * * * * * **

Where it remains still.

* * * * * * * * * * * * * * * * * **

Summary.

It is assumed that light is made up of quanta. It is shown that the Lorentz-Einstein transformation together with the Planck quantum relation leads us necessarily to associate particle motion and wave propagation, and that this idea gives a physical interpretation of Bohr's quantization condition. Diffraction is shown to be consistent with an extension of Newtonian dynamics. It is then possible to have both the particle and the wave character of light, and, by means of hypotheses suggested by electromagnetic theory and the correspondence principle, to give a plausible explanation of coherence and interference fringes. Finally, it is shown why quanta must take part in the kinetic theory of gases and how Planck's distribution is the limiting form of Maxwell's distribution for a gas of light quanta.

Many of these ideas may be criticized and perhaps modified, but little doubt can now remain of the real existence of light quanta. Moreover, if our ideas are correct, based as they are on the relativity of time, all the enormous experimental evidence of the "quantum" will support Einstein's conceptions.

1 October, 1923.

Note.– Since I wrote this paper, I have found a much more general result. The Principle of Least Action for a point particle can be written in the space-time notation as:–

$$\delta \int \sum_1^4 J_i dx^i = 0,$$

where the J_i are the four-dimensional energy-momentum vector of the particle.

Similarly, for the propagation of waves, we write:-

$$\delta \int \sum_1^4 \Theta_i dx^i = 0,$$

where the Θ_i are the covariant components of the four-dimensional vector whose time component is the frequency f/c, and whose space components are a vector along the ray, of magnitude $f/V_\phi = 1/\lambda$ (V_ϕ is the phase velocity). The quantum condition says that

$$J_4 = h\Theta_4.$$

More generally, I suggest putting

$$\vec{J} = h\vec{\Theta}.$$

From this statement, the identity of the Fermat and Maupertuis statements of the Principle of Least Action follows immediately, and it is possible to deduce rigorously the velocity of the phase wave in any electromagnetic field.

* – Communicated by R.H. Fowler, M.A.

Chapter V

KRAMERS and HEISENBERG:

Eine quantentheoretische Deuting $\cdots$ **bei der die Differentialquotienten durch durch h dividierte Differenzen, die Frequenzen durch quantentheoretische Frequenzen, und die Amplituden** Q **durch die charakteristischen Amplituden der quantentheoretischen Übergäng ersetzt werden, ergibt** $\cdots$

H.A. Kramers und **W. Heisenberg**, *Über die Streuung von Strahlung durch Atome*, Zeits. f. Phys. **31**, 681 (1925): A quantum interpretation of this expression – in which derivatives are replaced by differences divided by h, frequencies by quantum energy differences, and dipole moments Q by amplitudes characteristic of the quantum transitions – can be made in the following plausible way $\cdots$.

§V-1. Introduction.

Bohr's quantization of the Rutherford atom [1], leading for the first time to a theory of the discrete Balmer spectrum, provided the critical clue for further progress. The principle that the angular momentum of the bound electron orbits should be quantized in integral multiples of $\hbar$ was soon generalized by Ehrenfest [2] to the prescription that classical adiabatic invariants were logical candidates for quantization. Ehrenfest argued that for small disturbances of the mechanical system, quantized quantities could not change; they must change by a whole number at a jump, or they must remain constant. If the change in the system is sufficiently slow, the latter must be the case: quantized quantities must be *adiabatic invariants* and it is plausible to suppose that all adiabatic invariant quantities should be quantized..

The simple pendulum is an elementary example [3]. In an obvious notation, the pendulum has potential energy $V = mgl\phi^2/2$, kinetic energy $T = ml^2\dot\phi^2/2$, total oscillatory energy $E = T + V$, and frequency $\omega^2 = g/l$. As the pendulum is slowly pulled up reducing its length by $|dl|$, work is done on the system

$$dW = -mgdl + (\bar{V} - 2\bar{T})\frac{dl}{l} = dE_{cm} + dE.$$

The first term is the change in the center-of-mass potential energy; the second is the change of the oscillatory energy of the pendulum, averaged over many oscillations,

$$\frac{dE}{E} = \left(\frac{\bar{V} - 2\bar{T}}{E}\right)\frac{dl}{l}.$$

With $\bar{T} = \bar{V} = E/2$,

$$\frac{dE}{E} = -\frac{dl}{2l} = \frac{d\omega}{\omega}.$$

Born concludes that the ratio of the energy of oscillation E to oscillation frequency ω is an adiabatic invariant. Quantized according to Ehrenfest as

$$\frac{E}{\omega} = n\hbar,$$

it gives (almost) the energy levels of the quantum oscillator. The generalized coordinate $q = l\phi$ and its conjugate momentum $p = m\dot{q}$ describe an ellipse in phase space

$$\frac{p^2}{2mE} + \frac{m\omega^2}{2E}q^2 = 1$$

with semi-axes $a = \sqrt{2mE}$, $b = \sqrt{2E/m\omega^2}$ and area A swept out per cycle, where

$$\frac{A}{2\pi} = \frac{1}{2\pi}\oint pdq = \frac{ab}{2} = \frac{E}{\omega} \equiv J = n\hbar.$$

We include Karl Schwarzschild's pedagogical paper (here as **Paper V·1**) on 'normalizing' coordinates – which he renamed 'Action-Angle' variables – despite the fact that both Planck and Sommerfeld had priority, for two reasons: 1) Schwarzschild's work is still one of the clearest discussions of the subject, and 2) the intense human drama that simultaneously consumed Schwarzschild.

Kramers and Heisenberg used the Action-Angle variables in a canonical transformation involving the Poisson bracket to calculate the atomic dipole moment $\mathcal{M}$ induced by an incident electric field $\mathcal{E}$ interacting with the original atomic dipole moment $\mathcal{Q}$:

$$\mathcal{M} \sim \frac{\partial \mathcal{Q}}{\partial J} \frac{\partial(\mathcal{E}\mathcal{Q})}{\partial \Theta} - (J \leftrightarrow \Theta).$$

The Kramers-Heisenberg prescription for *inferring* a quantum interpretation of this classical result depends in an essential way on the choice of the action variable J which satisfies the quantization condition $J = m\hbar$. They make a 'plausible' interpretation of the classical derivative with respect to the now discrete quantized variable J as a difference divided by $\hbar$.

The deepest genesis of this step is discussed in detail by van der Waerden [4]. The idea had its origin in a paper by Born [5] already familiar to Heisenberg, and in which he as Born's assistant had played some role.

From Fig2 of their paper (here as **Paper V·2**), Kramers and Heisenberg replace

$$\frac{\partial \mathcal{Q}}{\partial J} \to \frac{(\mathcal{Q}_3 - \mathcal{Q}_1)}{\hbar}.$$

$\mathcal{Q}_3$ and $\mathcal{Q}_1$ are further interpreted as *transition amplitudes* (their original words!), and then in a *tour de force* of ingenuity, creativity and genius, Kramers and Heisenberg – after a series of cancellations followed by further interpretations of negative frequency contributions as various time orderings of events – obtain precisely the result of the soon to be discovered quantum mechanics.

Finally the stage was set for the ultimate crucial reflections of the young Heisenberg.

Footnotes and References:

1) Niels Bohr, Phil. Mag. **26**, 1-25 (1913) (here as **Paper III·1**).

2) P. Ehrenfest, Phys. Zeits. **15**, 657 (1914). See also Arnold Sommerfeld, *Atombau und Spektrallinien* (Vieweg & Sohn, 1924), 4th-Ed., Pp 397-407.

3) Max Born, *Problems of Atomic Dynamics* (Massachusetts Institute of Technology, Cambridge, Mass., 1926), 1st-Ed., Lectures 3,4,5; *Atomic Physics* (Blackie & Son, Ltd., Glasgow, London, 1958), 6th-Ed., Pp. 113-119, 338-341. Alternatively, see H. Goldstein, *Classical Mechanics* (Addison-Wesley, Reading, Mass., 1981), 2nd-Ed., Pp. 457-471; and

E.J. Saletan and A.H. Cromer, *Theoretical Mechanics* (Wiley, New York, 1971), Pp. 231-263.

4) B.L. van der Waerden, *Sources of Quantum Mechanics* (Classics of Science, Vol 5; Dover, New York, 1967) Gerald Holten, Ed., Pp 15-16. In van der Waerden's translation, Born's statement [5] is: "We are ⋯ forced to replace a classically calculated quantity ⋯

$$\sum_\tau \tau_k \frac{\partial \Phi}{\partial J_k} = \frac{1}{h}\frac{d\Phi}{d\mu}$$

by the linear average

$$\int_0^1 \sum_\tau \tau_k \frac{\partial \Phi}{\partial J_k} d\mu = \frac{1}{h}\left[\Phi(n+\tau) - \Phi(n)\right].$$

5) Max Born, Zeits. f. Phys. **26**, 379 (1924). Here Born coins the expression 'Quantum Mechanics'; and formulates the pre-matrix version of much of the paper: M. Born and P. Jordan, Zeits. f. Phys. **34**, 858 (1925), contained here as **Paper VII·1**.

Biographical Note on Schwarzschild.

Karl Schwarzschild (1873-1916) – From Eddington's eulogy in the Monthly Proceedings of the Royal Astronomical Society of February 1917 [1], we learn of the great distinction earned by Schwarzschild through his many varied accomplishments in Astronomy as well as in Physics and Mathematics, and most remarkably in experiment and instrumentation as well as in theory. Schwarzschild rose quickly from assistant at the Von Kuffner Observatory in Vienna to Director of the Observatory at Göttingen to – at 36 – Director of the Astrophysical Observatory at Potsdam.

Many of Schwarzschild's accomplishments were of the most practical kind possible in Astronomy but distinguished by their importance: determining the effective brightness of stars, correlating stellar brightness and color, and stellar color and motion, ellipsoidal stellar velocity distributions, diffractive measurements on double star separation, statistical analysis of stellar distributions. Closer to earth, Schwarzschild studied radiative equilibrium in solar energy transfer; designed reflecting telescopes, during the course of which he wrote a definitive work on the application of eikonal methods in practical physics.

But perhaps nothing in Schwarzschild's distinguished career foretold the great accomplishments of the dramatic final act.

Schwarzschild was descended from a family of prominent 17th-century Jewish bankers (the 'Black-shield', contemporaries and neighbors in the Frankfurt ghetto to the 'Red-shield' Rothschilds who went on to even greater prominence). Schwarzschild's way to combat anti-semitism was to be 'the best German of all'. So when WWI came, even though he was a 41 year old Professor of great prominence, he joined the army and served actively as an artillery officer in France and then in Russia, and was awarded the Iron Cross. In late 1915 he contracted an incurable disease and was hospitalized. It was in these awful circumstances that Schwarzschild wrote the three papers for which he is best remembered:

a) He solved the equations of Einstein's new General Theory of Relativity for the metric around a point mass [2]. This solution exhibited the Schwarzschild radius – the point of no return – characterizing Black Holes; but it also permits the calculation of the precession of the perihelion of elliptical planetary orbits characteristic of General Relativity. Einstein wrote him in the hospital: "My esteemed colleague, I had no idea $\cdots$" that an exact solution was possible. Einstein had calculated the precession in perturbation theory.

b) Then Schwarzschild solved for the metric inside an incompressible fluid sphere [3], a solution fundamental to gravitational collapse and black hole formation.

c) And finally, Schwarzschild wrote the pedagogic paper quoted here [4] (**Paper V·1**) where he explained beautifully the quantum significance of the canonical variables – familiar in Astronomy as Delaunay's variables [5] – which he gave the name 'Action-Angle Variables'.

And then he died a week after publication of the last paper. What killed him was 'phagedenos' (a form of acute necrotizing ulcerative gingivitis, bordering on gangrene, probably initially contracted as trench-mouth), a disease that is still quite dreadful and can be extremely resistant even to modern antibiotics, and of course for Schwarzschild there were none.

Schwarzschild left two sons. Martin Schwarzschild (1912-1997) fled Nazi Germany in 1935 for an outstanding career in Astrophysics at Princeton University [6]. His other son stayed in Nazi Germany and was murdered in the Holocaust [7].

Footnotes and References:

1) A.S. Eddington, Monthly Proceedings of the Royal Astronomical Society, Feb. 1917, Pp.314-319.

2) K. Schwarzschild, 'Über das Gravitationsfeld eines Massenpunktes nach der Einsteinschen Theorie,' *Sitz. d. Math. Phys. K. Deut. Akad. Wiss. Berlin* **16**, 189 (1916).

3) K. Schwarzschild, 'Über das Gravitationsfeld einer Kugel aus inkompressibler Flussigkeit nach der Einsteinschen Theorie,' *ibid.*, **16**, 424 (1916).

4) K. Schwarzschild, 'Zur Quantenhypothese,' *ibid.*, **16**, 548 (1916) (included here as **Paper V·1**).

5) C. Lanczos, 'The Variational Principles of Mechanics,' (University of Toronto Press, Toronto 1949), Pp. 243-254.

6) New York Times, 28:4, 12APR97; Physics Today, p.90, DEC97.

7) We are indebted to Professor Bertram Schwarzschild for information about this branch of his family tree.

Paper V·1: Excerpt from Sitz. d. Math.-Phys. K. Deut. Akad. Wiss. Berlin **16**, 548 (1916).

Zur Quantenhypothese

von K. SCHWARZSCHILD.

(Vorgelegt am 30. März 1916 [s. oben S.435].)

I.

§1. The Quantum hypothesis has recently been applied by Planck [1] and Sommerfeld [2] to the case of a mechanical system of many degrees of freedom. The problem is to give a principle for the partition of the phase space into elementary domains. I might point out, that for an important group of mechanical problems the evolution of the phase space takes place in a more understandable way when one uses a particular kind of canonical variable. The partition coincides in many, but not in all cases with that already suggested by Planck and Sommerfeld. I introduce two further applications, to the theory of the hydrogen Stark effect and to the theory of band spectra. Here, as for all earlier treated problems, the suggested partition is unquestionably practicable. A complication can occur in some cases but only at the boundary of phase space (see §5). I have so far not investigated the range of problems for which our partition is trouble free.

§2. I consider a mechanical system whose motion is quasi-periodic [3]. For these the following is understood: Let x_i be the coordinates, which uniquely specify the configuration of the system, p_i the corresponding momenta, and H the Hamiltonian, which yields the equations of motion for k degrees of freedom

$$\frac{dx_i}{dt} = \frac{\partial H}{\partial p_i}, \quad \frac{dp_i}{dt} = -\frac{\partial H}{\partial x_i}, \quad i = 1, 2, \cdots k.$$

Quasi-periodic systems are those whose solutions are of the form:

$$x_i = x_i(\alpha_1, \alpha_2 \cdots \alpha_k; \theta_1, \theta_2 \cdots \theta_k)$$
$$p_i = p_i(\alpha_1, \alpha_2 \cdots \alpha_k; \theta_1, \theta_2 \cdots \theta_k).$$

Here the α_λ ($\lambda = 1, 2 \cdots k$) are constants. The θ_λ are angle-variables proportional to the time:

$$\theta_\lambda = \omega_\lambda t + \beta_\lambda$$

(ω_λ and β_λ are constants, ω_λ the 'average velocity', and β_λ the initial value of the angle θ_λ), and the x_i, p_i are periodic with period 2π in the angles θ_λ.

One can always choose the α_λ so that they are canonically conjugate to the variables θ_λ. Then the equations of motion go over into:

$$\frac{d\alpha_\lambda}{dt} = -\frac{\partial H}{\partial \theta_\lambda}, \quad \frac{d\theta_\lambda}{dt} = +\frac{\partial H}{\partial \alpha_\lambda}.$$

Since the α_λ are constants, it follows that:

$$\frac{\partial H}{\partial \theta_\lambda} = 0, \quad \text{and} \quad H = H(\alpha_1, \alpha_2 \cdots, \alpha_k).$$

The Hamiltonian is a function only of the α_λ quantities. The equations for $d\theta_\lambda/dt$ give the average velocity of the angle θ_λ as a function of the quantities α_λ:

$$\omega_\lambda = \frac{\partial H}{\partial \alpha_\lambda}.$$

Since ω_λ is a reciprocal time, it follows that α_λ has the dimension of an Action (energy×time). We shall therefore designate the α_λ as 'action-variables' – J_λ – (variable in the sense of external perturbations of the system) in contrast to the 'angle-variables' θ_λ.

The volume of the phase space is:

$$\int dJ_1 d\theta_1 dJ_2 d\theta_2 \cdots dJ_k d\theta_k.$$

It permits the following natural evolution of the phase space: Each angle-variable θ_λ has the limits 0 and 2π. Each action-variable J_λ has bounds, such that integrated from limit to limit:

$$\int dJ_\lambda d\theta_\lambda = 2\pi \int dJ_\lambda = 2\pi \hbar \quad (\text{=Planck's Quantum of Action.})$$

It then follows that the limits on J_λ must be:

$$J_\lambda = \epsilon_\lambda + n_\lambda \hbar, \text{ with } n_\lambda \text{ an integer, and } \epsilon_\lambda \text{ a constant.}$$

In words: For quasi-periodic motions, a partition of the phase space into conjugate action-angle variables is given, such that the action-variables each increase by some integer multiple of $\hbar$ when the angle-variables increase by 2π.

§3. Degenerate case. When the general representation of the system is possible with fewer than k angle-variables – in other words, when one or more average velocities ω_λ vanishes for a given value of the J_λ, then the above rule holds only for pairs of variables 'with average velocity', whereas no evolution of the phase space occurs for those 'without average velocity'. These supplementary remarks are summarized by the remark that: For a vanishing average velocity ω_λ for which the Hamiltonian is independent of the variable J_λ, this pair of variables has no significance for the energy processes in the system.

§4. Normalizing the Variables. Boundaries of Phase Space. If instead of the θ_λ one introduces some linear functions of new variables θ'_λ with integer coefficients, then the x_i, p_i will also be periodic of period 2π in these new variables. The corresponding new canonical variables J'_λ will be linear functions of the J_λ with integer coefficients, since as is well known the two groups of canonical variables transform contragrediently under linear substitutions.

If for a particular mechanical problem one has found a representation of the motion by a number of angle-variables θ_λ proportional to the time, but if it can be shown that among the average velocities ω_λ there exist one or more commensurability relations of the form:

$$0 = n_1\omega_1 + n_2\omega_2 \cdots n_k\omega_k \quad \text{(with } n_\lambda \text{ integers)},$$

then one can introduce in place of some of the original θ_i, new variables θ'_i:

$$m_i\theta'_i = n_1\theta_1 + n_2\theta_2 \cdots n_k\theta_k \quad \text{(with } m_i \text{ integers)},$$

which have zero average velocity. If one chooses the m_i suitably (*e.g.*, so that they contain the determinant of the n_i as a factor), this gives a solution for which the

θ_i appear as integral multiples of the θ'_i. In this way, one can separate the angle-variables into two groups, the first (ignorable) group with zero average velocity, and the second (significant) group with angular velocities for which no commensurability relations exist. Such a 'normalizing' has already been tacitly assumed after the above discussion of the degenerate case.

There is now a particular definition to discuss. I will assume that one has found the angle-variables θ_λ which – by variation through 2π with the accompanying change of the J_λ – provide a simple over-view of the phase space. Then it should be clear that in the above described Normalization, the determinant of the old variables in terms of the new should be made as small as possible – in fact most conveniently equal to unity.

Without this provision one could *e.g.* choose $\theta_1 = n\theta'_1$ (with n some integer), and $J_1 = J'_1/n$, and one could, since J'_1 is some integer multiple of $\hbar$, go to an arbitrary time evolution of J_1, which is obviously not allowed physically.

There are also the boundaries of phase space to consider. Planck has noted that the elementary domains of phase space must be complete. The boundary of phase space will generally be defined by some inequalities between the variables J_λ. One needs the limits only for the variables with nonzero average velocities, since one leaves out all reference to the ignorable variables. I will now assume that for the remaining l significant variables the limiting conditions consist of l inequalities:

$$\sum_i n_{i\lambda} J_i > c_\lambda \quad \text{(with } c_\lambda \text{ constants and } n_{i\lambda} \text{ integers.)}$$

Then by the substitutions

$$J'_\lambda = \sum_i n_{i\lambda} J_i \quad \text{and} \quad \theta'_\lambda = \sum_\lambda n_{i\lambda} \theta_\lambda,$$

one can go over to new variables J'_λ and the corresponding θ'_λ, which retain the characteristic properties of our action-angle variables, and for which the limits of phase space are simply

$$J'_\lambda > c_\lambda.$$

Then a complete definition of the significant phase space variables results, if we take the c_λ for the initial values ϵ_λ. The ignorable variables will then obey their limits without question, when the whole phase space is considered.

The above prescription leads in a unique way to a consistent description of the phase space for many important problems. It is an interesting mathematical question for the future, to determine the most comprehensive range of problems for which its consistency and uniqueness holds.

§5. Thermal and Optical Properties of the System. As soon as the evolution of the phase space has been given, then the thermal properties of a body consisting of a large number of such systems in their stationary states are given in the way determined by Planck.

The optical properties give a very simple description of the frequencies ω emitted by the system, as determined by the Bohr's ansätz. The frequencies ω then follow from the equation:

$$\Delta E = \hbar\omega \;\; = \;\; H\left(\epsilon_1 + m_1\hbar, \epsilon_2 + m_2\hbar, \cdots \epsilon_\lambda + m_\lambda\hbar\right)$$
$$-H\left(\epsilon_1 + m_1'\hbar, \epsilon_2 + m_2'\hbar, \cdots \epsilon_\lambda + m_\lambda'\right),$$

where the m_λ, m_λ' are integers. There are just as many integers m_λ or m_λ' as there are incommensurable average velocities in the system (for a degenerate system this number is smaller than that of the degrees of freedom).

§6. Application of the Jacobi Integral Method. One seeks the connection between the original variables x_i, p_i and the action-angle variables $J_\lambda, \theta_\lambda$ from the Jacobi form of the canonical transformation:

$$p_i = \frac{\partial S}{\partial x_i}, \quad \theta_\lambda = \frac{\partial S}{\partial J_\lambda}, \quad S = S(x_1, x_2, \cdots x_k; J_1, J_2, \cdots J_k).$$

Here S must be a solution of the partial differential equation:

$$H(x_i, p_i) = H\left(x_i, \frac{\partial S}{\partial x_i}\right) = \text{const.}$$

When one has found a solution to this differential equation with k free parameters $\gamma_1, \gamma_2, \cdots \gamma_k$ which suffice for the integration of the problem, these constants in general will not be the sought-after constants J_λ and the quantity $\partial S/\partial \gamma_\lambda$ will not yet be the desired θ_λ. There is one case, which occurs for the most important problems, in which one does get the desired variables directly from the partial differential equation. One assumes that the variables x_i are representable as periodic functions with period 2π of the same number of auxiliary variables η_i, and that a solution of the partial differential equation can be found in the following form:

$$S = J_1\eta_1 + J_2\eta_2 + \cdots J_k\eta_k + T(J_i, \eta_i),$$

where the J_i are integration constants and T is a function of these constants and the variables η_i, which is periodic of period 2π in the η_i.

Then:

$$p_i = \frac{\partial S}{\partial x_i} = \sum_\lambda \left(J_\lambda + \frac{\partial T}{\partial \eta_\lambda}\right)\frac{\partial \eta_\lambda}{\partial x_i}, \text{ and } \theta_\lambda = \frac{\partial S}{\partial J_\lambda} = \eta_\lambda + \frac{\partial T}{\partial J_\lambda}.$$

From the first equation it follows that the p_i are periodic functions of the variables η_λ, just as we had prescribed for the x_i. From the second equation it follows that each η_λ and θ_λ simultaneously increase by 2π. If one inverts the equations to express the η_λ as functions of the θ_λ, their connection is of the form: $\eta_\lambda = \theta_\lambda$ plus a periodic function of the θ_i with period 2π.

The η_λ and also the x_i and p_i are periodic functions of the θ_i with period 2π. The θ_i are angle-variables of the required kind, and the canonically conjugate variables J_i are the action-variables – naturally apart from the normalization and any consideration of the boundaries of phase space.

In the following example we will show how simply the definition of the action variables can be chosen.

§7. **Example of Planck's Partition of Phase Space.** We analyze the motion of a particle of unit mass in the plane of rectangular coordinates x_1, x_2 under the effect of an anisotropic harmonic potential

$$V = \frac{1}{2}\left(A_1^2 x_1^2 + A_2^2 x_2^2\right).$$

As is well known:

$$x_1 = \gamma_1 \sin(A_1 t + \beta_1) \qquad x_2 = \gamma_2 \sin(A_2 t + \beta_2)$$

$$\dot{x}_1 = p_1 = \gamma_1 A_1 \cos(A_1 t + \beta_1) \qquad \dot{x}_2 = p_2 = \gamma_2 A_2 \cos(A_2 t + \beta_2),$$

with $\beta_1, \beta_2, \gamma_1, \gamma_2$ constants. Suitable angle-variables are

$$\theta_1 = A_1 t + \beta_1 \quad \text{and} \quad \theta_2 = A_2 t + \beta_2,$$

and we require their canonically conjugate action-variables. Since the two coordinates are obviously independent of one another, we can consider each coordinate by itself. The Hamiltonian of the motion in x_1 is:

$$H_1 = \frac{1}{2} \left(\dot{x}_1^2 + A_1^2 x_1^2 \right) = \frac{1}{2} A_1^2 \gamma_1^2.$$

The canonical variable J_1 must satisfy

$$\frac{d\theta_1}{dt} = A_1 = \frac{\partial H_1}{\partial J_1} = A_1^2 \gamma_1 \frac{\partial \gamma_1}{\partial J_1},$$

from which

$$1 = A_1 \gamma_1 \frac{\partial \gamma_1}{\partial J_1}, \quad \text{so} \quad J_1 = \frac{A_1 \gamma_1^2}{2},$$

and similarly $J_2 = A_2 \gamma_2^2 / 2$. The boundary conditions on the phase space are obviously $J_1, J_2 > 0$. Their elementary quantum domains are therefore given by

$$J_1 = m_1 \hbar, \quad \text{and} \quad J_2 = m_2 \hbar.$$

The corresponding value of the energy will be:

$$H = H_1 + H_2 = A_1 J_1 + A_2 J_2 = (A_1 m_1 + A_2 m_2) \hbar.$$

The volumes of the elementary domains in phase space are:

$$\int dJ_1 dJ_2 = m_1 \hbar \cdot m_2 \hbar.$$

We now look at the isotropic case $A_1 = A_2 = A$, where the two angles θ_1 and θ_2 have the same average velocity, and a degeneracy occurs. If we introduce a linear (integral) substitution of determinant 1:

$$\theta_1' = \theta_1, \quad \theta_2' = \theta_2 - \theta_1,$$

and θ_2' has average velocity zero, and θ_1' has average velocity A. The corresponding canonical variables become

$$J_1' = J_1 + J_2, \quad J_2' = J_2.$$

According to our rules only the subset of variables J_1' with average velocity nonzero has to be included. Since $J_1' = 0$ is a boundary of the phase space, we have to set $J_1' = m\hbar$, and the corresponding energy is

$$H = A(J_1 + J_2) = AJ_1' = Am\hbar.$$

The boundaries of the phase space are

$$J_1 = J_1' = J_2' \geq 0, \quad \text{and} \quad J_2 = J_2' \geq 0,$$

or $J_1' \geq J_2' \geq 0$. These limit only the range of J_2'. The volume of the elementary domains are

$$\int dJ_1' dJ_2' = \int dJ_1' J_1' = \int d\left(\frac{J_1'^2}{2}\right).$$

The sum of the first m elementary domains – integrated over J_1' from 0 to $m\hbar$ – is therefore $m^2\hbar^2/2$.

As one sees, according to our prescription the partition of the phase space changes essentially when one considers isotropy. Whereas for the anisotropic case, the elementary domains of the J_1, J_2-plane are squares, for the isotropic case they will be cut into diagonal bands. In the anisotropic case the orbital curves cover a two dimensional region everywhere densely. For isotropy however, they are closed ellipses. It should not be surprising that with this strong change in the character of the motion, even a change in the elementary domains results.

Naturally one can state that a certain anisotropy always exists and therefore the quadratic partition should always be retained even for the apparently isotropic case. Such an opinion is even more persuasive when one considers the case of the elastic force as the limiting case of another problem. For example, one adds to the elastic force a force proportional to the inverse third power of the separation, therefore assuming the potential $Ar^2 + B/r^2$ $(r^2 = x_1^2 + x_2^2)$. For $B = 0$ one gets back the isotropic elastic force. For $B \neq 0$ the form of the orbits is an ellipse with rotating major axis – precession of the perihelion. One has therefore two incommensurable periods in the motion, one for rotation within the ellipse itself, and a second for the precession of the perihelion. It is a problem without degeneracy. When one treats it according to the above rules and then goes to the limit $B \to 0$, without changing the phase space, then one gets exactly the treatment given by Planck and by Sommerfeld. According to the rules given here, one must use a new partition for the degenerate case $B = 0$. Planck's partition is more plausible, as would be the original choice of J_1, J_2, if one is treating the problem as an anisotropy rather than deviation from Hooke's Law.

One sees in general, that one can decide in this case – and analogously in many others – between Planck's choice and the partition described here, by replacing the given mechanical problem by a neighboring one of lesser degeneracy.

There is however another distinction to note with Planck's procedure. According to the above prescription, for a degenerate system the volume of the elementary domains in phase space depend on the boundary conditions of the ignorable variables (those without average motion). Planck, on the contrary, requires that the volumes should be an integer multiple of (powers of) $\hbar$. This requirement is already violated for the above example by the factor $\frac{1}{2}$ in the expression for the volume element $\frac{1}{2}m^2\hbar^2$.

As Planck himself has emphasized, only the results can decide the preference of one partition of phase space over another. $\cdots\cdots$

Gesamtsitzung vom 4. Mai 1916. – Mitteilung vom 30. März

Footnotes and References:
1) M. Planck, Verhandlungen der Deutschen Physikalischen Gesellschaft 1915, S.407 und 438.
2) A. Sommerfeld, Sitzungsber. der Kgl. Bayer. Akad. d. Wiss. 1915, S.425.
3) Vgl. z.B. C.V.L. Charlier, Mechanik des Himmels, Bd.I (Leipzig 1902).

Paper V·2: Excerpt from Zeitschrift für Physik **31**, 681 (1925).

On the Scattering of Radiation by Atoms

von **H.A. Kramers** und **W.Heisenberg**

in Kopenhagen. (Eingegangen am 5. Januar 1925.)

When an atom is exposed to radiation of angular frequency $\omega_\gamma = 2\pi f_\gamma$, it reradiates secondary spherical waves of the same frequency, coherent with the incoming radiation. But in general – as the correspondence principle requires – it also radiates other spherical waves of different frequencies. These angular frequencies are all of the form $|\omega_\gamma - \omega^*|$, where $\hbar\omega^*$ is the energy difference of the atom in the original and in some other state. The non-coherent radiation corresponds to certain processes recently found to demonstrate the role of light quanta. A wave-theoretic, correspondence principle analysis of the scattering interaction of atoms shows how this can take place in a natural and apparently unique way. The model clearly leads to a deeper understanding of the connection of the radiation of the atoms with the stationary states. This problem has been treated also by Bohr, Kramers and Slater, and their conclusions – if they are confirmed – would constitute interesting support for this interpretation.

§1. *Introduction.*

······ [Note added: We have skipped the preamble on classical electrodynamics and Rayleigh-Lorentz theory.] ······

The object of this work is to show how the careful application of correspondence ideas leads to the surprising result that the ansätz

$$\mathcal{E}, \mathcal{P}(t) \sim Re\left\{(\mathcal{E}_0, \mathcal{P}_0)\, e^{i\omega_\gamma t}\right\} \tag{1}$$

for the electric field vector $\mathcal{E}$ and the induced atomic electric dipole moment $\mathcal{P}$, is very useful for the description of the reaction of atoms to incident radiation of angular frequency ω_γ. The dipole moment will be generalized to a series of terms:

$$P(t) = Re \left\{ P_0 e^{i\omega_\gamma t} + \sum_j P_j e^{i(\omega_\gamma - \omega_j)t} \right\}. \tag{6}$$

Here $\hbar\omega_j = E_j - E_0$, is positive for states j with energy E_j higher than the energy E_0 of the original state, and negative for states of lower energy. The field $\mathcal{E}_0$ and the dipole moments P_0 and P_j are time-independent complex vectors, the latter of first order in $\mathcal{E}$ and ω_γ.

In words: Under irradiation by monochromatic light of frequency $f_\gamma = \omega_\gamma/2\pi$, an atom emits not only

a) coherent spherical waves of the same frequency, but also

b) non-coherent spherical waves whose frequencies correspond $\cdots\cdots$ to all possible transitions to other stationary states. $\cdots\cdots$ the average rate of radiation of energy is

$$S = \frac{(\omega)^4}{3c^3} (\bar{P} \cdot P)$$

where $\bar{P}$ is the vector complex conjugate to P $\cdots\cdots$

§2. *The Effect of External Radiation on a Classical Periodic System.*

We consider a non-degenerate quasi-periodic system, whose motion can be described by the canonical (action-angle) variables $J_1 \cdots J_n, \theta_1 \cdots \theta_n$. The electric moment of the system as a function of these variables can be represented by the following multiple Fourier series. [Note added: We suppress the explicit vector notation so $\mathcal{E} \cdot Q \to \mathcal{E}Q$, etc.]

$$\mathcal{M}(t) = \sum_{\tau's} \frac{1}{2} Q_{\tau_1 \cdots \tau_n} e^{i(\tau_1 \theta_1 \cdots \tau_n \theta_n)}. \tag{7}$$

The sum is to be taken over all positive and negative integer values of $\tau_1 \cdots \tau_n$. The coefficients Q are complex vectors whose components depend only on $J_1 \cdots J_n$. Their complex conjugates $\bar{Q}$ satisfy

$$\bar{Q}_{\tau_1 \cdots \tau_n} = Q_{-\tau_1 \cdots -\tau_n}. \tag{8}$$

The Hamiltonian H depends only on the J's. The fundamental angular frequencies are

$$\omega_k \equiv \dot{\theta}_k = \frac{\partial H}{\partial J_k} (k = 1 \cdots n). \tag{9}$$

We introduce a useful differential operator:

$$\frac{\partial}{\partial J} \equiv \sum_{r=1}^{n} \tau_r \frac{\partial}{\partial J_r}, \tag{10}$$

and also the angular frequency ω of the harmonic components of the motion:

$$\omega = \sum_{r=1}^{n} \tau_r \omega_r = \frac{\partial H}{\partial J}. \tag{11}$$

Our problem is to find the electric moment of the atom as a function of the time, for an atom exposed to a plane monochromatic beam of light whose wavelength is large compared to atomic dimensions. The electric vector of the light will be given by Eqn1. Define a new system of canonical variables J_r^*, θ_r^*, which evolve by an infinitesimal contact transformation

$$J_r^* - J_r = \frac{\partial K}{\partial \theta_r^*}, \quad \theta_r^* - \theta_r = -\frac{\partial K}{\partial J_r^*} \quad (r = 1 \cdots n) \tag{12}$$

from the old variables. It is possible to choose the function $K(J_r^*, \theta_r^*, t)$ so that in first approximation the J_r^* are independent of the time, and the θ_r^* increase linearly in the time, with $d\theta_r^*/dt = \omega_r$. The function K can be written

$$K = \text{Re} \left\{ \sum_{\tau's} \frac{i}{2} \frac{(\mathcal{E}\mathcal{Q}_{\tau_1 \cdots \tau_n})}{(\omega + \omega_\gamma)} e^{i(\theta^* + \omega_\gamma t)} \right\}, \tag{13}$$

with ω and θ^* given by the abbreviation (11). $\mathcal{Q}$ and ω here represent the same functions of J^* as they were of J. If we introduce into Eqn7 the new variables defined by Eqns12,13 and set $\theta^* = \omega t$, then we finally get the electric moment of the atom as a function of time to be:

$$M(t) = M_0(t) + M_1(t) \quad \text{where} \quad M_0(t) = \sum_{\tau's} \frac{1}{2} \mathcal{Q}_{\tau's} e^{i\omega t}. \tag{14}$$

Here and in the following, we omit the asterisk from the new variables. $\mathcal{M}_0$ corresponds to the motion of the undistorted atom. $\mathcal{M}_1$ can be written as the real part of a $2n$-fold sum over the τ and τ':

$$\mathcal{M}_1(t) = Re \sum_{\tau,\tau'} \frac{1}{4} e^{i(\omega+\omega'+\omega_\gamma)t} \left[\frac{\partial \mathcal{Q}}{\partial J'} \frac{(\mathcal{E}\mathcal{Q}')}{\omega'+\omega_\gamma} - \mathcal{Q} \frac{\partial}{\partial J} \frac{(\mathcal{E}\mathcal{Q}')}{\omega'+\omega_\gamma} \right]. \tag{15}$$

The sum is over all pairs of combinations of integer values of $\tau_1 \cdots \tau_n, \tau'_1 \cdots \tau'_n$; the abbreviated notation of Eqns10,11 is used throughout. We rearrange Eqn15 and sum first over τ_r, keeping fixed

$$\tau_r^0 = \tau_r + \tau'_r, \tag{16}$$

and then over τ_r^0 with

$$\omega^0 = \sum_r \tau_r^0 \omega_r. \tag{17}$$

We get

$$\mathcal{M}_1(t) = Re \sum_{\tau^0,\tau} \frac{1}{4} e^{i(\omega^0+\omega_\gamma)t} \left[\frac{\partial \mathcal{Q}}{\partial J'} \frac{(\mathcal{E}\mathcal{Q}')}{(\omega'+\omega_\gamma)} - \mathcal{Q} \frac{\partial}{\partial J} \frac{(\mathcal{E}\mathcal{Q}')}{\omega'+\omega_\gamma} \right]. \tag{18}$$

In the summation over τ at fixed τ^0, τ' is determined by Eqn16. Both ω' and ω^0 can take positive and negative values, since the summations are over all positive and negative integral values of the τ_r and τ_r^0. $\cdots\cdots$ it is necessary to assume that ω_γ does not coincide with any ω_r in the undisturbed motion.

Eqn18 says that the system – under the influence of the incident light – will emit radiation whose intensity is proportional to the intensity of the incident light; it splits up into harmonic components containing both the frequency ω_γ of the incident light, as well as frequencies which are the sums or differences of ω_γ and a frequency ω^0 of Eqn17. The frequency ω^0 itself need not occur in the undisturbed motion of the atom. Rather, ω^0 always has the form $|\omega| \pm |\omega'|$, where $|\omega|$ and $|\omega'|$ are both frequencies occurring in the undisturbed motion.

* *

In classical perturbation theory, the result can be recognized as the Poisson Bracket of the dipole moment with the generating function of the canonical transformation K which is the time integral of the perturbation Hamiltonian $\Delta H = -\vec{E} \cdot \vec{P}$:

$$K = \int^t \Delta H(t')dt'.$$

[See H. Goldstein, *Classical Mechanics* (Addison-Wesley, Reading, Mass., 1980), p450.]:

Then the induced dipole moment $\mathcal{M}_1$ is

$$
\begin{aligned}
\mathcal{M}_1 &= \sum_k \left(\frac{\partial \mathcal{M}_0}{\partial J_k} \delta J_k + \frac{\partial \mathcal{M}_0}{\partial \theta_k} \delta \theta_k \right) \\
&= \sum_k \left(-\frac{\partial \mathcal{M}_0}{\partial J_k} \frac{\partial K}{\partial \theta_k} + \frac{\partial \mathcal{M}_0}{\partial \theta_k} \frac{\partial K}{\partial J_k} \right) \\
&\sim \sum_{\tau's} \frac{\partial \mathcal{Q}}{\partial J'} \frac{(\mathcal{E}\mathcal{Q})}{\omega' + \omega_\gamma} - \mathcal{Q} \frac{\partial}{\partial J} \frac{(\mathcal{E}\mathcal{Q})}{\omega' + \omega_\gamma},
\end{aligned}
$$

where we have used

$$\int^t dt' e^{i(\omega' + \omega_\gamma)t'} \sim \frac{e^{i(\omega' + \omega_\gamma)t}}{i(\omega' + \omega_\gamma)},$$

which establishes Eqns15,18.

* *

§3. *Quantum Theory and the Coherent Radiation.*

······[Note added: We skip §3 in its entirety, and discuss instead §4 on the non-coherent processes, which leads directly to Heisenberg's starting point.]

§4. *The Non-Coherent Light.*

We now give a quantum interpretation for the terms in the classical Eqn18 which have ω^0 different from zero. For this we consider Fig2 for a system of two degrees of freedom. Let the points P, Q, R and S in the J_1, J_2-plane represent four stationary states, so that when the state P has the quantum number $n_k = (n_k)_P$, the states Q, R, S have the following quantum numbers:

$$(n_k)_Q = (n_k)_P + \tau_k + \tau_k' = (n_k)_R + \tau_k' = (n_k)_S + \tau_k$$
$$(n_k)_R = (n_k)_P + \tau_k, \quad (n_k)_S = (n_k)_P + \tau_k', \tag{30}$$

where the integers τ_k, τ_k' are defined in Eqns16,17. The different transitions labeled 1,2,3,4 are all possible. On the other hand, there is no possible transition which connects PQ. $\cdots\cdots$

* * * * * * * * * * * * * * * * * **

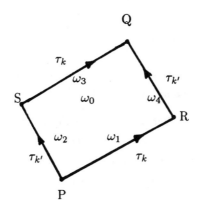

Fig.2: J_1, J_2-Plane of Four Stationary States.

* * * * * * * * * * * * * * * * * **

$\cdots\cdots$ A quantum interpretation of the expressions between the square brackets in Eqn18 – *in which the differentiations are replaced by differences divided* $\hbar$ [Note: Italics added. Note also that the use of Schwarzschild's Action-variable J is essential here, and that a Dipole Selection Rule $\Delta J = \pm1$ is implied.], the frequencies ω' by quantum energy differences, and the classical dipole moments by amplitudes Q characteristic of the quantum transitions – can be made in the following plausible way:

$$\frac{\partial Q}{\partial J'} \cdot \frac{(\mathcal{E}Q)}{\omega' + \omega_\gamma} \sim \frac{Q_3 - Q_1}{\hbar} \cdot \frac{1}{2} \left(\frac{(\mathcal{E}Q_4)}{\omega_4 + \omega_\gamma} + (4 \to 2) \right),$$

$$Q \cdot \frac{\partial}{\partial J} \left(\frac{(\mathcal{E}Q')}{\omega' + \omega_\gamma} \right) \sim \frac{Q_3 + Q_1}{2} \cdot \frac{1}{\hbar} \left(\frac{(\mathcal{E}Q_4)}{\omega_4 + \omega} - (4 \to 2) \right). \tag{32}$$

In the subtraction, a number of terms cancel and one gets as a quantum interpretation of the quantity in the brackets of Eqn18:

$$\{\tau_k, \tau_k'\} \sim \frac{1}{\hbar} \left\{ -\frac{\mathcal{Q}_1(\mathcal{E}\mathcal{Q}_4)}{\omega_4 + \omega_\gamma} + \frac{\mathcal{Q}_3(\mathcal{E}\mathcal{Q}_2)}{\omega_2 + \omega_\gamma} \right\}. \tag{33}$$

In this transformation it is tacitly assumed that the frequencies ω_γ, ω' and ω^0 are all positive, as are their quantum values:

$$\hbar\omega_0 = E(Q) - E(P),$$
$$\hbar\omega_1 = E(R) - E(P), \quad \hbar\omega_2 = E(S) - E(P),$$
$$\hbar\omega_3 = E(Q) - E(S), \quad \hbar\omega_4 = E(Q) - E(R). \tag{34}$$

When any one of these energies is negative, the corresponding vector Q in Eqn33 should be replaced by its complex conjugate $\bar{Q}$.

This must be summed over according to Eqn18, but first it is important to note that another term in (18), which is given by the exchange of τ_k and τ_k', can be specified by exactly the same quartet of stationary states P, Q, R, S. After this exchange, we get the expression (33) but with $1 \to 2$ and $3 \to 4$, which we write as $\{\tau_k', \tau_k\} \cdots\cdots$. We get finally:

$$\{\tau_k, \tau_k'\} + \{\tau_k', \tau_k\} \sim \frac{1}{\hbar} \left\{ \frac{\mathcal{Q}_4(\mathcal{E}\mathcal{Q}_1)}{\omega_1 + \omega_\gamma} - (1 \leftrightarrow 4) \right\}$$
$$+ \{1 \to 2 \text{ and } 4 \to 3\}. \tag{36}$$

Now we define a complex vector $\mathcal{M}(P, Q; R)$ which depends on the triplet of states P, Q, R:

$$\mathcal{M}(P, Q; R) = \frac{1}{4\hbar} \left\{ \frac{\mathcal{Q}_q(\mathcal{E}\mathcal{Q}_p)}{\omega_p + \omega_\gamma} - \frac{\mathcal{Q}_p(\mathcal{E}\mathcal{Q}_q)}{\omega_q + \omega_\gamma} \right\} e^{i(\omega_{qp} + \omega_\gamma)t}. \tag{37}$$

The small subscripts p, q, qp indicate the transitions $RP, QR; QP$. The frequencies satisfy:

$$\hbar\omega_p = E(R) - E(P), \quad \hbar\omega_q = E(Q) - E(R), \quad \hbar\omega_{qp} = E(Q) - E(P), \tag{38}$$

and can be negative where necessary. If ω_p, for example, should be negative, then Q_p should be replaced by its complex conjugate $\bar{Q}_p$, and so on.

Eqn36 can be expressed as the sum of $\mathcal{M}(P,Q;R)$ and $\mathcal{M}(P,Q;S)$, and from Eqn18 we see that the radiation moment of the atom consists of the sum of the real parts of such terms. Now a difficulty occurs; one does not know whether this expression corresponds to the reaction of the atom in state P or state Q (regardless of R). This question is resolved only by considerations which go beyond the correspondence principle. A somewhat similar question arises for spontaneous radiation from an undisturbed periodic system. $\cdots\cdots$ The correspondence principle tells us nothing about which of the two states emitted the radiation; only the conservation of energy allows us to conclude that the radiation originates from the state with the higher energy, in accord with Bohr's radiation postulate.

A similar problem confronts us now, and we will assume that the radiation moment (37) refers to the state Q when $\omega_{qp} + \omega_\gamma$ is positive, and to the state P when it is negative. This assumption can be justified in the following way: Radiation of the angular frequency $\omega_{qp} + \omega_\gamma$ means – as the basic process of energy exchange between the radiation field and the atom demands – that the atom in the original state has the possibility to change into another state with energy $\hbar|\omega_{qp} + \omega_\gamma|$. Irradiation of an atom with radiation of frequency ω_γ, on the other hand, always gives rise to the inherent possibility for the atom, of a change of state in which energy $\hbar\omega_\gamma$ either *appears* or *disappears*. [Note: Italics added.] An actual change of state of the atom will always consist of a transition to another stationary state; for radiation of frequency $|\omega_\gamma + \omega_{qp}|$ to be effective in bringing about such a transition, it must be accompanied by a reaction of frequency ω_γ. If $(\omega_\gamma + \omega_{qp})$ is positive, then this is only possible when the atom simultaneously gains an energy $\hbar\omega_\gamma$ and loses an energy $\hbar(\omega_\gamma + \omega_{qp})$; during the transition, the atom loses, in total, the energy $\hbar\omega_{qp}$, *i.e.*, the transition must be from the initial state Q, and *later* into the final state P. (This holds both for positive and negative values of ω_{qp}.) On the other hand, if $(\omega_\gamma + \omega_{qp})$ is negative (so that ω_{qp} is always negative), then the transition can only come about if the atom at the same time loses an energy $\hbar\omega_\gamma$ and an energy $\hbar(-[\omega_\gamma + \omega_{qp}])$, for a total loss of $-\hbar\omega_{qp} > 0$. For the transition to occur, the atom must initially be in state P and finally in state Q. $\cdots\cdots$ [1].

With the help of Eqns36,37, and the above understanding of the role of the atomic states in the interpretation of Eqn37, we are now able to give a quantum theoretic interpretation of Eqn18 in its full generality. For an atom in a stationary state P, the radiation moment induced by external radiation is

$$\mathcal{M}(t) = Re \sum_{Q,R} \{\mathcal{M}(Q,P;R) + \mathcal{M}(P,Q;R)\}, \tag{39}$$

where all stationary states R of the atom should be summed on, which are different from P and Q. The sum on Q should be separated into two separate sums, the first for which $E(Q) < E(P) + \hbar\omega$ and the second for which $E(Q) < E(P) - \hbar\omega$.

We now describe the results of the general formulas (37) and (39) in some special cases. For this we will always understand ω^* to be a positive quantity, whose contribution $\mathcal{M}(|\omega_\gamma \pm \omega^*|)$ to the total radiation moment of the atom in the state P will be given, assuming the existence of a state Q corresponding to that particular angular angular frequency $|\omega_\gamma \pm \omega^*|$ in the radiated light. It is necessary to distinguish various cases.

Case I. $E(Q) > E(P)$, $E(Q) - E(P) = \hbar\omega^*$. The state Q reacts to the scattered radiation when $\omega_\gamma > \omega^*$, and the frequency of the rescattered radiation is $\omega_\gamma - \omega^*$. The states R different from P and Q divide into three groups: R_a with $E(R_a) > E(Q) > E(P)$, R_b with $E(Q) > E(R_b) > E(P)$ and R_c with $E(Q) > E(P) > E(R_c)$. The absolute values of the frequencies corresponding to the transitions $R_{a,b,c} \to P,Q$ are $\omega_{1,3,5} = |E(R_{a,b,c}) - E(P)|$, and $\omega_{2,4,6} = |E(R_{a,b,c}) - E(Q)|$. With these, the expression for the induced atomic moment becomes

$$\mathcal{M}(\omega - \omega^*) = Re \frac{1}{4\hbar} e^{i(\omega-\omega^*)t} \times$$
$$\left\{ \sum_{R_a} \left(\frac{Q_2(\mathcal{E}\bar{Q}_1)}{\omega_1 - \omega_\gamma} + \frac{\bar{Q}_1(\mathcal{E}Q_2)}{\omega_2 + \omega_\gamma} \right) + (R_b) + (R_c) \right\}. \tag{40}$$

$\cdots\cdots$

* *

The first term in R_b

$$\sim \frac{\mathcal{Q}_4(\mathcal{E}\mathcal{Q}_1)}{\omega_1 - \omega_\gamma}$$

can be recognized in a familiar quantum mechanics context. In first order time independent perturbation theory, we calculate:

$$\psi_P \simeq \phi_P + \frac{1}{E_P - H_0}H'\phi_P.$$

The induced dipole moment between states Q, P is

$$
\begin{aligned}
\vec{d}_{QP} &= \langle\psi_Q|\vec{d}|\psi_P\rangle - \langle\phi_Q|\vec{d}|\phi_P\rangle \\
&\sim \langle\phi_Q|\vec{d}\frac{1}{E_P - H_0}\vec{\mathcal{E}}\cdot\vec{d}|\phi_P\rangle \\
&\sim \sum_R \langle\phi_Q|\vec{d}|\phi_R\rangle\frac{1}{E_R - E_P - \hbar\omega_\gamma}\vec{\mathcal{E}}\cdot\langle\phi_R|\vec{d}|\phi_P\rangle \\
&\sim \sum_R \vec{d}_{QR}\frac{1}{\hbar\omega_{RP} - \hbar\omega_\gamma}\vec{\mathcal{E}}\cdot\vec{d}_{RP} \\
&\sim \sum_R \frac{\mathcal{Q}_4}{\hbar}\frac{1}{\omega_{RP} - \omega_\gamma}\mathcal{E}\mathcal{Q}_1,
\end{aligned}
$$

in the Schrödinger-Dirac, the Heisenberg, and the original Kramers-Heisenberg notation. The classical prescription of '$\frac{1}{2}\times$ the Real part' does not survive, of course.

* *

Case II. $E(Q) < E(P)$ $\cdots\cdots$

Case III. $E(Q) = E(P)$ $\cdots$ [Note added: For the sake of brevity, we only discuss Case I, which results in expressions crucial to Heisenberg's insight leading to matrix mechanics.]

§5. *Concluding Remarks.*

One must require $\cdots\cdots$ both the energy distribution in the cavity radiation and the statistical equilibrium of the atoms be left unchanged. The ansätz by which this is achieved must be essentially that used by Pauli [2] to describe the thermal-equilibrium between free electrons and radiation. For example, in Case I

· · · · · ·

The results of this work show that it is *almost* possible from the classical theory, in accordance with the requirement of the correspondence principle in the limit of large numbers, to understand the radiation produced in an atomic system by external radiation. The theory is sufficient to lead to results of the form Eqn39 where $\mathcal{M}$ has the explicit character (37). · · · · · · difficulties when $\omega_\gamma + \omega_{rp} = 0$ or $\omega_\gamma + \omega_{qr} = 0$ · · · · · · spontaneous emission · · · · · · [Note added: The discussion devolves into the predicament of spontaneous radiation by an isolated atom, and refers to the earlier qualitative discussion of Bohr, Kramers and Slater. Our interest is to understand Eqn18 and the use of Action-Angle variables as the starting point of Heisenberg's invention of quantum mechanics from the requirement of matrix multiplication.] · · · · · ·

The fundamental idea of this work can be stated generally: the role of the atom in optical phenomena can always be expressed by the interaction between the radiation field and the atom in some one of its stationary states; the characteristic process, which we describe as a "transition from one stationary state to another", must then be regarded as being very short lived, and optical phenomena – so far as is yet known and analyzed – give no further insight into their nature.

Kopenhagen, Institut for teoretisk Fysik, Dezember 1924.

Footnotes and References:
1) N. Bohr, Naturw. **12**, 1115 (1924).
2) W. Pauli jr., Zeits. f. Phys. **18**, 272 (1923). · · · · · ·

PART TWO. Chapter VI

HEISENBERG:

Diese Frage hat nicht mit Electrodynamic zu tun, sondern sie ist $\cdots$ rein kinematischer Natur: $\cdots$ in einfachster Form $\cdots$: Gegeben an Stelle der klassischen Grösse x(t), tretende quantentheoretische Grösse; welche quantentheoretische Grösse tritt dann an Stelle von x(t)^2?

" This question has nothing to do with electrodynamics, but it is, it seems clear to us, of a purely kinematic nature; we can state it in the simplest form as follows: If one is given in place of the classical quantity $x(t)$ the corresponding quantum theoretic quantity – then what quantum theoretic quantity should take the place of $x^2(t)$?"
W. Heisenberg, Zeits. f. Phys. **33**, 879 (1925).

§VI-1. Introduction.

It was a time of giants, and among them Heisenberg appeared godlike. He recalls his first meeting at age 20 (at the end of his second university year) with Niels Bohr [1]. Sommerfeld had taken him along to the 'Bohr-festspiele' where " $\cdots$ Bohr concluded $\cdots$ one should assume Kramers' results were correct. I knew Kramers' work rather well $\cdots$ and dared to dissent $\cdots$ because the quadratic Stark-effect could be thought of as a limiting case of the scattering of light with very large wavelength $\cdots$ [Note added: with resonance at the electron's orbital frequency] $\cdots$ Bohr suggested that we should go for a walk $\cdots$ That $\cdots$ was the first thorough discussion of the fundamental $\cdots$ problems of modern atomic theory, and certainly had a decisive influence on my career." Imagine the precocity! And the profound qualities of both people to make such an exchange possible.

Heisenberg finished his hydrodynamics thesis with Sommerfeld at Munich at age 22 and was immediately involved as Born's assistant (along with Pauli and Jordan) in Born's attempts to modify classical perturbation theory to accommodate quantum transitions. Here he participated in Born's precursor to quantum mechanics where action-angle variables are used to introduce differences in place of derivatives [2], and collaborated with Kramers' [3] where this modification of the derivatives and the resulting reinterpretation of the classical perturbation terms in Kramers' dispersion theory seem to be his contribution [1].

June 1925 finds Heisenberg alone on the rocky island of Helgoland (to combat his allergies). He recounted years later to van der Waerden [1]: 'There was a moment when I had a revelation in which I saw that the energy was time independent. It was rather late at night. I calculated it with some difficulty, and it agreed. Then I climbed up on a cliff and saw the sun rise and felt very lucky.' [Note added: van der Waerden identifies this calculation as the derivation by matrix multiplication (not yet identified as such) of the harmonic oscillator Hamiltonian as a constant diagonal matrix with elements $(n + \frac{1}{2})\hbar\omega$.]

Heisenberg kept up a stream of letters to Pauli [1] reporting his discovery, his euphoria, his insecurities, and his optimism.

On June 21: 'My attempt to construct a new quantum mechanics progresses only slowly, but I am not really concerned, since it seems to me far-removed from the theory of quasi-periodic solutions. $\cdots$'

June 24: '$\cdots$ everything is still unclear and I presume only approximate $\cdots$ but perhaps the fundamental ideas are still correct. The basic principle is: In the calculation of any quantity such as energy, frequency etc, only relations between quantities controllable in principle should occur. (So far, the Bohr theory of hydrogen seems to me formally very much the same as Kramers' dispersion theory). $\cdots$' [Note added: Here Heisenberg reproduces the harmonic oscillator results of **Paper VI·1**.]

June 29: 'Meanwhile I have been pressing on somewhat, but not very much because in my heart I am convinced that this quantum mechanics is already correct, which is why Kramers called me the Optimist $\cdots$'

July 9: '$\cdots$ my interpretation has changed radically from day to day since Helgoland $\cdots$ It is actually my conviction that any interpretation of the Rydberg formula in terms of circular and elliptic orbits in classical geometry has not the slightest physical sense. My whole troubled efforts are intended to totally destroy the idea of the orbits, which one cannot even observe, and to replace them appropriately. Therefore I dare to send you this brief preliminary manuscript of my work because I believe that it – at least in the critical negative part – contains actual physics. In fact I have a very guilty conscience because I must beg you to return it to me in 2-3 days, since I must either make its existence known in the next few days or burn it. My own opinion of the paper, about which I am not really happy, is this: On the negative heuristic part I am firmly convinced, but on the positive more formal and speculative part, less so; but perhaps people can, someone can, make something sensible out of it. Therefore I beg you to read primarily the introduction ···'

van der Waerden does not record Pauli's responses, if any.

Heisenberg's discovery of Quantum Mechanics epitomizes and idealizes the creative act almost to the point of trivializing it. How does one isolate and formulate "in einfachster Form" the critical question which constitutes the crux of the mystery? Can such profound results ever again be achieved in such an elemental way? Feynman argued facetiously that they could not – 'nothing ever works the same way twice' – was his joking remark about such efforts. Feynman notwithstanding, can we understand what it was about Heisenberg – his intrinsic qualities, background, fundamental beliefs, elementary education, advanced education, teachers, colleagues, peers, thought processes – what was it that led Heisenberg in particular to ask 'How do you multiply x times x?'

The answer to this question is not an idle exercise. In fact it is fundamental to our whole educational philosophy from the neighborhood kindergarten to the most elite graduate schools and beyond to the great concentrations of intellectual superstars on such airless Everests as the Institute of Advanced Study. In a different area, it is the trillion-dollar computer software industry, where armies of very intelligent people are set to work in parallel on still very difficult problems which require abstract thinking and mastery of immense complexity, and with great success. How do these immensely successful people – in their hundreds of thousands

– differ from Heisenberg? Do their numbers, or those of any other modern cadre – for example the few thousand string theorists, or the lonely soul who proved Fermat's Last Theorem – include a modern Heisenberg?

What questions are being asked whose answers could be as profound that posed by Heisenberg? Are there any? Has the world passed on from a simpler time to the point where no relative neophyte – as Heisenberg was; after all, he apparently did not know about matrices – can hope to attain a broad but deep understanding of any subject without investing years in the complicated details, and make an immortal contribution whose purpose is to render such complications irrelevant.

It seems to us to be essential for any hope or aspiration to revolutionary advance that we take an optimistic attitude and act upon it *at all levels*. As physicists – not necessarily theoreticians or mathematical physicists – and teachers of physics and of aspiring physicists, we must rethink our education and research strategies to more nearly emulate the environment created by some of the founders of quantum mechanics. Our candidates for role models are two: Kramers and Born. We exclude Einstein, Dirac and de Broglie as too solitary and Bohr as too unsystematic. But Kramers and Born seemed to operate in a mode that can be adapted to all levels 'K through 22', which we will characterize (caricaturize?) by:
a) Small classes (like six) with an emphasis on dialog, discussion and debate.
b) A fanatical devotion to depth of understanding, and the ability to express it in multiple ways – verbally, orally, pictorially, intuitively, formally, mathematically, by analogy or contrast $\cdots$.
c) The exercise of exquisite taste to avoid irrelevant complications.
d) Practise in conjecture, validation and rejection of hypotheses, all part of the debates.
e) Lots of very expensive 'face'-time with able – hopefully gifted – next- to highest-level practitioners. It is important and difficult to make this face-time a significant professional activity for the higher level participants so that they will not feel that they are 'stealing' time away from their real (self-)interests.

The participants must – at every level – be self-selected. How they can gracefully exit is a problem that people rationalize every day and need not be an over-riding concern except among the very youngest.

We personally believe in programs that are narrowly focused, supported by simultaneous intense individual study and reading of assigned sources. Group meetings can alternate between formal seminar presentations as introductions to particular topics, and free-for-all round-table exchanges but always on the agreed upon topic. Superficiality and preciosity in the group or in any individual can be combated by requiring formal and even written presentations, which then become the starting point of the following discussion-debate. Just as superficiality can defeat the efforts of the junior participants, dogmatism or a personality cult around the leaders of the group can do mortal harm. The length of an individual conference can be usefully regulated by providing no place for the participants to sit. Standing also seems to encourage participation and an egalitarian feeling in the debate.

Hopefully such programs already exist. How might we judge their effectiveness? Certainly, no one has recently invented new ways of multiplying x times x. If there is a new Heisenberg among us, then there must be a hundred. So we need some intermediate criterion for the definition of success in such intellectual experiments: perhaps just deep interest in our discipline of choice, a lively joy in the revelation of understanding even if new only to us, an intimacy of understanding, an ability to articulate, to explain, to extrapolate, to question.

In the narrow confines of the subject of quantum mechanics, there must still be – it seems to us inevitable and unavoidable – a major investment involving the discipline of acquiring in the traditional way the fundamental foundation of mathematics, mechanics, modern physics and quantum mechanics. The small intimate discussion groups will hopefully enliven this traditional experience and better engage the minds and stimulate the creativity of the best students to even higher levels.

Footnotes and References:

1) B.L. van der Waerden, *Sources of Quantum Mechanics* (North Holland, Amsterdam, 1967), Pp.21-25; p.16; Pp.25-27.

2) M. Born, Zeits. f. Phys. **26**, 379 (1924).

3) H.A. Kramers and W. Heisenberg, Zeits. f. Phys. **31**, 681 (1925) (Here as **Paper V·2**.)

Biographical Note on Heisenberg.

Werner Karl Heisenberg (1901-1976) – was a student of Sommerfeld at Munich, with a PhD in hydrodynamics in 1923 [1]. From there, he became assistant to Born at Göttingen (1923-24); and a visitor at Copenhagen (1925-27). His discovery of the multiplication law of quantum mechanics was the first breakthrough into modern quantum mechanics; it was immediately recognized by Born and Jordan as matrix multiplication, and developed with them in the standard 'canonical' formulation paralleling the canonical formulation of classical mechanics (1925). In addition, Heisenberg was first to treat the effect of particle identity in quantum mechanics (1926); invented the Uncertainty Principle (1927); the theory of ferromagnetism (1928); formulated quantum electrodynamics with Pauli (1929); hole theory in solid state physics (1931); isotopic spin theory of charge independent nuclear force (1932); S-matrix theory and the bootstrap mechanism (1943); non-linear spinor field theory (denounced by Pauli) and spontaneous symmetry breaking (1947-). He was Professor at Leipzig (1927-41); at Berlin (1941-45); then Göttingen and Munich; and President of the Alexander von Humboldt Foundation (1953-).

Heisenberg has been much analyzed and criticized for his role in the Nazi atom bomb project [2]. The tape recordings of his reaction while in confinement at the time of the Hiroshima blast are interpreted by some as evidence of confusion and incompetence. There seems to be no doubt that his gross overestimate of the U235 critical mass effectively defeated the Nazi program for bomb development. Whether this was an error due to naivete or to calculated obstruction is still argued.

However culpable he was for his decision to remain a loyal German in spite of the Nazi regime, he paid with everything but his life. The full magnitude of the tragedy visited on him by the Nazis – from his early persecution including denunciation to Gestapo chief Himmler, to the ultimate degradation and intellectual destruction from which he never recovered his reputation or his creative gift – is described in detail in Cassidy's definitive biography [3]. His precocious and prolific contributions were all made before WWII, as were those of Bohr, Born, Dirac, Jordan, Pauli, and Schrödinger. So perhaps his only professional loss was his moral integrity, and that based on arguable evidence of the degree of his participation in Nazi atomic research.

What is unarguable is that Heisenberg did nothing to help Jewish Holocaust victims, in particular the parents of his 1925 colleague turned 1945 investigator and thereafter lifelong antagonist – Sam Goudsmit (co-discoverer with Uhlenbeck in 1925 of the electron spin of $\hbar/2$). Heisenberg's defense was that nothing could be done. The unspoken and somehow implied defense, which infuriated his critics, was that he did nothing because it would have endangered his most important mission, which was to subvert the Nazi atom bomb project.

Whatever the truth, we choose to remember him at his best, as Born first saw him [4]:"He looked like a simple peasant boy, with short, fair hair, clear bright eyes and a charming expression. $\cdots$ His incredible quickness and acuteness of apprehension has always enabled him to do a colossal amount of work without much effort $\cdots$", or as Kramers described both him and Bohr [5]:"tough, hard nosed, uncompromising, and indefatigable."

Footnotes and References:

1) E.P. Wigner, PHYSICS TODAY, AP76, p.86; The London Times 2FEB76, p.26.

2) Thomas Powers, *Heisenberg's War* (Little-Brown, New York, 1993), Pp. 326, 352, 365, 424, 553.

3) D.C. Cassidy, *Uncertainty – the Life and Science of WERNER HEISENBERG* (Freeman, NY, 1992), esp. Ch26.

4) B.L. van der Waerden, *Sources of Quantum Mechanics* (North Holland, Amsterdam, 1967), p.19.

5) A. Pais, *Niels Bohr's Times, In Physics, Philosophy, and Polity* (Clarendon Press, Oxford, 1991), Pp.21, 275-279.

Paper VI·1: Excerpt from Zeitschrift für Physik **33**, 879 (1925).

On the Quantum Reinterpretation
of Kinematical and Mechanical Relations

von **W. Heisenberg** in Göttingen.

(Eingegangen am 29. Juli 1925.)

This work contains the principle for constructing a quantum mechanics based exclusively on relations between observable quantities.

There is a serious objection to the formal rules in quantum theory for the calculation of observable quantities (e.g. the energy of the hydrogen atom). The calculational rules include essential relations between quantities which cannot be observed even in principle (e.g. position, orbital period of the electron). As a result, these rules lack the obvious physical requirement – which one could no longer hope to have – that any quantity could be verified experimentally. This hope can be maintained when the known rules are the result of and are applied to a particular limited range of quantum problems. However not even the hydrogen atom and its Stark effect follow the formal rules of such a quantum theory. Fundamental difficulties already appear in the problem of 'crossed fields' (the hydrogen atom in electric and magnetic fields of different directions), where the reaction of the atom to time-varying fields cannot be described by the known rules; and the extension of the quantum rules to atoms with more electrons has also proven impossible. This failure of the quantum rules, which were in fact constructed from classical mechanics, is presumed to indicate a deviation from classical mechanics. However this can hardly be seen as a minor deviation when one considers that the Einstein-Bohr frequency condition (which has general validity) is completely counter to classical mechanics; or even more convincing, from the standpoint of the wave theory which dictates the kinematics underlying this mechanics, where one concludes that classical mechanics absolutely cannot be imagined to have any validity even for the simplest quantum problems. In this state of affairs it seems advisable to abandon completely any hope of observation for the as yet unobservable quantities; and at the same time to concede that the piece-wise agreement of the existing quantum rules with experiments is more or less accidental; and to construct a quantum mechanics in which only relations between observable quantities occur. As the most important first Ansätz for such a quantum mechanics, one should look not only at the frequency condition but also at Kramers' dispersion theory [1] and the results following from it [2]. In the following we will construct such new quantum relations, and use them in the complete treatment of some special problems.

* * * * * * * * * * * * * * * * * * * **

In a 1963 letter to van der Waerden (loc. cit. p.29), Heisenberg expanded on his motivation: "I had always considered the wave theory of light to be reliable and believed that light waves must surely be generated by a source which had the same frequency and not some frequency of the electrons in the Bohr orbits of the atom. This point had already been emphasized by me (in my discussion with Bohr at the 1922 Bohr-Festspielen) $\cdots\cdots$ something in the atom must be oscillating with the correct frequency $\cdots$ one must somehow introduce some strange kinematics for the electrons which yields this frequency through the combination of two stationary states." The 'something oscillating with the correct frequency' had already been introduced as the 'virtual oscillators' of Bohr, Kramers and Slater.

* * * * * * * * * * * * * * * * * * * **

§1. In the classical theory, radiation from a moving electron (in the wave zone, i.e. $\mathcal{E}, \mathcal{B} \sim 1/r$) involves not only the dipole expressions:

$$\mathcal{E} = \frac{e}{r^3 c^2}[r[r\dot{v}]], \qquad \mathcal{B} = \frac{e}{r^2 c^2}[\dot{v}r],$$

but also quadrupole terms of the form, e.g. $\sim e\dot{v}v/rc^3$, and higher order dipole terms of the form, e.g. $\sim e\dot{v}v^2/rc^4$; and so on.

One can ask if each higher term must occur in the quantum theory. In the classical theory the higher terms can be calculated if the electron motion – i.e. its Fourier representation – is given. In quantum theory one would expect the same. This question has nothing to do with electrodynamics, but is of a purely kinematic nature. In its simplest form we can ask: if instead of the classical quantity $x(t)$, one is given its corresponding quantum quantity; then what quantum quantity should take the place of $x^2(t)$?

Before answering this question, we should note that in quantum theory it is not possible to correlate the position of the electron as a function of time with any observable quantities. However even in quantum theory the electron can radiate; the radiation will be described first of all by its angular frequency $\omega \equiv 2\pi f$, which in quantum theory is the difference of two energies:

$$\omega(n, n-\alpha) = \frac{1}{\hbar}\{W(n) - W(n-\alpha)\},$$

or in the classical theory in the form:

$$\omega(n, \alpha) = \alpha \cdot \omega(n) = \alpha\frac{1}{\hbar}\frac{dW}{dn}.$$

(Here one is assuming $2\pi n \cdot \hbar = J$, one of the canonical invariants.) To compare the classical and quantum theories, we have the combination rules for the frequency:

Classical:

$$\omega(n, \alpha) + \omega(n, \beta) = \omega(n, \alpha + \beta)$$

and Quantum:

$$\omega(n, n-\alpha) + \omega(n-\alpha, n-\alpha-\beta) = \omega(n, n-\alpha-\beta).$$

The amplitudes are also necessary for the description of the radiation. The amplitudes for a given polarization and phase can be given as complex vectors (with six independent components which are functions of two variables n and α) represented by the following expression:

Quantum: $Re\left\{\mathcal{A}(n, n-\alpha)e^{i\omega(n,n-\alpha)t}\right\}.$ (1)

Classical: $Re\left\{\mathcal{A}_\alpha(n)e^{i\omega(n)\cdot\alpha t}\right\}.$ (2)

[Note added: The erroneous 'Re' – real part – immediately vanished from Heisenberg's work.]

The phase of $\mathcal{A}$ might seem to have no physical significance in the quantum theory, since the frequencies are generally not commensurable. However even in the quantum theory the phase has a definite significance analogous to that in the classical theory. If one considers a particular quantity $x(t)$ in the classical theory, then one can think of quantities of the form

$$\mathcal{A}_\alpha(n)e^{i\omega(n)\cdot\alpha t},$$

which – whether or not the motion is periodic – combine in a sum or integral to produce $x(t)$:

$$x(n,t) = \sum_\alpha \mathcal{A}_\alpha(n)e^{i\omega(n)\cdot\alpha t} \text{ or or } x(n,t) = \int d\alpha \mathcal{A}_\alpha(n)e^{i\omega(n)\cdot\alpha t}.$$

Such a combination of the corresponding quantum quantities – because of the symmetry between n and $(n-\alpha)$ – seems somewhat arbitrary and meaningless; but if one can view the quantities

$$A(n, n-\alpha)e^{i\omega(n,n-\alpha)t}$$

as representing $x(t)$, then the question still needs to be answered: How should the quantity $x^2(t)$ be represented?

Classically the answer is obvious:

$$\mathcal{B}_\beta(n)e^{i\omega(n)\cdot\beta t} = \sum_\alpha \mathcal{A}_\alpha \mathcal{A}_{\beta-\alpha}e^{i\omega(n)\cdot(\alpha+\beta-\alpha)t}, \tag{3}$$

or the corresponding integral. Then

$$x^2(t) = \sum_\beta \mathcal{B}_\beta(n)e^{i\omega(n)\cdot\beta t}. \tag{5}$$

In quantum mechanics the simplest and most natural assumption is to replace Eqn (3) by the following:

$$B(n, n-\beta)e^{i\omega(n,n-\beta)t} = \sum_\alpha A(n, n-\alpha)A(n-\alpha, n-\beta)e^{i\omega(n,n-\beta)t}. \tag{7}$$

This gives the kind of structure already found in the rule for combining frequencies. If one makes the assumption (7), then the phases of the quantum amplitudes A have just as great a physical significance as in the classical theory: only the starting point of time – and therefore a common phase for all A – is arbitrary and without physical significance; but the phase of the individual A enters essentially in the quantity B [3]. A geometric interpretation of such quantum phase relations analogous to the classical theory at first seemed hardly possible.

This substitution is already motivated by Born's observation (see van der Waerden, loc. cit., p.21, p.30) that Einstein's emission probabilities A_{PQ} correspond to squares of classical amplitudes. Born, in a paper where Heisenberg had an acknowledged but unspecified role (Zeits. f. Phys. **26**, 379 (1924)), had coined the expression 'Quantum Mechanics' and the term 'transition amplitude' whose square was Einsteins transition probability, and later recalled speculating (in a paper with Jordan in Zeits. f. Phys. **33**, 479 (1925)) in "daily meetings in which Heisenberg often took part $\cdots$ that these amplitudes might be the central quantities and be handled by some kind of symbolic multiplication." One might wish that the noble Born had left this self-serving remark unsaid.

For the quantity $x^3(t)$, we easily find:

Classical:

$$\mathcal{C}(\eta, \gamma) = \sum_{\alpha, \beta} \mathcal{A}_\alpha(n) \mathcal{A}_\beta(n) \mathcal{A}_{\gamma - \alpha - \beta}(n); \tag{9}$$

and Quantum:

$$\mathcal{C}(n, n - \gamma) = \sum_{\alpha, \beta} \mathcal{A}(n, n - \alpha) \mathcal{A}(n - \alpha, n - \alpha - \beta) \mathcal{A}(n - \alpha - \beta, n - \gamma), \tag{10}$$

or the corresponding integrals. In a similar way all the amplitudes $x^n(t)$ can be represented in quantum theory. For any function $f[x(t)]$, one can obviously find the quantum analog if the function can be expanded as a power series.

Kramers and Heisenberg (see here in **Paper V·2**, Fig2 and Eqns 37,38) introduced the notation defining transition dipole moments as, e.g.,

$$\mathcal{Q}_p \sim \mathcal{Q}_{P \to R}$$

with time dependence $e^{-i(E_P - E_R)t}$,

and products

$$Q_q Q_p e^{i\omega_{qp} t} \sim Q_{R \to Q} Q_{P \to R} e^{i(E_Q - E_P)t}.$$

This notation and the concept of replacing the classical dipole moment P by the "charakteristischen Amplituden Q der quantentheoretischen Ubergäng" provided Heisenberg with the direct and compelling path to his discovery of quantum multiplication. Without the essential idea of giving equal billing to the initial and final states in the transition amplitude, the form of the product is obscure. Heisenberg's brilliance in isolating the essence of the problem 'in einfachster Form' – in the simplest way – was characteristic.

* * * * * * * * * * * * * * * * * * **

An essential difficulty does arise when we consider two quantities $x(t), y(t)$, and their product $x(t) \cdot y(t)$. If x is represented by $\mathcal{A}$ and y by $\mathcal{B}$, then the representation of $x \cdot y$ is:

Classical:

$$\mathcal{C}_\beta(n) = \sum_\alpha \mathcal{A}_\alpha(n) \mathcal{B}_{\beta-\alpha}(n),$$

and Quantum:

$$\mathcal{C}(n, n - \beta) = \sum_\alpha \mathcal{A}(n, n - \alpha) \mathcal{B}(n - \alpha, n - \beta).$$

Whereas classically $x(t) \cdot y(t)$ is always equal to $y(t) \cdot x(t)$, this need not be the case in quantum theory. In special cases, e.g. for the structure $x(t) \cdot x^2(t)$, this problem does not occur. The form $v\dot{v}$ in quantum theory could be replaced by

$$v\dot{v} \to \frac{d}{dt} \frac{v^2}{2} \equiv \frac{v\dot{v} + \dot{v}v}{2}.$$

Perhaps in similar ways some natural quantum average can be defined like Eqn (7).

Apart from the above difficulty, formulas like (7) might be useful to describe the interaction of electrons in an atom.

§ **2.** From these ideas – which concern the quantum theory of kinematics – we turn to the mechanics problem of determining the $\mathcal{A}, f, W$ of the system from the given forces. Previously this problem has been solved in two steps:

1. Integration of the equation of motion:

$$\ddot{x} + F(x) = 0. \tag{11}$$

2. Determination of the constants of the (periodic) motion by:

$$\oint pdq = \oint m\dot{x}dx = J(= 2\pi n \cdot \hbar). \tag{12}$$

To construct a quantum mechanics problem analogous to a particular classical problem, it seems very natural to take the equation of motion (11) directly over into the quantum theory, where it is only necessary - in order to maintain the fundamental principle of observable quantities – to replace the quantities $\ddot{x}, F(x)$ by their quantum representations known from §1. In the classical theory it is possible to solve (11) by the Ansätz of x as a Fourier series (or integral) with undetermined coefficients; we then get infinitely many equations with infinitely many unknowns, which can be reduced only in special cases to simple recursion relations for the $\mathcal{A}$. In quantum theory however we are restricted to this kind of solution of (11), because no quantum function can be defined which is directly analogous to the function $x(t)$.

This has the result that the quantum solution of (11) can be carried out only in the simplest cases. Before we do such a simple example however, the quantum determination of the constants of the motion must be derived from (12). We assume that the motion (classically) is periodic:

$$x = \sum_{\alpha=-\infty}^{+\infty} A_\alpha(n)e^{i\alpha\omega_n t}; \tag{13}$$

then

$$m\dot{x} = m\sum A \cdot i\alpha\omega_n e^{i\alpha\omega_n t}$$

and

$$\oint m\dot{x}dx = \oint m\dot{x}^2 dt = 2\pi m\sum A_\alpha(n)A_{-\alpha}(n)\alpha^2\omega_n.$$

Since x is real, then

$$\mathcal{A}_\alpha(n) = (\mathcal{A}_{-\alpha}(n))^*$$

and

$$\oint m\dot{x}^2 dt = 2\pi m \sum |\mathcal{A}_\alpha(n)|^2 \alpha^2 \omega_n. \tag{14}$$

This phase-integral is set equal to $2\pi n \cdot \hbar$. However this condition had to be forced on the mechanical calculation; in fact it seemed arbitrary even from the original standpoint of the correspondence principle which only determines $J = 2\pi n \cdot \hbar$ within an additive constant. So in place of (14) it is more natural to take:

$$\frac{d}{dn}(2\pi n \cdot \hbar) = \frac{d}{dn} \cdot \oint m\dot{x}^2 dt,$$

which says

$$\hbar = m \sum_{\alpha=-\infty}^{+\infty} \alpha \frac{d}{dn} \left\{ \alpha \omega_n \cdot |\mathcal{A}_\alpha(n)|^2 \right\}. \tag{15}$$

This restriction only defines the $\mathcal{A}_\alpha$ within a constant, an indeterminacy which has given rise to the possibility of empirically using half-integral quantum numbers.

If we ask whether (15) corresponds to some quantum relation between observable quantities, then one with the most remarkable uniqueness does occur to me. In fact, Eqn (15) assumes a simple quantum form connected to Kramers' dispersion theory [4]:

$$\hbar = 2m \sum_{\alpha=0}^{\infty} \left\{ |\mathcal{A}(n+\alpha, n)|^2 \omega(n+\alpha, n) - |\mathcal{A}(n, n-\alpha)|^2 \omega(n, n-\alpha) \right\}. \tag{16}$$

[Note: Recall the classical $\Rightarrow$ quantum prescription $\alpha \frac{d}{dn} F(n)|_{cl} \Rightarrow F(n+\alpha) - F(n)|_{qu}$.] Of course this expression uses a particular definition of the $\mathcal{A}$: the undetermined constant in $\mathcal{A}$ is chosen so that no radiation occurs in the groundstate. With n_0 labeling the groundstate, then for all $\alpha > 0$

$$\mathcal{A}(n_0, n_0 - \alpha) = 0.$$

The question of half-integral or integral quantization in a quantum mechanics limited to relations between observable quantities therefore can never occur.

Eqns (11) and (16) together contain a complete specification not only of the frequencies or energies, but also of the quantum transition probabilities. The actual mathematical derivations have been carried out so far only in the simplest cases; a complication arises in many systems, e.g. the hydrogen atom, because some solutions are periodic and some aperiodic. The result is that the quantum series (7) and (16) contain both a sum and an integral. In quantum mechanics, the separation into periodic and aperiodic motions cannot be carried out in general.

Nevertheless, Eqns (11) and (16) provide a solution – in principle – of the mechanical problem, and if this solution is not overdetermined or in contradiction with the given quantum relations, then a small perturbation of the mechanical problem by additional terms in the energy gives the expression already found by Kramers and Born – in contrast to what the classical theory would give. It must also be investigated whether Eqn (11) – in the suggested quantum form – corresponds to an energy integral

$$m\frac{\dot{x}^2}{2} + U(x) = \text{const}$$

and whether this energy satisfies the relation: $\Delta W = \omega \cdot \hbar$. A general answer to these questions would be found in the internal consistency of the proposed quantum mechanics and would lead to a quantum theory based only on observable quantities. Apart from the general connection between Kramers' dispersion formula and Eqns (11) and (16), we can only answer the above question in those special cases completely soluble by a simple recursion relation.

The general connection between Kramers' dispersion theory and our Eqns (11), (16) arises because according to Eqn (11) (i.e. its quantum analog) – just as in the classical theory – the oscillating electron behaves just like a free electron $\cdots$. This result follows from Kramers' theory if one keeps in mind Eqn (16). In fact Kramers finds for the moment induced by the wave $E\cos\omega t$:

$$M = 2e^2 E\cos\omega t \cdot \frac{1}{\hbar} \times$$

$$\sum_\alpha \left[\frac{|\mathcal{A}(n+\alpha, n)|^2 \omega(n+\alpha, n)}{\omega^2(n+\alpha, n) - \omega^2} - \frac{|\mathcal{A}(n, n-\alpha)|^2 \omega(n, n-\alpha)}{\omega^2(n, n-\alpha) - \omega^2} \right],$$

and for $\omega >> \omega(n + \alpha, n)$

$$M = -\frac{2e^2 E \cos \omega t}{\omega^2 \cdot \hbar} \times$$
$$\sum_{\alpha} \left[|A(n + \alpha, n)|^2 \omega(n + \alpha, n) - |A(n, n - \alpha)|^2 \omega(n, n - \alpha) \right],$$

which from (16) goes over into

$$M = -\frac{e^2 E \cos \omega t}{\omega^2 m}.$$

§3. Next we consider the simple example of the anharmonic oscillator:

$$\ddot{x} + \omega_0^2 x + \lambda x^2 = 0. \tag{17}$$

Classically this equation can be satisfied by the Ansätz

$$x = \lambda a_0 + a_1 \cos \omega t + \lambda a_2 \cos 2\omega t + \cdots \lambda^{\tau-1} a_\tau \cos \tau \omega t,$$

where the a are power series in λ which begin with a term free of λ. We investigate a quantum Ansätz with terms of the form

$$\lambda a(n, n); \quad a(n, n - 1) \cos \omega(n, n - 1)t; \quad \lambda a(n, n - 2) \cos \omega(n, n - 2)t \cdots \cdots;$$

The recursion formulas for the a and ω given from Eqns (3) or (7) (up to terms of order λ) are:
Classical:

$$2\omega_0^2 a_0 + a_1^2(n) = 0;$$
$$-\omega^2 + \omega_0^2 = 0;$$
$$2(-4\omega^2 + \omega_0^2)a_2(n) + a_1^2 = 0;$$
$$(-9\omega^2 + \omega_0^2)a_3(n) + a_1 a_2 = 0;$$
$$\cdots\cdots\cdots \tag{18}$$

and Quantum:

$$4\omega_0^2 a_0(n) + a^2(n + 1, n) + a^2(n, n - 1) = 0;$$

$$-\omega^2(n, n-1) + \omega_0^2 = 0;$$
$$2(-\omega^2(n, n-2) + \omega_0^2)a(n, n-2) + a(n, n-1)a(n-1, n-2) = 0;$$
$$2(-\omega^2(n, n-3) + \omega_0^2)a(n, n-3)$$
$$+a(n, n-1)a(n-1, n-3) + a(n, n-2)a(n-2, n-3) = 0;$$
$$\cdots\cdots\cdots \quad (19)$$

Here the quantization condition (with mass μ) is:

Classical ($J = 2\pi n\hbar$):

$$2\pi\mu\frac{d}{dJ}\sum_{-\infty}^{+\infty}\tau^2\frac{|a_\tau|^2\omega}{4} = 1,$$

and Quantum:

$$\frac{\mu}{2}\sum_0^\infty\left\{|a(n+\tau, n)|^2\omega(n+\tau, n) - |a(n, n-\tau)|^2\omega(n, n-\tau)\right\} = \hbar.$$

The first approximation in both classical and quantum theory is:

$$a_1^2(n) \text{ or } a^2(n, n-1) = \frac{2(n + \text{const})\hbar}{\mu\omega_0}. \quad (20)$$

The constant in (20) is determined by the condition that $a(n_0, n_0 - 1)$ should be zero in the groundstate labeled $n_0 \equiv 0$, so

$$a^2(n, n-1) = \frac{2n\hbar}{\mu\omega_0}.$$

From the classical recursion relation (18) (in first approximation in λ)

$$a_\tau = \eta(\tau)n^{\tau/2},$$

where $\eta(\tau)$ is independent of n. In quantum theory, from (19)

$$a(n, n-\tau) = \eta(\tau)\sqrt{\frac{n!}{(n-\tau)!}}, \quad (21)$$

where again $\eta(\tau)$ is independent of n. For large n the quantum result goes asymptotically into the classical value.

For the energy the classical Ansätz

$$\frac{\mu \dot{x}^2}{2} + \mu \omega_0{}^2 \frac{x^2}{2} + \frac{\mu \lambda}{3} x^3 = W,$$

in the present approximation is actually constant quantum mechanically and from (19), (20) and (21) has the value (up to order λ^2):

$$\text{Classical:} \quad W = n \hbar \omega_0. \tag{22}$$

$$\text{Quantum [from (7)]:} \quad W = (n + \frac{1}{2}) \hbar \omega_0. \tag{23}$$

According to this approach, even for the harmonic oscillator the energy is not given correctly by the 'classical mechanics', i.e. by (22), but has the form (23).

A more exact calculation for W, A, ω can be carried out for the simpler example of an anharmonic oscillator of the type:

$$\ddot{x} + \omega_0{}^2 x + \lambda x^3 = 0.$$

Classically one can set:

$$x = a_1 \cos \omega t + \lambda a_3 \cos 3\omega t + \lambda^2 a_5 \cos 5\omega t + \cdots;$$

analogously we investigate the quantum Ansätz

$$a(n, n-1) \cos \omega(n, n-1)t; \quad \lambda a(n, n-3) \cos \omega(n, n-3)t; \cdots.$$

The quantities a are again power series in λ, whose first term – as in (21) – has the form:

$$a(n, n-\tau) = \eta(\tau) \sqrt{\frac{n!}{(n-\tau)!}},$$

using (18),(19). The calculation of ω, a from (18),(19) up to order λ^2 or λ gives:

$$\omega(n, n-1) = \omega_0 + \lambda \cdot \frac{3n\hbar}{4\mu\omega_0{}^2} - \lambda^2 \cdot \frac{3\hbar^2}{64\mu\omega_0{}^2}(17n^2 + 7) \tag{24}$$

$$a(n, n-1) = \sqrt{\frac{2n\hbar}{\mu\omega_0}} \left(1 - \lambda \cdot \frac{3n\hbar}{8\mu\omega_0{}^3}\right). \tag{25}$$

$$a(n, n-3) = 2\sqrt{\frac{\hbar^3}{\mu^3\omega_0{}^7} n(n-1)(n-2)} \left(1 - \lambda \cdot \frac{39(n-1)\hbar}{16\mu\omega_0{}^3}\right). \tag{26}$$

The energy is defined as the constant term in

$$\mu \frac{\dot{x}^2}{2} + \mu\omega_0{}^2 \frac{x^2}{2} + \frac{\mu\lambda}{4} x^4.$$

(I cannot prove in general that the periodic terms are all actually zero, but it was the case in the terms calculated.) It turns out to be

$$W = (n + \frac{1}{2})\hbar\omega_0 + \lambda \cdot \frac{3(n^2 + n + \frac{1}{2})\hbar^2}{8\mu\omega_0{}^2}$$

$$- \lambda^2 \cdot \frac{\hbar^3}{64\mu^2\omega_0{}^5} \left(17n^3 + \frac{51}{2}n^2 + \frac{59}{2}n + \frac{21}{2} \right). \tag{27}$$

This energy can also be calculated using the Kramers-Born procedure in which the $\mu\lambda x^4/4$ term is a perturbation on the harmonic oscillator. The result is exactly the same as (27), which is remarkable support for the underlying quantum equations. The energies in (27) satisfy the formula [see(24)]:

$$\hbar\omega(n, n-1) = W(n) - W(n-1),$$

which also is a necessary requirement for the possibility of calculating the transition probability corresponding to Eqn (11) and (16).

In conclusion, the rotor should be mentioned, and reference made to the relation of Eqn (7) to the intensity formulas for the Zeeman effect [5] and for the multiplets [6].

The rotor can be thought of as an electron of mass μ circling a nucleus at a constant distance a. The equations of motion – both classical as well as quantum – say only that the electron describes a uniform constant rotation around the nucleus at a constant distance a and with an angular velocity ω. The quantization condition from (12) is:

$$\hbar = \frac{d}{dn}(\mu a^2 \omega),$$

or from (16): $\hbar = \mu \{a^2\omega(n+1, n) - a^2\omega(n, n-1)\}$, from which follows in both cases:

$$\omega(n, n-1) = \frac{\hbar \cdot (n + const)}{\mu a^2}.$$

The condition that the radiation vanish in the groundstate $(n_0 = 0)$, leads to

$$\omega(n, n - 1) = \frac{\hbar \cdot n}{\mu a^2}. \tag{28}$$

The energy is $W = \mu v^2 / 2$ or from (7)

$$W = \frac{\mu}{2} a^2 \cdot \frac{\omega^2(n, n - 1) + \omega^2(n + 1, n)}{2} = \frac{\hbar^2}{2\mu a^2} \left(n^2 + n + \frac{1}{2}\right), \tag{29}$$

which satisfies $\hbar\omega(n, n - 1) = W(n) - W(n - 1)$. In support of Eqns (28) and (29) as derived from the above theory, it should be emphasized that many band spectra (even those for which the existence of an electron motion is improbable) demand [7] formulas of the type (28), (29) (which one previously sought in the classical-mechanical theory with half-integral quantum numbers).

In order for the rotor to satisfy the Goudsmit-Kronig-Hönl formula, we must give up the spherical motion for the problem with one degree of freedom. We assume that the rotor precesses very slowly around the axis z of an external field fixed in space. The quantum number corresponding to this precession is called m. Then the motion can be represented by the quantities:

$$z: \qquad a(n, n - 1; m, m) \cos\omega(n, n - 1)t;$$
$$x + iy: \qquad b(n, n - 1 : m, m - 1)e^{i[\omega(n,n-1)+\sigma]t};$$
$$b(n, n - 1; m - 1, m)e^{i[-\omega(n,n-1)+\sigma]t}.$$

The equation of motion gives simply: $x^2 + y^2 + z^2 = a^2$, which from (7) gives rise to the equations [8]:

$$a^2(n, n - 1; m, m) + 2b^2(n, n - 1; m, m - 1) + 2b^2(n, n - 1; m, m + 1)$$
$$+ a^2(n + 1, n; m, m)$$
$$+ 2b^2(n + 1, n; m - 1, m) + 2b^2(n + 1, n; m + 1, m) = 4a^2, \tag{30}$$

and

$$a(n, n - 1; m, m)a(n - 1, n - 2; m, m)$$
$$= 2b(n, n - 1; m, m + 1)b(n - 1, n - 2; m + 1, m)$$
$$+ 2b(n, n - 1; m, m - 1)b(n - 1, n - 2; m - 1, m). \tag{31}$$

From (16) the quantization condition is:

$$b^2(n, n-1; m, m-1)\omega(n, n-1)$$
$$-b^2(n, n-1; m-1, m)\omega(n, n-1) \;=\; \frac{(m+\text{const})\hbar}{\mu}. \tag{32}$$

These equations correspond classically to the relations:

$$a_0{}^2 + 2b_1{}^2 + 2b_{-1}{}^2 \;=\; 2a^2;$$
$$a_0{}^2 \;=\; 4b_1 b_{-1}$$
$$(b_1{}^2 - b_{-1}{}^2)\omega \;=\; \frac{(m+\text{const})\hbar}{\mu}, \tag{33}$$

using a_0, b_1, b_{-1} for a unique (up to the undetermined constant for m) description.

The simplest solution of the quantum equations (30), (31), (32) is:

$$b(n, n-1; m, m-1) \;=\; a\sqrt{\frac{(n+m+1)(n+m)}{4n(n+\frac{1}{2})}};$$
$$b(n, n-1; m-1, m) \;=\; a\sqrt{\frac{(n-m+1)(n-m)}{4n(n+\frac{1}{2})}};$$
$$a(n, n-1; m, m) \;=\; a\sqrt{\frac{(n+m+1)(n-m)}{n(n+\frac{1}{2})}}.$$

These expressions agree with the formulas of Goudsmit, Kronig and Hönl; one cannot easily understand how these expressions represent a single solution of (30), (31), (32) – but it does seem plausible for one choice of boundary conditions (vanishing of the a, b at the 'boundary'; see the above cited work of Kronig, Sommerfeld and Hönl, Russell).

A similar approach leads to the intensity formula for multiplets, with the result that the known intensity rules agree with Eqns (7) and (16). This result is further support for the correctness of the kinematical equation (7).

Whether quantum mechanics can be based, as proposed here, on relations between observables – which can already be seen as successful in many respects – or whether this might only represent a crude approximation to the physical, and above all often very complicated problem of a quantum mechanics, can only be known through a deeper mathematical understanding of the very introductory methods used here.

Göttingen, Institut für theoretische Physik.

Footnotes and References:

1) H.Λ. Kramers, Nature **113**, 673 (1924).

2) M. Born, Zeits. f. Phys. **26**, 379 (1924); H.A. Kramers und W. Heisenberg, Zeits. f. Phys. **31**, 681 (1925); M. Born und P. Jordan, Zeits. f. Phys. (in press).

3) See also H.A. Kramers und W. Heisenberg, loc. cit. The phase agrees essentially with their expression for the induced moment.

4) This relation was already found from dispersion considerations by W. Kuhn, Zeits. f. Phys. **33**, 408 (1925); and by W.Thomas, Naturw. **13** (1925).

5) S. Goudsmit und R. de L. Kronig, Naturw. **13**, 90 (1925); H. Hönl, Zeits. f. Phys. **31**, 340 (1925).

6) R. de L. Kronig, Zeits. f. Phys. **31**, 885 (1925); A. Sommerfeld und H. Hönl, Sitzungsber. d. Preuss. Akad. d. Wiss. (1925) S.141; H.N. Russell, Nature **115**, 835 (1925).

7) See e.g. B.A. Kratzer, Sitzungsber. d. Bayr. Akad. (1922) S.107.

8) Eqn (30) is essentially identical to the Ornstein-Burger sum rule.

Chapter VII

BORN and JORDAN:

HEISENBERGschen Form der Quantenbedingung

$$pq - qp = h/2\pi i$$

die wir die "verschafte Quantenbedingung"
nennen und auf der alle weiteren Schlüsse beruhen.

"··· Heisenberg's form of the quantization condition

$$pq - qp = \frac{\hbar}{i} \cdot 1,$$

which we call the 'Canonical Quantization Condition'. From it all further results follow."
M. Born and P. Jordan, *Zeits. f. Phys.* **34**, 858 (1925).

§VII-1. Introduction.

Born recalls [1] receiving Heisenberg's manuscript on July 11, 1925 at the end of the summer term, as he was leaving Göttingen for a much needed rest: "··· I was fascinated. Heisenberg had taken up the idea of transition amplitudes and developed a calculus for them ··· one has a multiplication rule ···. His most audacious step was the suggestion of introducing the transition amplitudes of the coordinates q and momenta p in the formulae of mechanics ···.

I was deeply impressed by Heisenberg's considerations, which were a great step forward in the program which we had pursued."

Again, sadly, Born could not restrain himself from inserting remarks which – however true they might be – were too little, too late, self-serving and demeaning to a man who would have been better served by simply praising Heisenberg's success without trying to encroach on it.

In a letter to Einstein on July 15, "Heisenberg's new work, which will appear soon, looks very mysterious but is surely correct and deep."

Born further recalls that before July 19: " $\cdots$ one morning $\cdots$ I suddenly saw the light: Heisenberg's symbolic multiplication was nothing but matrix calculus $\cdots$ It meant that two matrix products **pq** and **qp** are not identical. $\cdots$ Closer inspection showed that Heisenberg's formula gave only the diagonal elements of the matrix **pq** $-$ **qp**: they were all equal to $h/2\pi i$. $\cdots$ I soon convinced myself that the only reasonable value of the non-diagonal elements was zero, and I wrote down the strange equation

$$pq - qp = \frac{h}{2\pi i},$$

$\cdots$ but this was only a guess $\cdots$."

On July 19, Born ran into Pauli and told him about the matrix-multiplication, and suggested that they collaborate on Heisenberg's new quantum mechanics. The response was shocking. It is difficult to imagine Pauli, the 25 year old assistant, – however distinguished – being so outrageously rude to Born, the 43 year old professor, – with his own share of distinction (according to Pais [2] ' a renowned physicist and teacher, $\cdots$ more than a hundred research papers, $\cdots$ six books.'). Pauli's outburst was recalled by Born as 'Yes, I know you are fond of tedious and complicated formalism. You are only going to spoil Heisenberg's physical ideas by your futile mathematics $\cdots$'

On July 20, Born extended the same invitation to Pascual Jordan, the third and most junior of his assistants with Pauli and Heisenberg, who accepted. After only a few days, Jordan was able to show that the matrix **pq** $-$ **qp** must be time independent and therefore presumably diagonal (from Heisenberg's prescription $a_{nm} \sim e^{i(\omega_n - \omega_m)t}$). "Then we began to write a joint paper $\cdots$."

In a 1964 letter to van der Waerden [3], Jordan recalls: "During Born's stay in Silvaplana I was in Hanover at my parents house, and I had some more thoughts on the problem, which were later presented in our work. It led to a correspondence with Born in which naturally I told him about my progress. I seem to recall a delay of some time in his response, necessitated because the two-fold strain of a somewhat stressful sanatorium treatment and the difficulty of our written communication on this exciting subject, became too much for him. It may in fact however have been as you say – namely that when we saw each other again in Göttingen, I had already extensively outlined the work in a first draft, ···."

Heisenberg and the even younger Jordan had a very genial correspondence [4]. From Munich, on August 20, Heisenberg writes: "I hear from Born that he has made great progress in the Quantum mechanics and was naturally very interested to learn about your calculations. Born wrote that you would like to have a copy of my work, so I am sending you the manuscript – unfortunately I have no more proofs. I would be very grateful whenever you could tell me a little about your own calculations ···". Again on September 10: " Dear Jordan! It has been very pleasing to hear that you have found the proof of the frequency condition and I beg you to write me in Copenhagen as soon as possible; because every day I want to press on further in physics ···". And on Sept 13: "Dear Jordan! Many thanks for your letter and the proof. Your work with Born seems to me a very great advance; the central point is obviously Bohr's frequency condition. I hope you will soon publish your beautiful results with Born, they are certainly ready ···". On Sept 21: "Dear Jordan! ··· I sent Born a sketch of the perturbation theory ···".

On Oct 23, Heisenberg to Pauli: "At the moment I am exclusively occupied with quantum mechanics and I very much doubt if the problem of writing this 3-Männerarbeit has a solution in any finite time. But I would like to hear your opinion about the outline, which terrible as it seems is only about 1/3 of the whole. The most important part of the work, in spite of its mathematical character, seems to be the Principal Axis Transformation; it is only an integration technique and so far helps only a little, but it is still formally very beautiful."

And finally on Nov 16, after submission of their paper, Heisenberg again to Pauli: " I have taken great pains to make the work as physical as possible but I am only half-satisfied. I am still rather unhappy about the whole theory but am glad that you – with your insight into mathematics and physics – are so completely on my side. Here I am in an environment which thinks and feels exactly the opposite, and I wonder if I am just too stupid to appreciate the mathematics. Göttingen is divided into two camps: one, like Hilbert (or Weyl in a letter to Jordan), which speaks of the great progress which should follow from the introduction of matrix considerations into physics; and the other, like Franck, which says that no one will ever really understand matrices."

Pauli regained his composure somewhat in a letter [5] to Kronig on Oct 9: "Heisenberg's mechanics has given me new enthusiasm and hope. The solution of the puzzle is not complete, but it is now again possible to make progress. One must first however try to rescue Heisenberg's mechanics from yet another flood of Göttingen's formal scholarship and better understand its physical meaning."

One has a strange vision of the central participants of this immortal drama rushing around on the blackened stage of some Wagnerian opera torn by sturm und drang. Why so peripatetic? Were these three ever under the same roof together? Why so emotional? Born in a sanatorium? Pauli of all people complaining of an excess of formalism? The naive vision of Professor Born leading his young assistants Heisenberg and Jordan in deep discussions at a blackboard, and evolving results which are then agreed upon and carefully transcribed for publication seems nowhere to have been realized, at least not in the creation phase. It was the end of the summer term, Born – for whatever reason – had been complaining of fatigue. What Jordan meant when he referred 39 years later to a 'strapaziøse Sanatoriums-Kur' was more than a vacation at a spa but apparently a month-long treatment for fatigue.

It must have been a traumatic time for Pauli [6]. His reign as the preeminent 'wunderkind' of physics was over. His accomplishments continued – including almost immediately the solution of Heisenberg's matrix mechanics for the discrete spectrum of the hydrogen atom, ironically – after his outburst to Born – surely the most mathematically complicated development so far of Heisenberg's new theory. There followed many contributions to field theory especially with Heisenberg and with Weisskopf, the neutrino hypothesis, the spin-statistics theorem – but he did not play a key role in this first wave of the creation of quantum mechanics, and he was no longer the brightest boy in the class. He cannot have been unscarred by the sudden explosion of creativity of a profound and fundamental kind never attained by him, in the very people closest to him while he himself was silent. Nor can the instantaneous ascendence of his junior colleague Heisenberg to a higher position in the pantheon of quantum theorists have been easy to accept.

Jordan was the youngest at 22 of Born's three assistants. His principal distinctions [7] were his role while still an undergraduate as assistant to Courant in the editing of *Methods of Mathematical Physics* (1924) which gave him an unusual sophistication in mathematics; and a PhD thesis – written under Born's supervision but immediately contradicted by Einstein himself! – which claimed a continuous spectrum of radiation in Compton scattering. Jordan was acknowledged as *a* major contributor in the papers with Born and with Born and Heisenberg, but he got very little recognition compared to his co-authors, and seems to have developed some resentment from about 1930 on. In those early years, he made major contributions including the first suggestion of field quantization (second quantization), a proof (simultaneous with others) of the equivalence of the Heisenberg and Schrödinger forms of quantum mechanics, and the introduction (with Wigner) of anti-commutation for particles satisfying the Pauli Exclusion Principle. His claims in later life of being *the* major contributor have to be taken with a certain reserve, although van der Waerden seems to support them [8]. van der Waerden identifies seven new ideas in the Born-Jordan paper. He credits 1) the recognition of matrix multiplication and 2) the basic commutation relation to Born. The proof of 3) the commutation relation, 4) energy conservation, and 5) the Bohr frequency condition, he credits to Jordan; although the 'proofs' are heuristic. More significantly, van der Waerden credits Jordan with the recognition that 6) the electromagnetic field is also to be regarded as a matrix, and 7) the absolute squares of the transition amplitudes determine the transition probabilities.

None the less, at this early phase of his career, Jordan was on the ascendency and seems to have enjoyed his participation without reservation.

Heisenberg was and remained different. He always had a serene certainty that his initial contribution defining the multiplication rule of amplitudes guaranteed his primacy. The source of such serenity could perhaps be found, as with Bohr, in a largeness of view in which the person is totally self-identified with an issue, in both their cases the issue being the purest understanding of quantum physics. He seemed genuinely pleased – without a hint of possessiveness or jealousy or envy, or (as so sadly evident in Born's case) of the need to lay claim even to that which was truly his – to share his creation with colleagues. But he was in no way ethereal

– in Kramers' words [9] he was "tough, hard nosed ⋯". There is no indication that Pauli at any stage tried on Heisenberg the bullying tactics that he used on so many others [10].

We turn now to the 'parsing' of the Born-Jordan paper and of the renowned 'Dreimännerarbeit', so-called by Heisenberg – the 'Three men's paper', of Born-Heisenberg-Jordan. For a full and authoritative analysis of who did precisely what, we commend the reader to van der Waerden [11]. Our objective is primarily to make the physics transparent by means of explanatory notes inserted directly into their abridged texts, and only coincidentally to apportion credit to the individual authors. For this latter task, we rely entirely on van der Waerden's judgements.

Footnotes and References:

1) B.L. van der Waerden, *Sources of Quantum Mechanics* (North Holland, Amsterdam, 1967), p.36.

2) A. Pais, *Inward Bound* (Clarendon Press, Oxford, 1986), p.252.

3) van der Waerden, loc. cit., p.40.

4) van der Waerden, loc. cit., Pp.40,49.

5) van der Waerden, loc. cit., p.37.

6) H. Atmanspacher and H. Primas, Journal of Consciousness Studies **3**(2), 112 (1996), for an account of his treatment for depression by Jung starting in 1932, which evolved to friendship and a shared interest in the nature of creativity.

7) I. Duck and E.C.G. Sudarshan, *Pauli and the Spin-Statistics Theorem* (World Scientific, Singapore, 1997), p.174. See also, I. Duck and E.C.G. Sudarshan, Am. J. Phys. **66**(4), 284 (1998).

8) van der Waerden, loc. cit., Pp.38,39.

9) A. Pais, *Niels Bohr's Times* (Clarendon Press, Oxford, 1991), p.21, from M. Dresden, *H.A. Kramers* (Springer, New York, 1987), p.481.

10) E. Segrè, *From X-rays to Quarks* (Freeman, New York, 1980), Pp.153-157, for an objective and favorable view of their relationship.

11) van der Waerden, loc. cit., Pp.38-40 and Pp.42-58.

Biographical Note on E.P. Jordan. (†, ††)

Ernst Pascual Jordan (1902-80), graduate of the Technische Hochschule in Hanover

(1921) and moved to Göttingen where his exceptional talents were immediately recognized by Born, who introduced him to the current problems of quantum physics. After a flawed PhD thesis (set right immediately by Einstein), he immediately joined Born in the development of Heisenberg's ideas into canonical quantum mechanics (1925; work which he was to claim in 1964 as substantially his own, due to Born's illness at the critical time); then with Born and Heisenberg (1926), credited with the idea of quantum field theory (1925); with Heisenberg, the quantum theory of electron spin (1926); the relation between Heisenberg and Schrödinger forms of quantum mechanics (1926); stated the indeterminacy principle simultaneous with Heisenberg's statement of the Uncertainty Principle (1927); with Klein, *second*-quantization for the Bose-Einstein gas (1927); with Wigner, anticommutation relations for the Pauli Exclusion Principle (1928); with Pauli, relativistic quantum electrodynamics (1928); invented nonassociative *Jordan*-algebras (1932); later work on biological, psychological, physiological, philosophical, political, and cosmological subjects which gained little acceptance. Professor at Rostock (1929-44), Berlin (1944-45), Hamburg (1947-71); member of the Bundestag (1957-61). He was a Nazi sympathizer, and accused as an informant (1936). He was unsuccessfully promoted by Pauli for a Nobel Prize, but was for some reason never recognized for his many early accomplishments, perhaps because they all seemed to coincide with his association with Born, Heisenberg, and Pauli.

Schucking was Jordan's student in 1952 and appears to have been as close to Jordan as anyone ever got and remembers him as a shy and kind man. He points out that Jordan had a persistent stutter which he covered in discussion with a nervous laugh. Schucking recalls Born's deep regret for delaying until too late a manuscript in which Jordan anticipated Fermi-Dirac statistics; that Pauli disdained him as not a true physicist but only a mathematician, " always a formalist"; and that Pauli only reluctantly accepted Jordan's creation-annihilation operators in quantum field theory.

One point is worth deeper investigation than we have given it: and that is which of Jordan and Born *really* possessed the sophisticated understanding of the theory of Hilbert spaces. Jordan had worked as a precocious undergraduate on the preparation of Courant and Hilbert's *Methods of Mathematical Physics* and quite possibly did first recognize the profound structure underlying Heisenberg's multiplication rule.

† – K. von Meyern in *Dictionary of Scientific Biography* (Scribners, New York, 1994), Ed: F.L. Holmes, Pp.448-454.
†† – E.L. Schucking, Physics Today **52**-10, 26 (1999).

Paper VII·1: Excerpt from Zeitschrift für Physik **34**, 858 (1925).

Zur Quantenmechanik

von **M. Born** und **P. Jordan** in Göttingen.

(Eingegangen am 27. September 1925.)

The Ansätz recently proposed by Heisenberg is developed into a systematic theory of quantum mechanics. The mathematical language is that of matrix algebra. After a brief introduction, the mechanical equations of motion are derived from a variation principle, and the proof of the energy law and the Bohr frequency condition from the Heisenberg quantization condition. The uniqueness of the solution and the significance of the phases in the partial waves is discussed in the example of the anharmonic oscillator. Finally, we investigate how the laws of the electromagnetic field fit into the new theory.

Introduction. Heisenberg's [1] recent ansätz for a new kinematical and mechanical foundation of Quantum Theory, is of great significance. He constructs a new theory – by a more or less artificial and forced modification of old ideas – by the postulate of a completely measurable system of dynamical variables. Heisenberg's physical idea has been carried through and explained in such a clear way that any additional remarks seem superfluous. However in a formal mathematical sense his observations – as he himself emphasized – are only in the initial stages. He has applied his hypothesis only to simple examples and not as a general theory. Encouraged by the hope that we might understand his work from its beginning, we have tried to clarify the mathematical content of his ansätz, and present our results here. It turns out that it is possible – based on Heisenberg's prescription – to construct a mathematical theory of quantum mechanics in remarkably close analogy to classical mechanics, and even to calculate characteristic features of quantum phenomena.

We confine ourselves at first to systems with one degree of freedom which we assume to be – classically speaking – periodic $\cdots\cdots$ non-relativistic $\cdots$ cartesian coordinates $\cdots\cdots$

The mathematical foundation of Heisenberg's ansätz is the multiplication rule for quantum variables, which he inferred by an ingenious correspondence argument.

··· this rule is nothing more than the multiplication of matrices. [Note: Italics added.] A two-index form (with discrete or continuous indices) – the so-called matrix – represents each physical variable which in the classical theory is a function of time. The mathematics of the new quantum mechanics is therefore characterized by matrix-analysis in place of the usual number-analysis.

We now have to understand the simplest questions of mechanics and electrodynamics in these terms. A variation principle based on the correspondence principle gives the Hamiltonian equations of motion in close analogy to the classical canonical equations. The quantization condition together with the equations of motion give a simple matrix expression. From this follows the proof of the energy law and Bohr's frequency condition in the sense postulated by Heisenberg. From one of these examples we then return to a detailed discussion of the role played in the new theory by the phases of the partial waves. To conclude, we show that the fundamental laws of the electromagnetic field in vacuum easily accommodate the new methods, and give a basis for Heisenberg's assumption that *the square of the magnitude of the element of the matrix representing the electric moment of the atom is a measure of the transition probability.* [Note added: Born was to get the Nobel Prize for his later interpretation of the wave function as a probability amplitude (here in **Paper XXX**), but the essential idea was present in Heisenberg's first paper, and again more explicitly in a footnote in a paper by Pauli which predated Born's remarks.]

Chapter I. Matrix Calculus.

§ 1. Elementary Operations. ······
§ 2. Symbolic Differentiation. ······
[Note added: In the interest of focusing on quantum mechanics, we have to assume that our readers are familiar with matrix manipulations and forgo Born and Jordan's introduction to the subject, much of which was soon rendered superfluous anyway. We go directly to quantum matrix dynamics.] ······

Chapter II. Dynamics.

§ 3. The Basic Relation. The dynamical system is described by the coordinate

q and the momentum **p** which are matrices

$$\mathbf{q} = \left(q(nm)e^{i\omega(nm)t}\right); \quad \mathbf{p} = \left(p(nm)e^{i\omega(nm)t}\right). \tag{24}$$

Here $f(nm) = \omega(nm)/2\pi$ is the quantum frequency corresponding to the transition between the states with quantum numbers n and m. The matrices (24) must be hermitian $q(nm) = q(mn)^*$ for all real t so

$$q(nm)q(mn) = |q(nm)|^2 \text{ and} \tag{25}$$

$$\omega(nm) = -\omega(mn). \tag{26}$$

If **q** is a cartesian coordinate, then (25) is a measure of the probability for the transition $n \leftrightarrow m$.

We will also require that

$$\omega(jk) + \omega(kl) + \omega(lj) = 0. \tag{27}$$

With (26) this can be expressed in terms of energy variables W_n for the individual states n, so that

$$\hbar\omega(nm) = W_n - W_m. \tag{28}$$

Then it follows that any function $\mathbf{g}(\mathbf{p}, \mathbf{q})$ always has the form

$$\mathbf{g} = \left(g(nm)e^{i\omega(nm)t}\right), \tag{29}$$

so the matrix $(g(nm))$ is the same function of the matrices $(q(nm))$, $(p(nm))$ as is g of **q** and **p**. We will always write the shorter form of the representation (24) as

$$\mathbf{q} = (q(nm)), \quad \mathbf{p} = (p(nm)) . \cdots \cdots \tag{30}$$

For a Hamiltonian function of the form

$$\mathbf{H} = \frac{1}{2m}\mathbf{p}^2 + \mathbf{V}(\mathbf{q})$$

we will assume with Heisenberg that the equations of motion are the same as the classical ones:

$$\dot{\mathbf{q}} = \frac{\partial \mathbf{H}}{\partial \mathbf{p}} = \frac{1}{m}\mathbf{p}, \text{ and } \dot{\mathbf{p}} = -\frac{\partial \mathbf{H}}{\partial \mathbf{q}} = -\frac{\partial \mathbf{V}}{\partial \mathbf{q}}. \tag{32}$$

$\cdots\cdots$ general Hamiltonians, relativistic mechanics, magnetic fields $\cdots$. [Note added: Following Heisenberg and Dirac, we can get these elementary differentiations by calculating $\delta\mathbf{H}$ for a change in $\mathbf{p}$, for example, which is proportional to the unit matrix $\delta\mathbf{p} \sim \mathbf{1}$. More general Hamiltonians of the form $\mathbf{p}^j\mathbf{q}^k\mathbf{p}^l$, etc, require a symmetrization constructed by Jordan designed solely to reproduce Eqns32.]

Classically the equations of motion are derivable from the Action Principle:

$$\int_{t_0}^{t_1} L\,dt = \int_{t_0}^{t_1} \{p\dot{q} - H(pq)\}\,dt = \text{Extremum}. \tag{33}$$

If we insert the Fourier transform of L and assume the time interval $(t_1 - t_0)$ to be very large, then only the constant term of L will contribute to the integral. The same form obviously follows in quantum mechanics – the diagonal sum $D(L) = \sum_k L(kk)$ should be an extremum:

$$D(L) = D(\mathbf{p}\dot{\mathbf{q}} - \mathbf{H}(\mathbf{pq})) = \text{Extremum}, \tag{34}$$

by a suitable choice of $\mathbf{p}$ and $\mathbf{q}$ for fixed $\omega(nm)$. By setting to zero the derivatives of $D(L)$ with respect to the elements of $\mathbf{p}$ and $\mathbf{q}$, one gets the equations of motion

$$i\omega(nm)q(nm) = \frac{\delta D(\mathbf{H})}{\delta p(nm)} \quad \text{and} \quad i\omega(nm)p(nm) = -\frac{\delta D(\mathbf{H})}{\delta q(nm)}.$$

From (26), (31) and (16) one finds equations of motion in the canonical form

$$\dot{\mathbf{q}} = \frac{\partial\mathbf{H}}{\partial\mathbf{p}} \quad \text{and} \quad \dot{\mathbf{p}} = -\frac{\partial\mathbf{H}}{\partial\mathbf{q}}. \tag{35}$$

For the quantization condition Heisenberg uses a relation deduced by Thomas [2] and Kuhn [3]. The equation

$$J = \oint p\,dq = \int_0^T p\dot{q}\,dt$$

can be recognized as the "classical" quantum theory. With the expansions

$$p = \sum_{\tau=-\infty}^{+\infty} p_\tau e^{2i\pi\tau t/T} \quad \text{and} \quad q = \sum_{\tau=-\infty}^{+\infty} q_\tau e^{2i\pi\tau t/T},$$

one gets the form

$$1 = 2\pi i \sum_{\tau=-\infty}^{+\infty} \tau \frac{\partial}{\partial J} (q_\tau p_{-\tau}). \tag{36}$$

If $p = m\dot{q}$, then the p_τ can be expressed in terms of the q_τ and one gets the classical equations in the form of a difference equation given by the Thomas-Kuhn relation. Since here $p \neq m\dot{q}$, we must transform Eqn (36) directly into a difference equation.

∗ ∗ ∗ ∗ ∗ ∗ ∗ ∗ ∗ ∗ ∗ ∗ ∗ ∗ ∗ ∗ ∗ ∗∗

This follows from

$$J = \int_0^T p\dot{q}\,dt = \frac{2\pi i}{T} \int_0^T dt \sum_{\tau=-\infty}^{+\infty} \tau(q_\tau p_{-\tau}),$$

where only $\tau_q = -\tau_p$ contributes to the integral. Following Heisenberg, we differentiate with respect to $J = nh$ and interpret the product $\sum_\tau q_\tau p_{-\tau}$ as the diagonal sum of the matrix product

$$D(qp) = \sum_{n',\tau} q(n' + \tau, n')p(n', n' + \tau).$$

Differentiation with respect to n picks out the term $n' = n$, and is interpreted following Kramers and Heisenberg as the difference

$$\tau \frac{\partial}{\partial n}(qp) \Rightarrow q(n + \tau, n)p(n, n + \tau) - q(n, n - \tau)p(n - \tau, n).$$

∗ ∗ ∗ ∗ ∗ ∗ ∗ ∗ ∗ ∗ ∗ ∗ ∗ ∗ ∗ ∗ ∗ ∗∗

The correspondence is

$$\sum_{\tau=-\infty}^{+\infty} \tau \frac{\partial}{\partial J} (q_\tau p_{-\tau}) \quad \text{with}$$

$$\frac{1}{h} \sum_{\tau=-\infty}^{+\infty} (q(n + \tau, n)p(n, n + \tau) - q(n, n - \tau)p(n - \tau, n));$$

where the $q(nm), p(nm)$ with negative indices are set to zero. Then corresponding to Eqn (36), *we get the quantization condition* [Note added: Recognized here by Born for the first time.]

$$\sum_k (p(nk)q(kn) - q(nk)p(kn)) = \frac{\hbar}{i}. \tag{37}$$

These are infinitely many equations, one for each n. For $\mathbf{p} = m\dot{\mathbf{q}}$ it gives

$$\sum_k \omega(kn)|q(nk)|^2 = \frac{\hbar}{2m},$$

in agreement with the Heisenberg quantization condition or the Thomas-Kuhn equation. Eqn (37) is the appropriate generalization of this equation.

However from (37) the diagonal sum $[D(\mathbf{pq})]$ will necessarily be infinite. Usually it would follow that $[D(\mathbf{pq}) - D(\mathbf{qp})] = 0$, whereas (37) leads to $[D(\mathbf{pq}) - D(\mathbf{qp})] = \infty$. The matrices involved are therefore no longer finite.

§ 4. Conclusions. Energy- and Frequency-law. With the formulation of the above paragraphs, the basic laws of the new quantum mechanics are given completely. All other laws of quantum mechanics must be derivable from these. First to be proven are the energy law and the Bohr frequency condition. The energy law says that if H is the energy, then $\dot{H} = 0$, or since

$$\mathbf{H} = \left(H(nm)e^{i\omega(nm)t} \right)$$

that H is a diagonal matrix. The diagonal terms $H(nn)$ of $\dot{H}$ – according to Heisenberg – are the energies of the different states of the system, and the Bohr frequency condition requires $\hbar\omega(nm) = H(nn) - H(mm)$, or $W_n = H(nn) + $const.

Next we examine the quantity $\mathbf{d} = \mathbf{pq} - \mathbf{qp}$. From (35)

$$\begin{aligned}
\dot{\mathbf{d}} &= \dot{\mathbf{p}}\mathbf{q} + \mathbf{p}\dot{\mathbf{q}} - \dot{\mathbf{q}}\mathbf{p} - \mathbf{q}\dot{\mathbf{p}} \\
&= \mathbf{q}\frac{\partial \mathbf{H}}{\partial \mathbf{q}} - \frac{\partial \mathbf{H}}{\partial \mathbf{q}}\mathbf{q} + \mathbf{p}\frac{\partial \mathbf{H}}{\partial \mathbf{p}} - \frac{\partial \mathbf{H}}{\partial \mathbf{p}}\mathbf{p}.
\end{aligned}$$

This has the general form

$$\sum_r \left(\mathbf{x}_r \frac{\partial \mathbf{f}}{\partial \mathbf{x}_r} - \frac{\partial \mathbf{f}}{\partial \mathbf{x}_r}\mathbf{x}_r \right)$$

which vanishes for $\mathbf{f}$ a function of the independent variables $\mathbf{x}_r$.

* * * * * * * * * * * * * * * * * * * **

Jordan's initial proof involves the expansion of functions **f** as symmetrized products of the individual **x**'s, which are then differentiated factor by factor. It is better to observe that the diagonal elements of **pq** − **qp** deduced by Heisenberg's argument are the unit matrix, and argue that this result is *necessary* for all elements in order that the result be a canonical invariant independent of the choice of q's and p's, which - as Hermitian matrices − must transform by unitary transformations. Ultimately, the quantization *prescription* must be *postulated* with the assistance of the correspondence principle on the basis of the canonical Poisson bracket formulation of classical mechanics.

* * * * * * * * * * * * * * * * * * * **

Therefore $\dot{\mathbf{d}} = 0$ and **d** is a diagonal matrix. The diagonal terms of **d** are given by the quantization condition (37). In summary, we obtain the matrix equation

$$\mathbf{pq} - \mathbf{qp} = \frac{\hbar}{i} \cdot \mathbf{1} \tag{38}$$

with **1** the unit matrix, which we call the "canonical quantization condition". All further results follow from this fundamental relation. [Note added: This is a startling generalization to make at such an early stage, and it remains the foundation of theoretical physics. The commutation relation is due to Born, but whether he alone also recognized its vast generality is not clear.]

From the form of this equation we conclude: If any equation (A) is derived from (38), then (A) remains correct when **p** and **q** are exchanged and $\hbar$ is replaced by $-\hbar$. This requires, e.g., only one proof for the equations

$$\mathbf{p}^n\mathbf{q} = \mathbf{qp}^n + n\frac{\hbar}{i}\mathbf{p}^{n-1} \text{ and } \mathbf{q}^n\mathbf{p} = \mathbf{pq}^n - n\frac{\hbar}{i}\mathbf{q}^{n-1}, \tag{39}$$

which is easily done by induction using (38).

To prove the energy law and the Bohr frequency condition, consider first the case

$$\mathbf{H} = \mathbf{H}_1(\mathbf{p}) + \mathbf{H}_2(\mathbf{q}).$$

H_1 and H_2 can be formally replaced by power series

$$H_1 = \sum_s a_s p^s, \quad \text{and} \quad H_2 = \sum_s b_s q^s,$$

and from (39)

$$Hq - qH = \frac{\hbar}{i} \frac{\partial H}{\partial p} \quad \text{and} \quad Hp - pH = -\frac{\hbar}{i} \frac{\partial H}{\partial q}. \tag{40}$$

Comparison with the equations of motion (35) gives

$$\dot{q} = \frac{i}{\hbar}(Hq - qH) \quad \text{and} \quad \dot{p} = \frac{i}{\hbar}(Hp - pH). \tag{41}$$

If the matrix $Hg - gH$ is abbreviated $[H, g]$, for which

$$[H, ab] = [H, a]b + a[H, b]; \tag{42}$$

then for $g = g(pq)$

$$\dot{g} = \frac{i}{\hbar}[H, g]. \tag{43}$$

If $g = H$, this gives

$$\dot{H} = 0. \tag{44}$$

So the energy law has been proven and H obtained as a diagonal matrix. Then from (41):

$$\begin{aligned}
\hbar\omega(nm)q(nm) &= (H(nn) - H(mm))\,q(nm) \quad \text{and} \\
\hbar\omega(nm)p(nm) &= (H(nn) - H(mm))\,p(nm),
\end{aligned}$$

which proves the Bohr frequency condition. [Note added: this development is solely due to Jordan, according to van der Waerden (loc. cit.)]

For general Hamiltonian functions, as in the simple example $H^* = p^2 q$, then $\dot{H}^* \neq 0$. However, the Hamiltonian

$$H = \frac{1}{2}\left(p^2 q + q p^2\right)$$

gives the same equations of motion as H^* and does satisfy $\dot{H} = 0$. So we can establish the energy law and the Bohr frequency condition in the following way:

for each $\mathbf{H}^*(\mathbf{pq})$ there is an $\mathbf{H}(\mathbf{pq})$ which gives the same equations of motion and for which $\dot{\mathbf{H}} = 0$, which does satisfy the energy-frequency conditions. $\cdots\cdots$

Chapter III. The Harmonic Oscillator.

The anharmonic oscillator $\cdots\cdots$ [Note added: We confine ourselves to the discussion of the original *a priori* solution of the harmonic oscillator.]

§ 5. **The Harmonic Oscillator.** The starting point of our discussion is the theory of the harmonic oscillator $\cdots\cdots$ with the energy

$$\mathbf{H} = \frac{1}{2}\mathbf{p}^2 + \frac{\omega_0^2}{2}\mathbf{q}^2. \tag{52}$$

Even for this simple problem, a supplement to Heisenberg's considerations is necessary. This involves a correspondence principle statement about the form of the solution. Since classically only first harmonic terms exist, Heisenberg tries a matrix with transitions only between neighboring states, with the form

$$\mathbf{q} = \begin{pmatrix} 0 & q_{01} & 0 & 0 & 0 & \cdots \\ q_{10} & 0 & q_{12} & 0 & 0 & \cdots \\ 0 & q_{21} & 0 & q_{23} & 0 & \cdots \\ \cdots & \cdots & \cdots & & & \end{pmatrix}. \tag{53}$$

Our aim is to construct the whole theory independently, without correspondence principle help from the classical theory. We ask if the form (53) of the matrix can be derived from the basic equations themselves or if not, what additional requirements are needed.

One sees immediately from the invariance against permutation of rows and columns stated in § 3, that the exact form of the matrix (53) can never be deduced from the basic equations; if one exchanges rows and columns in the same way, the canonical equations and the quantization condition remain invariant and therefore one has found an apparently different solution. But all these solutions are only apparent, i.e. they differ only by the numbering of the elements. We want to show that by a complete renumbering of the elements, any solution can be brought to the form (53). The equation of motion

$$\ddot{\mathbf{q}} + \omega_0^2 \mathbf{q} = 0 \tag{54}$$

gives

$$\left(\omega^2(nm) - \omega_0^2\right) q(nm) = 0, \tag{55}$$

where $\hbar\omega(nm) = W_n - W_m$. From the canonical quantization condition

$$\mathbf{pq} - \mathbf{qp} = \frac{\hbar}{i}\mathbf{1}, \tag{56}$$

it follows that for each n one n' must exist such that $q(nn') \neq 0$; because if an n were given, for which all $q(nn')$ were equal to zero, then the nth diagonal element of $\mathbf{pq} - \mathbf{qp}$ would be zero, which contradicts the quantization condition. Eqn (55) requires that there exist an n' for which $|W_n - W_{n'}| = \hbar\omega_0$. However in our basic postulate we have assumed that for $n \neq m$, then always $W_n \neq W_m$, so at most two such indices, n and n', exist; because the corresponding $W_{n'}, W_{n''}$ are solutions of the quadratic equation

$$(W_n - x)^2 = (\hbar\omega_0)^2;$$

if two such indices actually exist, it follows for the corresponding frequencies that

$$w(nn') = -w(nn''). \tag{57}$$

Then from (56)

$$\sum \omega(kn)|q(nk)|^2 = \omega(n'n)\left\{|q(nn')|^2 - |q(nn'')|^2\right\} = \frac{\hbar}{2}, \tag{58}$$

and the energy (52) is

$$
\begin{aligned}
H(nm) &= \frac{1}{2}\sum_k \left\{-\omega(nk)\omega(km)q(nk)q(km) + \omega_0^2 q(nk)q(km)\right\} \\
&= \frac{1}{2}\sum_k q(nk)q(km)\left\{\omega_0^2 - \omega(nk)\omega(km)\right\}.
\end{aligned}
$$

In particular for $m = n$:

$$H(nn) = W_n = \omega_0^2\left(|q(nn')|^2 + |q(nn'')|^2\right). \tag{59}$$

There are three possible cases:
a) There is no n'' and if $W_{n'} > W_n$;
b) There is no n'' and if $W_{n'} < W_n$;

c) There is an n''.

In case b) we consider n instead of n'; to this there are at most two possibilities $(n')'$ and $(n')''$, and of these one must equal n. Thus we come back to either case a) or case c) and can ignore case b).

In case a) $\omega(n'n) = +\omega_0$, and it follows from (58):

$$\omega_0 |q(nn')|^2 = \frac{\hbar}{2}, \tag{60}$$

so according to (59):

$$W_n = H(nn) = \omega_0^2 |q(nn')|^2 = \frac{1}{2}\omega_0 \hbar.$$

For the assumption $W_n \neq W_m$ for $n \neq m$ there is at most one index $n = n_0$ for which case a) occurs. When such an n_0 exists, we can construct a series of numbers $n_0, n_1, n_2, n_3 \cdots$ with $(n_k)' = n_{k+1}$ and $W_{k+1} > W_k$, and in every case $(n_{k+1})'' = n_k$. Therefore from (58) and (59), for $k > 0$:

$$H(n_k n_k) = \omega_0^2 \left\{ |q(n_k, n_{k+1})|^2 + |q(n_k, n_{k-1})|^2 \right\}, \tag{61}$$

$$\frac{1}{2}\hbar = \omega_0 \left\{ |q(n_k, n_{k+1})|^2 - |q(n_k, n_{k-1})|^2 \right\}. \tag{62}$$

From (60) and (62) follow

$$|q(n_k, n_{k+1})|^2 = \frac{\hbar}{2\omega_0}(k+1), \tag{63}$$

and then from (61)

$$W_{n_k} = H(n_k, n_k) = \hbar\omega_0 \left(k + \frac{1}{2}\right). \tag{64}$$

Next we ask if it is possible that there is no n for which case a) holds. We could then begin with arbitrary n_0, and take $n_0' = n_1$ and $n_0'' = n_{-1}$; to each of these again $n_1' = n_2$, $n_1'' = n_0$ and $n_{-1}' = n_0$, $n_{-1}'' = n_{-2}$ etc. In this way we would obtain the sequence of numbers

$$\cdots n_{-2} n_{-1} n_0 n_1 n_2 \cdots \tag{65}$$

and Eqns (61), (62) would hold for each k between $-\infty$ and $+\infty$. However that is impossible; because according to (62) the quantities $x_k = |q(n_{k+1}, n_k)|^2$ form a series of equally spaced numbers, and since they are positive, there must be a smallest. We label the corresponding index n_0 and return to the original case; therefore Eqns (63), (64) hold here also.

One sees further: each number n must be included in the n_k; because one could form a new series with n as the starting term, for which again Eqn (60) holds. The starting term of each series then has the same value $W_n = H(nn)$, which is impossible.

That completes the proof that the indices $0, 1, 2, \cdots$ can be reordered into a new series $n_0, n_1, n_2, \cdots$ for which Eqns (63),(64) hold; in these new indices the solution has the Heisenberg form (53). This appears therefore as a "normal-form" of the general solution. It has according to (64) the property that $W_{n_k+1} > W_{n_k}$. Conversely if one requires that $W_n = H(nn)$ with n always increasing, then necessarily $n_k = k$; this principle therefore defines the normal form uniquely. But this only fixed the notation and clearly defined the calculation; physically nothing new was added.

The deep distinction with the semi-classical determination of the stationary states is that the classically calculated orbits are continuously related to one another and occur from the outset in a definite order, to which the quantum orbits must be subsequently reordered. The new mechanics is indeed a discontinuous theory in that here there is no mention of physical processes to define the order of quantum states. But the quantum numbers are actually not to be understood as indices which one can order and regulate according to some practical viewpoint (e.g. by increasing energy W_n).

§ 6. Anharmonic Oscillator. ······

Chapter IV. Remarks on Electrodynamics.

According to Heisenberg the square of the absolute magnitude of the matrix element $|q(nm)|^2$, for $\mathbf{q}$ a cartesian coordinate, is a measure of the transition probability. We conclude with an *explanation of this assumption* based on general considerations. [Note: Italics added.] For that it is necessary to introduce the representation of the fundamental laws of electrodynamics in the new theory. Our understanding here has a preliminary character; but it should allow us to deduce the fundamental position taken on the problem. A further discussion will be given later with all the implications of the present theory for the theory of light quanta.

We only mention certain points, without going into quantization conditions for systems with many degrees of freedom. Even so one already comes fairly far in electrodynamics, as one can see by the following considerations. Electromagnetic oscillations are a system of infinitely many degrees of freedom. Nevertheless, the basic laws already developed for one degree of freedom apply to this case also, since it is a system of uncoupled oscillators when analyzed in terms of its eigen-oscillations. No doubt one can handle this situation also. The fact that the basic equations of electrodynamics are linear is of critical importance; because then the equivalent oscillators – in contrast to most other systems – are already harmonic oscillators, and the validity of the energy law follows independent of the quantization condition: From

$$\mathbf{H} = \frac{1}{2}\left(\mathbf{p}^2 + \omega_0{}^2 \mathbf{q}^2\right)$$

follows

$$
\begin{aligned}
\dot{\mathbf{H}} &= \frac{1}{2}\left(\dot{\mathbf{p}}\mathbf{p} + \mathbf{p}\dot{\mathbf{p}} + \omega_0{}^2\dot{\mathbf{q}}\mathbf{q} + \omega_0{}^2\mathbf{q}\dot{\mathbf{q}}\right) \\
&= \frac{1}{2}\omega_0{}^2\left(-\mathbf{qp} - \mathbf{pq} + \mathbf{qp} + \mathbf{pq}\right) = 0.
\end{aligned}
$$

In a corresponding way, without going into the specific quantization conditions, one might expect that the laws of electrodynamics of the vacuum should be generalized as matrix representations of Maxwell's equations. On this basis, we do obtain Heisenberg's postulated result for the significance of the $|q(nm)|^2$.

§ 7. Maxwell's Equations, Energy- and Momentum-Laws. We will represent matrices as boldface. $\cdots$ The electric and magnetic fields $\vec{\mathbf{E}}, \vec{\mathbf{B}}$ will be regarded as matrices whose elements are harmonic plane wave components, e.g.,

$$\vec{\mathbf{E}} = \left(E(nm)e^{i\omega(nm)(t - x/c)}\right). \tag{89}$$

Of course, it must be understood that the indices n, m are no longer restricted to a discrete range, and that they may also include vector indices.

One will keep Maxwell's equations as matrix equations:

$$\vec{\nabla} \times \vec{B} = \dot{\vec{E}}, \quad \vec{\nabla} \times \vec{E} = -\dot{\vec{B}}. \tag{90}$$

The differentiations are carried out on each individual element of the matrix. We now derive the energy-momentum law; for that it is necessary to make a few obvious remarks about the multiplication of matrix-vectors. We define the scalar product as, e.g.,

$$\vec{F} \cdot \vec{G} = F_x G_x + (y) + (z), \tag{91}$$

and the vector product as

$$\left(\vec{F} \times \vec{G}\right)_x = F_y G_z - F_z G_y. \tag{92}$$

Since matrix multiplication is not commutative, it is not generally true that

$$\vec{F} \cdot \vec{G} = \vec{G} \cdot \vec{F}, \quad \vec{F} \times \vec{G} = -\vec{G} \times \vec{F}.$$

On the otherhand, it is still true that

$$\vec{\nabla} \cdot \left(\vec{F} \times \vec{G}\right) = \left(\vec{\nabla} \times \vec{F} \cdot\right) \vec{G} - \vec{F} \cdot \left(\vec{\nabla} \times \vec{G}\right). \tag{93}$$

We define the energy density $\mathbf{\Phi}$ as the (scalar) matrix

$$\mathbf{\Phi} = \frac{1}{8\pi} \left(\vec{E}^2 + \vec{B}^2\right). \tag{94}$$

Then

$$8\pi \dot{\mathbf{\Phi}} = \vec{E} \cdot \dot{\vec{E}} + \dot{\vec{E}} \cdot \vec{E} + \vec{B} \cdot \dot{\vec{B}} + \dot{\vec{B}} \cdot \vec{B},$$

and from (90):

$$8\pi \dot{\mathbf{\Phi}} = \vec{E} \cdot \vec{\nabla} \times \vec{B} + \vec{\nabla} \times \vec{B} \cdot \vec{E} - \vec{B} \cdot \vec{\nabla} \times \vec{E} - \vec{\nabla} \times \vec{E} \cdot \vec{B},$$

and from (93)

$$\dot{\mathbf{\Phi}} + \vec{\nabla} \cdot \vec{S} = 0, \tag{95}$$

where

$$\vec{S} = \frac{1}{8\pi} \left(\vec{E} \times \vec{B} - \vec{B} \times \vec{E}\right). \tag{96}$$

This is Poynting's law for matrix electrodynamics; $\vec{S}$ is Poynting's vector.

In a similar way, we derive the momentum law $\cdots\cdots$ four dimensional vector analysis $\cdots$ relativity $\cdots$

§ 8. **Spherical Waves. Dipole Radiation.** Since our goal is to calculate the radiation from an oscillator, we must review spherical waves. For that we introduce Hertz's vector $\vec{Z}$ as a matrix vector from which we get

$$\vec{E} = \vec{\nabla}\left(\vec{\nabla}\cdot\vec{Z}\right) - \ddot{\vec{Z}} \quad , \quad \vec{B} = \vec{\nabla}\times\dot{\vec{Z}}. \tag{100}$$

In the classical theory $\vec{Z}$ is proportional to $e^{i\omega(t-r/c)}/r$. This expression can be expanded in plane waves using the identity [4]

$$\frac{e^{ikr}}{r} = \frac{ik}{2\pi}\int e^{ik(\vec{r}\cdot\hat{s})}d\omega_{\hat{s}}; \tag{101}$$

where $\vec{r}$ is the radius vector from the center of the spherical wave to the field point, and $\int d\omega_{\hat{s}}$ is an integration over the directions of the unit vector $\hat{s}$. In this way, we get the matrix representation for spherical waves in the form (89):

$$\vec{Z} = \left(eq(nm)\frac{e^{i\omega(nm)(t-r/c)}}{r}\right); \tag{102}$$

here the matrix $e\mathbf{q} = (eq(nm))$ is the electric moment which gives rise to the wave.

From here the calculation of the electromagnetic fields and the radiation is the same as in the classical theory, where $\vec{r}$ is the radius vector associated with each matrix. One gets

$$\vec{B} = -\frac{e}{c^2 r}\hat{r}\times\ddot{\mathbf{q}}, \quad \vec{E} = \frac{e}{c^2 r}\hat{r}\times\left(\hat{r}\times\ddot{\mathbf{q}}\right) \tag{103}$$

and finally

$$\vec{S} = \frac{e^2}{4\pi c^3 r^2}\hat{r}\left(\hat{r}\times\ddot{\mathbf{q}}\right)^2. \tag{104}$$

The integration over all space directions follows in the same way as in the classical theory. The result for the energy radiated per second is:

$$\int\vec{S}\cdot\hat{r}r^2 d\omega_r = \frac{2e^2}{3c^3}\ddot{\mathbf{q}}^2. \tag{105}$$

Averaged over time, the result is the diagonal matrix

$$\frac{2e^2}{3c^3}\langle \ddot{\vec{q}}^2 \rangle. \tag{106}$$

For oscillations in a fixed direction, we can replace the matrix vector $\vec{q}$ by the matrix scalar $\mathbf{q} = q(nm)$; then we get the radiation

$$\frac{2e^2}{3c^3}\langle \ddot{\mathbf{q}}^2 \rangle = \frac{2e^2}{3c^3}\left(\sum_k \omega(nk)^4 |q(nk)|^2\right). \tag{107}$$

We cannot give a complete theory of radiation $\cdots\cdots$. Here we will only prove that the radiation is effectively given by the quantities $|q(nk)|^2$; the result (107) shows that is the case, but we see at once that there are terms not appropriate for a stationary state emitting spontaneous radiation: because the spontaneous transitions always go only to a state of lower energy, i.e. for appropriate enumeration to states of smaller quantum numbers. But we can in a completely formal way impose this condition in our theory; for that we take – not the mean value – but the diagonal sum of the radiation matrix (105); that gives

$$\mathrm{Diag}\left(\frac{2e^2}{3c^3}\ddot{\mathbf{q}}^2\right) = \frac{2e^2}{3c^3}\sum_{nk}\omega(nk)^4 \cdot |q(nk)|^2. \tag{108}$$

Here we can re-sum the right side and write:

$$\frac{4e^2}{3c^3}\sum_n\left(\sum_{k<n}\omega(nk)^4 \cdot |q(nk)|^2\right). \tag{109}$$

In this way the desired ordering is obtained: now for each state n, there is radiation from transitions to all states $k < n$, with the classically known intensity. This agrees with experience if the indices n are ordered with increasing energy W_n. Therefore Heisenberg's assumption is made correct in the above sense.

* * * * * * * * * * * * * * * * * * **

Born and Jordan's *very first* quantum mechanical discussion of the theory of radiation has never been excelled in its economy, clarity and beauty. It is next immediately made *completely* explicit that the matrix indices of the fields include the photon occupation number or the second quantized representation of the field operators. Here certainly was the genesis of quantum field theory, for which credit was later explicitly given by Born to Jordan (see van der Waerden, loc. cit.).

＊＊＊＊＊＊＊＊＊＊＊＊＊＊＊＊＊＊**

In conclusion, we emphasize that the transition probabilities include the statistical weights of the states, and in fact each row and column, i.e. each diagonal term of **W** designates a state to which we must ascribe the same statistical weight. This result (in its generalization to systems of many degrees of freedom) leads to the fundamental principle of Bose-Einstein statistics for light quanta, as will be shown later.

Comment Added in Proof: The generalization of the theory to many degrees of freedom has been completed with Herrn W. Heisenberg, and will be included in the continuation of this work. There also, other points met here will be explored further.

Footnotes and References:
1) W. Heisenberg, Zeits. f. Phys. **33**, 879 (1925).
2) W. Thomas, Naturw. **13**, 627 (1925).
3) W. Kuhn, Zeits. f. Phys. **33**, 408 (1925).
4) P. Debye, Ann. d. Phys. **30**, 705 (1909); Eqn (7"), p.758.

Paper VII·2: Excerpt from Zeitschrift für Physik **35**, 557 (1926).

Zur Quantenmechanik II

von **M. Born, W. Heisenberg** und **P. Jordan**

in Göttingen (Eingegangen am 16. November 1925.)

The quantum mechanics developed in Part I from Heisenberg's Ansätz is applied to systems of many degrees of freedom. Perturbation theory is developed for non-interacting and for a large class of interacting systems, and its connection with the eigenvalue theory of Hermitian forms is established. The results include the derivation of the laws of momentum and angular momentum, and of selection rules and intensity formulas. Finally, the ansätz is applied to the statistics of the eigenmodes of radiation in a cavity.

Introduction. This work applies the new theory of quantum mechanics, whose physical and mathematical foundation was developed in two earlier works by the authors [1]. It extends the new theory [I; Ch2] to systems of many degrees of freedom and introduces 'canonical transformations' to reduce the integration of the equations of motion to well-known mathematical questions. The new theory gives first: a perturbation theory (Ch1 §4) closely similar to the classical theory, and second: the connection of quantum mechanics with the mathematical theory of quadratic forms of infinitely many variables (Ch3). Before we go on to these developments of the theory, we will try to define precisely its physical content.

The starting point of the new theory was the conviction that it was impossible to overcome the difficulties of a step by step approach to quantum theory. Instead, for the mechanics of atoms and electrons, there should be a mathematical system of relations between observable quantities only, based on simplicity and consistency, like classical mechanics. Such a system of quantum relations would contrast with the old quantum theory: the problem is that the new theory cannot be directly interpreted geometrically, because the electron motion cannot be described by the concepts of space and time available to us; the characteristic feature of the new theory is that it requires a profound change of the existing kinematics and mechanics; the new quantum mechanics does however constitute an important improvement, in that the basic postulates make the theory a complete whole.

For example, the existence of discrete stationary states in the new theory is just as natural as the existence of discrete eigenfrequency modes for the old classical theory (see Ch1).

If one asks for the fundamental difference required by the basic quantum postulates, between the new quantum theory and the classical theory, then it seems to us that our formalism – if it proves correct – is a quantum mechanics as similar to the classical mechanics as one could hope. We retain the energy and momentum conservation laws and the form of the equations of motion (see Ch1 §2). This similarity of the new theory with the classical constitutes a self-contained correspondence principle which is surely closer than could have been imagined. In fact, the new theory can be viewed as an exact formulation of Bohr's correspondence ideas. It is important for the further development of the new theory, that this correspondence can be realized exactly and used to describe the transition from the symbolic quantum geometry to the observable classical geometry. It seems to us a particularly essential feature of the new theory, that in it the continuous- and the line-spectra of atoms occur on equal footing, as solutions of the same equation of motion which are mathematically very close to one another (see Ch3 §3); any difference between 'quantized' and 'unquantized' motions appears in the theory only in the sense that specific motions are possible only for large quantum numbers. This condition appears in the fundamental quantum mechanical equation (Ch1 §1), which is valid for all possible motions and which gives a specific definition of the motion in general.

We would like to emphasize the mathematical unity and simplicity of the new theory, which is essential for progress in the problem of the atom. It is clear that the new theory solves the principal difficulties of the quantum theory. For example, the classical radiation-resistance force is still not included in the theory, but there are only indirect indications of any consistency problem for the new quantum mechanics (see Ch1 §5). It seems that this fundamental difficulty is reduced from before, and that the new theory also provides a better hope for its solution in collision processes. In the old theory, these difficulties required a change of the basic postulates of quantum theory, and violated energy conservation in collisions, as has been pointed out by Bohr [3]. In the new theory, the basic postulate gives energy conservation as a consequence of the quantum mechanical equations. The result is that the Franck-Hertz momentum requirement appears as a natural mathematical consequence of the new theory. The hope is that in a treatment of collisions in the new quantum mechanics, such difficulties will not appear.

The $\cdots\cdots$ anomalous Zeeman effect $\cdots\cdots$ after a recent Note of Uhlenbeck and Goudsmit [4] has entered a new phase $\cdots\cdots$.

Finally, we have a new way to handle an old statistical problem$\cdots\cdots$ by quantizing the eigenmodes of a $\cdots$ cavity $\cdots$ to get Planck's formula. However, Einstein [5] found a false value for the fluctuations $\cdots\cdots$ We have found that the new theory leads to the correct value, which is important support for the new quantum mechanics.

Chapter 1. Systems of One Degree of Freedom.

§ 1. **Basic Principles.** I: A quantum variable – it could be a coordinate or momentum or some function of both – will be represented by an array of quantities

$$a(nm)e^{i\omega(nm)t} \tag{1}$$

or suppressing all factors, on an array of quantities

$$a(nm) \tag{2}$$

dependent only on the indices n, m. We can speak of a (usually infinite) 'matrix'.

II: The computational operations such as addition and multiplication of quantum variables are defined by the corresponding rules for matrices.

III: Differentiation $\cdots$ We are led to two kinds of differentiations of a matrix function $\mathbf{f}$ with respect to a matrix argument $\mathbf{x}$:
a)Derivatives of the first kind:

$$\frac{\partial \mathbf{f}}{\partial \mathbf{x}} = \lim_{\alpha \to 0} \frac{\mathbf{f}(\mathbf{x} + \alpha \cdot \mathbf{1}) - \mathbf{f}(\mathbf{x})}{\alpha} \tag{3}$$

b)$\cdots$ [Note added: We have suppressed most of the earlier discussion due to Jordan as superfluous. See details in Born and Jordan [6] and in **Paper VII·1**. Jordan acknowledged that Heisenberg's definition was "surely an appropriate simplification." (See van der Waerden, loc. cit., p.55, citing letters of Oct5,7 from Heisenberg in Copenhagen to Born and to Jordan in Göttingen.)]

IV: Calculations with quantum variables – because of the violation of the commutative law of multiplication – remain somewhat undefined, until the value of **pq** − **qp** has been written down. We introduce as the fundamental quantum relation:

$$\mathbf{pq} - \mathbf{qp} = \frac{\hbar}{i}\mathbf{1}. \tag{5}$$

We will return later to the physical interpretation of this relation. It is important to emphasize that Eqn5 is the only fundamental equation of the new quantum mechanics in which Planck's constant $\hbar$ occurs. The constant $\hbar$ enters into the basic laws in such a simple way, that one can see from Eqn5 that in the limit $\hbar = 0$ the new theory goes over into the classical one, as it must do physically.

An important relation can be derived from Eqn5: Let $\mathbf{f(pq)}$ be some function of **p** and **q**, then

$$\mathbf{fq} - \mathbf{qf} = \frac{\hbar}{i}\frac{\partial \mathbf{f}}{\partial \mathbf{p}}, \quad \mathbf{pf} - \mathbf{fp} = \frac{\hbar}{i}\frac{\partial \mathbf{f}}{\partial \mathbf{q}} \cdots\cdots \tag{6}$$

§ 2. The Canonical Equations, Energy Conservation and Frequency Condition. For the Hamiltonian $\mathbf{H(pq)}$ the canonical equations are

$$\dot{\mathbf{p}} = -\frac{\partial \mathbf{H}}{\partial \mathbf{q}}; \quad \dot{\mathbf{q}} = \frac{\partial \mathbf{H}}{\partial \mathbf{p}}. \tag{7}$$

From the combination principle

$$\omega(nm) + \omega(mk) = \omega(nk), \tag{8}$$

it follows that ω can be represented in the form

$$\omega_{nm} = \frac{W_n - W_m}{\hbar} \tag{9}$$

$\cdots\cdots$[Note added: There follows a long argument identifying **W** with **H** in diagonal form, satisfying $\dot{\mathbf{H}} = 0$. This ends Heisenberg's introduction and review of the basic results of Born and Jordan's Part I (here as **Paper VII·1**).]

§ 3. Canonical Transformations. By a 'canonical transformation' one means a transformation of the variables $\mathbf{p}, \mathbf{q}$ into new variables $\mathbf{P}, \mathbf{Q}$ for which

$$\mathbf{pq} - \mathbf{qp} = \mathbf{PQ} - \mathbf{QP} = \frac{\hbar}{i}\mathbf{1}; \tag{16}$$

then the canonical equations (7) hold for $\mathbf{P}, \mathbf{Q}$ as they do for $\mathbf{p}, \mathbf{q}$. A general transformation which satisfies this condition is

$$\mathbf{P} = \mathbf{S}\mathbf{p}\mathbf{S}^{-1}, \quad \mathbf{Q} = \mathbf{S}\mathbf{q}\mathbf{S}^{-1}, \tag{17}$$

where $\mathbf{S}$ is an arbitrary quantum theoretic quantity; we might note $\cdots\cdots$

$$\mathbf{f}(\mathbf{P}, \mathbf{Q}) = \mathbf{S}\mathbf{f}(\mathbf{p}, \mathbf{q})\mathbf{S}^{-1} \tag{18}$$

The importance of the canonical transformation is the following: given any set $\mathbf{p}_0, \mathbf{q}_0$ which satisfies Eqn16, then one can reduce the problem of the integration of the canonical equations for a Hamiltonian $\mathbf{H}(\mathbf{pq})$ to the problem of finding a canonical transformation $\mathbf{S}$ such that with

$$\mathbf{p} = \mathbf{S}\mathbf{p}_0\mathbf{S}^{-1}, \quad \mathbf{q} = \mathbf{S}\mathbf{q}_0\mathbf{S}^{-1}, \tag{19}$$

the Hamiltonian

$$\mathbf{H}(\mathbf{pq}) = \mathbf{S}\mathbf{H}(\mathbf{p}_0\mathbf{q}_0)\mathbf{S}^{-1} = \mathbf{W} \tag{20}$$

will be a diagonal matrix. Eqn20 is the analog of Hamilton's partial differential equation; $\mathbf{S}$ corresponds to the action function. [Note added: This is Born's reformulation of Heisenberg's earlier perturbative development in terms of an operator $\mathcal{G}$ which Heisenberg now recognized as $S = e^{\mathcal{G}}$. In a Sept 29 letter to Jordan (van der Waerden, loc. cit., Pp.49-50), Heisenberg says: "Born's form of the canonical transformation is certainly the heart of the new theory $\cdots$ since it is the *most general.*"]

§ 4. Perturbation Theory. Consider the mechanical problem defined by the Hamiltonian:

$$\mathbf{H} = \mathbf{H}_0(\mathbf{pq}) + \lambda\mathbf{H}_1(\mathbf{pq}). \tag{21}$$

We assume the mechanical problem defined by $\mathbf{H}_0(\mathbf{pq})$ to have been solved; therefore the solutions $\mathbf{p}_0, \mathbf{q}_0$ are known which satisfy the canonical commutation relation (Eqn5), and make $\mathbf{H}_0(\mathbf{p}_0\mathbf{q}_0) = \mathbf{W}_0$ a diagonal matrix. We then seek a transformation **S** such that

$$\mathbf{p} = \mathbf{S}\mathbf{p}_0\mathbf{S}^{-1}, \quad \mathbf{q} = \mathbf{S}\mathbf{q}_0\mathbf{S}^{-1}, \tag{22}$$

and $\mathbf{H}(\mathbf{pq}) = \mathbf{S}\mathbf{H}(\mathbf{p}_0\mathbf{q}_0)\mathbf{S}^{-1} = \mathbf{W}$, i.e., **H** is a diagonal matrix. To construct a solution, we use the ansätz

$$\mathbf{S} = 1 + \lambda\mathbf{S}_1 + \lambda^2\mathbf{S}_2 + \cdots \tag{23}$$

for which

$$\mathbf{S}^{-1} = 1 - \lambda\mathbf{S}_1 + \lambda^2(\mathbf{S}_1^2 - \mathbf{S}_2) + \lambda^3 \cdots\cdots \tag{24}$$

If we assume for **H** the form Eqn21, then we can expand in powers of λ and obtain the iteration equations

$$
\begin{aligned}
\mathbf{W}_0 &= \mathbf{H}_0(\mathbf{p}_0\mathbf{q}_0) \\
\mathbf{W}_1 &= \mathbf{H}_1 + \mathbf{S}_1\mathbf{H}_0 - \mathbf{H}_0\mathbf{S}_1 \\
\mathbf{W}_2 &= \mathbf{H}_2 + \mathbf{S}_1\mathbf{H}_1 - \mathbf{H}_1\mathbf{S}_1 \\
&\quad + \mathbf{H}_0\mathbf{S}_1^2 - \mathbf{S}_1\mathbf{H}_0\mathbf{S}_1 + \mathbf{S}_2\mathbf{H}_0 - \mathbf{H}_0\mathbf{S}_2 \text{ usw,}
\end{aligned} \tag{25}
$$

where $\mathbf{H}_0, \mathbf{H}_1$ have the arguments $\mathbf{p}_0, \mathbf{q}_0$. $\cdots\cdots$ The solution satisfies $\mathbf{S}\mathbf{S}^\dagger = 1$ and $\mathbf{q} = \mathbf{q}^\dagger, \mathbf{p} = \mathbf{p}^\dagger$ are hermitian.

In the first approximation

$$\mathbf{W}_1 = \bar{\mathbf{H}}_1, \tag{29}$$

where $\bar{\mathbf{H}}_1$ is the diagonal part of $\mathbf{H}_1$; and

$$S_1(mn) = \frac{H_1(mn)}{\hbar\omega_0(mn)}(1 - \delta_{mn}). \tag{30}$$

$\cdots\cdots$ In second approximation

$$W_2 = \bar{H}_2 + \sum{}' \frac{H_1(nk)H_1(kn)}{\hbar\omega_0(nk)}, \tag{31}$$

where the sum $\sum'$ omits the term with $k = n$. $\cdots\cdots$

§ 5. Systems for which the time occurs explicitly in the 'Hamiltonian'. The treatment of the quantum mechanical effect of an external force which explicitly

depends on the time, seems to us of particular interest, because for these problems some differences between quantum and classical theories become evident. $\cdots\cdots$
[Note added: In the interest of space and time, we must omit this section.]

Chapter 2. Fundamentals of the Theory of Systems of Arbitrarily Many Degrees of Freedom.

§ 1. The Canonical Equations of Motion; Perturbation Theory. $\cdots\cdots$

It is clear that the general rules of matrix analysis described in Ref. [6] and in Ch1 are applicable in the theory of systems of many degrees of freedom. The equations of motion follow directly from the variation principle, so that we can write immediately:

$$\dot{\mathbf{q}}_k = \frac{\partial \mathbf{H}}{\partial \mathbf{p}_k}; \quad \dot{\mathbf{p}}_k = -\frac{\partial \mathbf{H}}{\partial \mathbf{q}_k}. \tag{2}$$

What is essentially new is the generalization of the commutation relations to:

$$\begin{aligned}
\mathbf{p}_k \mathbf{q}_j - \mathbf{q}_j \mathbf{p}_k &= \frac{\hbar}{i}\delta_{kj}, \quad k,j = 1,\cdots, \\
\mathbf{p}_k \mathbf{p}_j - \mathbf{p}_j \mathbf{p}_k &= 0, \\
\mathbf{q}_k \mathbf{q}_j - \mathbf{q}_j \mathbf{q}_k &= 0.
\end{aligned} \tag{3}$$

$\cdots\cdots$[Note added: Pauli, Weyl, and Dirac were quick to pick up the trail with similar developments (see van der Waerden, loc. cit., p.53.]

§ 2. Interacting Systems. We will now examine interacting systems. $\cdots\cdots$

Chapter 3. Connection with the Theory of the Eigenvalues of Hermitian Forms.

§ 1. General Methods. $\cdots\cdots$

§ 2. Applications to Perturbation Theory. $\cdots\cdots$

§ 3. Continuous Spectra. $\cdots\cdots$ [Note added: Ch3 is Born's mathematical summary of the theory. He left Göttingen on Oct28 for a US visit and did not participate further in the production of the Dreimännerarbeit.]

Chapter 4. Practical Applications of the Theory.

§ 1. Laws of Momentum and Angular Momentum; Intensity Formulas and Generalizations. $\cdots\cdots$

§ 2. The Zeeman Effect. $\cdots\cdots$

§ 3. Coupled Harmonic Oscillators. Statistics of the Wave Fields. A system of coupled harmonic oscillators given by

$$\mathbf{H} = \frac{1}{2} \sum_{k=1}^{N} \frac{\mathbf{p}_k^2}{m_k} + \mathbf{V}(\mathbf{q}), \tag{35}$$

with a potential $\mathbf{V}(\mathbf{q})$ quadratic in the coordinates (with numerical coefficients) represents the simplest imaginable system of many degrees of freedom. As in Ch2 §1, the commutation relations remain invariant under orthogonal transformations of the coordinates and momenta. It can – as in the classical theory – be transformed into a system of uncoupled oscillators. For example, the vibrations of a crystal lattice can be analyzed in terms of their eigenmodes, just as in the classical theory. Each individual eigenmode is treated as a single linear oscillator, and the uncoupled eigenmodes are equivalent to the original system. The same conclusion holds for: a) the uncountably many degrees of freedom needed to understand the vibrations of a continuum of idealized elastic-coupled bodies; or b) the countably many of an electromagnetic cavity.

In the old quantum theory, the oscillations of an electromagnetic cavity have been extensively investigated. This simplest imaginable problem of coupled oscillators can be handled here too. The prescription that the energy of an eigenmode should be an integral multiple of $\hbar\omega$ bears a formal similarity to the theory of light-quanta. One hopes therefore to understand light-quanta from a consideration of cavity oscillations. Of course, it is clear that such an attack on the problem of light-quanta can give no insight into one most essential aspect of the problem: the coupling of separated atoms. This problem does not even enter the question of the cavity oscillations. Nevertheless, a close connection can be established between the cavity eigenmodes and light-quanta: the statistics of the eigenmodes of the cavity must correspond with the statistics of the light-quanta, and conversely.

Debye [7] showed that the partition of individual light-quanta over the eigenmodes of a cavity gives such statistics, and he derived Planck's formula in this way. Such a mixture of wave-theoretic and light-quantum concepts cannot be the essence of the problem, however. It would have the result of completely separating the wave-theoretic side of the problem from the light-quantum theory, and would treat the cavity radiation by general statistical laws, e.g. as a quantum theoretic atomic system. The corresponding light-quantum statistics is then shown to be Bose statistics [8]. This result seems natural, but has nothing to do with the assumption of independent light-quanta. It is to be understood as the assignment of the statistics to the eigenmodes – and only shows that the assumption of statistically independent light-quanta would not be consistent. [Note added: This section is credited by van der Waerden (loc. cit., p.55) entirely to Jordan. It does seem however to be strongly influenced by Heisenberg's initial fundamental philosophical premise – 'that a mixture of wave-theoretic and light-quantum concepts cannot be the essence $\cdots$.']

Treatment of the cavity radiation by the old quantum theory resulted in the basic difficulty that it did indeed lead to the Planck Radiation Law, but not to the correct mean-square fluctuation of the energy in any partial volume. This result shows that any treatment of the eigenmodes of a mechanical system or of an electromagnetic cavity by the old theory must lead to serious contradictions. We conjectured that the modified kinematics of the new theory would yield the correct value for the fluctuations, eliminate the contradictions and make possible the expected statistics for the cavity radiation.

The state of the oscillators is specified by the 'quantum-numbers' $n_1, n_2, n_3, \cdots$ of the individual oscillators, and the energy of the individual states – up to an additive constant – is

$$E_n = \hbar \sum_k \omega_k n_k. \tag{36}$$

The additive constant, the 'zero-point energy', is $E_0 = \frac{1}{2}\hbar \sum_k \omega_k$, which would be infinite for infinitely many degrees of freedom. E_n will be the thermal energy. Each state of the system designated by a distinct set $n_1, n_2, n_3, \cdots$ will be ascribed the same statistical weight. The result is: For waves in an s-dimensional isotropic volume of magnitude $V = L^s$, propagating with velocity v, the number of eigenmodes with frequencies in the interval $df = d\omega/2\pi$ is equal to the number of 'Bose-Einstein cells' in df; this holds for arbitrary s, and includes the vibrations of membranes or strings. The number of eigenmodes for df is determined by the number of ways positive integers $n_1, \cdots, n_s$ can be chosen so that

$$f = \frac{v}{2L}\sqrt{n_1^2 + \cdots + n_s^2}$$

is in df. If $K_s(a)$ is the volume of an s-dimensional sphere of radius a, then there are $K_s(f) \times V/v^s$ eigenmodes of frequency less than f. Equivalently, the number of cells in df can be determined by: the momentum components $p_1, p_2, \cdots, p_s$ satisfy

$$\frac{E}{v} \equiv p \to \frac{hf}{v} = \sqrt{p_1^2 + \cdots + p_s^2},$$

and the volume of the cells in the $2s$-dimensional phase-space is $(2\pi\hbar)^s$. Therefore the number of cells corresponding to frequencies less than f is $K_s(f) \times V/v^s$, as before. [Note added: the artifice of doubling the dimension of phase-space accounts for complex or left- and right-moving modes.]

A well-defined reversible pairing of light-quanta to the eigenmodes can be constructed such that the individual pairs belong to the same df. This pairing can also be done so that the direction of an eigenmode and the corresponding light-quantum are the same. From Eqn36 then, the quantum number of each oscillator is the same as the number of quanta in the corresponding light-quantum cell. The statistics of the light-quanta gives the statistics for the corresponding eigenmodes and conversely. One sees that the statistical-weights of the states of the oscillator are determined directly by the basic assumption of the Bose-Einstein statistics for the light-quanta. The equal probability complexions are defined to have as many quanta as there are in each state [9].

According to Debye's statistics, $\cdots\cdots$ [Note added: We abandon this diversion and return to the usual Bose-Einstein statistics.]

The inadequacy of the classical theory for the fluctuations in radiation fields appears in the following way: Suppose a volume V is in thermal equilibrium with a very large volume of the same kind, and can freely exchange waves in the frequency interval f to $f + df$, but no others, and let E be the energy of waves of frequency f in V. Then – following Einstein – the mean square fluctuation $\overline{\Delta^2} = \overline{(E - \overline{E})^2}$ can be calculated by Boltzmann's Principle. If z_f is the number of eigenmodes (cells) per unit volume per unit frequency, then

$$\overline{E} = \frac{z_f \hbar \omega}{e^{\hbar\omega/kT} - 1} \cdot V, \tag{39}$$

which gives

$$\overline{\Delta^2} = \hbar\omega\overline{E} + \frac{\overline{E^2}}{z_f V}. \tag{40}$$

* * * * * * * * * * * * * * * * * * * *

Using $\overline{E^2} = (z_f \hbar\omega \cdot V)^2 \times \overline{n^2}$ and $\overline{n^2} = \dfrac{\sum_{n=0}^{\infty} n^2 e^{-n\hbar\omega/kT}}{\sum_{n=0}^{\infty} e^{-n\hbar\omega/kT}}$.

* * * * * * * * * * * * * * * * * * * *

In classical wave theory – as Lorentz [10] has shown – only the second term in Eqn40 occurs. This contradiction occurs also for the waves in a crystal lattice or an elastic continuum. The problem arises [11] because additivity of the entropy of the volume V and of the very large volume was assumed in the Einstein derivation. This additivity exists in the classical theory only in the validity range of the Rayleigh-Jeans Law. The statistical independence of the sub-volume is therefore not a rigorous result of the old theory, and any failure of this assumption must also affect the simple problem of the harmonic oscillator.

We will now calculate the fluctuations according to quantum mechanics. To avoid irrelevant complications, we consider the simplest case of a vibrating string, first in the classical theory.

The length of the string is $L = \pi$ [Note added: This choice is made here to tidy up the original equations.] and its sideways displacement is $u(x,t)$. With the expansion

$$u(x,t) = \sum_{k=1}^{\infty} q_k(t) \sin kx \tag{41}$$

or

$$q_k(t) = \frac{2}{\pi} \int_0^\pi dx \sin kx \cdot u(x, t),$$

the energy of the string becomes a quadratic sum over the Fourier coefficients $q_k(t)$ as coordinates. It has the familiar form

$$H = \frac{1}{2} \int_0^\pi dx \left\{ u'^2 + \dot{u}^2 \right\} = \frac{\pi}{4} \sum_{k=1}^\infty \left\{ \dot{q}_k(t)^2 + k^2 q_k(t)^2 \right\}. \tag{42}$$

For the energy of the string in the interval $(0, a)$ we get

$$E(a) = \frac{1}{2} \int_0^a dx \cdot \sum_{j,k=1}^\infty \times$$
$$\left\{ \dot{q}_j \dot{q}_k \sin jx \sin kx + q_j q_k jk \cos jx \cos kx \right\}. \tag{43}$$

If we assume in Eqn43 $\cdots\cdots$ that all wavelengths are small compared to a, then we get directly $E(a) = \frac{a}{\pi} H$. One sees that the difference

$$\Delta = E - \overline{E},$$

where $\overline{E}$ is the average over the phase ϕ_k in

$$q_k = a_k \cos(\omega_k t + \phi_k); \quad \omega_k = k, \tag{44}$$

eliminates the $j = k$ terms in Eqn43. This phase average is identical with the time average. Carrying out the integration

$$\Delta = \frac{1}{4} \sum_{j \neq k} \left\{ \dot{q}_j \dot{q}_k K_{jk}(-) + jk q_j q_k K_{jk}(+) \right\} \tag{45}$$

with

$$K_{jk}(\mp) = \frac{\sin(j-k)a}{(j-k)a} \mp \frac{\sin(j+k)a}{(j+k)a}.$$

We also need

$$\Delta^2 = (\Delta_1 + \Delta_2)^2 = \Delta_1^2 + \Delta_2^2 + \Delta_1 \Delta_2 + \Delta_2 \Delta_1 \tag{46}$$

with

$$\Delta_1^2 + \Delta_2^2 = \frac{1}{16} \sum_{j \neq k} \sum_{m \neq n} \times$$
$$\left\{ \dot{q}_j \dot{q}_k \dot{q}_m \dot{q}_n K_{jk}(-) K_{mn}(-) + jkmn q_j q_k q_m q_n K_{jk}(+) K_{mn}(+) \right\}$$

and

$$\Delta_1\Delta_2 + \Delta_2\Delta_1 = \frac{1}{16}\sum_{j\neq k}\sum_{m\neq n} \times$$

$$\{jk(q_j q_k \dot{q}_m \dot{q}_n + \dot{q}_m \dot{q}_n q_j q_k)K_{jk}(+)K_{mn}(-)\}.$$

From Eqn44, $\overline{\Delta_1\Delta_2 + \Delta_2\Delta_1} = 0$ and

$$\begin{aligned}
\overline{\Delta^2} &= \overline{\Delta_1^2} + \overline{\Delta_2^2} \\
&= \frac{1}{8}\sum_{j\neq k}\left\{\overline{\dot{q}_j^2\dot{q}_k^2}K_{jk}(-)^2 + j^2k^2\overline{q_j^2 q_k^2}K_{jk}(+)^2\right\}.
\end{aligned} \tag{47}$$

[Note added: Again, if all wavelengths are small compared to a] $\cdots$, the sums over j,k go into integrals over ω_j,ω_k:

$$\overline{\Delta^2} = \frac{1}{8}\int_0^\infty\int_0^\infty d\omega_j d\omega_k \left\{\overline{\dot{q}_j^2\dot{q}_k^2}K_{jk}(-)^2 + j^2k^2\overline{q_j^2 q_k^2}K_{jk}(+)^2\right\}.$$

Finally, assume that a is large and use the limit

$$\lim_{a\to\infty}\frac{1}{a}\int_-^+\frac{\sin^2\omega a}{\omega^2}f(\omega)d\omega = \pi f(0). \tag{48}$$

Then only the first term in

$$K_{jk}(\mp) \sim \frac{\sin(\omega_j - \omega_k)a}{\omega_j - \omega_k}$$

contributes in Eqn45 and gives in Eqn47:

$$\overline{\Delta^2} = \frac{a}{8}\int_0^\infty d\omega\left\{\left(\overline{\dot{q}_\omega^2}\right)^2 + \left(\omega^2\overline{q_\omega^2}\right)^2\right\}. \tag{49}$$

The average energy in a is

$$\overline{E} = \frac{a}{4}\int_0^\infty d\omega\left\{\overline{\dot{q}_\omega^2} + \omega^2\overline{q_\omega^2}\right\}. \tag{50}$$

The relation

$$\overline{\dot{q}_\omega^2} = \omega^2\overline{q_\omega^2} \tag{51}$$

which is true here remains valid in quantum mechanics, as seen in Ch1. In order to get the quantities used in Eqns39,40, we use $z_f = 2$ from Eqn44 [Note added:

These conclusions are no doubt correct, but BHJ are being cavalier in their arguments. The doubling of the number of degrees of freedom (corresponding to a complex Fourier expansion for a quantum string) - giving $z_f = 2$ can be invoked in a number of ways – here by averaging over the phase.] and get $d\omega_k = 2\pi df_k = dk$. Then with $V = a$:

$$\overline{\Delta^2} = \frac{\overline{E^2}}{2V}, \tag{52}$$

we get the second – classical – term in Eqn40.

For the transition to quantum mechanics, Eqns41,42,43 are to be interpreted as matrix equations for $\mathbf{u}, \mathbf{q}, \mathbf{H}, \mathbf{E}$. x is still a number; when we consider the continuous string as a series of point masses, x (multiplied by a lattice constant) represents the number of a particular point mass.

The matrix $\mathbf{q}_k$ has $2N$ dimensions, where N is the number of eigenmodes (possibly infinite). The matrix-elements $q_k(nm)$ of $\mathbf{q}_k$ all vanish except those with

$$n - m = \pm 1. \tag{53}$$

The phase average of a matrix is just the diagonal matrix of its diagonal elements. From Eqn53, conclusions similar to Eqn44 can be found. The derivation which led to Eqn46 remains valid for quantum mechanics. Also Eqn47 holds for the diagonal matrix $\overline{\Delta_1^2 + \Delta_2^2}$ with matrices $\mathbf{q}_k$; and finally, when we calculate the part of $\overline{\Delta^2}$ designated as $\overline{\Delta^2}$:

$$\overline{\Delta_1^2 + \Delta_2^2} = \frac{\overline{E^{*2}}}{2V}.$$

Here E^* is no longer the average thermal energy of Eqns49,50,51, but the sum of this and the zero-point energy. From elementary oscillator formulas

$$E^* = \hbar\omega \cdot V + E,$$

so

$$\overline{\Delta_1^2 + \Delta_2^2} = \frac{1}{2}(\hbar\omega)^2 V + \hbar\omega\overline{E} + \frac{\overline{E^2}}{2V}; \tag{54}$$

then the zero-point energy for df will equal

$$\frac{V}{L} \cdot \frac{\hbar\omega}{2} \cdot Lz_f df = \hbar\omega \cdot V df.$$

We still have to calculate $\overline{\Delta_1\Delta_2 + \Delta_2\Delta_1}$. Corresponding to Eqn49 we get

$$\overline{\Delta_1\Delta_2 + \Delta_2\Delta_1} = \frac{a}{8}\int_0^\infty d\omega \cdot \omega^2 \left\{ (\mathbf{q}_\omega\dot{\mathbf{q}}_\omega)^2 + (\dot{\mathbf{q}}_\omega\mathbf{q}_\omega)^2 \right\}.$$

From the commutation relations – with $L/2$ as the 'mass' in Eqn42 –

$$-\mathbf{q}_j\dot{\mathbf{q}}_j(nn) = \dot{\mathbf{q}}_j\mathbf{q}_j = \frac{\hbar}{Li}.$$

The contribution to $\overline{\Delta_1\Delta_2 + \Delta_2\Delta_1}$ after division by df is

$$\overline{\Delta_1\Delta_2 + \Delta_2\Delta_1} = -\frac{1}{2}(\hbar\omega)^2 V,$$

and with Eqn54 follows

$$\overline{\Delta^2} = \hbar\omega\overline{E} + \frac{\overline{E^2}}{z_f V} \tag{55}$$

in agreement with Eqn40.

* * * * * * * * * * * * * * * * * * *

The choice $z_f = 2$ is unnecessary. Making z_f explicit we get

$$E^* = z_f \cdot \frac{\hbar\omega}{2} \cdot V + E \quad \text{and}$$

$$\overline{\Delta_1{}^2 + \Delta_2{}^2} = \frac{\overline{E^{*2}}}{z_f V} \Rightarrow \frac{z_f V}{4}(\hbar\omega)^2 + \hbar\omega \cdot \overline{E} + \frac{\overline{E^2}}{z_f V}.$$

The contribution from the commutation relation is

$$\overline{\Delta_1\Delta_2 + \Delta_2\Delta_1} = -\frac{1}{2}(\hbar\omega)^2 \frac{z_f}{2} \cdot V,$$

with $z_f/2$ separate commutation relations for z_f (real) generalized coordinates, each counted as a degree of freedom.

* * * * * * * * * * * * * * * * * * *

When one remembers that this problem is far removed from those for which quantum mechanics was created, the result appears particularly encouraging for the further possibilities of the new theory. One is spared the wave-interference calculation by which Ehrenfest obtained the above result, and could conclude directly from the additivity of the entropies of the partial volumes in the quantum mechanics that no contradictions can arise for any similar problem. We conjecture from our above result that the additivity actually holds in general.

The terms in Eqn55 not found in the classical theory are closely related to the zero-point energy. The essential difference of the new theory from the old is not a difference of their mechanical laws, but of the kinematics of the new theory. One can already see in Eqn55, in which no mechanical principles enter at all, an obvious difference of the quantum kinematics from the old theory.

If the new quantum mechanics should prove correct, then its most important advance over the old theory might be this: kinematics and mechanics are brought into close connection, as existed in the classical theory. The fundamentally new point of view, which follows from the basic postulates of the new quantum theory, for the mechanical concepts and the notion of space and time, finds its proper expression in the kinematics as well as in the mechanics and in the connection between them.

Footnotes and References:

1) W. Heisenberg, Zeits. f. Phys. **33**, 879 (1925) (here as **Paper VI·1**); M. Born and P. Jordan, Zeits. f. Phys. **34**, 858 (1925) (here as **Paper VII·1**); cited in the following as (Part) I.

2) Added in Proof: In a recent paper by P. Dirac (Proc. Roy. Soc. London **109**, 642 (1925)) (here as **Paper VIII·1**), some of the conclusions contained in Part I and in this work, and also further new consequences of the Theory have been set forth.

3) N. Bohr, Zeits. f. Phys. **34**, 142 (1925).

4) G. Uhlenbeck and S. Goudsmit, Naturwiss. **13**, 953 (1925).

5) A. Einstein, Phys. Zeits. **10**, 185, 817 (1909).

6) See Part I. [Note added: here most of these details have been left to the original.]

7) P. Debye, Ann. d. Phys. **33**, 1427 (1910). See also P. Ehrenfest, Phys. Zeits. **7**, 528 (1906).

8) S.N. Bose, Zeits. f. Phys. **26**, 178 (1924).

9) A. Einstein, Sitzungsber. d. Preuss. Akad. d. Wiss. 1925, p.3.

10) H.A. Lorentz, Les Theories Statistiques en Thermodynamique (Leipzig, 1916), p.59.

11) P. Ehrenfest, Report in the Göttingen Seminar on the Structure of Matter, Summer 1925. It has since been published, Zeits. f. Phys. **34**, 362 (1925).

Chapter VIII

DIRAC:
"Thus the most general differential operation on a quantum variable is the difference of its HEISENBERG products with some other quantum variable."

P.A.M. Dirac, Proc. Roy. Soc. A**109**, 642 (1925). "··· I saw that it provided the key to the problem of quantum mechanics. ···"

§VIII-1. Introduction

Heisenberg lectured on his new quantum mechanics at Cambridge less than three weeks after giving his first paper to Born at Göttingen, and actually a day before it was submitted for publication to *Zeitschrift für Physik*. On July 28, Heisenberg spoke at the Kapitza Club in front of an audience that included R.H. Fowler (but NOT Dirac – perhaps deterred by Heisenberg's title 'Termzoologie und Zeemanbotanik'). Fowler was intrigued and asked for proofs of the paper. Not until early September did Dirac become aware of Heisenberg's work, when Fowler suggested that he think about it. In a 1961 interview with van der Waerden [1] Dirac recalled: "At first I could not make much of it, but after two weeks I saw that it provided the key to the problem of quantum mechanics. I proceeded to work it out for myself ··· (based on) Transformation Theory of Hamiltonian Mechanics from lectures by R.H. Fowler and from Sommerfeld's book ···."

Dirac's later recollection [Segre 2] demonstrates his solitary and reflective nature which seemed to produce the most profound ideas from the most ethereal of beginnings. He independently discovered the canonical commutation relations (already known to Born and Jordan [3]): "··· I used to take long walks on Sunday alone ··· the idea occurred to me that the commutator $AB - BA$ was very similar to the Poisson bracket of classical mechanics in the Hamiltonian form. That was an idea that I just jumped at ···. But ··· I did not know very well what was a Poisson bracket ··· it had slipped out of my mind ··· when the library opened, I checked what a Poisson bracket really is and found that it was as I had thought $\cdots$. This provided a very close connection between ordinary classical mechanics and the new mechanics involving the noncommuting quantities introduced by Heisenberg. $\cdots\cdots$ I continued developing this work and Heisenberg and the people working with him developed the matrix point of view in Göttingen independently; we had some correspondence but we were working essentially independently."[4]

Dirac's work was directly based on Heisenberg's rule for the multiplication of quantum variables, already (unknown to Dirac) recognized by Born and soon to be published in great detail by Born and Jordan [3]. But Dirac in what became typical fashion cut through their many detailed arguments, especially those of Jordan, to recognize the basic operation of differentiation of quantum variables to be the commutator of the original quantum variable with another quantum variable. Then he reversed the argument of Born [3] (immediately preceding Eqn37 of Born and Jordan) to prove the connection between the commutator product and the classical Poisson bracket (immediately preceding Dirac's Eqn10) in the correspondence limit of large quantum numbers. Dirac immediately generalized this result by "the fundamental assumption that *the* $\cdots$ *(commutator)* $\cdots$ *of two quantum quantities is equal to $i\hbar$ times their Poisson bracket*" $\cdots$. In this, Dirac demonstrated a complete mastery of Action-angle variables (from Fowler's lectures and Sommerfeld's *Atombau und Spektrallinien*) and their most recent application by Kramers and Heisenberg [5].

In this and a follow-up paper [6], Dirac replicated many of the results of Born, Heisenberg and Jordan on the connection of canonical Heisenberg quantum mechanics to classical mechanics. In all of these, and in another great result on the hydrogen boundstate spectrum in which he was scooped by Pauli [7] by a matter of five days, Dirac made great contributions – always in his inimitably elegant and succinct way – but always finished second to the Göttingen cohort. Whether this was a motivation or not, Dirac's career of singular and incomparable contributions was only beginning.

Footnotes and References:

1) B.L. van der Waerden, *Sources of Quantum Mechanics* (North Holland, Amsterdam, 1967), Pp.40-43, Pp.58-59.

2) E. Segre, *From X-Rays to Quarks* (Freeman, New York, 1980), p.158.

3) M. Born and P. Jordan, Zeits. f. Phys. **34**, 858 (1925), here as **Paper VII·1**. Born and Jordan use the result of Kramers and Heisenberg (Eqn32 of [5] below), which in turn has its roots in an earlier paper by Born (M. Born, Zeits. f. Physik, **26**, 379 (1924), Eqns30,33; available in Segre (loc. cit. p.191,192)) in which Heisenberg was an acknowledged assistant but not a co-author.

4) See Segre's (loc. cit., p.322) reference to P.A.M. Dirac's "Recollections of an exciting era," in *History of 20th Century Physics, Proceedings of the International School of Physics, Course 57* (Academic Press, New York, 1977), and many other primary sources also referenced here.

5) H.A. Kramers and W. Heisenberg, Zeits. f. Phys. **31**, 681 (1925) (here as **Paper V·2**).

6) P.A.M. Dirac, Proc. Roy. Soc. **A110**, 561 (1926). (Available in Segre, loc. cit., p.417.)

7) W. Pauli, Zeits. f. Phys. **36**, 336 (1926). (Available in Segre, loc. cit., p.387.)

Biographical Note on Dirac.(†)

Paul Adrien Maurice Dirac (1902-84) is a godlike figure whose name permeates modern physics. His contributions to the genesis of quantum mechanics were accomplished in complete isolation compared to the community of Bohr, Heisenberg, Born, Jordan, and Pauli. He worked in the solitary mode of a Schrödinger. His earliest contributions were prompted by Heisenberg's initial breakthrough but thereafter were certainly independent, although they were anticipated by a matter of days or weeks by parallel achievements of Heisenberg, of Born and Jordan, of Pauli, and of Fermi. However Dirac's insights were distinguished by such clarity, originality, and beyond everything by such elegance that he was immediately recognized as a founder of the quantum theory. His first completely unique contribution was his creation of the Dirac equation for the relativistic electron.

Dirac studied electrical engineering (1919-21) and applied mathematics at Bristol; was an Exhibitioner and research student at St John's College, Cambridge (1923-25); made his reputation with his first papers establishing the fundamental principles of quantum mechanics, and demonstrating the equivalence of the Schrödinger and the Heisenberg formulations (1925,1926); he was also (and in every case independently) cofounder of Fermi-Dirac statistics (1926); coinventor with Heisenberg of wave function symmetrization for identical particles (1926); coinventor, with Born, Heisenberg, and – in particular – Jordan, of quantum field theory (1927); he discovered the Dirac equation (1928); invented hole theory (1931); predicted the positron and antiproton (1933). Dirac wrote *The Principles of Quantum Mechanics* (1930); became Fellow of the Royal Society (1930); Lucasian

Professor of Mathematics (1932-69), the chair once held by Newton; won the Nobel Prize (1933); Royal Medal of the Royal Society (1939); Copley Medal (1952); Max Planck Medal (1952), previously awarded only to Planck and Einstein; Order of Merit (1973); became Professor of Physics, Florida State University (1971-84).

Dirac's later work, although not popularly known, has had a deep impact on modern theoretical physics. Goddard and Taylor[††] cite the quantum theory of magnetic monopoles, the dynamics of constrained systems, extended models of elementary particles, and steps toward a quantum theory of gravity as all having great importance fifty years after their publication. Dirac's speculative idea on the role of the Lagrangian in quantum mechanics was brought to fruition only after fifteen years as Feynman's Path Integral formulation of quantum mechanics. This is the most powerful formulation of quantum field theory, and is the only viable way to handle the complexities and constraints of modern theories from non-abelian gauge theories through super-symmetric theories to super-string theories.

Dirac's elegance manifested itself in an economy of words which made his written work difficult in an opposite sense to that of Bohr. But his writing was practically effusive compared to his conversation, where he usually seemed content with his own thoughts and internalized those. One story has him on the occasion of leaving on a voyage to Japan, refusing the gift of a book with the words: "Reading precludes thought."

(†– The London Times, 23 Oct 1984, p10e; 25 Oct 1984, p20g; ††– P. Goddard and J.C. Taylor, The London Times, 15 Nov 1984, p18g.)

Paper VIII·1: Excerpt from Proc. Roy. Soc. A**109**, 642 (1925).

The Fundamental Equations of Quantum Mechanics

P.A.M. Dirac, 1851 Exhibition Senior Research Student,
St. John's College, Cambridge.

(Communicated by R.H. Fowler, F.R.S. – Received November 7th, 1925.)

§ 1. Introduction.

The experimental facts of atomic physics require a change from classical theory for their description. In Bohr's theory, this change requires: a) the special assumption of stationary states of the atom, in which it does not radiate; and

b) certain 'quantum conditions', rules which fix the stationary states and the frequencies of the radiation during transitions between them. These assumptions are foreign to classical theory, but have been very successful in the interpretation of many atomic phenomena. The only way the classical theory enters is: a) through the assumption that the classical laws hold for the motion in the stationary states, although they fail completely during transitions; and b) by the assumption of the Correspondence Principle, that classical theory gives the right results in limiting cases when Planck's constant $\hbar$ can be considered sufficiently small.

Recently Heisenberg [1] has proposed a new theory based on the premise that it is not the equations of classical mechanics that are at fault, but that the mathematical operations by which physical results are deduced from them require modification. *All* the information supplied by classical theory can then be made use of in the new theory.

§ 2. Quantum Algebra.

Consider a multiply periodic non-degenerate dynamical system of n degrees of freedom, defined by equations connecting the coordinates and their time derivatives. We solve the problem in classical theory in the following way: Assume that each of the coordinates x can be expanded in the form of a multiple Fourier series in the time, thus

$$x = \sum_{\alpha_1 \cdots \alpha_n} x(\alpha_1 \cdots \alpha_n) \cdot e^{i(\alpha_1 \omega_1 + \alpha_2 \omega_2 + \cdots + \alpha_n \omega_n)t} \equiv \sum_\alpha x_\alpha e^{i(\alpha \cdot \omega)t}.$$

Substitute these values in the equations of motion, and equate to zero the total coefficient of each harmonic term. The resulting equations (the A equations) determine the amplitudes x_α and the frequencies $(\alpha \cdot \omega)$ (measured in radians per unit time). The solution is not unique. There is an n-fold infinity of solutions labeled by n constants $\kappa_1 \cdots \kappa_n$. Each x_α and $(\alpha \cdot \omega)$ is now a function of two sets of numbers, the α's and the κ's, and can be written $x_{\alpha\kappa}, (\alpha \cdot \omega)_\kappa$.

In the quantum problem, Heisenberg assumes that each coordinate can be represented by harmonic components of the form $\exp \cdot i\omega t$, whose amplitude and frequency each depend on two sets of integers $j_1 \cdots j_n$ and $k_1 \cdots k_n$, and written $x(jk), \omega(jk)$. The differences $j_r - k_r$ correspond to the previous α_r, but neither the

j's nor any function of the j's and k's play the part of the previous κ's in specifying to which solution each particular harmonic component belongs. For example, we cannot take together all the components for which all the j's have a given set of values, and say that these by themselves form a single complete solution of the equations of motion. The quantum solutions are all interlocked, and must be considered as a single whole. The result of this is that, although in classical theory each of the A equations relates amplitudes and frequencies having one particular set of κ's, the amplitudes and frequencies occurring in the quantum A equation do not have one particular set of values for the j's, or for any function of the j's and k's, but have their j's and k's related in a special way, which will appear later. [Note added: This is a not very illuminating description of the seemingly impossible Gordian knot which Heisenberg succeeded in untangling. Dirac reiterates Heisenberg's prescription before beginning his own new considerations in §3.]

In classical theory we have the obvious relation

$$(\alpha \cdot \omega)_\kappa + (\beta \cdot \omega)_\kappa = (\{\alpha + \beta\} \cdot \omega)_\kappa.$$

Following Heisenberg, the corresponding relation in quantum theory is

$$\omega(j, j - \alpha) + \omega(j - \alpha, j - \alpha - \beta) = \omega(j, j - \alpha - \beta)$$

$$\text{or } \omega(jm) + \omega(mk) = \omega(jk). \tag{1}$$

This means that $\omega(jk)$ has the form $\Omega(j) - \Omega(k)$, the Ω's being frequency levels. In Bohr's theory these would be $1/\hbar$ times the energy of the levels, but we do not need to assume this.

In classical theory two harmonic components with the same κ's multiply as:

$$a_{\alpha,\kappa} \cdot e^{i(\alpha \cdot \omega)_\kappa t} \times b_{\beta,\kappa} \cdot e^{i(\beta \cdot \omega)_\kappa t} = (ab)_{\alpha+\beta,\kappa} \cdot e^{i(\{\alpha+\beta\} \cdot \omega)_\kappa t} \text{ where}$$

$$(ab)_{\alpha+\beta,\kappa} = a_{\alpha,\kappa} b_{\beta,\kappa} \quad \text{(classical)}.$$

In quantum theory an (nm) and an (mk) component multiply as:

$$a(nm) \cdot e^{i\omega(nm)t} \times b(mk) \cdot e^{i\omega(mk)t} = ab(nk) \cdot e^{i\omega(nk)t} \text{ where}$$

$$ab(nk) = a(nm)b(mk) \quad \text{(quantum)}.$$

We are led to consider the product of the amplitudes of an (nm) and an (mk) component as an (nk) amplitude. This, together with the rule that only amplitudes labeled by the same pair of numbers can be added in an A equation, replaces the classical rule that all amplitudes in an A equation must have the same set of κ's.

We now define the ordinary algebraic operations on quantum variables. The sum of x and y is

$$\{x + y\}(nm) = x(nm) + y(nm) \quad \text{(quantum or classical)}$$

and the product is

$$xy(nm) = \sum_k x(nk)y(km) \quad \text{(quantum)} \tag{2}$$

similar to the classical product

$$(\alpha \cdot xy)_\kappa = (xy)_{\alpha,\kappa} = \sum_r x_{r,\kappa} y_{\alpha-r,\kappa} \quad \text{(classical)}.$$

An important difference between the two algebras is that, in general,

$$xy(nm) \neq yx(nm) \quad \text{(quantum)}$$

and quantum multiplication is not commutative, although it is associative and distributive. The quantity (2) with components $xy(nm)$ we call the Heisenberg product of x and y, and write simply as xy. Whenever two quantum quantities are multiplied, the Heisenberg product is understood. Ordinary multiplication is, of course, implied in the products of other quantities that are related to sets of n's which are explicitly stated.

The reciprocal of a quantum quantity x is defined by either of the equivalent relations $\{1/x\} \cdot x$ or $x \cdot \{1/x\} = 1$. In a similar way the square root of x is defined by $\sqrt{x} \cdot \sqrt{x} = x$. It is not obvious that there are always solutions $\cdots\cdots$.

We can now generalize the classical equations of motion to quantum theory provided we can decide the correct order of the factors in each of the products.

Any equation derivable from the equations of motion by algebraic processes not involving the interchange of the factors of a product, and by differentiation and integration with respect to t, can also be taken over into quantum theory. In particular, the conservation of energy equation remains valid.

The equations of motion do not suffice to solve the quantum problem. Even in classical theory the equations of motion do not determine the $x_{\alpha\kappa}$ and $(\alpha \cdot \omega)_\kappa$ until we define the κ's. We could choose the κ's such that $\frac{\partial E}{\partial \kappa_r} = w_r$, where E is the energy of the system. This would identify the κ_r as the action variables J_r. The corresponding equations in quantum theory are the quantization conditions.

§ 3. Quantum Differentiation.

So far the only differentiation we have considered in quantum theory is with respect to the time t. We now determine the most general quantum operation d/dv that satisfies the laws

$$(\text{I}) \quad \frac{d}{dv}(x + y) = \frac{d}{dv}x + \frac{d}{dv}y, \text{ and (II) } \frac{d}{dv}(xy) = \frac{d}{dv}x \cdot y + x \cdot \frac{d}{dv}y.$$

(Note that the order of x and y is preserved in the last equation.)

Law (I) requires the components of dx/dv to be linear functions of x, *i.e.*,

$$\frac{dx}{dv}(nm) = \sum_{n'm'} a(nm; n'm')x(n'm'). \tag{5}$$

There is one coefficient $a(nm; n'm')$ for each set of integral values of n, m, n', m'. Law (II) imposes conditions on the a's. Substituting from (5) into (II) and equating the nm components on either side, gives

$$\sum_{n'm'k} a(nm; n'm')x(n'k)y(km') = \sum_{kn'k'} a(nk; n'k')x(n'k')y(km)$$
$$+ \sum_{kk'm'} x(nk)a(km; k'm')y(k'm').$$

This must be true for all x and y, so equating the coefficients of $x(n'k)y(k'm')$ on either side, we get

$$\delta_{kk'}a(nm; n'm') = \delta_{mm'}a(nk'; n'k) + \delta_{nn'}a(km; k'm').$$

* * * * * * * * * * * * * * * * * * **

Setting $k = k'$, we already see that the $a(nm; n'm')$ have $m = m'$ or $n = n'$ or both. Matrix differentiation thus defined has the form of an *ordinary* matrix product

$$\frac{dx}{dv}(nm) = \sum_k A(nk)x(km) + x(nk)B(km)$$

with one (unsummed) index carried by the matrix x being differentiated. We have yet to see that $A = -B$, and that the derivative of a quantum variable x is the commutator x with some other quantum variable corresponding to v. v itself is not yet specified, nor is there yet a prescription to find the corresponding A. Dirac made the purest prescription that *all* operations must be definable as matrix operations of addition and multiplication, as distinct, for example, from ordinary differentiation. He was including differentiation of one quantum variable with respect to another. For example, if v is a space coordinate, then $A = -B = iP$, the momentum operator which generate the corresponding translations. He included ordinary differentiation with respect to the non-dynamical non-quantum quantity *time* as corresponding to $A = H = $ diagonal. The distinction between the time as a non-dynamical variable and the space coordinates as dynamical variables is not made.

* * * * * * * * * * * * * * * * * * **

To proceed further, consider the various cases:

1) $k = k', m = m', n \neq n'$. This gives $a(nm; n'm) = a(nk; n'k)$, so $a(nm; n'm)$ is independent of m for $n \neq n'$. Similarly

2) $k = k', m \neq m', n = n'$ gives $a(nm; nm')$ independent of n for $m \neq m'$. Next

3) $k \neq k', m = m', n = n'$ gives

$$a(nk'; nk) = -a(km; k'm) \equiv a(kk') \tag{6}$$

depending only on k, k', for $k \neq k'$. The last case

4) $k = k', m = m', n = n'$ gives $a(nm; nm) = a(nk; nk) + a(km; km)$. This means

we can put

$$a(nm; nm) = a(mm) - a(nn), \tag{7}$$

which completes (6) by defining $a(kk')$ when $k = k'$. Eqn 5 now reduces to

$$
\begin{aligned}
\frac{dx}{dv}(nm) &= \sum_{m' \neq m} a(nm; nm')x(nm') + \sum_{n' \neq n} a(nm; n'm)x(n'm) \\
&\qquad + a(nm; nm)x(nm) \\
&= \sum_{m' \neq m} a(m'm)x(nm') - \sum_{n' \neq n} a(nn')x(n'm) \\
&\qquad + \{a(mm) - a(nn)\} x(nm) \\
&= \sum_{k} \{x(nk)a(km) - a(nk)x(km)\},
\end{aligned}
$$

and finally

$$\frac{dx}{dv} = xa - ax. \tag{8}$$

Thus the most general differentiation operation satisfying laws (*I*) and (*II*) that can be performed on a quantum variable is that of taking the difference of its Heisenberg products with some other quantum variable. It is easily seen that one cannot in general change the order of differentiation, *i.e.*,

$$\frac{d^2x}{du\,dv} \neq \frac{d^2x}{dv\,du}.$$

As an example in quantum differentiation, consider the case when a is a constant, so $a(nm) = 0$ unless $n = m$. Then

$$\frac{dx}{dv} = x(nm)a(mm) - a(nn)x(nm).$$

If $ia(mm) = \Omega(m)$, the frequency level previously introduced, then

$$\frac{dx}{dv}(nm) = i\omega(nm)x(nm),$$

and the quantum differentiation becomes ordinary differentiation with respect to t. [Note added: With $x(nm) \sim \exp \cdot i\omega(nm)t$.]

* * * * * * * * * * * * * * * * * * **

Dirac seems to have misspoken here. Instead of '··· consider the case when a is a *constant* ···', he should have said '······ when a is *diagonal* ···'.

* * * * * * * * * * * * * * * * * **

§ 4. The Quantum Conditions.

We shall now consider what the expression $(xy - yx)$ corresponds to in classical theory. To do this we suppose that $x(n, n - \alpha)$ varies only slowly with the n's, the n's being large numbers and the α's small ones, so that we can put $x(n, n-\alpha) \simeq x_{\alpha\kappa}$ where $\kappa_r = n_r \hbar$ or $(n_r + \alpha_r)\hbar$, these being practically equivalent. We now have

$$
\begin{aligned}
(xy - yx) &= \\
&x(n, n - \alpha)y(n - \alpha, n - \alpha - \beta) - y(n, n - \beta)x(n - \beta, n - \alpha - \beta) \\
&= \{x(n, n - \alpha) - x(n - \beta, n - \beta - \alpha)\}\, y(n - \alpha, n - \alpha - \beta) \\
&\quad - \{y(n, n - \beta) - y(n - \alpha, n - \alpha - \beta)\}\, x(n - \beta, n - \alpha - \beta), \\
&\simeq \qquad \hbar \sum_r \left\{ \beta_r \frac{\partial x_{\alpha\kappa}}{\partial \kappa_r} y_{\beta\kappa} - \alpha_r \frac{\partial y_{\beta\kappa}}{\partial \kappa_r} x_{\alpha\kappa} \right\}.
\end{aligned}
\tag{9}
$$

* * * * * * * * * * * * * * * * * * * **

Here Dirac borrows the rules (in reverse) from Kramers and Heisenberg [2] (who in turn had borrowed them unacknowledged from Born (1924)) for replacing quantum differences by classical derivatives in a correspondence principle limit. Next – again following Kramers and Heisenberg – Dirac uses the Action-Angle variable representation from the classical theory of light-atom scattering.

* * * * * * * * * * * * * * * * * * **

Now

$$
i\beta_r \left\{ y_\beta \cdot e^{i(\beta \cdot \omega)t} \right\} = \frac{\partial}{\partial \theta_r} \left\{ y_\beta \cdot e^{i(\beta \cdot \omega)t} \right\}
$$

where the θ_r are angle variables, equal to $\omega_r t$. The (nm) component of $(xy - yx)$ corresponds in classical theory to

$$
-i\hbar \sum_{\alpha+\beta=n-m} \sum_r \times
$$

$$\left\{ \frac{\partial}{\partial \kappa_r} \left\{ x_\alpha e^{i(\alpha \cdot \omega)t} \right\} \frac{\partial}{\partial \theta_r} \left\{ y_\beta e^{i(\beta \cdot \omega)t} \right\} - \frac{\partial}{\partial \kappa_r} \left\{ y_\beta e^{i(\beta \cdot \omega)t} \right\} \frac{\partial}{\partial \theta_r} \left\{ x_\alpha e^{i(\alpha \cdot \omega)t} \right\} \right\},$$

so $(xy - yx)$ corresponds to

$$(xy - yx) \simeq -i\hbar \sum_r \left\{ \frac{\partial x}{\partial \kappa_r} \frac{\partial y}{\partial \theta_r} - \frac{\partial y}{\partial \kappa_r} \frac{\partial x}{\partial \theta_r} \right\}.$$

If we take κ_r to be the action variable J_r, this is $i\hbar$ times the Poisson bracket

$$[x, y] = \sum_r \left\{ \frac{\partial x}{\partial \theta_r} \frac{\partial y}{\partial J_r} - \frac{\partial y}{\partial \theta_r} \frac{\partial x}{\partial J_r} \right\} = \sum_r \left\{ \frac{\partial x}{\partial q_r} \frac{\partial y}{\partial p_r} - \frac{\partial y}{\partial q_r} \frac{\partial x}{\partial p_r} \right\}$$

where the p's and q's are any set of canonical variables of the system.

The fundamental Poisson brackets for various combinations of the p's and q's are

$$[q_r, q_s] = [p_r, p_s] = 0, \quad [q_r, p_s] = \delta_{rs} (= 1 \text{ if } r = s, = 0 \text{ if } r \neq s). \tag{10}$$

The general bracket expressions satisfy the laws *I* and *II*, which now read

$$(IA): [x, z] + [y, z] = [x + y, z], \quad (IIA): [xy, z] = [x, z]y + x[y, z].$$

$\cdots$ with $[x, y] = -[y, x]$. If x and y are given as algebraic functions of the p_r and q_r, $[x, y]$ can be expressed in terms of the $[q_r, q_s]$, $[p_r, p_s]$ and $[q_r, p_s]$, and thus evaluated without using the commutative law of multiplication $\cdots$. The bracket $[x, y]$ still has a meaning in quantum theory when x and y are quantum variables, provided the fundamental brackets are still given by (10).

We make the fundamental assumption that *the difference between the Heisenberg products of two quantum quantities is equal to $i\hbar$ times their Poisson bracket:*

$$(xy - yx) = i\hbar[x, y]. \tag{11}$$

This is equivalent – in the limiting case of classical theory – to taking the arbitrary quantities κ_r that label a solution to be the J_r. It seems reasonable to take (11) as the general quantum conditions.

It is not obvious that all the information in (11) is consistent. Because the quantities on either side of (11) satisfy the laws IA and IIA, the only independent conditions given by (11) are those for which x and y are p's and q's, namely

$$q_r q_s - q_s q_r = 0, \quad p_r p_s - p_s p_r = 0, \text{ and } q_r p_s - p_s q_r = i\hbar \delta_{rs}. \tag{12}$$

If the only grounds for believing that Eqns12 were consistent with each other and with the equations of motion were that they are known to be consistent in the limit when $\hbar \to 0$, the case would not be very strong, since one might be able to deduce that $\hbar = 0$, which would not be an inconsistency in the limit. There is much stronger evidence than this, however, because the classical operations obey the same rules as the quantum ones. If one can get an inconsistency by applying the quantum operations, then by applying the classical operations in the same way one must also get an inconsistency. If a series of classical operations leads to the equation $0 = 0$, the corresponding series of quantum operations must also lead to the equation $0 = 0$, and not $\hbar = 0$. There is no way of obtaining a quantity that *does not vanish* by a quantum operation with quantum variables if the corresponding classical operation with the corresponding classical variables gives a quantity that *does vanish*. The possibility of deducing by quantum operations the inconsistency $\hbar = 0$ therefore cannot occur. The *correspondence between quantum and classical theories lies not so much in the limiting agreement when $\hbar \to 0$ as in the fact that the fundamental mathematical operations in the two theories obey the same laws of (10,12).*

For a system of one degree of freedom, if we take $p = m\dot{q}$, the only quantum condition is $m(q\dot{q} - \dot{q}q) = i\hbar$. Equating the constant part of the left-hand side to $i\hbar$, we get $2m \sum_k q(nk)q(kn)\omega(nk) = \hbar$. This is equivalent to Heisenberg's quantum condition. By equating the remaining components on the left-hand side to zero we get further relations not given by Heisenberg's theory.

The quantum conditions (12) overcome – in many cases – the difficulty of the order in which factors in products in the equations are to be taken. The order does not matter except when a p_r and a q_r are multiplied together, and this never occurs in a system describable by a potential energy function that depends only on the q's and a kinetic energy function that depends only on the p's.

It should be pointed out that the classical quantity in Kramers and Heisenberg's theory of scattering by atoms [2] has components of the form (8) (with $\kappa_r = J_r$) which are interpreted in their theory in agreement with the present theory. No classical expression involving differential coefficients can be interpreted in quantum theory unless it can be put in this form.

§ 5. Properties of the Quantum Poisson Bracket.

In this section we deduce certain results that are independent of the assumption of the quantum conditions (11) or (12).

In classical theory, the Poisson brackets satisfy the identity

$$[x, y, z] \equiv [[x, y], z] + [[y, z], x] + [[z, x], y] = 0. \tag{13}$$

In quantum theory, this result is obviously true when x, y, and z are expressible in any way as sums and products of p's and q's, so it must be generally true. Note that the identity (13) with the Poisson brackets replaced by the difference of Heisenberg products $(xy - yx)$ is obviously true, so there is no unconsistency with (11).

If H is the Hamiltonian of the system, the equations of motion can be written classically $\dot{p}_r = [p_r, H] \quad \dot{q} = [x, H]$. These equations will be true in quantum theory if the orders of factors in products in the equations of motion are unimportant. They can be taken to be true if these orders are important, but one can decide the orders of the factors in H. From IA and IIA it follows that

$$\dot{x} = [x, H] \tag{14}$$

in quantum theory for any x.

If A is an integral of the equations of motion in quantum theory, then $[A, H] = 0$. The action variables J_r must, of course, satisfy this condition. If A_1 and A_2 are two such integrals, then from (13) $[A_1, A_2] = \text{const}$ as in classical theory.

The classical conditions that a set of variables P_r, Q_r be canonical are

$$[Q_r, Q_s] = [P_r, P_s] = 0 \quad \text{and} \quad [Q_r, P_s] = \delta_{rs}.$$

These conditions can be taken over into quantum theory as the conditions for the quantum variables P_r, Q_r to be canonical.

In classical theory we can introduce the set of canonical variables ξ_r, η_r related to the uniformizing variables J_r, θ_r by $\xi_r = \sqrt{J_r} \cdot e^{i\theta_r}$, and $\eta_r = -i\sqrt{J_r}e^{-i\theta_r}$.

Presumably there will be a corresponding set of canonical variables in quantum theory, each containing only one kind of component, so that $\xi_r(nm) = 0$ unless $m_r = n_r - 1$ and $m_s = n_s(s \neq r)$, and $\eta_r(nm) = 0$ except when $m_r = n_r + 1$ and $m_s = n_s(s \neq r)$. One can consider the existence of such variables as the condition for the system to be multiply periodic in quantum theory. The components of the Heisenberg products of ξ_r and η_r satisfy the relation

$$\xi_r\eta_r(nn) = \xi_r(nm)\eta_r(mn) = \eta_r(mn)\xi_r(nm) = \eta_r\xi_r(mm) \tag{15}$$

where the m's are related to the n's by $m_r = n_r - 1$, $m_s = n_s(s \neq r)$.

The classical ξ's and η's satisfy $\xi_r\eta_r = -i \cdot J_r$. This relation need not hold between the quantum ξ's and η's. The quantum relation could be, *e.g.*, $-i \cdot J_r = \eta_r\xi_r$, or $= (\xi_r\eta_r + \eta_r\xi_r)/2$. A detailed investigation of any particular dynamical system is necessary in order to decide what it is. If the last relation is true, we can introduce the canonical variables

$$\xi_r' = (\xi_r + i\eta_r)/\sqrt{2}, \quad \eta_r' = (i\xi_r + \eta_r)/\sqrt{2},$$

and then $J_r = (\xi_r'^2 + \eta_r'^2)/2$. This is the case that actually occurs in the harmonic oscillator. In general J_r is not necessarily even a rational function of the ξ_r and η_r, an example of this being the rigid rotor considered by Heisenberg.

§ 6. The Stationary States.

A quantity C that does not vary with the time, has all its (nm) components zero except those with $n = m$. It is convenient to assign each set of n's to a definite state of the atom, as in Bohr's theory. Each $C(nn)$ belongs to that certain state in precisely the same way that *every* quantity occurring in classical theory belongs to that certain configuration. The components of a varying quantum quantity are so interlocked, however, that it is impossible to associate them with a given state.

When all the quantities are constants, a relation between quantum quantities reduces to a relation between $C(nn)'s$ belonging to a definite stationary state n. This relation will be the same as the classical theory relation, assuming the classical laws hold for stationary states; in particular, the energy will be the same function of the J's as in classical theory. This is justification for Bohr's assumption of the mechanical nature of the stationary states. It should be noted though, that the variable quantities associated with a stationary state in Bohr's theory – the amplitudes and frequencies of orbital motion – have no physical meaning and are of no mathematical importance.

With x and H in (11), we get using (14)

$$x(nm)H(mm) - H(nn)x(nm) = i\hbar \cdot \dot{x}(nm) = -\hbar \cdot \omega(nm)x(nm),$$

or $H(nn) - H(mm) = \hbar \cdot \omega(nm)$. This is just Bohr's relation connecting the frequencies with the energy differences.

The quantum condition (11) applied to the canonical variables ξ_r, η_r gives $(\xi_r\eta_r - \eta_r\xi_r)(nn) = i\hbar \cdot [\xi_r, \eta_r] = i\hbar$. This equation combined with (15) shows that

$$(\xi_r\eta_r)(nn) = -n_r i\hbar + \text{ const.} \tag{16}$$

It is known physically that an atom has a ground state which does not radiate. This is taken into account by Heisenberg's assumption that all amplitudes $C(nm)$ having a negative n_r or m_r vanish, or rather do not exist, if we take the ground state to have every n_r equal zero. This makes $\xi_r\eta_r(nn) = 0$ when $n_r = 0$ on account of (15); so the constant in (16) must be zero in general.

If $\xi_r\eta_r = -i \cdot J_r$, then $J_r = n_r\hbar$ – just the ordinary rule for quantizing the stationary states – so in this case the frequencies are the same as in Bohr's theory. If $\frac{1}{2}(\xi_r\eta_r + \eta_r\xi_r) = -i \cdot J_r$, then $J_r = (n_r + \frac{1}{2})\hbar$. In this case, half-integral quantum numbers would have to be used to get the correct frequencies by Bohr's theory.

So far we have considered only multiply periodic systems. There does not seem to be any reason why the fundamental equations (11) and (12) should not apply even to non-periodic systems $\cdots$ such as a general atom. One might not expect the stationary states of such a system to be simply classifiable, except perhaps when there are pronounced periodic motions, so one would have to assign a separate number n to each stationary state according to some plan. Our quantum variables would still have harmonic components each related to two n's, and Heisenberg multiplication could be carried out exactly as before. There would then be no ambiguity in the interpretation of (12) or of the equations of motion.

I would like to express my thanks to Mr R.H. Fowler, F.R.S., for many valuable suggestions in the writing of this paper.

Footnotes and References:

1) W. Heisenberg, Zeits. f. Phys. **33**, 879 (1925).
2) H. Kramers and W. Heisenberg, Zeits. f. Phys. **31**, 681 (1925), Eqn18.

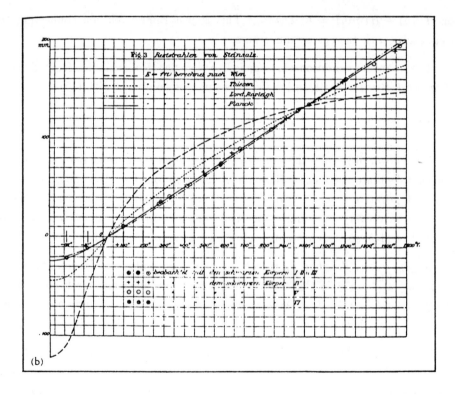

(b)

Original data analysis of their Blackbody data by Rubens and Kurlbaum validating Planck's equation. Emission vs. temperature at fixed frequency. [From F. Rubens and F. Kurlbaum, *Ann. d. Phys.* **4**, 649 (1901) in E. Segrè, From X-Rays to Quarks (Freeman, NY, 1980), p.69.]

Chapter IX

SCHRÖDINGER:

"Dann soll ψ das 'Hamiltonsche Integral'

$$\int d\tau \left\{ K^2 T \left(q, \frac{\partial \psi}{\partial q} \right) + \psi^2 V \right\}$$

stationär machen unter der normierenden Nebenbedingung

$$\int \psi^2 d\tau = 1.$$

Die Eigenwerte dieses Variationproblems liefern die Quanten-niveaus der Energie."

"Then ψ should make *stationary* 'Hamilton's Integral' $\cdots$ subject to the *normalization restriction* $\cdots$. The *eigenvalues* of this variational problem $\cdots$ give $\cdots$ the *quantum energy-levels*." *Ann. d. Phys.* **79**, *361 (1926); p.376.*

§ IX-1. Introduction.

Schrödinger's invention of wave mechanics was the saving grace which made the whole subject of quantum mechanics intuitively accessible for the vast majority even of physicists; to say nothing of practitioners in other fields such as chemistry, biology, engineering; and especially, for all non-scientists who want only a cultural glimpse – not necessarily an understanding – of the subject.

At the time, Schrödinger's great achievement was immediately hailed by the older generation of physicists – led by Planck and Einstein – who welcomed a formulation of quantum mechanics in the familiar language of differential equations and continuous ψ-fields whose square somehow corresponded to the electron charge distribution. Even Max Born (see §9 of **Paper X·2**) paid tribute to Schrödinger's formulation as "the easiest way by far $\cdots$ to follow $\cdots$ an event". In his own work – and he had just finished the 'Dreimännerarbeit' – Born found Schrödinger's wave mechanics essential for a solution of the scattering problem, and was led to rudiments of the 'Copenhagen'-interpretation of the ψ-function.

Schrödinger believed (in the way many still think for 'practical' quantum mechanics) in the wave function as a classical wave describing an actual continuous distribution of the electron in a stationary state. Transition amplitudes – the dipole matrix elements – were somehow 'beats' between mixed wave forms which lead continuously from one to the other, with continuous radiation at the beat frequency. This was disputed by Heisenberg in a seminar given by Schrödinger at Munich (late summer, 1926) [1]: "I pointed out that Schrödinger's conception would not even explain Planck's radiation law. $\cdots$ I was taken to task by Wien – '$\cdots$ he understood my regrets that quantum mechanics was finished $\cdots$ all such nonsense as quantum jumps $\cdots$ would soon be solved by Schrödinger'. "

After the dramatic success of his wave equation and the praise from such as Planck and Einstein who saw this as a return to a form of classical objective reality, Schrödinger was immediately confronted with an intense debate from Bohr who rejected his semi-classical intuitive interpretation of the square of the wave function as the electron's charge distribution, as at best misleading. Segrè [2] describes the October 1926 exchange in Copenhagen, between Schrödinger alone on one side, and Bohr and Heisenberg – battling to save matrix mechanics from being supplanted by the too facile wave mechanics – on the other: "The protracted discussions were so laborious that Schrödinger fell ill from exhaustion and went to bed $\cdots$. This did not stop Bohr from continuing the argument with Schrödinger in his bedroom. $\cdots$ Bohr was a kind soul, polite and seriously concerned $\cdots$ but on such a vital problem of physics he could not control himself." They never reached an accord.

Heisenberg recalls [1] "$\cdots$ Bohr was normally most considerate and friendly $\cdots$ he now struck me as an almost remorseless fanatic $\cdots$ not prepared to make the least concession or to grant that he could ever be mistaken. $\cdots$ how passionate the discussions $\cdots$ how deeply rooted the convictions of each man." Heisenberg reconstructs the exchanges between Bohr and Schrödinger, beginning with Schrödinger: "$\cdots$ the whole idea of quantum jumps is $\cdots$ nonsense. $\cdots$ is this jump gradual or sudden? $\cdots$ we must ask ourselves precisely how the electron behaves during the jump. $\cdots$ what laws govern its motion during the jump? $\cdots$ sheer fantasy." and Bohr's fabulous response, reflecting his whole philosophy of far ranging speculation driven by incontrovertible facts, without regard to human emotional ties to any status quo: "What you say is absolutely correct. But it does not prove that there are no quantum jumps $\cdots$ only that we cannot imagine them $\cdots$"; and Schrödinger: "$\cdots$ leave that to the philosophers." And culminating at last with Schrödinger's outburst "If all this damned quantum jumping were really here to stay, I should be sorry I ever got involved with quantum theory." and Bohr's equally famed reply " But the rest of us are extremely glad that you did $\cdots$".

Bohr [3] remembers it somewhat differently: "$\cdots$ a special opportunity for a lively exchange of views. $\cdots$ Heisenberg and I tried to convince him that his beautiful treatment $\cdots$ could not be brought into conformity with Planck's Law $\cdots$" evidently he meant without the emission of discrete quanta.

Heisenberg again: "$\cdots$ we fully realized how difficult it would to persuade even leading physicists that they must abandon all attempts to construct perceptual models of atomic processes."

Schödinger himself, sad to say, was not psychologically prepared to continue the immense intellectual struggle that was now undertaken - led by Bohr and Heisenberg in a debate that is lively to this day. He spent the rest of his life trying to argue away "these damned quantum jumps", a subject which shows no sign of going away. And eventually he withdrew to less contentious subjects where he did not have to engage in the intellectually-bruising debates with the implacable Copenhagen school and their abstract argumentation.

§ IX-2. So Where Did Schrödinger Go Wrong?

If indeed he did go wrong. Because the debate has never been completely settled, and seems even to be increasing in volume. Over the years many impassioned pleas [e.g., 4], and cogent arguments [e.g., 5] have maintained that not only is the issue still open, but that with sufficient ingenuity even such standard quantum mechanical devices as the collapse of the wave function accompanying a quantum jump (and more; see Cramer, Cushing [5]) can be produced in a continuous classical view of emission and absorption. The prescriptions may seem excessively artful and contrived (involving for Cramer [5]: advanced and retarded radiation; the elimination of fields in favor of action at a distance; the return to concepts like Slater's 1923 virtual oscillators; and much more); but they are in some sense an alternative to standard quantum mechanics. Not ever – in our judgement, or probably even in the judgement of their proponents – a preferable alternative, but simply *an* alternative; which renders the abstract 'Copenhagen' interpretation of standard quantum mechanics – originating with Bohr and Heisenberg's domination of Schrödinger in their 1926 debate – as sufficient but *not necessary*.

The principal *bête noire* of this school of thought – and there are many – is the 'collapse of the wave function' during a measurement. For Schrödinger – who believed the wavefunction to be a classical field whose square (appropriately normalized) gives, say, the real physical charge distribution of an individual electron in an individual atom – there was no place for such an uncontrolled collapse. There was no quantum jump. A $2P$ electron emitted electromagnetic radiation in a $(2P \to 1S)$-transition in smooth continuous way at the appropriate beat frequency $\hbar\omega = E_{2P} - E_{1S}$, and the wave function evolved continuously from ψ_{2P} to ψ_{1S}. This is in fact, *in its conclusion*, the correct calculation that one makes – semi-classically, wave mechanically, quantum mechanically, or field theoretically – to describe the transition, and it gives all the required numbers.

There *was* and *is* no criticism of Schrödinger's ability to calculate. The criticism leveled at Schrödinger by Bohr and Heisenberg was not for what he *did*, but for what he *believed*, or rather for what he did *not* believe.

Because Schrödinger did not believe that the photon was emitted in 'one piece', and that the emission process had only a 'before' and an 'after', and that the system 'existed' in only one of the two states, '$2P$-electron' or '$1S$-electron plus photon'. Rather he believed that there was an 'individual reality' to his description of the atom as being in some 'in-between' state with the emission being only partially completed. All this a result of Schrödinger insisting that his ψ be a classical 'physical' field.

Bohr and Heisenberg could not tolerate Schrödinger's semi-classical and continuous view of the photon emission process, because in their judgement – which obviously prevailed at the time, and has survived *every* test *ever* devised in generation after generation of refinements of experimental techniques – it contradicted the quantum hypothesis of Planck, and the photon hypothesis of Einstein. They therefore insisted on a 'before' or an 'after', but no 'in-between' [6]. So what *did* Schrödinger's correct calculation *mean* for the interpretation of Schrödinger's ψ?

Here then in brief is the orthodox (von Neumann) *Copenhagen Interpretation* of the wave function, with all its quibbles and qualifications which have resulted in its history of disbelief and rejection, amounting almost to an ongoing religious civil war between the unconverted and the (unthinking??) faithful true-believers. We recite the litany:

The Schrödinger wave function ψ_N is a probability amplitude referring to a *statistical ensemble* of N ($>> 1$) identical independent quantum systems identically prepared in the state $\psi(t = 0)$. As time progresses, $\psi(t)$ evolves according to the Schrödinger wave equation. At a later time T, the fraction F of the states in the ensemble which will be observed to be in some state ϕ is

$$F(T) = | \int \phi^* \psi(T)|^2$$

(with suitable normalizations). This fraction is the transition probability produced by Schrödinger's equation. Now we can repeat the process with the FN (still $>> 1$) systems in a new ensemble which evolve from the initial state $\psi_{FN}(t = T) = \phi$.

The evidently objectionable expression 'collapse of the wave function' might better be replaced by something like 'selection from the ensemble'.

The construction of the ensemble (the preparation of the N systems in the state ψ_N and the selection of those FN in some particular state ϕ) is *outside* the purview of the Schrödinger equation and in a some never-never land of 'measurement', yet to be discussed.

§ IX-3. An Expert Comments.

How should we feel about this circuitous and mysterious interpretation as a physical law?

Let's hear what E.P. Wigner had to say [7]: "··· desirable to scrutinize the orthodox view ··· look for loopholes ··· avoid the conclusions of the orthodox view" and even though he himself accepts the orthodox theory, he confesses that it "makes one uneasy." His conclusion is that orthodox quantum mechanics only gives probabilistic connections between successive observations. Anything else would require a change in the equations of quantum mechanics. He goes on "The assumption of two types of changes of the state vector is a strange ··· true dualism" and "the uncertainty ··· does not increase in time if the system is left alone··· chance enters when a measurement is carried out".

Wigner finally concludes that it is necessary to *postulate* a macroscopic apparatus for which " measurements which leave the system object-plus-apparatus in one of the states with a definite position of the pointer cannot be described by the linear laws of quantum mechanics." Wigner's key point seems to be that projection of the object-plus-apparatus wave function onto a 'state' of the apparatus characterized only by its macroscopic position, involves a nonlinear incoherent trace over a sensibly infinite number of apparatus quantum states, that is to say: an operation outside the linear Schrödinger evolution.

Wigner ultimately maintains that "The only possible question which can yet be asked is whether such a reduction must be postulated also when a measurement with a macroscopic apparatus is carried out. ··· this is true if the validity of quantum mechanics is admitted for all systems."

We leave the situation in this unsatisfactory state with the promise that we will revisit it, but the warning that we will not resolve it.

Footnotes and References.

1) W. Heisenberg, *Physics and Beyond* (Harper and Row, New York, 1971) as excerpted in *NIELS BOHR A Centenary Volume* (Harvard Press, Cambridge, 1985) A.P. French and P.J. Kennedy, Eds., Pp.163-166.
2) E. Segrè, *From X-rays to Quarks* (W.H. Freeman, New York, 1980), Pp.160-165.
3) N. Bohr, See Heisenberg, loc. cit., p.165. See also A. Pais, *Niels Bohr's Times* (Clarendon Press, Oxford, 1991), p.298, for reference to Schrödinger's immediate reaction (in

a letter to Wien) "Bohr, and especially Heisenberg, ⋯ totally, cloudlessly, amiable and cordial ⋯ Bohr talks ⋯ in a dreamlike, visionary and really quite unclear manner ⋯ full of consideration ⋯".

4) J. Dorling, *Schrödinger's original interpretation of the Schrödinger equation: a rescue attempt*, in SCHRÖDINGER (Cambridge University Press, Cambridge, 1987) C.W. Kilmister, Ed., p.16-41.

5) J.G. Cramer, Phys. Rev. **D22**, 362 (1980), and many discussions in Physics Today which can be traced from references in S. Goldstein, Physics Today **51**-3, 42 (1998). In this regard, see comments by W. Heisenberg, *The Development of the Interpretation of the Quantum Theory*, in NIELS BOHR and the Development of Physics (Pergamon Press, London, 1955) W.Pauli, Ed., Pp.12-29;298-385 (excerpted here as **Paper XI·2**).

6) The frequent resort to single quotation marks 'thus', is to remind us that each word must be carefully and meticulously defined. Here it *is* true that: "It all depends on what the meaning of 'is' is."

7) E.P. Wigner, *The Problem of Measurement*, Am.J.Phys. **31**-1, 6-15 (1963). Wigner was a junior member of the first generation of quantum theorists as co-inventor (with Jordan) of anticommutation relations for fermions, with many other achievements culminating in the 1963 Nobel Prize.

Paper IX·1: Excerpt from Annalen der Physik **79**, 361 (1926).

Quantization as an Eigenvalue Problem

von **E. Schrödinger**.

Eingegangen 27. Januar 1926.

Chapter One.

§ 1. In the simplest case of the hydrogen atom, I show that the usual quantization prescription arises from another requirement, in which the mention of 'whole numbers' no longer occurs. Furthermore, this new requirement produces discreteness in a completely natural way, just like the discreteness of the number of nodes of a vibrating string. The new understanding is generalizable and strikes very deeply at the real essence of the quantum prescription.

The customary form of this prescription is related to Hamilton's partial differential equation:

$$H\left(q, \frac{\partial S}{\partial q}\right) = E. \tag{1}$$

One usually looks for a solution for S in the form of a <u>sum</u> of functions, each of one individual independent variable q.

Instead, we introduce for S a new unknown ψ, where ψ is a <u>*product*</u> of well-behaved functions of the individual coordinates. That is, we set

$$S = K \log \psi, \tag{2}$$

where K must have the dimension of an *action*. Eqn1 becomes

$$H\left(q, \frac{K}{\psi} \frac{\partial \psi}{\partial q}\right) = E. \tag{1}$$

We do *not* solve Eqn1 directly, but consider the following problem: Eqn1 can always – at least for the *one*-electron problem – be put in the form of a quadratic function of ψ and its first derivatives, which equals zero. We seek those real functions ψ which are unique, finite, and twice differentiable, and which produce an *extremum* of the integral of the quadratic form [1] over the whole configuration space. *We replace the quantum conditions by this variation problem.*

First we choose for H the Kepler Hamiltonian, and show that the given requirements can be fulfilled for all positive E but only for a discrete set of negative

E-values. In this case then, the variation problem has both a continuous and a discrete eigenvalue spectrum. The discrete spectrum corresponds exactly to the Balmer series, the continuous spectrum to the hyperbolic orbits. For numerical agreement, K must be set equal to $\hbar$.

* * * * * * * * * * * * * * * * * * **

Schrödinger's paper contains the wave-mechanic solution of the hydrogen atom familiar to everyone. The derivation is based on the generalization of the Hamilton-Jacobi equation of classical mechanics. Schrödinger's wave-mechanical solution for the Balmer series followed by a mere ten days parallel work of Pauli (Zeits. f. Phys. **36**, 336 (1926)), and by five that of Dirac (Proc. Roy. Soc. A**110**, 561 (1926)). The invention of Schrödinger's wave equation was welcomed even by Max Born who acknowledged that it made calculations much easier than Heisenberg's approach. The *formal* equivalence of the two *theories* was immediately assumed and quickly proved by Schrödinger himself (here as **Paper IX·3**), among others.

* * * * * * * * * * * * * * * * * * **

In the Kepler problem, with e and m equal to the charge and mass of the electron, the quadratic form corresponding to (1) is:

$$\mathcal{L} \equiv \left(\frac{\partial \psi}{\partial x}\right)^2 + \left(\frac{\partial \psi}{\partial y}\right)^2 + \left(\frac{\partial \psi}{\partial z}\right)^2 - \frac{2m}{\hbar^2}\left(E + \frac{e^2}{r}\right)\psi^2 = 0, \qquad (1)$$

with $r = \sqrt{x^2 + y^2 + z^2}$.

* * * * * * * * * * * * * * * * * * **

Easily recognized as (the negative of) the Lagrangian density with a Lagrange multiplier E included to impose the normalization restriction, but not yet generalized to complex ψ.

* * * * * * * * * * * * * * * * * * **

The variation problem requires

$$\delta J = \delta \int \int \int dx\,dy\,dz\,\mathcal{L} = 0, \qquad (3)$$

where the integral extends over all space. In the usual way, one finds

$$
\frac{1}{2}\delta J = \int dS \cdot \frac{\partial \psi}{\partial n}\delta\psi
$$
$$
- \int\int\int dx\,dy\,dz \left[\nabla^2\psi + \frac{2m}{\hbar^2}\left(E + \frac{e^2}{r} \right)\psi \right]\delta\psi = 0. \tag{4}
$$

So first we must have

$$
\nabla^2\psi + \frac{2m}{\hbar^2}\left(E + \frac{e^2}{r} \right)\psi = 0, \tag{5}
$$

and second, the surface integral at infinity must vanish

$$
\int dS \cdot \frac{\partial \psi}{\partial n}\delta\psi = 0. \tag{6}
$$

(If it turns that we have to supplement this condition with a further restriction on the behavior of $\delta\psi$ at infinity, then the *continuous* eigenvalue actually exists. More on this later.)

The solution of (5) can be found in spherical coordinates r, θ, ϕ, by assuming ψ to be a *product* of functions of r, of θ, and of ϕ – the method is well known. The dependence on the polar angles is expressed by a *spherical harmonic function*; for the dependence on r – the function we call χ – we get the differential equation:

$$
\frac{d^2\chi}{dr^2} + \frac{2}{r}\frac{d\chi}{dr} + \left(\frac{2mE}{\hbar^2} + \frac{2me^2}{\hbar^2 r} - \frac{L(L+1)}{r^2} \right)\chi = 0, \tag{7}
$$

where $L = 0, 1, 2, 3 \cdots$. The restriction of L to an integer is necessary for the dependence on the polar angles to be well defined. – We need the solutions of (7) which are finite for all real non-negative r-values. Eqn7 has *two* singularities in the complex r-plane [2], at $r = 0$ and $r = \infty$, of which the second is an 'essential singularity' for all solutions, whereas the first is not. These two singularities are the endpoints of our interval. The requirement of *remaining finite* at the endpoint is equivalent to a boundary condition on the function χ. The equation *in general* has no solution which remains finite at *both* endpoints. Such solutions do exist, but only for special values of the constants appearing in the equation. The important determination of these *special* values is the *essential point* of this whole investigation.

* * * * * * * * * * * * * * * * * * * **

Schrödinger appears to have had all the requisite skills right at his fingertips. The situation has never been more clearly defined than it was at this first instance. Schrödinger acknowledges his debt to de Broglie (see [5], and here as **Paper IV·1**), but there was more to his development than simply plunking down a wave equation. He also gives the greatest thanks (in his Footnote [2]) to Herman Weyl "Für die Anleitung zur behandlung"– For the (instruction, guide, leading, conducting of (?!!)) *introduction to the analysis* – of Eqn7.

* * * * * * * * * * * * * * * * * * * **

First we examine the singular point $r = 0$. The *fundamental equation* which determines the behavior of the solution $\psi \sim r^\rho$ at this point is

$$\rho(\rho - 1) + 2\rho - L(L+1) = 0 \tag{8}$$

with the roots $\rho_1 = L$, $\rho_2 = -(L+1)$. The two solutions at $r = 0$ behave like r^L and $r^{-(L+1)}$, but only the first is useful because L is never negative. It will correspond to an ordinary power series of *increasing* exponents, which begins with r^L. (The other solution, which does not interest us, $\cdots$ can also contain a logarithm.) Since the next singular point is at $r = \infty$, the series must converge and produce a *transcendental function*. We have therefore established that: *The desired solution is a uniquely determined transcendental function which behaves at $r = 0$ like r^L.*

Next we investigate the behavior of this function at $r \to \infty$ on the positive real axis. We simplify (7) by the substitution

$$\chi = r^\alpha U, \tag{9}$$

where α is chosen so that the $1/r^2$ term vanishes. For this, α must be either L or $-(L+1)$. Then Eqn7 becomes

$$\frac{d^2 U}{dr^2} + \frac{2(\alpha + 1)}{r} \frac{dU}{dr} + \frac{2m}{\hbar^2} \left(E + \frac{e^2}{r} \right) U = 0. \tag{7}$$

Its solutions at $r = 0$ behave like r^0 and $r^{-(2\alpha+1)}$. For $\alpha = L$, it is the first of these $\sim r^0$; and for $\alpha = -(L+1)$, it is the second $\sim r^{(2L-1)}$, which becomes a transcendental function and according to (9) leads to the desired solution, which is indeed unique. We choose $\alpha = L$, so our solution at $r = 0$ goes like $U \sim r^0$. Eqn7 – Laplace's equation – has the general form

$$U'' + \left(\delta_0 + \frac{\delta_1}{r}\right)U' + \left(\epsilon_0 + \frac{\epsilon_1}{r}\right)U = 0. \tag{7}$$

In our case, the constants have the value

$$\delta_0 = 0, \quad \delta_1 = 2(L+1), \quad \epsilon_0 = \frac{2mE}{\hbar^2}, \quad \epsilon_1 = \frac{2me^2}{\hbar^2}. \tag{11}$$

This type of equation is basically quite easy to handle, because the so-called Laplace transformation, which in general gives back *again* an equation of *second* order, *here* leads to the *first* order, which is solvable by quadratures. This allows a representation of the solution of (7) by a complex integral. Here I simply state the final result [3]. The integral

$$U = \int_C e^{zr}(z - c_1)^{\alpha_1 - 1}(z - c_2)^{\alpha_2 - 1}dz \tag{12}$$

is a solution of (7) for any integration contour C for which

$$\int_C \frac{d}{dz}\left[e^{zr}(z - c_1)^{\alpha_1}(z - c_2)^{\alpha_2}\right]dz = 0. \tag{13}$$

The constants $c_1, c_2, \alpha_1, \alpha_2$ have the following values: c_1 and c_2 are the roots of the quadratic equation $z^2 + \delta_0 z + \epsilon_0 = 0$ and

$$\alpha_1 = \frac{\epsilon_1 + \delta_1 c_1}{c_1 - c_2}, \quad \alpha_2 = \frac{\epsilon_1 + \delta_1 c_2}{c_2 - c_1}. \tag{14}$$

From (11) we get $c_1 = -c_2 = +\frac{\sqrt{-2mE}}{\hbar}$; and

$$\alpha_1 = \frac{me^2}{\hbar\sqrt{-2mE}} + L + 1, \quad \alpha_2 = -\frac{me^2}{\hbar\sqrt{-2mE}} + L + 1.$$

The integral representation (12) not only allows one to see the asymptotic behavior of the totality of solutions, when r goes to infinity in a particular way, but also the behavior of a *particular* solution, which is otherwise very difficult.

We will at first *exclude* the case when α_1 and α_2 are real whole numbers. That occurs for both cases simultaneously when

$$\frac{me^2}{\hbar\sqrt{-2mE}} = \text{ a real whole number.} \qquad (15)$$

We therefore assume at present, that (15) does not hold.

* * * * * * * * * * * * * * * * * * **

This is a verbatim translation of Schrödinger's original, which is uncharacteristically tortuous. Presumably, he is giving great emphasis to the completeness and uniqueness of his exact characterization of the solutions of the Coulomb radial differential equation subject to physical boundary conditions. He is not content to just find the Balmer states, but in their *very first* exploration in wave mechanics wants to employ full rigor.

* * * * * * * * * * * * * * * * * * **

The behavior of all solutions as $r \to \infty$ – we will always think of real positive limits – is characterized [4] by the behavior of the two linearly independent solutions – which we call U_1 and U_2 – obtained by following two special integration paths C. *Both* contours come in from infinity, and go back on the same paths, such that

$$\lim_{z \to \infty} e^{zr} = 0, \qquad (16)$$

i.e., the real part of zr should be infinitely negative. Here the requirement (13) is obeyed. For the solution U_1, the point c_1 is encircled once by the contour C_1; and for the solution U_2, the point c_2 is encircled once by the contour C_2.

These two solutions asymptotically (in the sense of Poincaré) for large real positive r-values are given by $(j = 1, 2)$

$$U_j \sim e^{c_j r} r^{-\alpha_j} (-1)^{\alpha_j} (e^{2\pi i \alpha_j} - 1)\Gamma(\alpha_j)/(-2c_j)^{\alpha_j - 1}, \qquad (17)$$

We show the first term in the asymptotic series of increasing integer powers of $1/r$.

We now need to understand the two cases $E \gtrless 0$.

1. $\underline{E > 0}$. We first note that the restriction (15) to integer values is certainly violated here because this quantity is pure imaginary. Furthermore, from (14),

both c_1 and c_2 are pure imaginary. Since r is real, the exponential functions in (17) are periodic functions of infinite extent. The values of α_1 and α_2 in (14) show that U_1 and U_2 *both* go like $r^{-(L+1)}$ near zero. *The same must also hold for the complete transcendental function U, since it is a linear combination of U_1 and U_2.* Still further, (9) and (10) show that the function χ, i.e. the whole transcendental solution of the *original* Eqn7, always goes like $1/r$ at zero, since it consists of U multiplied by r^L. We can therefore conclude:

The Euler differential Eqn5 of our variation problem has – for each positive E – two solutions which are unique, finite and stable and diverge at zero like $1/r$ multiplied by a standing wave. – More remains to be said about the surface boundary condition (6).

2. $\underline{E < 0}$. In this case, the restriction (15) to integral values is not automatically violated, so we must keep its effect in mind. Then from (14) and (17), at $r \to \infty$ U_1 grows beyond all bounds, whereas U_2 vanishes *exponentially*. Our complete transcendental function U (and the same holds for χ) will remain finite when and only when U_1 is identical with U_2 within a numerical factor. *But this is never the case.* One knows that: if in (12) one chooses for the integration contour C a path closing around *both* points c_1 and c_2, then – because the sum $\alpha_1 + \alpha_2$ is an integer – that path is *actually closed* in the Riemann sense of the integrand. In this case the restriction (13) is sufficient to show that the integral (12) actually represents *our complete transcendental function U*. It can be expanded in a power series in r which converges in any case for r sufficiently small, satisfies the differential equation (7), and must coincide with that one U. Therefore: U is represented by (12), where the contour C is a closed path around both points c_1 and c_2. This closed path can still be distorted so that it breaks up into the two earlier integration paths corresponding to U_1 and U_2, which therefore appear in U in a *linear combination* and in fact with *non-vanishing coefficients* such as 1 and $e^{2\pi i \alpha_1}$. That means U cannot be identified with U_2, but must also contain U_1.

The complete transcendental function U, which alone satisfies (7) for the problem under consideration, does *not* – under the assumptions made – remain finite at large r. As to the *completeness*-question, i.e. the proof that our method has found *all* linearly independent solutions, we assert: *For any negative E which does not satisfy the restriction (15), our variation problem has NO solution.*

We still have to investigate the family of discrete negative E-values which satisfy the restriction (15). Then α_1 and α_2 are both integers. Of the two integration paths which earlier gave the fundamental solutions U_1 and U_2, the first must be changed in order to give a nonvanishing result. Since $\alpha_1 - 1$ is certainly positive, the point c_1 is now neither a branch point nor a pole of the integrand, but a simple zero. The point c_2 can also be regular, namely when $\alpha_2 - 1$ is not negative. In *every* case, two suitable integration paths can be specified and the integration on them even carried out in closed form with known functions, and the behavior of the solutions made completely explicit. [Note added: See Schrödinger's Footnote [2] where he acknowledges the mathematical expertise of Herman Weyl in the analysis of the solutions of Eqn7 using the complex integral prescription of Eqn12. The Reference [3] is to a 1900 textbook by Schlesinger, and the original theory is due to Poincaré and Horn. Just what elements of the analysis are due to Weyl, and what would have been possible to Schrödinger unaided is of course not clear. What is clear, is that the level of mathematical sophistication in this first analysis of the solutions of the Schrödinger Coulomb wave equation is remarkable.]

$$\text{Consider now } \frac{me^2}{\hbar\sqrt{-2mE}} = n; \quad n = 1, 2, 3, 4 \cdots . \tag{15}$$

$$\text{Then from (14) above } \alpha_1 - 1 = n + L, \qquad \alpha_2 - 1 = -n + L. \tag{14}$$

We must investigate the two cases, $n \leq L$ and $n > L$. Consider first:

a) $n \leq L$. Then c_1 and c_2 lose their singular character, acquiring instead the role as beginning- and end-point of the integration path, in order to fulfill the requirement (13). A third point with significance here is negative real infinity. Each path between two of the three points yields a solution and of these three solutions, two are linearly independent, as one can easily confirm by calculating the integrals in closed form. In particular, the *complete transcendental solution* is produced by the integration path from c_1 to c_2. *This* integral remains regular for $r = 0$ as can be shown even without calculation. I simply state this fact because the actual calculation is likely to obscure it. On the other hand, it can be shown that this integral grows without bound for positive, infinitely large r. One of the other two integrals remains finite for large r, but is infinite for $r = 0$. So in the case $n \leq L$ we get *no* solutions. Now consider:

b) $n > L$. According to (14), c_1 is a zero point, and c_2 is a pole of at least first order. Two independent integrals are then produced: one by the path from $z = -\infty$, carefully avoiding the pole, which leads to the zero point; the other by the *residue* at the pole. The *latter* is the complete transcendental function. We need it multiplied by r^L to get the solution χ of the original Eqn7 (within an adjustable constant factor). We get

$$\chi(r) = f(x) = x^L e^{-x} \sum_{k=0}^{n-L-1} \frac{(-2x)^k}{k!} \binom{n+L}{n-L-1-k}, \tag{18}$$

where $x = r \times \sqrt{-2mE}/\hbar$. The criterion for a usable solution is that it remain finite for all real non-negative r. The boundary condition (6) is satisfied by its exponential vanishing at infinity.

We summarize the results for negative E: *For negative E, our variation problem has solutions only when E satisfies the condition (15). The integer L, which gives the order of the spherical harmonic appearing in the solution, can only take values smaller than n (which is always possible). The part of the solution dependent on r is given by (18).*

One finds further: *The solutions for a permissible combination of (n, L) have exactly $(2L + 1)$ arbitrary constants; for a given n-value there are n^2 arbitrary constants.* [Note added: Schrödinger does not explicitly mention 'Laguerre polynomials' or other special functions until his next paper (here as **Paper IX·2**).]

We have thereby confirmed the important properties of the eigenvalue spectrum of our variation problem, at least in its principal feature of discreteness. In this note I will not concern myself with the proof of the completeness of the eigenfunctions. Based on far ranging experiments, one may conjecture that no eigenvalues have been missed.

It should be pointed out that the eigenfunctions for positive E are not well defined by the variation problem as originally stated, since ψ goes to zero at infinity only as $1/r$, and $\partial\psi/\partial r$ goes as $1/r^2$. The surface integral (6) therefore is of order $\delta\psi$ at infinity. To actually derive the positive spectrum from the variation problem, one must add another restriction to the *problem*: perhaps that $\delta\psi$ should vanish at infinity, or that it should tend to a constant value, independent of the direction in which one goes to spacelike infinity; in the latter case the spherical harmonics make the surface integral vanish.

§ 2. The condition (15) gives

$$E_n = -\frac{me^4}{2\hbar^2 n^2},\tag{19}$$

just the Bohr energy levels for the Balmer terms, with $K = \hbar$. Our n is the principal quantum number. $L + 1$ is analogous to the azimuthal quantum number which, with another number to completely specify the spherical harmonics, can be assigned the significance of 'azimuthal' and 'polar' quantum numbers. These numbers determine *here* the system of nodal-lines on the sphere. The 'radial quantum number' $n - L - 1$ is the number of 'nodal spheres', i.e. the number of real positive zeros of the function $f(x)$ in (18). – The positive E-values correspond to the continuum of hyperbolic orbits, which one could describe in the sense of the radial quantum number $\to \infty$. This corresponds to the radial function having *standing* oscillations out to infinity.

It is interesting to estimate the domain where the functions (18) are appreciably different from zero and where their oscillations take place, and the *general order-of-magnitude* of the axis of the corresponding ellipse. The factor multiplying the radius r in the argument of the constant-free function f is – obviously – the inverse of the length

$$\frac{a_n}{n} = \frac{\hbar n}{\sqrt{-2mE_n}} = \frac{\hbar^2 n}{me^2},\tag{21}$$

where a_n is the semi-major axis of the nth elliptical orbit. (This follows from (19) and the relation $E_n = -e^2/2a_n$.) The quantity (21) is the order-of-magnitude of the domain for small integers n and L, because it can be safely assumed that the roots of $f(x)$ are then of order unity. That is not obviously the case when n is large, but the above assertion is confirmed there fairly accurately.

§ 3. It is natural to relate the ψ-function to an oscillation-mode in the atom, to which the electron orbits of today's much questioned reality correspond in some sense. I originally had the intention to establish the new formulation of the quantum prescription in this more obvious way, but have instead presented the above new mathematical structure because in it the essentials appear more clearly. It appears to me that the essential features of the quantum prescription do not require the mysterious postulate of 'matrix ordering', but that this is in fact a later step: it has its basis in the finiteness and uniqueness of a certain spatial function.

I would prefer not to debate the meaning of these oscillation-modes until some more complicated case has been successfully calculated. It is not yet certain that its results should be a perfect copy of the customary quantum theory. For example, the relativistic Kepler problem – solved exactly, following the above prescription – leads surprisingly to *half-integral fractional* quanta (radial and azimuthal).

Before leaving the discussion of the oscillations, a few remarks are in order. Above all it should be emphasized, that for the impetus of these ideas in the first place, I have to thank the inspired thesis of Louis de Broglie [5] and his discussion of the spatial distribution of the 'phase waves', of which he showed there was always a 'whole number' in the orbit for each period of the electron. The principal difference is that de Broglie thought of traveling waves, whereas we, when we apply our formula to the oscillation-modes, are led to eigen-oscillations. I have recently shown [6] that one can base the Einstein gas law on the idea of such standing eigen-oscillations, assuming the dispersion law of de Broglie's phase waves. The above derivation for the atom can be viewed as a generalization of these ideas from the ideal gas model.

If one takes the particular function (18), multiplied by a spherical harmonic of order L, as the description of the eigen-oscillation process, then the quantity E must have something to do with the *frequency* of that process. One recalls that for a familiar vibration process the corresponding 'parameter' is proportional to the *square* of the frequency. Such a prescription in the present case for negative E-values would lead to *imaginary* frequencies. Moreover, the quantum theorists feel that the energy should be proportional to the frequency itself and not its square.

[Note added: Schrödinger's conjectures in the remaining paragraphs have had no lasting value and foretell a divergence of his views from the mainstream of quantum theory as dictated primarily by Bohr. Schrödinger remained marvelously creative in quantum mechanics but more and more in a contrarion mode.]

The paradox is resolved in the following way: There is a priori *no natural zero-level* for the 'parameter' E of the variational equation (5), since the unknown function ψ appears multiplied not only with E but also with a function of r, which can be changed up to a constant compensating any change in the zero-level of E. The 'resolution of the vibration theorists' for this problem is to suppose that not E itself – which we continue to call the energy – but (E added to a fixed constant) is proportional to the square of the frequency. This constant should be *very large* compared to the negative E-values [which are limited by (15)]. Then first, the frequencies will be *real*, and second our E-values, since only relatively small frequency *differences* occur, will actually be very nearly proportional to these frequency differences. That is all the quantum theory can demand, at least so long as the zero level of the *energy* is not specified.

The idea that the oscillation frequency should perhaps be given by

$$\omega = C'\sqrt{C + E} = C'\sqrt{C} + \frac{C'}{2\sqrt{C}}E + \cdots, \tag{22}$$

where $C >> E$, has a very important significance. *It provides an understanding of the Bohr frequency condition.* According to Bohr, the *emission frequencies* are proportional to the *energy differences*, that is – from (22) – the differences of the eigenfrequencies ω of each hypothetical oscillation-mode of the system. And in fact the eigenfrequencies are all very large compared to the emission frequencies, in accord with the above approximation. The emission frequency therefore appears as low 'difference tones' which with many higher frequencies result from the eigen-oscillations themselves. In the propagation of energy from one to another normal mode, that *some* lightwaves occur at frequencies equal to each eigenfrequency *difference*, is very understandable; one needs only imagine that the lightwave is causally connected with the *beats* necessarily occurring at each space point during the transition, and that the frequency of the light is the frequency of the intensity maximum of the beat process.

It might create doubt because this conclusion is based on the approximate form (22) (expansion of the square root), where the Bohr frequency condition it-self apparently has the character of an approximation. This is only apparent and is completely avoided when one uses the *relativistic* theory in which a deeper un-derstanding is possible. The large additive constant C corresponds naturally with the rest energy mc^2 of the electron. Also the apparently *repeated* and *independent* appearance of the constant $\hbar$ (which was introduced in (20)) will be clarified or even avoided by the relativistic theory. But unfortunately its unobjectionable derivation at present still contains the above difficulties.

It is hardly necessary to ask how much faith we should put in the idea, that for a quantum transition the energy from one oscillator goes over to another, like the notion of electrons jumping. The change in the oscillator can take place continuously in space and time, it can continue happily as long as the emission process lasts, according to experiments (canal ray experiment of W. Wien); and yet when the atom is exposed to an electric field for a relatively short time during this transition, the eigenfrequencies change, and the beat frequencies are at once detuned, but only as long as the field acts. These experimentally established facts are most difficult to comprehend, as one perhaps realizes from the discussion in the attempted solution of Bohr-Kramers-Slater.

As for the rest, in the euphoria over the intellectual speculation on all these things, one should not forget that the idea of the atom oscillating but not radiating when in the particular form of *one* eigen-oscillation – while I admit that it must be closely examined – is still very far removed from the *natural* description of an oscillating system. A macroscopic system is known not to behave this way, but occurs in general in a potpourri of its eigen-oscillations. However one should not anticipate clarifying his insight on this point. Even a potpourri of eigen-oscillations would not make a difference if no other beat frequencies occur than those the atom *usually* radiates. Even the simultaneous emission of each of these spectral lines by the same atom contradicts no experiment. One might well think that only in the groundstate (and approximately in certain 'metastable' states), can the atom oscillate with *one* eigenfrequency and *not* radiate, since in this case no beats occur. The *excitation* would consist of the simultaneous stimulation of one or more different eigenfrequencies, whose beats produce the light emission.

It seems possible that in general all eigenfunctions belonging to the same frequency are excited at the same time. The multiplicity of the eigenvalues corresponds to a *degeneracy*. The removal of the quantization of a degenerate system leads to the random partition of the energy among all the eigenfunctions belonging to each eigenvalue.

Added in Proof on 28.Febr.1926.

For a conservative system in classical mechanics the variation problem can be expressed more elegantly than was done above, without explicit reference to Hamilton's partial differential equation: Let $T(q,p)$ be the kinetic energy as a function of the coordinates and momenta, V the potential energy, $d\tau$ the 'rational measure' of the volume element of configuration space, i.e. not simply the product $dq_1 dq_2 \cdots dq_n$, but this product divided by the square-root of the discriminant of the quadratic form appearing in $T(q,p)$. (See Gibbs, *Statistical Mechanics.*) Then ψ is required to make *stationary* 'Hamilton's Integral'

$$\int d\tau \left\{ \hbar^2 T \left(q, \frac{\partial \psi}{\partial q} \right) + V \psi^2 \right\}, \tag{23}$$

subject to the *normalization restriction* $\int d\tau \psi^2 = 1$. As is well-known, the *eigenvalues* of this variation problem are the *stationary values* of the integral (23) and for our purpose give the *quantum energy-levels*.

Zürich, Physikalisches Institut der Universität.

Footnotes and References:

1) These solutions are not completely well-defined.

2) For the introduction to the analysis of Eqn7, I owe the greatest thanks to Hermann Weyl. For the unsupported assertions in the following, I refer to L. Schlesinger, *Differentialgleichungen* (Sammlung Schubert Nr. 13, Göschen 1900, especially Ch. 3 and 5.)

3) See L. Schlesinger, loc. cit. The theory is due to H. Poincaré and J. Horn. [Note added: An admittedly cursory examination of the 1908 and 1922 editions of Schlesinger's text reveals that they are replete with references to Poincaré and to Horn, and are even generally relevant, but does not turn up Schrödinger's explicit development.]

4) When (15) is fulfilled, at least one of the two integration paths described in the text is unsuitable, since it leads to a vanishing result. [Note added: This seems to be the essential point of departure of Schrödinger's (and Weyl's (?)) development from anything known before.]

5) L. de Broglie, Ann. d. Physique **3**, 22 (1925) (Thesis Paris 1924).

6) Soon to appear in Zeits. f. Phys.

Paper IX·1b: Excerpt from Annalen der Physik **79**, 489 (1926).

Quantization as an Eigenvalue Problem

von **E. Schrödinger**.

Eingegangen 23. Februar 1926.

Chapter Two.

§ 1. Hamilton's Analogy Between Mechanics and Optics.

Before we analyze the eigenvalue problem of the quantum theory for other special systems, we examine the *general* consistency between the Hamilton partial differential equation of a mechanical problem and the *wave equation* 'belonging' to it, as in the Kepler problem of Eqn (5) in the first chapter. We considered this consistency only briefly as it was described by the – unexplained – transformation Eqn (2), and the – also unexplained – transition from the *setting-to-zero* of an expression, to the requirement that the *space integral* of that expression should be *stationary* [2].

The *internal* consistency of Hamilton's theory with the occurrence of wave propagation is not new. It was known to Hamilton himself, and was the starting point of his Theory of Mechanics as an outgrowth of his *Optics in Inhomogeneous Media* [3]. Hamilton's Variation Principle can be understood as Fermat's Principle for wave propagation in configuration space (q-space), and Hamilton's equation expresses Huygen's Principle for this wave propagation. This powerful and momentous idea of Hamilton has been omitted from most modern versions of mechanics for the sake of a discussion of its analytic consistency [4].

We consider general conservative classical mechanical systems. Hamilton's equation is

$$\frac{\partial W}{\partial t} + T\left(q_k, \frac{\partial W}{\partial q_k}\right) + V(q_k) = 0. \tag{1}$$

W is the action function, i.e. the time-integral of the Lagrangian $T - V$ along a system orbit as a function of the endpoint q_k and the time t. q_k is a generalized coordinate with $k = 1 \cdots n$ and n the dimension of the configuration space. T is the kinetic energy as a function of q_k and the momentum p_k (a quadratic form of the latter), for which the prescription is $p_k \equiv \partial W / \partial q_k$. V is the potential energy. To solve the equation, make the ansatz

$$W = -Et + S(q_k), \tag{2}$$

to get $T(q_k, p_k) = E - V(q_k). \cdots (1')$ The energy E is an integration constant.

Eqn $(1')$ is most simply expressed as a geometric statement in q-space, following Hertz. It is particularly simple and clear when the kinetic energy of the system is expressed in terms of a non-euclidian metric. If $\bar{T}$ is the kinetic energy in terms of the *velocities* $\dot{q}_k$, then the line-element is defined as

$$ds^2 = \bar{T}(q_k, \dot{q}_k)\, dt^2. \tag{3}$$

The right side contains dt only superficially, since it means (with $\dot{q}_k dt = dq_k$) a quadratic form in dq_k.

From this expression one can construct: the angle between line-elements, perpendicular directions, divergence and curl of a vector, gradient of a scalar, Laplacian and so on. One can think of the operations just as in three dimensional euclidian space, although the actual expressions are a little more complicated because the line-element (3) must be used throughout in place of the euclidian one. *In the following all geometrical expressions in q-space are to be understood in this non-euclidian sense.* In the calculations a most important distinction must be made: one must distinguish between *covariant* and *contravariant* components of vectors and tensors.

The dq_k are the prototype *contravariant* vector, for which the q_k-dependent coefficients in $\bar{T}$ are the fundamental *covariant* metric-tensor. T is the contravariant form corresponding to $\bar{T}$, formed by the naturally covariant vector $p_k = \partial W / \partial q_k \equiv \mathrm{grad}_k W$ which corresponds to the contravariant velocity vector $\dot{q}_k$. (T and $\bar{T}$ have the invariant value of the kinetic energy, expressed in terms of the covariant momenta, or the contravariant velocities.)

Eqn $(1')$ can be expressed simply as
$(\mathrm{grad} W)^2 = 2(E - V)$ or $|\mathrm{grad} W| = \sqrt{2(E - V)}. \cdots (1'')$

This requirement is easy to understand. Assume that a function W (of the form (2)) has been found, which satisfies all the requirements. Then this function represents a group of surfaces in q-space labeled by $W = $ const., with a single W-value assigned to each surface.

On the one hand, Eqn (1″) gives the prescription for the exact construction of any surface of this group: *when one with its W-value is known,* then successively all others and their W-values can be constructed. On the other hand, if just the data necessary for this construction is given, that is if *only* the surface and its W-value *are given completely arbitrarily,* then there *is* an ambiguity in the exact construction. In all this, we are supposing the time to be fixed. The construction prescription therefore *exhausts* the content of the differential equation: one can obtain *each* of its solutions from a suitably assumed function plus the W-value.

Now the construction prescription: Consider an arbitrary surface with the value W_0. To find the surface which corresponds to the value $W_0 + dW_0$, label the sides of the given surface arbitrarily as positive or negative. Erect at each point of the surface a perpendicular to it with the above sign and the magnitude

$$ds = \frac{dW_0}{\sqrt{2(E - V)}}. \tag{4}$$

The endpoints of the perpendiculars define the surface $W_0 + dW_0$. By continuing, one can construct the family of surfaces on each side side of W_0 for all $W \gtrless W_0$. The construction is *ambiguous*, since one could at the first step take the *other* side as the positive one. For the later steps one no longer meets this double valuedness, i.e. one cannot later arbitrarily change the designation of the surface one has just reached without involving an instability in dW. Moreover, the two possibilities one reaches are not identical unless the W-values are.

Next we consider the most simple dependence on the *time*. Eqn (2) shows that at some later time $t + \Delta t$ the original surface for a particular value W_0 evolves into the surface originally assigned to $W = W_0 + \Delta W$. The W-value changes according to a definite simple law from surface to surface, and for positive E in the direction of increasing W-values. Instead of this one can imagine that the surfaces move forward, in that each assumes the form and position of the one it follows, and also the W-value *which it had*. The law of evolution of the surfaces is therefore given, that e.g. the surface W_0 at time $t + \Delta t$ must have reached the position which at time t the surface $W_0 + E\Delta t$ had. That would result – from (4) – if one moved each point of the surface W_0 forward in the direction of the positive side by

$$ds = \frac{E\Delta t}{\sqrt{2(E - V)}}. \tag{5}$$

That is, the surfaces propagate with a normal velocity

$$u = \frac{ds}{dt} = \frac{E}{\sqrt{2(E - V)}}, \tag{6}$$

which for constant E is a function of position only.

Now one recognizes that our system of surfaces $W = $ const. can be interpreted as the system of wave-fronts of a forward advancing but stationary wave motion in q-space, for which the magnitude of the phase velocity at each point of the space is given by (6). Then the 'perpendicular'-construction can often be replaced by Huygen's construction for elementary waves [with the radius (5)] and their envelope. The 'index of refraction' is proportional to the reciprocal of (6), dependent on position but not on direction. q-space is 'optically inhomogeneous but isotropic'. The elementary waves are spheres, but of course – as we emphasize once again – spheres in the sense of the non-euclidian line-element (3).

The action function W plays the role of the *phase* for our wave-system. The Hamilton equation is the expression of Huygen's Principle. If one writes Fermat's Principle as

$$0 = \delta \int_{P_1}^{P_2} \frac{ds}{u} = \delta \int_{P_1}^{P_2} \frac{2T}{E} dt = \frac{1}{E} \delta \int_{P_1}^{P_2} 2T \, dt, \tag{7}$$

then one is led directly to Hamilton's Principle in Maupertuis' form (where the time integrals are to be understood with the usual grain of salt, i.e. $T + V = E = $ const., even during the variation). The 'rays', i.e. the trajectories orthogonal to the wave-fronts, are trajectories of the system with the energy E, in agreement with the well-known equation

$$p_k = \frac{\partial W}{\partial q_k}. \tag{8}$$

This states that for each particular action function a family of orbits can be derived as the gradient of a velocity-potential [5]. (The momentum p_k is simply a covariant velocity vector, which Eqn (8) says is the gradient of the action function.)

In spite of all the talk in the above discussion of wave-fronts, propagation velocity, Huygen's Principle, etc., one has really not looked at the analogy of mechanics to *wave*-optics, but to *geometric*-optics. Although the idea of *rays*, which is of primary importance in mechanics, belongs to *geometric*-optics, it is scarcely one of its more refined products. Even Fermat's Principle can be expressed purely in geometrical optics just by use of an appropriate index of refraction. And the system of W-surfaces interpreted as wave-fronts represents the mechanical motion in a somewhat looser interpretation, as the system-point of a mechanical system with rays advancing not at the phase velocity u but with a velocity (at constant E) proportional to $1/u$; directly from (3)

$$v = \frac{ds}{dt} = \sqrt{2T} = \sqrt{2(E - V)}. \tag{9}$$

This disparity is easily understood. First, from (8): the system-point velocity is *large* where grad$W$ is large, i.e. where the W-surfaces are close together, i.e. where u is small. Second, from the representation of W as the time-integral of the Lagrangian, this changes during the motion [by $(T - V)dt$ during dt], so the system-point *cannot* remain in contact with the same W-surface.

This does not include in the analogy such important ideas from the wave theory as amplitude, wavelength, frequency – all generally referred to as wave-*like*. There seems for some a mechanical parallel; but for the wave function itself there is none. W has the role of *phase* of the waves – which remains somewhat ill-defined because of the indefiniteness of the wave-*form*.

For a qualitative analogy, this defect is not disturbing. The analogy *persists* even with the *geometric*-optics or when one presses forward with a very primitive wave optics, but not with the complete structure of wave optics. That geometric optics for *light* is only a gross approximation does not change this. For the further construction of the q-space optics in the wave-theoretic sense one must take care not to violate the geometrical-optics restrictions too far; one must assume [6] the *wavelength* to be small compared to all orbital dimensions. One learns nothing new.

But even this first attempt at a wave-theoretic explanation leads to a disturbing thing, that a completely different suspicion arises: *We now conclude that our classical mechanics must fail for very small orbits and very strong curvature.* This failure is a complete analog of the failure of geometric optics, i.e. the 'optics with infinitely small wavelength', which occurs when the 'object' or 'opening' is no longer large compared to the actual *finite* wavelength. Perhaps our classical mechanics *is* the *perfect* analog of geometric optics but even as such it is false and not in agreement with reality. It fails when the radius of curvature and the dimensions of the orbit are no longer large compared to the relevant wavelength in q-space. Then a 'wave mechanics' must be found [7] – and the place to look is surely the wave-theoretic development of Hamilton's ideas.

§ 2. 'Geometric' and 'Wave' Mechanics.

We next assume that a construction expressing this analogy is to represent the above wave-system as *sine*-waves. This is the simplest and most obvious choice, yet the *arbitrariness* of this assumption must be emphasized, since on it depends an account of fundamental significance. From this the wave function should contain the time only through a factor of the form $\sin(\cdots)$, whose phase argument is a linear function of W. The coefficient of the action W must be the inverse of an action, since the phase itself is dimensionless. We further assume that this factor should be universal, i.e. not depend on E, but also not depend on the nature of the mechanical system. We designate it as $1/\hbar$. The time factor is

$$\sin\left(\frac{W}{\hbar} + \text{const}\right) = \sin\left(-\frac{Et}{\hbar} + \frac{S(q_k)}{\hbar} + \text{const}\right). \tag{10}$$

This gives the *frequency* f of the waves from

$$2\pi f t(\equiv \omega t) = \frac{Et}{\hbar}. \tag{11}$$

It also gives the frequency of the q-space waves without appreciable artificiality as proportional to the energy of the system [8]. That makes sense only when E is absolute, not – as in classical mechanics – where it is only determined up to an additive constant. The *wavelength* according to (6) and (11) is *independent* of this additive constant and given by

$$\frac{\lambda}{2\pi} = \frac{u}{\omega} = \frac{\hbar}{\sqrt{2(E-V)}}, \tag{12}$$

where the radical is just twice the kinetic energy. If we make a rough comparison of this wavelength with the orbital dimension a of a of a hydrogen electron as it would be in classical mechanics, we get the order of magnitude $\hbar/mva$, where m is the electron mass and v its orbital velocity. The denominator mva is the magnitude of the angular momentum in the Kepler orbit of atomic dimensions, which all quantum theories agree to be $\mathcal{O}(\hbar)$. We obtain the correct order of magnitude when we identify our dimensional action constant with Planck's quantum of action.

If in (6) one replaces E by ω, then one gets $u = \hbar\omega/\sqrt{2(\hbar\omega - V)}. \cdots (6')$ The dependence of the wave-velocity on the system-energy, or equivalently on the *frequency*, is a *dispersion relation* for the waves. This dispersion relation is of great interest. We have pointed out in §1 that the moving wave-front has only a loose correspondence with the motion of the system-point, because their velocities are not equal and cannot be equal. From (9), (11) and (6), the velocity of the system-point has a very direct relation to the wave-velocity. One gets

$$v = \frac{df}{d(f/u)}, \tag{13}$$

i.e. the velocity of the system-point is that of a *wave-group* which fills a small frequency interval (the signal velocity). In his remarkable and original thesis [9], to which I am indebted for the inspiration of this work, de Broglie has already derived this relation for the 'phase-waves' of electrons based essentially on assumptions from relativity theory. One sees that it is a theorem of greater generality which springs not from relativity but from the usual mechanics of a conservative system.

This fact can be used to establish a very intimate connection between wave propagation and system-point motion. One imagines a wave-packet with relatively small dimensions in all directions. Such a wave-packet will presumably follow the same law of motion as a single system-point of the mechanical system. If the wave-packet is to be *equivalent* to the system-point, and one is to see it as approximately point-like, its dimensions must be small compared to the dimensions of the orbit. That would only be the case if the orbit dimensions – in particular its radius of curvature – were very large compared to the wavelength. Then by analogy with the usual optics it would be clear that the dimensions of the wave-packet not only could not be compressed under the magnitude of a wavelength, but conversely that the packet must spread out in all directions over a large number of wavelengths unless it was *approximately monochromatic*. But then we must require – since the wave-packet as a whole moves with a well defined group-velocity – that it should correspond to a mechanical system *of definite energy* (see Eqn (11)).

Such wave-packets have been constructed in the same way in optics by Debye [10] and von Laue [11], where the exact analytical representation of a spherical wave or a ray-bundle was given. They gave a connection to a part of Hamilton-Jacobi Theory not mentioned in §1: the derivation of the equations of motion in integral form by differentiation of a complete integral of Hamilton's equation with respect to the constants of integration. This is the Jacobi equation with the interpretation: the system-point of the mechanical system coincides with *that one* point where a continuum of wave-trains meet *with equal phase*.

In optics one obtains the rigorous wave-theoretic representation of a 'ray-bundle' of 'sharply' defined finite cross section, which according to Debye meet at a focal point and then fly away from one another, in the following way: One superposes a *continuum* of *even* waves, each of which individually would fill the *whole* space, in such a way that the wave directions vary within a given spatial angle. Outside a given cone the waves cancel one another almost exactly through interference. They represent the desired limited ray-bundle exactly wave-theoretically by diffraction phenomena throughout the stipulated region. – One can represent in this way not only a *finite* but also an *infinitesimal* ray, simply by allowing the wave directions to vary only over an infinitesimal angle. von Laue used this in his famous treatment of degrees of freedom of a ray-bundle [11]. Finally, one can work not with a monochromatic wave, but with the frequency allowed to vary over an infinitesimal range. Then by an appropriate distribution of amplitudes and phases, an excitation can be confined to a restricted domain which is relatively small even in the longitudinal direction. In this way, one gets the analytic representation of an 'energy-packet' of relatively small dimension propagating with the group-velocity. The actual *position* of the energy-packet – when its detailed structure is not important – is given in a very plausible way as that one space-point where *all* the superposed waves have *exactly* the same phase.

We will now extend these ideas to q-space waves. We choose a wave-packet at time t, at point P in q-space, moving in direction R, and with average frequency f corresponding to the average E-value $E = 2\pi\hbar f = \hbar\omega$. For the mechanical system these data correspond to a given initial position and initial velocity.

To carry out the optical construction, we need *one* family of wave-fronts for the given frequency, i.e. *one* solution, which we call W, of Hamilton's equation for the given E-value, with the following property: at time t the family of wave-fronts going through the point P, say

$$W = W_0, \tag{14}$$

should have their perpendiculars in the prescribed direction R. However this is still not enough. The wave-group W must be specified in an n-fold ($n=$ number of degrees of freedom) infinity of ways, such that the wave-normal at the point P can spread in an $(n-1)$-dimensional infinitesimal cone and with the frequency $f = E/2\pi\hbar$ in an infinitesimal interval. All members of this continuum of waves must have the same phase at time t at point P. Then *somewhere*, at a given later time, some point exists where all these phases coincide again.

It is sufficient to have a solution W of Hamilton's equation which depends on n constants α_k. We take $\alpha_1 \equiv E$, and choose $\alpha_2 \cdots \alpha_n$ so that the family of surfaces goes through P with the direction R. We understand $\alpha_1, \alpha_2 \cdots \alpha_n$ to have these values and from (14) the surfaces going through P at t is *this* group. Also we consider a *continuous group*, whose α_k-values are in an infinitesimally neighboring α_k-range. Any member of this continuum, i.e. therefore *a group*, will be given by

$$W + \sum_{k=1}^{n} \frac{\partial W}{\partial \alpha_k} d\alpha_k = \text{Const.} \tag{15}$$

for a *fixed* value of $d\alpha_1 \cdots d\alpha_n$ and *variable* Const. Each member of *this group* will be determined by the following choice of constant:

$W_0 + \sum_{k=1}^{n} (\partial W / \partial \alpha_k) \, d\alpha_k|_0 = \text{Const.}, \cdots (15')$

The surfaces (15) for all possible values of $d\alpha_1 \cdots d\alpha_n$ constitute *a group*. They all go through P at t, their normals fill an infinitesimal $(n-1)$-dimensional cone, and their E-parameters vary in a small range. The family of surfaces (15') is that one which goes through P at t. The phase of the wave functions which belong to the group (15), are determined directly from this by substitution using (15').

We now ask: *for a particular time* is there always some point at which all surfaces of the class (15') intersect *and at which* all wave functions belonging to the class (15) coincide in phase? The answer is: *it does give* a point of corresponding phases, but it is *not* the intersection point of the class of surfaces (15'), because these *no longer* correspond to a given time. The point of definite phase is such that the representative of (15) which arises from (15') *changes continuously*.

One sees this in the following way: for the intersection point, all terms of (15') must at some time simultaneously satisfy

$$W = W_0 \quad \text{and} \quad \frac{\partial W}{\partial \alpha_k} = \left(\frac{\partial W}{\partial \alpha_k}\right)_0 \quad \text{for all } k, \tag{16}$$

where the $d\alpha_k$ are arbitrary within a small range. In these $(n+1)$ equations, the right sides are constants and the left sides are functions of $(n+1)$ quantities $q_1 \cdots q_n, t$. The equations are satisfied identically for the initial time and the n-coordinates of P. For any arbitrary different time there will be *no* solution for the $q_1 \cdots q_n$, because the system *overdetermines* these n quantities.

One can still proceed in the following way: leave aside the first equation $W = W_0$ for the moment and determine the q_k as functions of the time and the constants according to the above n equations. This point is called Q. For it naturally, the *first* equation is *not* satisfied, but gives an action W different from W_0. If one goes back to the derivation of (16) from (15'), this means that point Q is a member not of the group (15'), but of a class of surfaces arising from it by allowing the action to increase from W_0 to W, which we call (15''). For it Q is an intersection point. So each of the class (15) evolves according to the displacements given by (15'). This evolution requires changing the Const. in (15) by *the same amount* for all displacements. In that way the *phase angle* is changed the same amount for all displacements. As just stated, this determines the new surfaces, i.e. the members of the class which we call (15″), and they intersect at the point Q, all with the same phase angle. That can be stated as: the n equations

$$\frac{\partial W}{\partial \alpha_k} = \left(\frac{\partial W}{\partial \alpha_k}\right)_0 \quad k = 1 \cdots n, \tag{17}$$

determine the point Q as a function of the time, as a point of common phase for the whole group of waves (15).

Of the n surfaces whose intersection point Q is determined by (17), only the first is significant, which satisfies the above (only the first of Eqns (17) contains the time). The $(n-1)$ stationary surfaces determine the *trajectory* of the point Q on its intersection-line. It can be shown that this intersection-line is the trajectory orthogonal to the class $W = $ Const. W therefore satisfies the requirement of Hamilton's equation identically in $\alpha_1 \cdots \alpha_n$. If one differentiates Hamilton's equation with respect to α_k for $k = 2 \cdots n$, then one obtains the statement that all the perpendiculars at each point of a surface $\partial W / \partial \alpha_k = $ Const. are perpendicular to each other, i.e. that each surface *contains* the perpendicular of the others. If the intersection line of the $(n-1)$ fixed surfaces (17) does not branch – which is usually the case – then each element of the intersection-line must be a *unique common* line-element on the $(n-1)$ surfaces, with the perpendicular going through the same point coinciding with the W-surface, so the intersection line is orthogonal to the W-surface, as we set out to show.

Very briefly, one can infer for Eqn (17) the following properties: $W / \hbar$ is the phase of the wave function. If one considers not just *one*, but a fixed manifold of wave-systems and orders them in a continuous way by some continuous parameters α_i, and introduces the equations $\partial W / \partial \alpha_i = $ Const., then all infinitely close individual wave-systems of this manifold have corresponding phase. These equations determine the the geometric locus of the point of definite phase. If the equations are sufficient to confine this locus to a point, the equations then specify *this point* as a definite function of the time.

Equations (17) are just the Jacobi equations, so we have also shown:
The point of definite phase for a given n-parameter infinitesimal manifold of wave-

systems propagates according to the same law as the system-point of the correspond-
ing mechanical system.

The superposition of these wave-systems actually gives an appreciable excitation only in a relatively small region of the point of definite phase, whereas
everywhere else it is almost totally destroyed by interference; this is true at least
for the choice of suitable amplitudes and even for special choices of the *form* of the
wave-system; it turns out to be exact, and holds even for very difficult situations. I
will use this as a physical hypothesis in the examples, without pursuing the proof.
That will only be worth the trouble when the hypothesis has been tested and when
its application will justify such analysis.

One might suppose that the region in which the excitation could be confined
would measure at least a large number of wavelengths in each direction. First, that
is obviously incorrect, since even at a distance from the point of definite phase of
only *a few* wavelengths the phase coherence no longer holds, and the interference
is already so strong that each point is independent. Secondly, one can use the hint
from the three-dimensional euclidian case of ordinary optics, to see that at least
it generally behaves in this way.

What I now conjecture with greater certainty is the following:
The actual mechanical event is described by the corresponding *wave-process* in
q-space and not by the motion of a *system-point* in this space. The study of the
system-point motion, which is the subject of classical mechanics, is only an approximation and has the same significance as geometric or ray-optics compared to
the actual optical processes. A macroscopic mechanical process will be represented
as a wave-process of the above kind, which with appropriate approximations can
be regarded as pointlike in agreement with the geometric structure of the curved
orbits. We have seen that exactly the same laws of motion actually hold for such
a wave-packet, as classical mechanics ascribes to the system-point. This method
of treating the problem loses its validity when the orbit is no longer large compared to the wavelength. Then the *rigorous* wave-theoretic treatment *must* be
used, i.e. in order to make clear the possible behavior, one *must* proceed from the
wave equation and not from basic equations of mechanics. Ultimately, these are as

unsuitable for the explanation of the microscopic structure of mechanical events as is geometric optics for the explanation of *diffraction phenomena.*

When one does seek the representation of this microstructure in the context of classical mechanics, of course with additional very artificial assumptions it has been possible and practical results of the highest significance have been obtained. Thus it seems to me very significant, that this theory – I mean the quantum theory of Sommerfeld, Schwarzschild, Epstein, et al – has the most intimate connection with Hamilton's equation and the Hamilton-Jacobi theory, i.e. with that form of classical mechanics which already contains the clearest hint of the real wave-character of mechanical events. Hamilton's equation corresponds to Huygen's Principle (in its old naive form, not in the rigorous Kirchoff form). And like geometric optics supplemented by a completely universal prescription (Fresnel's zone construction), which in far-reaching measure does justice to the diffraction phenomena, the Theory of the Action Function can be borrowed from light for the processes in atoms. Alternatively one must engage in insoluble contradictions, as one did when one attempted to use for these atomic processes the idea of the *system-point*; just as one gets lost in the incomprehensibility of trying to follow the *light rays* in the realm of diffraction phenomena.

One's immediate conclusions are: I will give no definite results for actual events, about which absolutely nothing has been obtained so far, but which can only be understood by investigating the wave equation; I will merely illustrate the state of affairs qualitatively. For that, we think of a wave-packet having somehow a small, perhaps closed 'orbit', whose dimensions are only of the order of magnitude of the wavelength and therefore *small* compared to the dimensions of the wave-packet itself. It is clear that the 'system-orbit' in the sense of classical mechanics – that is the orbit of the point of definite phase - will lose its distinctive role completely, because around this point a whole continuum of points is spread out which also have a completely definite phase but which describe a completely different orbit. Alternatively we might say: the wave-packet not only fills the whole orbital region but moves over it in all directions.

In *this* sense, let me point out de Broglie's 'phase-waves' accompanying the orbit of the electron in the sense that the electron orbit itself, just as in the atomic binding, no longer has any particular significance, and certainly not that of the locus of the electron in its orbit. In this sense the now more and more prevalent conviction is *first of all*: that the *phase* of the electron's motion in the atom should be ascribed the real significance; *secondly*: that one cannot simply say that the electron at a definite moment *is in a definite quantum-orbit* determined by the quantum considerations; *thirdly*: that the real laws of quantum mechanics require not a definite prescription for the *detailed orbit*, but the elements of the whole orbit-manifold of the system, so that any possible particular connection between the different orbits can be made [12].

It is not inconceivable that a careful analysis of the experimental evidence might lead to the kind of proposition described here, in which the experimental evidence is the expression of the structure of the actual events. All these conclusions lead systematically to abandoning the concept of 'electron position' or of 'electron orbit'. If one chooses not to abandon these, then one is left with gross contradictions. These contradictions are so strongly evident, that one has even doubted whether atomic phenomena can be analyzed atall in the space-timelike mode of thinking. I would regard such a philosophical standpoint as a definitive decision similar in its effect to a complete 'striking of arms'. Then we would *never* change our thought-patterns, and we would *never* understand that which we cannot understand now. There are such things – but I do not think that atomic structure is among them. – From our point of view, such doubt provides no answers, although – or better stated – *because* its emergence is certainly understandable. In the same way, geometric optics, by its continued pursuit – trying to understand diffraction phenomena by the ray-ideas of macroscopic optics, and constantly failing – must eventually lead to the conclusion that *the Laws of Geometry* are not applicable, since it constantly concludes that the *rectilinear* and *independent* light-rays used in it must exhibit the most remarkable *curvatures* and be *visibly affected* even in a homogeneous medium. I consider this analogy to be *very* apt. As far as the unmotivated *curvatures* are concerned, the atomic analogy is not lacking – one only needs to think of the explanation of the anomalous Zeeman effect as arising from 'non-mechanical forces'.

How should one best make progress for the wave-like generalization of mechanics? Instead of starting from the fundamental equations of mechanics, one must suppose progress to be more probable starting from the wave equation for q-space and its manifold of solutions. So far in this work, the wave equation has not been explicitly used, and in fact not yet even formulated. The only purpose for its introduction so far was to get the *wave velocity* as a function of the energy (or frequency) by Eqn (6), and the wave equation itself is not uniquely determined by this result. Indeed, it is not even certain that it must be of second order, nor that the striving for simplicity is correct. These are just the next things to try. So one assumes that the wave function ψ satisfies

$$\nabla^2 \psi - \frac{1}{u^2}\ddot{\psi} = 0 \qquad (18)$$

for processes which depend on the time only through a factor $e^{i\omega t}$. With Eqns (6) and (11), this gives

$$\nabla^2 \psi + \frac{2}{\hbar^2}(E - V)\psi = 0 \cdots (18')$$

with $E = \hbar\omega$. The differential operations are to be understood as being with respect to the line-element (3). – But even under the assumption of second order this is not uniquely connected to (6). A possible generalization would be to replace $\nabla^2 \psi$ by

$$f(q_k)\vec{\nabla}\cdot\left(\frac{1}{f(q_k)}\vec{\nabla}\psi\right), \qquad (19)$$

where $f(q_k)$ could be any arbitrary function of the q_k, but in any plausible example must be free of $E, V(q_k)$ and the coefficients of the line element (3) (one might think for example of $f = u$). Our ansatz is further dictated by the desire for simplicity, since I believe this to be a *sine qua non* for success.

The substitution of a *partial* differential equation as equivalent to the basic dynamical equations of the atom-problem might at first seem extremely dubious because of the enormous manifold of solutions which satisfy such equations. Even classical dynamics had no way to limit the possibilities, but led to an extensive manifold of solutions, in fact to a continuous family whereas according to all expectations only a discrete number of these solutions actually seem to exist. The task of quantum theory is to understand the rule by which the discrete class of actually appearing solutions is selected by the 'quantum restrictions' from the continuous class of orbits permitted by the classical mechanics. It seems to be an unpromising start for a new attempt in this direction, when it begins by *increasing* the number of solutions transcendentally, instead of decreasing them.

In fact the problem in classical dynamics can also be presented as a *partial* differential equation, just Hamilton's equation. But the manifold of solutions of the *problem* does not correspond to the manifold of solutions of Hamilton's equation. An arbitrary 'complete' solution of Hamilton's equation solves the problem *totally.* Some *other* complete solution yields the same orbits, but in another representation.

Concerning the doubts directed against Eqn (18) as the basis of atomic dynamics, I will stipulate auxiliary conditions must be added to this equation. These will however have a character which is completely foreign and unintelligible in terms of the former 'quantum conditions'. Instead, they are of the type we are used to in the physics of partial differential equations: initial- or boundary-conditions which are in no way *analogous* to the quantum conditions. In all cases in classical mechanics which I have investigated so far, Eqn (18) *carries within itself the quantum conditions.* These appear in such a way that determination of frequency or energy levels for a stationary process is generally possible *automatically* – without any further physical assumptions – from nearly obvious requirements on ψ: that it should be unique, finite and stable in all configuration space.

The manifest doubts are therefore changed into the opposite, in any case where energy levels or frequencies occur. (Just what it is that 'oscillates' is a question that can be answered directly only for the one-body problem in ordinary three-dimensional space.) The determination of the quantum levels *no longer follows* in two separate steps: 1) Determination of all dynamically possible orbits, and 2) *Rejection* of most of the solutions found in 1) and selection of a few by special requirements. The quantum levels are determined *at once* as the *eigenvalues* of Eqn (18) *which carry in themselves the natural boundary conditions.*

Just how far the analytic investigation can be taken in complicated cases is beyond my judgement so far, but I will conjecture on it. In most problems one has the feeling that for the above two-step procedure, step 1 must be done as the final result: Energy – usually as a very simple function of the quantum numbers – is looked for. The use of the Hamilton-Jacobi method results in a great simplification, in that the actual calculation of the mechanical solution is turned around. It uses the integral which represents the momentum merely to evaluate a complex closed integration, rather than as the overall variable, which makes much more trouble. Even so, the complete solution of Hamilton's equation must be actually known, i.e. represented by quadrature. Therefore the integration of the mechanical problem must be done in principle for arbitrary initial conditions. – For the determination of the eigenvalues of a differential equation one first seeks the solution without regard to the boundary- or continuity conditions and picks out those parameter values for which the solution has the correct behavior. An example of this is given in our first Chapter. One sees in this example – what is typical for eigenvalue problems – that the solution, which *generally* is given only in very difficult inaccessible analytic form (Eqn (12), Ch.1), corresponds to an eigenvalue of extraordinary simplicity. I do not know if all *direct* methods of calculation of the eigenvalues are perfected. For the distribution of the eigenvalues, higher order numbers are the known case. But this limiting case is *not* very interesting here, it corresponds to the classical macroscopic mechanics. For spectroscopy and atomic physics, only the *first* 5 or 10 eigenvalues are of interest, although *the most interesting of all* would determine the *ionization threshold.* For a sharper wording: each eigenvalue problem can be given as a Mini-Max problem without any direct reference to the differential equation. It seems likely that direct methods for an approximate calculation of the eigenvalues can be found, so that the *urgent* need is met. It should be possible to determine in each case whether the eigenvalues numerically *known* from spectroscopy *satisfy* the problem or not.–

I should acknowledge explicitly the fact that Heisenberg, Born, Jordan, and several other prominent researchers [12], have put forth an attempt at removing the quantum difficulties which has had noteworthy success. It is very difficult to doubt that it contains a part of the truth. In its goals, Heisenberg's research is remarkably close to what we have just described. Its method however is so completely different, that I have not been able to find the connecting link. I do believe that these two approaches are not opposed to each other, but – from their extraordinary difference of view-points and methods – that they will supplement one another in that where one fails, the other will succeed. The strength of the Heisenberg program lies in its promise to give the *line intensities*, a question for which I myself so far have no answer. The strength of my research – if I am allowed to voice an opinion – lies in the physical viewpoint which the bridge built between the macroscopic and the microscopic mechanical events, and the outwardly different treatments which they require, make understandable. For me personally, a particular excitement occurred at the end of the previous chapter with the appearance of the emitted frequencies as 'beats', from which I also saw that they provide a clear understanding of the intensity formula.

§3. Examples of Applications.

······ [Note added: It is with sincere regret that we omit the discussion of these classic original developments. We give priority to Schrödinger's more qualitative and general arguments on the grounds that they are almost universally disregarded. We simply list the sub-section titles.]

1) The Planck Oscillator. The Degeneracy Problem.
2) The Rotor with Space-fixed Axis.
3) The Rigid Rotor with Free Axis.
4) The Non-rigid Rotor (Diatomic Molecule).

······ Perturbation theory expands the analytic possibilities of the new theory by an extraordinary amount. As a result of the greatest practical importance, I might mention that the first order *Stark Effect* formula, already deduced by Epstein and unassailably verified experimentally, is actually found completely correctly.

Zürich, Physikalisches Institut der Universität.

Footnotes and References:
1) See this Ann. d. Phys. **79**, 361 (1926). It is *not* absolutely necessary for understanding to read the first Chapter *before* the second.
2) The calculational techniques in the previous chapter are *not pursued further*. They serve only for preliminary rough orientation about the superficial connection between the wave equation and Hamilton's equation. ψ does not represent the actual displacement of a definite motion described in the relations of Eqn (2) of the first chapter. On the other

hand, the connection between the wave equation and the variation problem is obviously of the utmost reality: the integrand of the stationary integral is the Lagrange function for the wave process.

3) See for e.g., E.T. Whittaker, Analytische Dynamik (Deutsche Ausgabe bei Springer 1924) Ch.11, p.306ff.

4) Felix Klein since the summer of 1891 in his Lectures on Mechanics repeatedly advocated the Jacobi Theory of quasi-optical Analyses in higher non-euclidian spaces. See F. Klein, Jahresber. d. Deutsch. Math. Ver. **1** (1891) and Zeits. f. Math. u. Phys. **46** (1901). (Ges.-Abh. II. pp. 601 and 603.) In the second Note, Klein reviewed his lecture at the Naturforscherversammlung in Halle, where ten years earlier he had established the connection with and the great significance of Hamilton's optical treatment. He emphasized that it "had not not found the general acceptance which I had expected for it." – For the reference to F. Klein I need to thank a brief friendly reminder from Herr Prof. Sommerfeld's 'Atombau', 4th-Ed., p.803.

5) See particularly A. Einstein, Verh. d. D. Physik. Ges. **19**, 77 (1917). The statement given here of the quantum conditions is closest among all the older statements to the foregoing. de Broglie has also referred back to this.

6) For the optical case, see A. Sommerfeld and Iris Runge, Ann. d. Phys. **35**, 290 (1911). There (in expansion of a comment by P. Debye) it is shown that the equation of *first* order and *second* degree for the *phase* ('Hamilton's equation') can be derived exactly from the equation of *second* order and *first* degree for the *wave function* ('wave equation') in the limit of vanishing wavelength.

7) See also A. Einstein, Berl. Ber. S. 9ff (1925).

8) In the first chapter this relation was treated in the framework of complete speculation merely as an approximate equation.

9) L. de Broglie, Annals de Physique (10) **3**. 22 (1925). (Thesis, Paris, 1924.)

10) P. Debye, Ann. d. Phys. **30**, 755 (1900).

11) M. von Laue, Ann. d. Phys. **44**, 1197 (1914).

12) W. Heisenberg, Zeits. f. Phys. **33**, 870 (1925); M. Born and P. Jordan, Zeits. f. Phys. **34**, 858 (1925); M. Born, W. Heisenberg and P. Jordan, Zeits. f. Phys. **35**, 557 (1926); P. Dirac, Proc. Roy. Soc. Lond. **A109**, 642 (1925).

Paper IX·2: Excerpt from Annalen der Physik **79**, 734 (1926).

On the Relation of the Heisenberg-Born-Jordan Quantum Mechanics to Mine

von **Erwin Schrödinger**.

Eingegangen 18. März 1926.

§ 1. Introduction and Outline.

With the most extraordinarily different starting points and calculational paths, the Heisenberg matrix mechanics [1] on the one hand, and the theory I recently presented here called the 'wave ' mechanics [2] on the other, it is truly remarkable that these two new quantum theories agree in all well-known special cases, even where they disagree with the old quantum theory. I mention in particular the peculiar 'half-integral' properties of the oscillator and the rotor. This detailed agreement is even more mysterious since their starting points, ideas, methods, and complete mathematical apparatus seem entirely different. Above all, the passage to classical mechanics in the two theories seems to lead in diametrically opposite directions. In the Heisenberg theory the classical continuous variables are represented by a system of discrete numerical matrices, labeled by a pair of integral indices, and determined by *algebraic* equations. The author himself characterizes it as 'a generally discontinuous theory' [1]. Wave mechanics, in contrast, proceeds to classical mechanics from the opposite foundation of *a continuous theory*. It begins by replacing an event usually described by a few dependent variables satisfying a few *total* differential equations, with a continuous *field-like* event in the configuration space, which for a single event is determined by a *partial* differential equation derivable from an Action Principle. This Action Principle and the corresponding differential equation replace the equations of motion *and* the quantization conditions of the older 'classical quantum theory' [3].

In the following the very most intimate connection between Heisenberg's matrix mechanics and my own wave mechanics will be explained. From the formal mathematical standpoint of course one has to show that the two theories are *identical*. The basic idea of the proof follows next.

The Heisenberg theory seeks the solution of a quantum mechanical problem as the solution of a system of infinitely many algebraic equations, whose unknowns – infinite matrices – are identified as the classical position and momentum variables of the mechanical system, and functions of them, which satisfy peculiar *multiplication rules. Each* position coordinate, *each* momentum coordinate, and *any* function of them, corresponds to *an* infinite matrix.

I will then show (§2 and §3) that any function of coordinate and momentum can be represented in such a way that these matrices *always satisfy* the Born-Heisenberg multiplication rules (the 'quantization condition' or the 'commutation relation'; see below). This assignment of matrices to functions is *general*, since it assumes nothing about the *particular* mechanical system under consideration, but is the same for all such systems. (In other words: the Hamiltonian itself does not affect the commutation relations.) The actual ordering of variables on the other hand, is *undetermined* to a high degree. This is made obvious *by changing* to an *arbitrary* complete orthogonal function system with the *same configuration space*. (Note: not '*pq*-space', but '*q*-space'.) The essential *indeterminacy* of the ordering is obvious, since one can transform to an arbitrary orthogonal system.

In this very general way, matrices are constructed which obey the general multiplication rule, as I show in §4. The *particular* problem of the characteristic algebraic equation which the *matrices* of the coordinate and momentum form with the *matrix* of the Hamiltonian function (the 'equations of motion'), is completely solved if one uses a *particular* orthogonal system: namely the *eigenfunctions* of the very partial differential equation which constitutes the basis of my wave mechanics. The solution of the naturally occurring *boundary value problem* of this differential equation is *completely equivalent* to the solution of Heisenberg's algebraic equations. *All* of Heisenberg's matrix elements – which one needs to determine the 'transition probabilities' or 'line intensities' – can be found *by differentiations and integrations* once the *boundary value problem* is solved. [Note added: Emphasis added here and similarly later.] In addition to these matrix elements, other quantities closely related to them occur in the wave mechanics with the obvious significance as partial wave amplitudes of the electric moments of the atom. The intensity and polarization of the emitted light can then be found entirely *on the basis of the Maxwell-Lorentz theory.* A brief sketch of these developments can be found in §5 [Note added: omitted here].

§ 2. The Connection of an Operator and its Matrix to a Well-Ordered Function; and the Confirmation of the Commutation Relations.

The starting point for the construction of matrices is the simple observation that the original Heisenberg multiplication rule for the $2n$ coordinates and conjugate momenta – $q_1 \cdots q_n; p_1 \cdots p_n$ - agrees exactly with the multiplication law which holds for the usual *linear differential operators* in the domain of the n variables $q_1 \cdots q_n$. The ordering has to be specified, because in a *function* each of the p_k's must be replaced by the operator $\partial/\partial q_k$. – In fact each operator $\partial/\partial q_k$ commutes with $\partial/\partial q_m$ for all m, and even with q_m when $m \neq k$. The operator obtained for $m = k$ by commutation and subtraction

$$\frac{\partial}{\partial q_k} q_k - q_k \frac{\partial}{\partial q_k}, \tag{1}$$

operating on any function of the q_k, *reproduces* the function. That is: this operator is the *identity*. This simple fact will appear in the matrix representation as the Heisenberg commutation relation. Now to the systematic proof.

Because of the 'non-commutativity', a particular 'function in the usual sense' of the q_k, p_k does not correspond to a uniquely determined operator, but to a 'function symbol described in a particular way'. Moreover, since we can define no other operations with the operators $\partial/\partial q_k$ than addition and multiplication, the functions of q_k, p_k must always be expressible as regular power series in the p_k, so that we can define the operations of p_k as those of $\partial/\partial q_k$. For the evaluation of any individual term in such a power series, the following form is sufficiently general:

$$F(q_k, p_k) = f(q_1 \cdots q_n) p_r p_s p_t g(q_1 \cdots q_n) p_{r'} h(q_1 \cdots q_n) p_{r''} p_{s''} \cdots. \tag{2}$$

[Note added: Here f is some function of the $q_1 \cdots q_n$. This discussion borrows heavily from Born and Jordan's construction of matrix differentiation in [1].] We define the 'well-ordered function' corresponding to F as the following operator:

$$[F, u] \equiv f(q_1 \cdots q_n) K^3 \frac{\partial^2}{\partial q_r \partial q_s \partial q_t} g(q_1 \cdots q_n) K \frac{\partial}{\partial q_{r'}} K^2 \frac{\partial^2}{\partial q_{r''} \partial q_s''} \cdots u, \tag{3}$$

where p_k is replaced by $K\partial/\partial q_k$. [Note added: $K \to \hbar/i$ in the following.] By $[F, u]$ I mean that function (in the usual sense) of $q_1 \cdots q_n$ which one gets when one carries out the operations F on any (ordinary) function $u(q_1 \cdots q_n)$. [Note added: NOT to be confused with the commutator bracket.] If G is another well-ordered function, then $[GF, u]$ is that function of u where *first* the operations of F are carried out, and *then* the operations of G; or equivalently the operation GF. Naturally, it is not in general the same as $[FG, u]$.

Next we construct a well-ordered function such as F by representing its operators (3) in a specific but arbitrary orthogonal system, with the basic objective: all q-space should correspond to a matrix in the following way. For abbreviation we will write the variables $q_1 \cdots q_n$ simply as x, and $\int dx$ as the integration measure of q-space. Now consider the functions

$$u_1(x)\sqrt{\rho(x)}, \cdots u_n(x)\sqrt{\rho(x)}, \cdots \quad \text{to infinity,} \tag{4}$$

$$\text{which satisfy} \int dx \rho(x) u_j(x) u_k(x) = 0, j \neq k; \quad = 1, j = k. \tag{5}$$

It will be assumed that these functions u_k vanish strongly enough on the natural boundary of q-space (in general at infinity), that the surface integrals appearing as by-products of integrations-by-parts which occur later, should all vanish.

The function F represented by (2) with the operator (3) will now be assigned the following matrix

$$F^{jk} = \int dx \rho(x) u_j(x)[F, u_k(x)]. \tag{6}$$

(the indices on the left do not imply any 'covariance' connection; we write the matrix indices *above*, since we later have to write the matrix elements of the q_l, p_l, for which we reserve the lower place.) – In words: the j, k-matrix element of an operator is calculated by multiplying the orthogonal function $u_j(x)$ with the *row* index j times the density function $\rho(x)$, then the operator acting on the orthogonal function $u_k(x)$ with the *column* index k, and integrating $\int dx$ over the whole domain [4].

It is easy to show that addition and multiplication of the well-ordered functions – that is, the corresponding operators – is the same as matrix addition and matrix multiplication of the corresponding matrices. For addition this is trivial. For multiplication the proof is as follows: let G be another well-ordered function like F and G^{lm} its corresponding matrix

$$G^{lm} = \int dx \rho(x) u_l[G, u_m]. \tag{7}$$

We want to show that the product obeys $(FG)^{km} = \sum_l F^{kl} G^{lm}$. First we transform the expression (6) for F^{kl} in the following way. By a series of integrations-by-parts, the operator F is 'shifted-over' from the function u_l to the function ρu_k. In the 'shifting', the *order* of the operations is clearly reversed. All the surface integrals should vanish as above. The 'shift' operation, with a sign change for an odd number of derivatives, will be $[\bar{F}, \rho(x) u_k(x)]$, $\cdots\cdots$. We get

$$F^{kl} = \int dx u_l[\bar{F}, \rho u_k]. \tag{6}$$

If one now calculates the matrix product, then

$$\sum_l F^{kl} G^{lm} = \sum_l \left\{ \int dx u_l(x)[\bar{F}, \rho u_k] \times \int dx' \rho u_l(x')[G, u_m] \right\}$$

$$= \int dx [\bar{F}, \rho u_k] \times [G, u_m]. \tag{8}$$

The last equation is the result of the 'completeness relation' of the orthogonal functions $u_l(x)$ [5], applied to the 'expansion coefficients' of the functions $[G, u_m]$ and $[\bar{F}, \rho u_k]/\rho$.

$*\,*\,*\,*\,*\,*\,*\,*\,*\,*\,*\,*\,*\,*\,*\,*\,*\,*\,*\,*$

All this is familiar in the *Dirac* delta-function notation for the completeness relation – $\sum_l \rho(x) u_l(x) u_l(x') = \delta(x - x')$ - with $\delta(x - x')$ defined by the integration over x' for *any* function f

$$\int_{<x}^{>x} dx' \delta(x - x') f(x') = f(x).$$

$*\,*\,*\,*\,*\,*\,*\,*\,*\,*\,*\,*\,*\,*\,*\,*\,*\,*\,*\,*$

Next, by new integrations-by-parts in (8), we shift the operators in $[\bar{F}, \rho u_k]$ over to the function $[G, u_m]$, where the operator again has its original order. Finally, one gets

$$(FG)^{km} = \sum_l F^{kl}G^{lm} = \int dx \rho u_k[FG, u_m]. \tag{9}$$

On the left stands the km^{th} element of the matrix product FG. On the right, the km^{th} element of the matrix corresponding to the well-ordered product FG, which is what we set out to prove.

§ 3. The Heisenberg Quantization Condition and the Rule for Partial Integration.

Since the operation (1) is the identity, then the well-ordered function for $pq - qp$ is the *operator* – multiplication by $\hbar/i$. The corresponding *matrix* is

$$(p_l q_l - q_l p_l)^{jk} = \hbar/i \int dx \rho u_j u_k = 0 \text{ for } j \neq k, \quad = \hbar/i \text{ for } j = k. \tag{11}$$

This is Heisenberg's quantization condition. – In a self-explanatory way, we find that the q_j and p_j correspond to the matrices:

$$(q_j)^{km} = \int dx \rho u_k q_j u_m \text{ and } (p_j)^{km} = \int dx \rho u_k \frac{\hbar}{i} \frac{\partial}{\partial q_j} u_m. \tag{12}$$

We turn next to the 'Rule for Partial Differentiation'. A well-ordered function such as (2) can be differentiated with respect to q_k [6], without changing the order of factors *at each step*, if q_k should be entered after each q_k differentiation and all these results added. Then it is easy to show that the following relation holds

$$\left[\frac{\partial F}{\partial q_k}, u\right] = \frac{i}{\hbar}[p_k F - F p_k, u]. \tag{13}$$

The idea is this: instead of directly differentiating with respect to q_k, I can make it easier for myself and simply write $ip_k/\hbar$ which is just $\partial/\partial q_k$. However, the operator $\partial/\partial q_k$ affects not only the q_k dependence in F (which it *should*), but also the q_k dependence of u (which it *should not*). *This error is corrected*, when I subtract the operation $[Fp_k, u]$!

Consider next partial differentiation with respect to a p_k. This is even simpler than that for $\partial/\partial q_k$, since the p_k occur only as products of powers, like p_k^3 which stands for individual factors $p_k p_k p_k$. Then we can say: for partial differentiation with respect to p_k, each individual p_k which occurs in F should be omitted just once (with all the remaining p_k's held fixed); and all the results added. How does this affect an operator like (3)? – 'Each individual $K\partial/\partial q_k$ is to be omitted once, and all the results added.' This prescription gives the operator equation:

$$\left[\frac{\partial F}{\partial p_k}, u\right] = \frac{i}{\hbar} \left[F q_k - q_k F, u\right]. \tag{15}$$

To demonstrate this, construct the operator $[F q_k, u]$, and then 'take q_k through F from right to left'; that is, by successive permutations, bring the operator $F q_k$ to $q_k F$. The exchanges meet a barrier whenever the q_k stands next to a $\partial/\partial q_k$. In this case, the exchange is not simply $q_k \partial/\partial q_k$, but $q_k \partial/\partial q_k + 1$. Each of these 'inserts' yields a by-product of the permutation, as one easily observes directly from the resulting 'partial derivative quotients'. At the end of the cross-overs there remains the operator $q_k F$, which is explicitly subtracted in (15), proving the result. Equations (14) and (15) hold also for matrices, since from (6) any linear operator corresponds to one and only one matrix (naturally for appropriate function systems $u_j(x)$) [7].

§ 4. The Solution of the Heisenberg Equations of Motion.

We have just shown that the matrices constructed from well-ordered functions using an arbitrary complete orthonormal system (4) with the definitions (3) and (6), satisfy all the Heisenberg relations including the commutation relation (11). Now we consider a particular mechanics problem characterized by a specific Hamiltonian $H(q_k, p_k)$. The authors of quantum mechanics [1] took this function directly from the *usual* mechanics, which naturally does *not* yield the result in 'well-ordered' form; in the usual analysis these do not differ except in the order of the factors. The 'ordering' or 'symmetrization' of the function in a particular way for quantum mechanics, for example by replacing the usual mechanical function qp^2 by $(p^2 q + q p^2)/2$ or even by $(p^2 q + p q p + q p^2)/3$, are all the same according to (11). This function is already 'well-ordered', that is, the order of the factors is irrelevant. I will not go into the general symmetrization rule here [1]. The main points are, if I understand things correctly: H^{kj} should be a diagonal matrix in which the symmetrized functions are identical with those originally given by the usual analysis [8]. We will comply with these requirements in a direct way.

Then the authors require that the matrices $(q_l)^{jk}$, $(p_l)^{jk}$ obey an infinite set of 'equations of motion', which they originally wrote as follows:

$$\left(\frac{dq_l}{dt}\right)^{jk} = \left(\frac{\partial H}{\partial p_l}\right)^{jk} \quad \text{and} \quad \left(\frac{dp_l}{dt}\right)^{jk} = -\left(\frac{\partial H}{\partial q_l}\right)^{jk}, \tag{18}$$

with $l = 1, 2, 3 \cdots n$ and $j, k = 1, 2, 3 \cdots \infty$. The upper index pair j, k again means the particular matrix-element corresponding to the well-ordered function. The significance of partial derivatives on the right needs explanation, since they do not appear to correspond to the total derivative d/dt on the left. By that the authors mean: there should be an infinite series of numbers $\omega_1, \omega_2, \omega_3, \omega_4 \cdots \infty$, such that the above equations are satisfied when one assigns to d/dt operating on the jk^{th} matrix element the effect of multiplication by $i(\omega_j - \omega_k)$.

The series of eigenfrequencies (19) however is not always known beforehand, but forms with the q, p matrix elements the numerical unknowns of the system of equations (18). With the time dependence (20), the commutation relations (14) and (15), and the definition (12), these have the form:

$$(\omega_j - \omega_k)q_l^{jk} = \frac{1}{\hbar}(Hq_l - q_lH)^{jk} \text{ and } (\omega_j - \omega_k)p_l^{jk} = \frac{1}{\hbar}(Hp_l - p_lH)^{jk}. \tag{18}$$

We must have enough information to solve these equations with nothing more than the eigenvalues of the matrices constructed from the orthogonal system (4). I now make the following assertions:

1.) The equations (18') are generally sufficient if one chooses as the orthogonal system the *eigenfunctions* of the naturally occurring boundary value problem of the differential equation:

$$E\psi = [H, \psi] \tag{21}$$

where ψ is an unknown function of $q_1 \cdots q_n$; E is the eigenvalue parameter. The density function is the same as that which must be used to make (21) self-adjoint. The angular frequencies ω_j are the eigenvalues E_j divided by $\hbar$. H_{jk} becomes the diagonal matrix $H^{kk} = E_k$.

2.) If one has symmetrized the function H in a suitable way – the symmetrizing process in my opinion is not uniquely defined *a priori* – then (21) is *identical with the wave equation of my own wave mechanics* [9].

The assertions 1) are almost immediately obvious when one first thinks about the question, whether or not (21) corresponds in general to a sensible boundary value problem in the domain of all q-space. This is necessary in order that multiplication by a suitable function can always make it self-adjoint, and so forth. This question is further resolved in assertion 2). – Since by (21) and the definition of the eigenvalue and eigenfunctions $[H, u_j] = E_j u_j$, then from (6)

$$H^{jk} = \int dx \rho u_j [H, u_k] = E_k \int dx \rho u_j u_k$$
$$= 0 \quad \text{for} \quad j \neq k, \qquad = E_k \quad \text{for} \quad j = k; \tag{23}$$

and for example, $(Hq_l)^{jk} = \sum_m H^{jm} q_l^{mk} = E_j q_l^{jk}$

$$(q_l H)^{jk} = \sum_m q_l^{jm} H^{mk} = E_k q_l^{jk}, \tag{24}$$

so the right side of the first equation (18') is $\frac{E_j - E_k}{\hbar} q_l^{jk}$, and similarly for the second equation. In this way the assertions 1) are proven.

Turning now to 2), the equivalence of the (suitably symmetrized) Hamiltonian with the wave operator of wave mechanics. Firstly, it seems to me that the symmetrization process is not unique. For example, for one degree of freedom, consider the *usual* Hamiltonian

$$H = \frac{1}{2}(p^2 + q^2). \tag{25}$$

This can be taken unchanged into quantum mechanics as a 'well-ordered' function. However, it seems to me that with equal justification one could use the well-ordered function

$$H = \frac{1}{2} \left(\frac{1}{f(q)} p f(q) p + q^2 \right), \tag{26}$$

where $f(q)$ would be some 'density function' $\rho(x)$. (26) is obviously just a special case of (27) and the question arises, how is it possible in general, especially for a complicated Hamiltonian, to choose the special case. $\cdots\cdots\cdots$

So the solution of the whole Heisenberg-Born-Jordan system of matrix equations reduces to the natural boundary value problem of a linear differential equation. One must solve the boundary value problem, then calculate the required matrix elements using (6).

To clarify what is meant by the *natural* boundary value problem, I refer to a particular example [10]. It turns out that the naturally infinite boundary of q-space constitutes a singularity of the differential equation and allows only one unique finite stable boundary condition. It seems likely that a general requirement for the applicability of the theory is that the problems should first of all be microscopic. If the domain of the length coordinate is imagined to be reduced (for example, particles in a 'box'), the reduction must be done in a fundamental way by introducing a suitable potential. The *vanishing* of the eigenfunctions at the boundary is usually sufficient, and indeed corresponds to the known behavior of the integrals of (6). $\cdots\cdots$

Footnotes and References:

1) W. Heisenberg, Zeits. f. Phys. **33**, 879 (1925); M. Born and P. Jordan, Zeits. f. Phys. **34**, 858 (1925) and **35**, 557 (1926) (the last with Heisenberg). $\cdots$ I occasionly refer to the last of these as 'Quantum Mechanics I and II'. See also P. Dirac, Proc. Roy. Soc. London **A109**, 642 (1925) and **A110**, 561 (1926). (See here as **Papers VI·1, VII·1,2** and **VIII·1**.)

2) E. Schrödinger, Ann. d. Phys. **79**, 361 (Ch.1); 489 (1926) (Ch.2). These reports are completely independent of this Note, which only provides the connecting link among them. (See here as **Paper IX·1**.)

3) My theory was motivated by L. de Broglie, Ann. de Physique **3**, 22 (1925) (Thesis, Paris, 1924) (see here as **Paper IV·1**); and by infinitely far-sighted remarks of A. Einstein, Berl. Ber. 1925, p.9. I myself was unaware throughout of any genetic connection with Heisenberg's work. I naturally knew of his theory, but felt deterred, to say nothing of repelled, by the seemingly very difficult methods of the transcendental algebra and by the lack of any physical insight.

* * * * * * * * * * * * * * * * * * * **

This aside is a marvelous glimpse of the dynamic personality of Schrödinger which is also present in his energetic technical writing. The "infinitely far-sighted remarks" of Einstein remain to be traced down. His reaction of being "repelled" by the "transcendental" formalism of the matrix mechanics with its "lack of any physical insight" was implicitly acknowledged by Born who immediately adopted the wave mechanics after failing to make progress on the scattering problem using the matrix formalism, and as a by-product was led to his Nobel Prize winning prescription for the interpretation of quantum mechanics (see the next chapter).

* * * * * * * * * * * * * * * * * * **

4) F^{jk} is the jth 'expansion coefficient' of the operator F acting on u_k.

5) See for example Courant-Hilbert, *Methods of Mathematical Physics I*, p.36. It is important to remember that the completeness relation for the 'expansion coefficients' holds even when the expansions themselves are *not* convergent. If they converge, then the validity of (8) is immediately completely obvious.

6) For all these definitions we follow Heisenberg. From the strictly logical standpoint, the resulting proof is actually too restrictive and we could write down the rules (14) and (15) directly, since they have been proved by Heisenberg and are needed only for the sum-, product- and for the commutation relation (11), which we have proven.

7) The *converse* also holds, in the sense that for a given orthogonal system and density function *no more* than *one* linear differential operator can correspond to a given *matrix* according to our prescription (6). For in (6) let F^{jk} be given, let $[F, u]$ be the required linear operator (whose *existence* we *assume*), and let $\phi(x)$ be piecewise continuous and differentiable as necessary, but otherwise a completely arbitrary function of $q_1 \cdots q_n$. Then the completeness relation applied to the functions ϕ and $[F, u_k]$ gives

$$\int dx \rho \phi [F, u_k] = \sum_j \left\{ \int dx \rho \phi u_j(x) \times \int dx' \rho u_j(x')[F, u_k] \right\}$$

The right side holds identically, since it involves only the expansion coefficients of ϕ and the given matrix F^{jk}. By 'shifting', one can transform the left side into the kth 'expansion coefficient' of the function $[\bar{F}, \rho\phi]/\rho$. So all the expansion coefficients of this function are uniquely determined, and so also the function itself (Courant-Hilbert, p.37). Since ρ is given and ϕ is arbitrary, then we can say: the action of the shift operator on an *arbitrary* function – that is, the *shift operator* itself – is *uniquely* defined by the matrix F^{jk}.

Note that the expandability of the functions has been assumed. – That there must always be a linear operator for any arbitrary matrix, has not been proved.

8) The *stronger* requirement: that these should yield the quantum mechanical equations of motion – I maintain to be too strong. In my opinion it led the authors to restrict themselves to polynomials in q, which is unnecessary.

9) Ann. d. Phys. **79**, 510 (1926). See Eqn 18.

10) Ann. d. Phys. **79**, 362, 510 (1926). See eqns 23,24.

Paper IX·3: Excerpt from Die Naturwiss. **14**, 664 (1926).

Der Stetig Übergang von Mikro- zur Makromechanik

von **E. Schrödinger**, Zurich.

Based on ideas of de Broglie [1] and Einstein [2], I have sought to show [3] that the usual differential equations of mechanics, which determine the coordinates of mechanical systems as functions of time, are no longer adequate for 'small' systems; one has instead *partial* differential equations which determine a quantity ψ (the 'wave function') as a function of the coordinates and the time. Just like the differential equation for a vibrating string or some other oscillating system, the displacement ψ is a superposition of pure time-harmonic (i.e., sine-like) vibrations, whose frequencies correspond exactly with the 'term-frequencies' of the microscopic system. For example, the Planck harmonic oscillator [4] with the energy

$$\frac{m}{2}\left(\frac{dq}{dt}\right)^2 + \frac{m\omega_0^2}{2}q^2, \tag{1}$$

when expressed in terms of the dimensionless length-variable

$$x = q\sqrt{\frac{m\omega_0}{\hbar}}, \tag{2}$$

has the eigensolutions [5]

$$\psi_n = e^{-x^2/2}H_n(x)e^{-i\omega_n t}, \tag{3}$$

with $\omega_n = \omega_0\left(n + 1/2\right)$; $n = 0, 1, 2, 3\cdots$. H_n are the well-known Hermite polynomials [6] (e.g., $H_0 = 1$, $H_1 = 2x, \cdots$). When multiplied by $e^{-x^2/2}$ and the 'normalization factor' $1/\sqrt{2^n n!}$, they become Hermite's orthogonal functions. These are the amplitudes of the eigen-oscillations. $\cdots$ The similarity with the usual string functions is clear.

I choose a number $A \gg 1$ and construct the following superposition of eigenfunctions:

$$\psi = \sum_{n=0}^{\infty}\left(\frac{A}{2}\right)^n\frac{\psi_n}{n!} = e^{-i\omega_0 t/2}\sum_{n=0}^{\infty}\left(\frac{A}{2}e^{-i\omega_0 t}\right)^n\frac{1}{n!}e^{-x^2/2}H_n(x). \tag{4}$$

These are the *normalized* eigenfunctions (see above) combined together with the coefficients

$$\frac{A^n}{\sqrt{2^n n!}},$$ (5)

which, as one easily shows [7], has the effect of singling out a relatively small group of eigenfunctions in the neighborhood of the mean n-value

$$\bar{n} = \frac{A^2}{2}.$$ (6)

The sum in (4) is evaluated using the identity [8]

$$\sum_{n=0}^{\infty} \frac{s^n}{n!} e^{-x^2/2} H_n(x) = e^{-s^2+2sx-x^2/2}.$$ (7)

Therefore

$$\psi = e^{-i\omega_0 t/2} e^{[-A^2/4 e^{-2i\omega_0 t}+Axe^{-i\omega_0 t}-x^2/2]}.$$ (8)

If one takes, as I presume we should, the real part of the right side, then

$$\psi = e^{\left(\frac{A^2}{4}-\frac{(x-A\cos\omega_0 t)^2}{2}\right)} \cos\left[\frac{\omega_0 t}{2} + A\sin\omega_0 t \cdot (x - \frac{A}{2}\cos\omega_0 t)\right].$$ (9)

In this result, the first factor is of special interest. It displays a relatively high and narrow 'peak' with the shape of a 'Gaussian error curve', centered on the moving point

$$x = A\cos\omega_0 t.$$ (10)

The width of the peak is of order 1, which is very small compared to A. According to (10) the peak oscillates in the same way as would a point mass with the same energy function (1) in the usual mechanics. The amplitude in x is A, or in terms of q is

$$a = A\sqrt{\frac{\hbar}{m\omega_0}}.$$ (11)

The energy E of a point mass m oscillating according to the usual mechanics with the same amplitude and frequency is

$$E = \frac{1}{2}ma^2\omega_0^2 = \frac{A^2}{2}\hbar\omega_0,$$ (12)

which, according to (6), is exactly $\bar{n}\hbar\omega_0$, where $\bar{n} = A^2/2$ is the mean quantum number of the given wave packet. The 'correspondence' is therefore complete in this respect also.

The *second* factor in (9) is a rapidly varying function of both x and t, of magnitude ≤ 1, which modulates the first factor into a wave packet such as sketched in Fig2. $\cdots$ A more exact analysis of (9) shows that Fig2 – at only *one* instant – does not fully represent the situation. The *number* and *width* of the 'oscillations' or 'waves' in the wave packet depend on time. The waves are most *numerous* and *smallest* near the origin $x = 0$; they are completely symmetric at the turning points $x = \pm A$ which occur at $\cos\omega_0 t = \pm 1$, when the second factor in (9) is independent of x. The extension of the wave packet ('density of mass points') always stays the same. The 'variation' of the 'curvature' can be interpreted as an indicator of the velocity, which can be clearly understood from the general wave-mechanical point of view – as I will not describe further here.

* * * * * * * * * * * * * * * * * * **

The previous paragraph is a red-herring introduced by Schrödinger's erroneous presumption of the significance of the real part of ψ. Things are actually easier and more clear if we take the absolute square $|\psi|^2$, which was soon found to be the correct prescription. Then we get the probability density of the wave packet (8) to be

$$|\psi|^2 = e^{A^2/2}e^{-(x-A\cos\omega_0 t)^2}.$$

* * * * * * * * * * * * * * * * * **

Our wave packet holds together permanently, and does *not* spread out with time, as one would expect from optics for example. That does not, of course, say much about the completely analogous behavior of a bump on a string. One easily shows, however, that by multiplication of two or three expressions like (3) – one in x, one in y, and the third in z – either an even- or an odd-dimension wave packet can be represented, which revolves on a periodic ellipse [9]. Such a wave packet always remains compact, in contrast to a wave packet in classical optics which spreads out with time. The difference may be due to the fact that our wave packet is constructed from individual *discrete* harmonic components, not from a *continuum* of them.

* * * * * * * * * * * * * * * * * * **

Heisenberg almost immediately showed that the harmonic oscillator was a completely misleading example of long-term coherence in a wavepacket. In any realistic example, the packet is constructed as a superposition of waves whose frequencies are not rational multiples of each other, and as a result they lose phase coherence rapidly. This did not deter Schrödinger and many after him from their belief in the interpretation of the wave function as a 'guiding' wave accompanying the classical particle along its trajectory.

* * * * * * * * * * * * * * * * * * **

It might be noted that a common additive constant C which could be added to each ω_n in (3) (corresponding to the 'rest energy' of the mass points) makes no essential change. It only adds a phase Ct in the angular bracket of (9). With that, the timelike oscillations within the wave packet are much faster, but the envelope of the packet (10) describing the pendulum on the whole, and even its shape, remain completely unchanged.

It is obvious that in a completely analogous way one could construct wave packets for the hydrogen atom which at high quantum numbers would revolve in Kepler ellipses; but these are computationally much more complicated than the especially simple pedagogic example considered above.

* * * * * * * * * * * * * * * * * * **

This has become the field of study known as Rydberg atoms, where the classical and quantum worlds do coexist with marvelous compatibility for hydrogenic orbits with principal quantum numbers of order 100's.

* * * * * * * * * * * * * * * * * * **

Footnotes and References:

1) L. de Broglie, Ann. de phys. **3**, 22 (1925) (Thesis, Paris, 1924).

2) A. Einstein, Berlin. Ber. 1925, p. 9ff.

3) E. Schrödinger, Ann. d. Phys. **79**, 361, 489, 734 (1926).

4) I. e., a point mass m on a line is attracted to a fixed point by a force which is proportional to its distance from the point; according to the usual mechanics, such a point mass executes sinusoidal motion of frequency $f_0 = \omega_0/2\pi$.

5) i is $\sqrt{-1}$. On the right it is customary to assume the real part.

6) See Courant-Hilbert, *Methoden der Mathematischen Physik* I, Kap. II, §10, 4 S.76 (Berlin:Springer 1924).

7) $z^n/n!$ for large z as a function of n has a very sharp maximum for $\bar{n} = z$. By squaring (5) we get $\bar{n} \sim A^2/2$.

8) Courant-Hilbert l.c., Eqn 58.

9) It is worth noting that for an even-dimension oscillator, the quantum levels are integral; for an odd-dimension oscillator, half-integral. The spectroscopically important half-integral representations are due to the odd dimensionality of space.

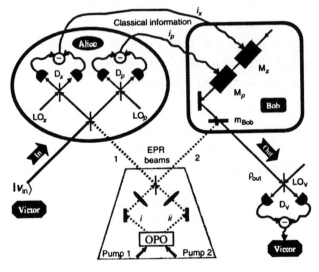

Fig. 1. Schematic of the experimental apparatus for teleportation of an unknown quantum state $|v_{in}\rangle$ from Alice's sending station to Bob's receiving terminal by way of the classical information (i_x, i_p) sent from Alice to Bob and the shared entanglement of the EPR beams (1, 2).

Schematic of Kimble's apparatus for teleportation of encrypted quantum state information from 'Alice to Bob'. See our ChXIV, p.460. [From A. Furusawa, J. Sorenson, S. Braunstein, C. Fuchs, H. Kimble, and E. Polzik, *Unconditional Quantum Teleportation* in Science **282**, 706 (1998).]

PART THREE. Chapter X

MAX BORN:

Schrödingersche Quantenmechanik

gibt eine ganz bestimmte antwort; aber es handelt sich um keine Kausalbeziehung. Hier erhebt sich der ganze Problematik des Determinismus.

"Schrödinger quantum mechanics provides $\cdots$ a completely definite answer; $\cdots$ but it is not a specific result. $\cdots$ Here arises the whole Problem of Determinism."
Zeits. f. Phys. **37**, *863 (1926); p.866.*

§X-1. Introduction.

From the first, much effort has been expended to apportion just the *right* amount of credit – never too much, usually as little as possible – to Max Born for his introduction of the probabilistic interpretation of quantum mechanics. Pais [1] gives many examples of mean (and usually self-serving) remarks: beginning with Bohr "the 'interpretation' of the wave function was obvious" $\cdots$ "We had never dreamed that it could be otherwise."; and 'Dirac had the same idea.'; and 'So did Wigner, who told me (Pais) $\cdots$ that he, too, had thought of identifying Φ_{mn} or $|\Phi_{mn}|$ with a probability. When Born's paper came out $\cdots$ "$\cdots$ I (Wigner) soon realized that Born was right."'; and Mott "$\cdots$ given Schrödinger, de Broglie, and the experimental results, this must have been quickly apparent to everyone $\cdots$ in 1928 it was already called the 'Copenhagen interpretation' – I do not think I ever realized that Born was the first $\cdots$"; and Casimir "I do not recall that Born was especially referred to. He was of course co-creator of matrix mechanics."; and Heisenberg – in the fourth reference in the third footnote of his 1927 paper announcing the uncertainty principle (see **Paper XI·1**) – says "$\cdots$ it is mathematically analyzed in the fundamental work of M. Born $\cdots$ and its interpretation used for scattering phenomena." Heisenberg precedes this with a reference to 'the statistical aspect of quantum mechanics $\cdots$ in Born, Heisenberg, and Jordan (Chapter 4, §3 of **Paper VII·2**)', but that development (the *only* reference to work earlier than Born's) – based on fluctuations of a quantum gas – is peripheral if not completely irrelevant to the *interpretation* of the wave function. Next, Bohr in his 1927 Como Lecture (see **Paper XII·1**) does not mention Born at all. Finally, Heisenberg – thirty years later – (see **Paper XI·2**) relented somewhat in grudging and still inadequate credit: "Born was able to recognize $\cdots$ the probability wave $\cdots$ contains statistical statements $\cdots$ as in Gibbs' thermodynamics $\cdots$" and " Born's hypothesis $\cdots$ is a particular case of this (Heisenberg's subsequent) more general conclusion."

The resolution of such manifestly unfair behavior is found in a principle enunciated by van der Waerden [2]. Paraphrased in briefest form, it states that: What counts is 'Who *published* What, and When.'

Born's interpretation of the wave function is contained in a few lines in his 'preliminary' version submitted in late June 1926. This predates by two weeks Schrödinger's lecture at Munich, to which Heisenberg took initial exception; and by about three months the beginning of the Schrödinger-Bohr debate in Copenhagen. His conclusions were general, in spite of the fact that they were based on a heuristic discussion of the asymptotic properties of the wave function of an electron elastically or inelastically scattered by an atom. He left no doubt that he understood the wave function to predict the result of repeated experiments on an ensemble of identical systems. In this respect he had already corrected Schrödinger's naive [3] semi-classical interpretation of the wave function in terms of a classical density or even probability wave.

Born's work was not complete, in the sense that he did not specify how to prepare the ensemble, or what happened to the wave function after the experiment. Nor did he correctly understand what the status (or rather the lack of it) of the Schrödinger equation was during these interventions. Here Born left the further development of the interpretation to the Copenhagen school, in particular to Heisenberg and Bohr. He did not mention the process of experiment, detection, measurement, preparation, etc.; nor did he mention the paradox of wave function collapse. Nonetheless, and perhaps only from our informed perspective, these steps are inevitable once he assigned the wave function the role of describing an ensemble; and detection the role of relegating the acted-upon-systems to a new sub-ensemble. It also seems that he did not correctly understand that these auxiliary steps were necessary to provide the logical completeness which he conjectured for quantum mechanics, nor that Schrödinger's equation by itself *did not* describe the quantum jumps.

The apportionment of credit along the strictly documented lines defined by van der Waerden, would seem to be transparent on the record in spite of the persistent refusal primarily by Bohr and Heisenberg to make proper acknowledgement of Born's priority. The failure to acknowledge cannot be forgiven by any of the usual modern excuses that the result was buried in some obscure journal, or lost in the flood of information, or not properly promoted, etc, etc. Born's result was published in the primary journal by one of the primary founders of quantum mechanics. Imperfect and incomplete though it was, Born's result explained the significance of the Schrödinger wave function in unusually transparent terms that made the Copenhagen interpretation logically inevitable.

§X-2. The Essence of Born's Interpretation.

Here we isolate the key points of Born's discovery papers. They are emphasized again in context in **Papers X·1,2**. First of all, Born's papers are sprinkled with misstatements, misconceptions, and possibly errors. We list a few in footnote [4], and will not refer to them again. This is not to say that they have done no harm. His reference to Einstein and 'ghost' or 'guiding' fields has given license to almost all the retrograde interpretations of quantum mechanics which abound to this day. And they do obscure the positive contribution which nevertheless was there to suggest the way for Bohr and Heisenberg.

And now to the positives in Born's papers. The essence of Born's first paper reduces to:
I have studied the collision process in the existing theory. Schrödinger's form is by far the most useful here. Both before and after the collision, the electron is

free and *a definite state of the atom and a definite momentum state of the electron must be defined.* When the two interact, there is a complicated wave function which determines the asymptotic behavior at infinity. The problem is to solve the Schrödinger wave equation with the boundary condition $\cdots$(the incoming electron). The 'scattered' wave at infinity describes the system after the collision. With a simple perturbation calculation, $\cdots$ there is a unique solution which for $z \to +\infty$ approaches (see the text of Born's first paper (**Paper X·1**) for details):

$$\psi_{\text{scatt}}(r \to \infty) \simeq \sum_{\vec{k}_f, m} |\vec{k}_f; m\rangle \langle m; \vec{k}_f | V | \vec{k}_i; n\rangle.$$

The energy of the scattered particle is given by

$$\frac{k_f{}^2}{2m} = E_n - E_m + \frac{k_i{}^2}{2m}.$$

Only one interpretation is possible. $\langle m; \vec{k}_f | V | n; \vec{k}_i\rangle$ is the probability amplitude that the incident electron is scattered from the incident momentum $\vec{k}_i$ into the outgoing momentum $\vec{k}_f$ with the energy loss $E_m - E_n$.

Quantum mechanics therefore provides a *definite* answer for the result of a collision; but not a *specific* result. One gets no answer to the question "What is the state after the collision?", but only to the question "What is the probability of a given effect of the collision?"

Here arises the whole Problem of Determinism. $\cdots$ The 'distribution function' $\Phi \simeq \langle f | V | i \rangle$ is now readily understood and calculable.

Born's second paper must be read selectively, sprinkled as it is in the introduction with references to 'ghost' fields etc, but culminating in still acceptable remarks beginning with 'Einstein's remark that "the waves $\cdots$ gave the probability that a light quantum travels in a particular direction."' And Born's conclusion that 'only the probability of a particular (result) is determined by the (scattering wave function)'; and 'The motion of the particle follows probability laws, but the probability (amplitude) propagates in accordance with (Schrödinger's equation).'

Here begin Born's two most instructive interpretational developments. The first identifies the wavefunction as a probability amplitude *applicable to an en-*

semble of identical systems: Any function ψ can be expanded in eigenfunctions ψ_n as $\psi = \sum_n c_n \psi_n$. Our idea explains that this coherent superposition gives the probability that in any such group of independent atoms, the state ψ_n occurs with a definite probability $|c_n|^2$.

The second supports his "only possible interpretation" of the scattered wave function: For aperiodic accelerated systems, it is natural to speak as follows: Choose the incoming wave appropriately; then the expansion coefficients $c(k)$ for the outgoing waves are uniquely determined. The result is the scattering amplitude for the collision.

These two remarks in §1&2 of Born's second paper are important supplements to the conclusions already strongly made in his first paper.

We skip over Born's original developments to his §9 Concluding Statement: "The Schrödinger formulation is the easiest way by far to discuss the problem of transitions. It becomes possible to follow the whole space-time development of an event. But it does not allow a unique prediction of individual events. The complete indeterminacy was mentioned previously."

§X-3. von Neumann's Measurement Postulate.

von Neumann in his classic and authoritative treatise *Mathematische Grundlagen der Quantenmechanik* [6] rigorously defined the original notion of 'strong' quantum measurement in which a quantum state is completely resolved into its separate orthogonal component amplitudes for a particular Hilbert space. These amplitudes (actually each absolute square for the corresponding probability) are then 'measured' in a catastrophic, incoherent, 'classical' detection which produces a cascade sensibly described – without recourse to quantum niceties – as macroscopic. The result is the irreversible 'collapse' of the wave function into the component detected during the macroscopic measurement process. The prescription is that the measurement and collapse must be treated outside the realm of quantum mechanics as an interruption of the unitary Hamiltonian evolution of the quantum mechanical state.

The Stern-Gerlach experiment is the prototypical 'strong' quantum measurement for detecting the component of spin-1/2 parallel or anti-parallel to the gradient of an external magnetic field. Even in this simple situation, paradoxical results can be found [7] resulting from a so-called 'weak' measurement in which the spin-components are not completely resolved along the z-direction, and then fully resolved along the x-direction by a 'strong' post-selection measurement. Aharanov, Albert, and Vaidman pose a paradox for such a sequence of Stern-Gerlach 'measurements' " $\cdots$ the result of a measurement of a component of the spin of a spin-1/2 particle can turn out to be 100." The resolution of this paradox is instructive [8] (and obviously does *not* involve outrageous spin components), and the paradoxical result has been observed [9] in a laser-optics analog of the original spin-1/2 situation.

The orthodox view was challenged most productively by the paradox of Einstein, Podolsky, and Rosen (see our Ch**XIII** and **Paper XIII·1**); but defended by Bohr (see **Paper XIII·2,3**) orthodoxy prevailed until Bohm and Bell sharpened the paradox, thereby leading to crucial new concepts of *coherent* manipulation of quantum states (see our Ch**XIV**) in Aspect experiments. Finally, 'measurements' have been developed – quantum non-demolition measurements – which do not involve the collapse of the wave packet, or any catastrophic, irreversible, decohering violence to the quantum state (see our Ch**XIV, XVIII**). Such QND-measurements go beyond Born-Bohr, but do not contradict them. The kind of information available in QND-measurements simply can not include anything constrained by the Heisenberg Uncertainty Principle and typically involves the coherent *comparison* of two quantum systems, without requiring specific knowledge of either one. There is a close analogy to the description of two oscillators in terms of center-of-mass and relative coordinates. The CM momentum $P = p_1 + p_2$ and the relative coordinate $q_r = q_1 - q_2$ are commuting operators and can be determined precisely and simultaneously without restriction by the uncertainty principle or disruption of the quantum state; but no precise knowledge of the sort p_1, q_1 can be obtained, and even that only at the usual price of decohering the original quantum state. These situations were described at length already in von Neumann's original work, but not of course anticipating any of the subtleties of modern quantum-optical experimental techniques.

§X-4. Concluding Remarks.

At this point, our purpose is not too defend the orthodox Copenhagen interpretation (whatever that is) with all it's associated baggage of paradox, debate, confusion, abstraction, alternatives, etc. Our point is that Born's idea – to interpret the Schrödinger wave function as a probability amplitude associated with an ensemble of identical quantum systems; which propagates according to the Schrödinger equation; and gives transition amplitudes and probabilities in agreement with experiment – leads *inevitably* to the steps taken primarily by Heisenberg and Bohr to physically interpret Born's mathematical prescriptions. Once Born wrote down the scattered wave function (here in modern notation, in 1st-Born approximation) $\psi_{scatt} \simeq \sum_{f'} |f'\rangle\langle f'|V|i\rangle$ the interpretation in terms of Hilbert space state vectors [7] was obvious; and the interpretation as a continuously evolving probability was clearly stated. What remained to be done was to make sense of it all, and this Born left to others. The interruption of the Schrödinger propagation and the collapse of the wave function; the quantum jumps; the nature of the detection and measurement process; the 'information theory' aspect of the wave function; the correspondence with classical expectations and prejudgements; $\cdots\cdots$ all spring naturally – if not easily or even understandably – from Born's first "only possible interpretation" of the scattered wave function.

Footnotes and References:

1) A. Pais, *Max Born's Statistical Interpretation of Quantum Mechanics"* in Science **218**, 1193 (1982); see also A. Pais, *Inward Bound* (Oxford University Press, New York, 1986), Pp.255-261; and A. Pais, *Niels Bohr's Times* (Clarendon Press, Oxford 1991), Pp.284-289.

2) B.L. van der Waerden, *Exclusion Principle and Spin* in Theoretical Physics in the Twentieth Century, A Memorial Volume to Wolfgang Pauli (Interscience, New York, 1960) M. Fierz and V.F. Weisskopf, Eds., p.199. We quote van der Waerden: " $\cdots$ the *historical method* follows, step by step, the development of ideas $\cdots$ as far as they can be traced in their publications. In my opinion, a thorough understanding of a physical theory can be reached only by the historical method. $\cdots$ This implies a severe restriction in the domain of investigation. $\cdots$ the essential ideas must be explained $\cdots$."

3) To avoid endless circumlocution (and also as a reflection of our own predilection for the traditional orthodox Copenhagen interpretation) we inevitably refer to such alternatives in somewhat negative terms. These alternatives are subjects of an ongoing debate which

will – and indeed *should* – never end.

4) A short list of Born's misconceptions includes the remarks:

a) the Schrödinger form $\cdots$ describe(s) $\cdots$ quantum jumps. Born erred here. The Schrödinger amplitude evolves continuously, the 'jumps' must be imposed by the 'interpretation'.

b) the theory is complete $\cdots$ include(s) the transition problem $\cdots$ I $\cdots$ demonstrate this. (Note: Again subject to the above caveat.) And more in his second paper:

c) Einstein $\cdots$ spoke of a 'ghost field' $\cdots$ no energy or momentum belonged to the field itself however.

d) the guiding field $\cdots$ just as when particles $\cdots$ really fly about. (Note: Born makes no use whatsoever of these 'ghost' and 'guiding' concepts, except apparently as some sort of intuitive crutch.)

5) In Ch9 of his second paper, Born makes an almost correct remark about the extent to which he is seeking to describe the transition process. There is no hint of describing any 'jump', but only of describing the time development of an 'event'. Reinterpreted as the 'probability of an event' renders this acceptable.

6) J. von Neumann, *Mathematische Grundlagen der Quantenmechanik* (Springer, Berlin, 1932). [English translation: *Mathematical Foundations of Quantum Mechanics* (Princeton, Princeton NJ, 1955) R.T. Beyer, Transl.] See especially Ch5 on entropy considerations and reversibility of macroscopic measurements. von Neumann (in 1932: Ch6, §2: *Composite Systems*) recognized the possibility of EPR-type complications [See Footnote 212, p.429 of Beyer's translation], and the concept of entanglement [so-called by Schrödinger] which he credits to Landau (Zeits. f. Phys. **45**, 1927).

7) Y. Aharanov, D. Albert, and L. Vaidman, Phys. Rev. Lett. **60**, 1351 (1988); Phys. Rev. Lett. **62**, 2325 (1989).

8) I. Duck, P. Stevenson, and E.C.G. Sudarshan, Phys. Rev. D **40**, 2112 (1989).

9) N. Ritchie, J. Story, and R. Hulet, Phys. Rev. Lett. **66**, 1107 (1991).

10) M. Born, W. Heisenberg and P. Jordan, Zeits. f. Phys. **35**, 557 (1926). See especially Ch3, p.581.

Biographical Note on Born.

Max Born (1892-1970) – won the 1954 Nobel Prize in Physics (with Walther Bothe) for his statistical interpretation of the wave function [a], work done 28 years earlier.

Born in Breslau (now in Poland), son of a professor of anatomy; a student at Berlin, Heidelberg and Zurich, PhD at Göttingen, then professor at Berlin and Frankfurt. After WWI service in the cavalry (where he reports writing papers on horseback [b]), he was professor and chair at Göttingen. After a career studying crystal structure, Born – inspired by the discovery of matrix mechanics by Heisenberg (one of his assistants, Pauli and Jordan were the other two) – turned to the development of quantum mechanics in the Heisenberg formulation, with great result. He was remembered by Maria Goeppert-Mayer [c] from her student days for his informal friendliness, unusual in German professors of the day.

Throughout the frenetic days of the development of matrix mechanics in the late summer and early fall of 1925, Born was held back by a chronic fatigue that required treatment at a sanatarium. After the development of his interpretation of quantum mechanics (see **Papers X·1,2**), Born was unwell [e], his three remarkable assistants – Heisenberg, Jordan, and Pauli – had moved on from Göttingen, to be replaced by a new generation. Born was fired from his Chair at Göttigen because of his Jewish origin [f] and left Nazi Germany immediately in 1933 for positions at Cambridge, Bangalore, and as professor of natural philosophy at Edinburgh (1936-53). He was the author of some 300 papers and 20 books, many concerned with the survival threat posed by militaristic science [b,d].

His grand-daughter Olivia Newton-John (1948-) emerged from Australia to become an international recording and film star (1965-) [g].

Footnotes and References:

a) The LONDON TIMES, 4NOV54 p.4; and NYTIMES, 4NOV54, p.33.

b) NYTIMES 6JAN70, p.1.

c) PHYSICS TODAY, MAR70, p.97.

d) The LONDON TIMES, 6JAN70, p.10.

e) N. Mott, in NIELS BOHR (Harvard University Press, Cambridge, 1985) A.P. French and J.P. Kennedy, Eds., p.174; described by Born himself as a 'nervous breakdown' (see M. Born, *My Life & My Views* (Scribner's, New York 1968), p.37).

f) A. Pais, Science **218**, 1193 (1982): on p.1194 we learn from Born's daughter (Footnote 1) that Born's Sephardic Jewish ancestral name was Abarbanel. The family name was changed to Born by a previous generation.

g) Ramzi Salti, *Olivia Newton-John Biography*, DISCoveries Magazine (JUN95).

Paper X·1: Excerpt from Zeits. f. Phys. **37**, 863 (1926).

Zur Quantenmechanik der Stossvorgänge

[Vorläufige Mitteiling.[1]]

von **Max Born**, Göttingen

(Eingangen am 25. Juni 1926.)

From an investigation of the collision process, we show that quantum mechanics in the Schrödinger form allows us to describe not only the stationary states but also the quantum jumps.

The quantum mechanics begun by Heisenberg has so far been applied only to the calculation of the stationary states and the oscillatory amplitudes corresponding to transitions among them (I intentionally avoid the expression 'transition probability'). It appears that the still evolving formalism has been confirmed. But these questions represent only one side of the quantum problem; what seems equally important is the question of the nature of the 'transition' itself. On this point opinion is divided; many assume that the question of the transition cannot be understood in the present form of quantum mechanics, but that new concepts will be necessary. I myself, from my impression of the completeness of the logical structure of quantum mechanics, conjecture that the theory is complete and that it must also include the transition problem. I believe that I have been able to demonstrate this. [Note added: There is a semantic distinction which Born is not making, between the *theory* (i.e., Schrödinger's calculation) and the *interpretation* (i.e., Bohr's auxiliary explanation of it).]

Bohr has pointed out that the basic difficulties of the quantum idea which occur for the emission and absorption of light by an atom, also occur for the interaction of atoms at short distances, and therefore for collisions. For these, one has to deal with the obscure wave fields associated with material particles which underlie the formalism of quantum mechanics. Accordingly, I have studied the problem of the interaction of a free particle (α-ray or electron) and a given atom in order to determine if a description of the collision process is possible within the framework of the existing theory. [Note added: Born's probabilistic thinking is the beginning of the transition from Schrödinger's classical field interpretation of ψ to the full orthodox Copenhagen abstraction of ψ.]

Of the different forms of the theory, that of Schrödinger proves most useful here, and on this basis I recommend it as the deepest formulation of the quantum laws. [Note added: The word 'deepest' was an ill-considered choice whose sole result was to annoy and alienate Heisenberg. Born had struggled at length and in vain to solve the scattering problem in the Heisenberg formulation.] The train of thought of my investigation now follows.

* * * * * * * * * * * * * * * * * * * **

This remark is generous of Born, who had exhibited a strong proprietary interest in the matrix-formulation of quantum mechanics which was invented by his assistant Heisenberg, and further developed by himself with Heisenberg and Jordan. The matrix formulation was, and to a large extent still is, viewed as being more fundamental than, but ultimately equivalent to, Schrödinger's more intuitive wave-mechanical formulation. The intuitive advantage of the Schrödinger wave-mechanics was warmly acknowledged by Bohr, in spite of heated exchanges with Schrödinger over its significance.

Born's papers of June and July 1926 predate the famous debate between Schrödinger and Bohr (allied with Heisenberg) which began with Schrödinger's lecture in Munich in late summer 1926 and ended in Copenhagen in early October 1926.

Schrödinger resisted Bohr's extremely abstract interpretation of his work, but both camps opposed Born's less abstract and less far reaching interpretation as being variously and simultaneously: unnecessary, incomplete, wrong, obvious, well-known, and on and on. Born was lastingly resentful of these slights. [Note added: For a full account, see A. Pais, *Niels Bohr's Times* (Clarendon, Oxford, 1991), Pp.284-289; also A. Pais, *Inward Bound* (Clarendon, Oxford, 1986), esp. p.260; and the essay by A. Pais, *Science*, **218**, 1193 (1982).]

* * * * * * * * * * * * * * * * * * * **

When one calculates in quantum mechanics the interaction of two systems, one cannot know – as in classical mechanics – how a particular state of one system will be influenced by any individual state of the other system. All states of both systems are coupled together in a complicated way. That holds also for a non-periodic process such as a collision where a particle – say an electron – comes in from infinity and then goes out to infinity. Here the picture must be that both before and after the collision, when the electron is essentially free and the coupling is small, *a definite state of the atom and a definite uniform motion of the electron must be definable.* [Note: Italics added.] It is then a question of describing the asymptotic behavior of the coupled particles. I was not able to do this with the matrix form of quantum mechanics, but with the Schrödinger form things went well. [Note added: Indoctrinated as we all are by Bohr and Wheeler, among others, it is now natural for us to recognize the full implication of Born's somewhat off-hand (here italicized) remark. Born stopped just short of the ultimate *definition* of the state of the scattered particle, and the necessity of the actual measurement and the consequent 'collapse of the wave function' insisted upon by Bohr.]

According to Schrödinger, if the atom is in the nth quantum state there is an oscillatory amplitude for any state variable in the whole space with a constant angular frequency $\omega = W_n^0/\hbar$. An electron moving in a straight line is a special case of such an amplitude corresponding to a plane wave. When the two systems come into interaction, there is a resulting complicated oscillation. However one easily sees that it can be specified by its asymptotic behavior at infinity. One is dealing not with a 'diffraction problem', for which an incoming plane wave will be bent or scattered at the atom; in place of the boundary conditions used in optics for the description of the image, one has here the potential energy of interaction between the atom and the electron.

The problem is to solve the Schrödinger wave equation for the electron-atom combination with the boundary condition that the solution – in a definite direction of the electron coordinates – goes over asymptotically into a plane wave in this direction (the incoming electron). For this solution, what is of interest is the behavior of the 'scattered' wave at infinity; this describes the behavior of the system after the collision. Let us explain this somewhat further. Let $\psi_1^0(q_k), \psi_2^0(q_k), \cdots$ be eigenfunctions of the undisturbed atom (we assume a discrete series); the undistorted (straight-line) motion of the electron corresponds to the eigenfunction $\exp(i\vec{k}\cdot\vec{r})$, which form a continuous manifold of plane waves whose de Broglie wavelength $\lambda = 2\pi/k$ is related to the kinetic energy of the translational motion by $\tau = \hbar^2 k^2/2\mu$. The eigenfunction of the undisturbed state for which the electron comes out in the $+z$-direction is therefore

$$\psi^0_{n,\tau}(q_k, z) = \psi^0_n(q_k)e^{ikz}.$$

Now let $V(q_k;\vec{r})$ be the interaction potential energy between the atom and the electron. With a simple perturbation calculation, one can show that there is a uniquely determined solution of the Schrödinger equation for a given interaction V, which asymptotically approaches the above function for $z \to +\infty$.

What is important now, is how the solution develops 'after the collision'.

Next follows the calculation: according to the perturbation calculation, the scattered wave has the asymptotic expression at infinity

$$\psi^1_{n\tau}(\vec{r};q_k) = \sum_m \int d\Omega(\hat{k}_{nm})\Phi_{n\tau;m}e^{i\vec{k}_{nm}\cdot\vec{r}}\psi^0_m(q_k).$$

* * * * * * * * * * * * * * * * * * * **

This is more easily recognized in modern (Dirac-)notation as the scattered wave in 1st Born approximation:

$$\psi_{\text{scatt}}(r \to \infty) \simeq \sum_{\vec{k}_f;m} e^{i\vec{k}_f\cdot\vec{r}}|m\rangle\langle m; \vec{k}_f|V|k_i\hat{z}; n\rangle.$$

* * * * * * * * * * * * * * * * * * * **

This means: the perturbation at infinity can be represented as a superposition of solutions for the undisturbed processes. The wavelength $\lambda_{n\tau;m}$ of the scattered particle is given by de Broglie's formula from its energy

$$W_{n\tau;m} = \hbar\omega^0_n - \hbar\omega^0_m + \tau,$$

where $f^0_{n,m} \equiv \omega^0_{n,m}/2\pi$ are the frequencies of the undisturbed atoms.

Only one particle-like interpretation of this result is possible: $\Phi_{n\tau;m}(\vec{k}_{n\tau;m})$ is the probability [2] that the electron incident from the z-direction is scattered into the direction of $\vec{k}_{n\tau;m}$ and its energy has decreased by the amount $E_m - E_n = \hbar\omega_m^0 - \hbar\omega_n^0$ corresponding to the quantized excitation of the atom from state n to state m.

Schrödinger quantum mechanics therefore provides a completely *definite* answer to the question of the effect of a collision; however it is not a *specific* result. One gets no answer to the question "What is the state after the collision?", but only to the question "What is the probability of a given effect of the collision?" (where naturally the quantum energy law must be obeyed). [Note: Italics added.]

* * * * * * * * * * * * * * * * * * **

In this brief, direct, simple, unequivocal statement begins the whole subject of the probabilistic interpretation of quantum mechanics.

* * * * * * * * * * * * * * * * * * **

Here arises the whole Problem of Determinism. From the standpoint of quantum theory, there is no quantity which is uniquely specified in an individual case as the result of a collision; but even experimentally we have no indication that there exists one among all the eigenproperties of the atom which is guaranteed a specific value as a definite result of a collision. Should we still hope to discover such properties (perhaps phases of the internal atomic motion) and to specify them in individual cases? Or should we believe that the agreement between theory and experiment on their absence, rather than indicating any restrictions on specific results, is some precise balance on which the non-existence of such restrictions depends? I myself believe that determinism in the atomic world must be given up. But that is a philosophical question, for which physical arguments are not solely decisive.

* * * * * * * * * * * * * * * * * * **

Here Born anticipates and disposes of hidden variable complications, and 'pilot wave' proposals such as those of Schrödinger and Bohm.

In this last regard, his own language was not completely specific and whether or not he had some reservations about the 'actual' meaning of the wave function is not clear. In the expanded version of this paper (see the following **Paper VI·2**), he does refer to Einstein's initial description of the 'ghost field' showing the way for the light quantum which Einstein had visualized as a co-moving singularity.

✳ ✳ ✳ ✳ ✳ ✳ ✳ ✳ ✳ ✳ ✳ ✳ ✳ ✳ ✳ ✳ ✳ ✳ ✳✳

In any case, the indeterminism exists in a practical sense for the experimental as well as for the theoretical physicist. The 'distribution function' Φ widely used by the experimenters is now also readily understood theoretically. One can calculate it from the interaction potential $V(\vec{r}; q_k)$; however the calculations are too complicated to describe here. I will only explain the properties of the function Φ in a few words. If, for example, the atom before the collision is in the groundstate $n = 1$, then it follows that

$$W_{1\tau;m} = \tau - \hbar\omega_m^0 + \hbar\omega_1^0,$$

so that for an electron with kinetic energy smaller than the smallest excitation energy of the atom, then necessarily only $m = 1$, and $W_{11} = \tau$ can occur; the result is 'elastic scattering' of the electron with the distribution function Φ_{11}. If τ exceeds the first excitation energy, then there is also inelastic scattering with the distribution Φ_{12}, and so on. $\cdots\cdots$

I imagine further that the problem of the absorption and emission of light must be handled in an entirely analogous way as 'boundary-value' solutions of the wave equation $\cdots\cdots$ A more detailed discussion will soon appear $\cdots$

Footnotes and References:
1) This report was originally prepared for the "Naturwissenschaften", but could not be accepted there because of its length. I hope that its publication here does not appear superfluous.
2) Correction added in Proof: Exact considerations show that the probability is proportional to the square of the quantity $\Phi_{n,\tau;m}$.

Paper X·2: Excerpt from Zeits. f. Phys. **38**, 803 (1926).

Quantenmechanik der Stossvorgange

von **Max Born** in Göttingen.

(Eingegangen am 21.Juli 1926.)

The Schrödinger form of quantum mechanics allows us to define in a natural way the probability of any state in terms of the amplitude of the corresponding normal mode. This leads to a theory of collision processes in which the transition amplitudes can be determined by the asymptotic behavior of aperiodic solutions.

Introduction: Collision processes have been the most productive experimental source for the fundamental assumptions of quantum theory, and are also best suited to illustrate the physical consequences of the formal laws of 'quantum mechanics'. These laws appear to yield the correct term values of the stationary states and the correct amplitudes for the radiation emitted in the transitions. But on the physical interpretation of the formalism, opinions are divided. Heisenberg's view, based on the matrix-mechanics form of quantum mechanics [2] and developed by him jointly with Jordan and myself, proceeds from the idea that an exact representation of the process in space-time is not generally possible, and instead formulates relations between observable quantities which can only be understood in the classical sense as eigenproperties of the motion. Schrödinger on the other hand [3] seems to ascribe to waves, which he – like de Broglie – pictures as the vehicle of atomic processes, a reality of the same kind as light waves; he explains "Wave packets build up, which are relatively localized in all dimensions" and which obviously are to represent the particle motion directly.

Neither of these two concepts seems satisfactory to me. I try a third interpretation and explore here its usefulness in collision processes. For it, I refer to a remark of Einstein's about the behavior of wave fields and light quanta; he said something to the effect that the waves only exist to show the light quanta the way, and he spoke in this sense of a 'ghost field'. This gave the probability that a light quantum – the carrier of energy and momentum – travels in a particular direction; no energy or momentum belonged to the field itself however.

These ideas construct quantum mechanics – as it was then – on the model of the electromagnetic field. By analogy between the light quantum and the electron, one would imagine the laws of the electron motion to be formulated in a similar way. And so here it is natural to think of the de Broglie - Schrödinger waves as the 'ghost field', or better as the 'guiding field'.

I tentatively suggest the following idea: the guiding field ψ, a scalar function of the coordinates of all involved particles and the time, propagates according to Schrödinger's equation. Momentum and energy however will be carried along, just as when particles (electrons) really fly about. The paths of these particles are defined only in so far as their energy and momentum are specified; however, only the probability of a particular path will be determined by the value of the distribution function of their impacts. One could summarize, somewhat paradoxically, something like: The motion of the particle follows probability laws, but the probability itself propagates in accordance with the law of causality [4].

* **

Born's idea is slowly and in his word, tentatively, taking form in an unrefined and ill-defined language being used for the first time. The last sentence should be "the probability *wave* propagates in accordance with the *Schrödinger wave equation.*" Born used the terms 'probability' and 'probability amplitude' interchangeably at first (although he corrected this in footnote [2] of **Paper VI·1**). His use of the term 'causality' here is non-relativistic, as explained in footnote [4]. In his first version (see **Paper X·1**), Born was very close to the eventual Copenhagen interpretation of the wave function. This expanded version of his ideas is unfortunately somewhat blemished by the introduction of unnecessary furbelows (a bit of elaborate trimming, a useless flourish) on that contribution. He bases his excursions on a *very* early remark of Einstein's about his conception of the photon vis-á-vis the classical field (see **Paper II·2**), which was never pursued by Einstein himself; and on an erroneous conception of Schrödinger (quickly disposed of by Heisenberg as untenable (see **Paper XI·1**)) identifying quantum wave-packets as classical particle 'singularities'.

There is a small but vigorous, vocal, very determined, and very ingenious subset of quantum theorists who take *very* seriously these unorthodox appendages of 'ghost fields' 'guiding' the *real* particles on their probabilistically determined classical trajectories. Their challenge is to construct an alternative to the perceived paradoxes and excessive abstractions of the orthodox Copenhagen interpretation without introducing any observable differences. The price that must be paid is reckoned by the proponents of these alternatives to be worth the elimination of perceived ugliness in the orthodox theory. This is largely a matter of taste, or rather lack of it.

* * * * * * * * * * * * * * * * * * **

If one considers the two stages of the development of the quantum theory, then the first dealt with periodic processes and was completely unsuited to demonstrate the usefulness of such a concept; the second, which considered aperiodic stationary processes, was somewhat more suggestive of this interpretation. We will concern ourselves with this in the following work. In fact, a third stage could be mentioned, which considers non-stationary events which apparently originate from a space-time independent coupling.

A precise description of the idea is only now possible on the basis of a new mathematical development [5]; therefore we turn our attention to that, and later return to the hypothesis.

§ **1.** Definition of the Density and Probability for a Periodic System: We begin with a complete formal analysis of the discrete stationary states of a non-degenerate system. These may be characterized by Schrödinger's differential equation

$$[H - W]\psi = 0. \tag{1}$$

The eigenfunctions should be normalized to 1:

$$\int \psi_m^*(q)\psi_n(q)dq = \delta_{nm}. \tag{2}$$

Any given function $\psi(q)$ can be expanded in the eigenfunctions:

$$\psi(q) = \sum_n c_n \psi_n(q). \tag{3}$$

Previously we have directed our attention only to the eigenstates ψ_n and their eigenvalues W_n. As explained in the introduction, our idea explains that the coherent superposition of functions in (3) yields the probability that in any such group of independent atoms, the states occur with a definite probability.

* * * * * * * * * * * * * * * * * *

Here is the first interpretation of the wave function in terms of the distribution of an *ensemble* of identically prepared systems $\cdots$ " in any such group of independent atoms, the states occur with a definite probability."

* * * * * * * * * * * * * * * * * *

The completeness relation

$$\int |\psi(q)|^2 dq = \sum |c_n|^2 \tag{4}$$

leads us to interpret this integral as the number of atoms. For the individual eigenfunctions normalized to 1 (or: the a priori weight of the states are 1), $|c_n|^2$ represents the probability of the state n and the total of these parts combines additively.

* * * * * * * * * * * * * * * * * * **

$\cdots$ leads us to interpret this integral as the number of *identical but independent systems in the ensemble* $\cdots\cdots |c_n|^2$ represents the *probable number of systems in the ensemble* in the state n $\cdots$

* * * * * * * * * * * * * * * * * * **

In order to justify this interpretation we consider the motion of a mass point in three dimensional space under the action of the potential energy $U(\vec{r})$; this gives the differential equation

$$\nabla^2 \psi + \frac{2\mu}{\hbar^2} (W - U) \psi = 0. \tag{5}$$

If one substitutes for W, ψ an eigenvalue W_n and an eigenfunction ψ_n, multiplies the equation by ψ_m^* and integrates over the space $(dS = dx\,dy\,dz)$, then one gets:

$$\int dS \left\{ \psi_m^* \nabla^2 \psi_n + \frac{2\mu}{\hbar^2} \left(W_n - U \right) \psi_m^* \psi_n \right\} = 0.$$

From Green's theorem with the help of the orthogonality relation (2) we get

$$\delta_{mn} W_n = \int dS \left\{ \frac{\hbar^2}{2\mu} \left(\vec{\nabla}\psi_m^* \cdot \vec{\nabla}\psi_n \right) + \psi_m^* U \psi_n \right\}. \tag{6}$$

Each energy level can thus be expressed as a space integral of the energy density of the eigenfunctions.

If one now constructs the corresponding integral for any function

$$W = \int dS \left\{ \frac{\hbar^2}{2\mu} |\vec{\nabla}\psi|^2 + U|\psi|^2 \right\}, \tag{7}$$

then using the expansion (3) one gets the expression

$$W = \sum_n |c_n|^2 W_n. \tag{8}$$

In agreement with our interpretation of the $|c_n|^2$, the right side is the mean value of the total energy of a system of atoms; this mean value can be represented as the space integral of the energy density of the function ψ.

But in other respects the essence of our idea cannot be illustrated as long as we consider only periodic processes.

§ **2.** Aperiodic Systems. We go therefore to aperiodic processes and consider for simplicity the case of uniform straight-line motion along the x-axis. Here the differential equation is

$$\frac{d^2\psi}{dx^2} + k^2\psi = 0, \qquad k^2 = \frac{2\mu}{\hbar^2} W; \tag{1}$$

which has as eigenvalues all positive real W and as eigenfunctions $\psi = c e^{\pm ikx}$. In order to define the density and the probability, one must normalize all the eigenfunctions. Since for (2) the corresponding integral diverges, it is natural to use instead the 'mean value':

$$\lim_{a \to \infty} \frac{1}{2a} \int_{-a}^{+a} dx |\psi(k, x)|^2 = 1, \tag{2}$$

from which $c = 1$ and the normalized eigenfunctions are

$$\psi(k, x) = e^{\pm ikx}. \tag{3}$$

Any function of x can be constructed from these. For that, it is necessary to choose the interval of the k-scale which corresponds to unit weight. Here one considers the free motion as the limiting case of a periodic one, namely the eigenfunctions on a finite section of the x-axis. The number of modes per unit length in the interval $(k, k + \Delta k)$ is

$$\frac{\Delta k}{2\pi} = \Delta \left(\frac{1}{\lambda} \right),$$

where λ is the wavelength. One sets

$$\psi(x) = \int_{-\infty}^{+\infty} \frac{dk}{2\pi} c(k) e^{ikx}, \tag{4}$$

where $|c(k)|^2$ is the probability in the interval $dk/2\pi$.

For a mixture of atoms for which the eigenfunctions have the distribution $c(k)$, the number analogous to (4) of §1 is given by the integral

$$\int_{-\infty}^{+\infty} dx |\psi(x)|^2 = \frac{1}{(2\pi)^2} \int_{-\infty}^{+\infty} dx \left| \int_{-\infty}^{+\infty} dk c(k) e^{ikx} \right|^2. \tag{5}$$

If only the small interval $k_1 \le k \le k_2$ is occupied, then

$$\int_{-\infty}^{+\infty} dk c(k) e^{ikx} = \bar{c} \int_{k_1}^{k_2} dk e^{ikx} = \frac{\bar{c}}{ix} \left(e^{ik_2 x} - e^{ik_1 x} \right),$$

where $\bar{c}$ is an average value. Then one has

$$
\begin{aligned}
\int_{-\infty}^{+\infty} dx |\psi(x)|^2 &= \frac{|\bar{c}|^2}{(2\pi)^2} \int_{-\infty}^{+\infty} \frac{dx}{x^2} |e^{ik_2 x} - e^{ik_1 x}|^2 \\
&= \frac{|\bar{c}|^2}{(2\pi)^2} 4 \int_{-\infty}^{+\infty} \frac{dx}{x^2} \sin^2 \frac{k_2 - k_1}{2} x = \frac{1}{2\pi} |\bar{c}|^2 (k_2 - k_1).
\end{aligned}
$$

The momentum of the translational motion corresponding to the eigenfunction (3) is according to de Broglie, equal to

$$p = \frac{h}{\lambda} = \hbar k. \tag{6}$$

One can also interpret this as a 'Matrix' by defining the matrix in the continuous spectrum not by an integral but by a mean value

$$
\begin{aligned}
p(k, k') &= \frac{\hbar}{i} \lim_{a \to \infty} \frac{1}{2a} \int_{-a}^{+a} dx\, \psi^*(k, x) \frac{\partial \psi(k', x)}{\partial x} \\
&= \frac{\hbar}{i} \lim_{a \to \infty} \frac{1}{2a} \int_{-a}^{+a} dx\, e^{-ikx} ik' e^{ik'x}.
\end{aligned}
$$

In this way

$$p(k, k') = \hbar k \quad \text{for} \quad k = k', \qquad = 0 \quad \text{for} \quad k \neq k'. \tag{7}$$

If one replaces $\Delta k = k_2 - k_1$ by $\Delta p/\hbar$, then finally

$$\int_{-\infty}^{+\infty} dx\, |\psi(x)|^2 = |\bar{c}|^2 \frac{\Delta p}{2\pi\hbar}. \tag{8}$$

Finally one has the result that a cell of length $\Delta x = 1$ and of momentum spread $\Delta p = 2\pi\hbar$ has weight 1, in agreement with the well-known law of Sackur and Tetrode [6], and that $|\bar{c}|^2$ is the number of particles with momentum $p = \hbar k$.

Now we go on to accelerated motions. Here one can define in an analogous way a given distribution of moving particles. But for a collision process this is not the meaningful question. For these processes each motion before and after the collision has a straight-line asymptote. At very long times (compared to the duration of the collision) before and after the collision the particle is in an essentially free state described by the following prescription: For the asymptotic behavior of the wave function before the collision the expansion coefficient $|c(k)|^2$ is known; can one calculate from that the expansion coefficients after the collision?

For this it is natural to speak of a stationary particle current. Mathematically the problem can be expressed as follows: The stationary wave function must be broken up into ingoing and outgoing waves; these are asymptotically plane waves. One represents both by Fourier integrals of the form (4) and chooses the expansion coefficients $c(k)$ for the incoming waves appropriately; then it can be seen that the $c(k)$ for the outgoing waves are uniquely determined. The result is the distribution into which a given incident particle is transformed by the collision.

To visualize the situation clearly, first consider the one-dimensional case.

§ **3.** The Asymptotic Behavior of the Eigenfunctions in the Continuous Spectrum for One Degree of Freedom. The Schrödinger equation is

$$\frac{d^2\psi}{dx^2} + \frac{2\mu}{\hbar^2}\left(W - U(x)\right)\psi = 0, \tag{1}$$

where $U(x)$ is the scattering potential. With the abbreviations

$$\frac{2\mu}{\hbar^2}W = k^2, \qquad \frac{2\mu}{\hbar^2}U(x) = V(x); \tag{2}$$

then we have

$$\frac{d^2\psi}{dx^2} + k^2\psi = V\psi. \tag{3}$$

We investigate the asymptotic behavior of the solution at infinity. For that, in order to have a simple situation, we assume that $V(x)$ vanishes at infinity faster than $1/x^2$, that is $V(x) < K/x^2$, where K is a positive number [7].

We now determine $\psi(x)$ by an iterative procedure; it should have

$$u_0(x) = e^{ikx}$$

and $u_1(x), u_2(x), \cdots$ should be the solutions of the successive iterative equations

$$\frac{d^2u_n}{dx^2} + k^2u_n = V u_{n-1}$$

which vanish for $x \to +\infty$.

These satisfy

$$u_n(x) = \frac{1}{k}\int_{-\infty}^{+\infty} d\xi\, u_{n-1}(\xi)V(\xi)\Theta(\xi - x)\sin k(\xi - x),$$

as one can easily verify (here $\Theta(x) = 0, 1$ for $x < 0, > 0$).

········· Consequently the series

$$\psi(x) = \sum_{n=0}^{\infty} u_n(x) \tag{4}$$

converges absolutely ··· and is the required solution of our differential equation.

* * * * * * * * * * * * * * * * * * **

Here unfortunately Born's discussion wanders and becomes confusing, if not confused, by an unconventional choice of phase of his plane-waves. He never actually makes use of his one dimensional iterative solution but instead observes that there are two linearly independent solutions for each value of the energy $E = \hbar\omega = \hbar^2 k^2/2\mu$, which are asymptotic to

$$\psi \sim e^{ikx-i\omega t} \quad \text{and} \quad \sim e^{-ikx-i\omega t},$$

using accepted phase conventions. From these he constructs ψ^L at $x \to -\infty$ and ψ^R at $x \to +\infty$ as

$$\psi^{L,R}(x,t) = A^{L,R} e^{ikx-i\omega t} + B^{L,R} e^{-ikx-i\omega t},$$

which he separates into waves (more properly wave packets) 'incoming' at times $t \to -\infty$ from $x \sim \omega t/k \to -\infty$ and 'outgoing' at times $t \to +\infty$. This allows him to identify

$$\psi^L = 1 \cdot e^{ikx-i\omega t} + \mathcal{R} \cdot e^{-ikx-i\omega t} \quad \text{and} \quad \psi^R = \mathcal{T} \cdot e^{ikx-i\omega t},$$

for incident amplitude unity, reflected amplitude $\mathcal{R}$, and transmitted amplitude $\mathcal{T}$. Born's choice of iterative solution actually produces 1 instead of $\mathcal{T}$ on the right, and $1/\mathcal{T}$ and $\mathcal{R}/\mathcal{T}$ on the left. The choice

$$\begin{aligned}
u_n &= \frac{1}{k}\int_{-\infty}^{+\infty} d\xi\, u_{n-1}(\xi) V(\xi) \times \\
&\quad \frac{1}{2}\left\{\Theta(x-\xi)e^{ik(x-\xi)} - \Theta(\xi-x)e^{-ik(x-\xi)}\right\}
\end{aligned}$$

is suitable for our purpose.

Born goes on to show that Schrödinger's equation implies energy conservation and also the time independence of

$$\frac{\partial \psi}{\partial x} \frac{\partial \psi*}{\partial t}, \text{ which leads to } |\mathcal{R}|^2 + |\mathcal{T}|^2 = 1.$$

* * * * * * * * * * * * * * * * * * **

§ **4.** Conservation of Energy. $\cdots\cdots$

§ **5.** Generalization to Three Dimensions. $\cdots\cdots$

§ **6.** Elastic Scattering. $\cdots\cdots$

§ **7.** Inelastic Electron Scattering. $\cdots\cdots$

§ **8.** Physical Consequences. $\cdots$ These sections have become standard fare in all wave mechanics classes, but are too purely technical to include here.

§ **9.** Concluding Remarks. On the basis of the preceding remarks it is my opinion that quantum mechanics allows us to formulate and solve not only the problem of the stationary state, but also that of the transition process. The Schrödinger formulation seems to be the easiest way by far to do justice to these problems; moreover it makes it possible to follow in a completely natural way the whole space-time development of an event. Even so, just like the earlier theory, it does not allow the unique prediction of individual events. I have made particular mention in my earlier report of this complete indeterminacy, since it seems to me to be in close agreement with the result of experiment. Moreover, it is important to emphasize that there is no parameter present in the theory which would permit the theory to select of itself a unique single-valued result. In classical mechanics there are the 'phases' of the motion, for example the coordinates of the particles at a particular instant. It seems to me extremely unlikely that any such quantities corresponding to these phases, could be freely introduced into the new theory; however, Herr Frankel has assured me that it is in fact possible. Nonetheless, this possibility would not change the practical indeterminacy of the collision process where one cannot define the value of the phases; it would lead back to the same formulas as the 'phaseless' theory just discussed.

* * * * * * * * * * * * * * * * * * * **

After introductory references to 'ghost fields' 'guiding' the classical point particles on their probabilistically determined trajectories, Born never mentions the concept again. What he *wrote* above has been *completely absorbed* into the standard development and the orthodox Copenhagen interpretation of quantum mechanics.

* * * * * * * * * * * * * * * * * * * **

I foresee the possibility that the laws of motion of light quanta can be handled in a completely analogous way [8]. There one has the same basic problem of aperiodic emission processes for the free radiation, and a boundary-condition problem, or an initial-condition problem, for the coupled wave equations of the Schrödinger ψ-quantity and the electromagnetic field. The law for this coupling is one of the most important problems; it is, I know, treated in other places [9]. When this law is formulated, it will be possible to derive a rational theory for the lifetime of states, the transition probability for radiation processes, of damping and of linewidth.

Footnotes and References:
1) See my preliminary version, Zeits. f. Phys. **37**, 863 (1926).
2) W. Heisenberg, Zeits. f. Phys. **33**, 879 (1925)); M. Born and P. Jordan, ebenda **34**, 858 (1925); M. Born, W. Heisenberg and P. Jordan, ebenda **35**, 557 (1926). See also P.A.M. Dirac, Proc. Roy. Soc. **109**, 642 (1925); **110**, 561 (1926).
3) E. Schrödinger, Ann. d. Phys. **78**, 361, 489, 734 (1926). See in particular the second paper on p.499. Also Die Naturw. **14**, 664 (1926).
4) This shows that the knowledge of the state at all points at an instant determines the distribution of the state at all later times.
5) In the mathematical derivation of this work, Herr Prof. N. Wiener of Cambridge, Mass., has helped in the friendliest way; I wish to express and acknowledge my thanks, for without him I would not have reached this goal.
6) A. Sackur, Ann. d. Phys. **36**, 958 (1911); **40**, 67 (1913); H. Tetrode, Phys. Zeits. **14**, 212 (1913); Ann. d. Phys. **38**, 434 (1912).
7) This assumption rules out the case of the pure Coulomb potential and the dipole potential.
8) The difficulties found so far in optics, seem to me in part due to the tacit assumption that the center of the wave and the emitted particle must be in the same place. But this is clearly not the case for the Compton Effect and will indeed prove to be untrue in general.
9) O. Klein, Zeits. f. Phys. **37**, 895 (1926).

Chapter XI

HEISENBERG:

··· so muss man bestimmte Experimente angeben, mit deren Hilfe man den "Ort des Elektrons" zu messen gedenkt; anders hat dieses Wort keinen Sinn. ··· ist kein Mangel, z.B.: Man beleuchte das Elektron und betrachte es unter einem Mikroskop. ··· Man wird aber in Prinzip etwa ein Γ-Strahl-Mikroskop bauen ···

"··· one must specify a definite experiment to measure the "position of the electron"; otherwise these words have no meaning. ··· there is no mystery, for example: Simply illuminate the electron and observe it under a microscope. ··· At least in principle one can build a γ-ray microscope ···"

W. Heisenberg, Zeits. f. Phys. **43**, 172 (1927).

§XI-1. Introduction.

It was a time of giants, and among them Heisenberg appeared godlike [1]. In this one great paper [2] Heisenberg reaffirmed his monumental and unique genius. Without this paper his stature might have rested on his "lucky" conjecture recognized by Born and Jordan as matrix multiplication of quantum variables, and later exploited with them in the 'Dreimännerarbeit' [3]. In this paper, published alone and some say in haste, Heisenberg articulated the interpretation of quantum mechanics in a series of steps which remain the fundamental pedagogical foundations of the orthodox 'Copenhagen' interpretation. If nothing else had ever been written on the subject, it is fair to say that all the now known results of quantum mechanics are *implicit* in this paper of Heisenberg's.

The accusation exists that the paper was published in haste and created a rift between Bohr and Heisenberg, eventually patched up by Pauli [4], but ended the nirvana which had existed between them. Perhaps all the ideas contained here had existed in some incomplete and unarticulated form before this synthesis by Heisenberg, and certainly Heisenberg could not have created this train of thought in isolation. In this regard, we once again invoke van der Waerden's principle [5]: "What counts is who *published* what, and when."; tempered at last by a more complete set of acknowledgements by Heisenberg to Bohr and Born, and especially to his old junior colleagues Jordan and Pauli.

Heisenberg defined the foundation for the interpretation of quantum mechanics in a number of simple intuitive logically compelling steps:
1) by defining the pragmatic *empirically possible* meaning of the very words that express quantum mechanics; and
2) the fundamental limitations on measurements imposed by the basic concept of the Planck-Einstein-de Broglie quantum process; generally expressed in his
3) Heisenberg Uncertainty Principle $\Delta p \Delta q \sim \hbar$, which he deduced from qualitative arguments summarized in his
4) The Gedanken-experiment with his mythological γ-ray microscope [6] to measure simultaneously the position and momentum of an electron; and then
5) derived formally from the Dirac-Jordan-Schrödinger formalism.

But Heisenberg went much further in pursuit of his goal of *understanding* quantum mechanics: "$\cdots$ we understand a physical theory only when we can qualitatively explain $\cdots$ all simple cases, and can show that the theory involves no internal contradictions." This pragmatism is what distinguished Heisenberg from Dirac, Pauli, and Jordan who had comparably sophisticated formal skills. Only Bohr had a similar instinct for the fundamental, but his formal skills were of a heuristic nature and his communication skills, of course, were notorious. Max Born had a similar pragmatic plus formal combination of talents, but seems to have passed the baton and dropped out of the competition at an early stage.

Heisenberg applied his position measurement Gedanken-experiment first to the question of defining the "1S-orbit of the electron in the hydrogen atom." He immediately concluded that a γ-ray of energy $E_\gamma > 10 \text{keV}$ was necessary to get a measurement of precision $< 10^{-8} \text{cm}$, inevitably ejecting the electron from the "orbit". The result is that no orbit can be defined experimentally. The only quantity accessible to the position measurement Gedanken-experiment is the electron position probability distribution over an ensemble of identical 1S-hydrogen atoms.

Heisenberg had the foresight already at this point to preclude any kind of classically describable orbit for the electron, entirely removing any Bohmian orbits from the realm of the observable. Nor is it possible to define precisely the electron velocity without losing corresponding sight of its position, in accord with Eqn1. The Stern-Gerlach experiment replaces the microscope in a measurement of the energy, which Heisenberg shows by another Gedanken-experiment to satisfy an energy-time uncertainty relation $\Delta E \Delta T \sim \hbar$.

More Gedanken-experiments are constructed. One of particular interest is to understand the concept of the classical orbit. The provocative remark is made that *"The orbit exists only because we observe it."* Before we turn to the well-known paradoxes based on this remark, let's explore Heisenberg's applications of it. What is involved is repeated measurements on a single system, which can be shown to be equivalent to single measurements on repeated systems, i.e., an ensemble of systems, with the ensemble defined by the result of the earlier measurement. There is a successive *collapse of the wave function* after each measurement to correspond to the appropriate new ensemble. The new wave packet has the dimensions appropriate to the new uncertainties, so the wave packet keeps its size rather than dispersing very quickly as it would if it were to continue propagating. An analogous phenomenon is pointed out to occur even in successive measurements of a classical orbit, where the successive possible errors are reduced to the instrumental uncertainties at each observation, and not allowed to propagate endlessly.

A particularly simple Gedanken-experiment illustrates the gist of the *wave function collapse*. It is a two level system with the upper state $\psi_2(E_2)$ unstable against quantum jumps to the stable ground state $\psi_1(E_1)$, with a mean lifetime $\tau = 1/(2\alpha)$. Some provision is made to start the system in state 2 at time $t = 0$ and to measure the energy at various later times $t_1, t_2, \cdots$. How does quantum mechanics describe this system?

Heisenberg writes the wave function as

$$\chi(t) = e^{-\alpha t}\psi(E_2)e^{-iE_2t/\hbar} + \left(1 - e^{-2\alpha t}\right)^{1/2}\psi(E_1)e^{-iE_1t/\hbar},$$

which describes an ensemble $\mathcal{E}_2(t)$ of a large number of identical systems known to have E_2 at $t = 0$. At time t_n (a large number of) energy measurement(s) produces the result E_2 (with probability $P_2(t_n) = \exp(-2\alpha t_n)$) or the result E_1 (with probability $P_1(t_n) = 1 - P_2(t_n)$). Systems whose energies are determined at t_n to be E_2 are assigned the wave function $\chi(t - t_n)$ (within a phase) and relegated to a new ensemble $\mathcal{E}_2(t - t_n)$ where they begin life anew, exponentially decaying from state 2 to state 1. Systems whose energies are determined to be E_1, are assigned the wave function $\psi(E_1)e^{-iE_1t}/\hbar$ (again, within a phase) and relegated to an ensemble $\mathcal{E}_1(t)$ of stable ground state systems.

Here we witness the much reviled: "wave function collapse" following a measurement corresponding to relegation to a new, possibly different, ensemble. Many authors – with Heisenberg the first – have demonstrated the equivalence of repeated measurements of this kind on a single system at a succession of times to be equivalent to an ensemble. Heisenberg also demonstrates from his just invented Uncertainty Principle that the energy determination destroys all phase coherence.

§XI-2. Bohr-Heisenberg Personality Conflicts.

The stress involved in the intense creativity at the genesis of quantum mechanics is nowhere better documented than here in the conflict that arose between Heisenberg and Bohr in their struggle from opposite poles toward an understanding. MacKinnon [7] writes: "For months, Heisenberg and Bohr had protracted, often heated, discussions concerning idealized experiments, physical interpretations, and the meaning of the concepts involved. Their dialog eventually broke down because of two points $\cdots$ Heisenberg completely rejected any wave interpretation of electrons $\cdots$ one should begin with the formalism, work out its consequences, then relate these to experiments, both real and gedanken. Bohr insisted that a consistent physical interpretation is more basic than the mathematical formalism that builds on it [Surely an error on Bohr's part. It is inconceivable that any clarity of understanding of quantum mechanics could have been achieved by any purely heuristic hand-waving arguments without the concepts of state-vectors and superposition. At the very least, Heisenberg's and Bohr's modes of thought have to be co-equal, maybe (to coin a phrase) complementary.] ··· Schrödinger's wave interpretation must have a significant role ··· Cooperative effort proved extremely difficult. Bohr went to Norway for a skiing trip and ··· Heisenberg remained in Copenhagen and within days wrote the first draft of his Uncertainty Principle paper which Pauli approved." Now the rift between Bohr and Heisenberg became critical. Bohr evidently thought Heisenberg's paper was the draft of a joint paper, but Heisenberg rushed it into print on his own. Pauli eventually negotiated a face-saving Addendum to Heisenberg's paper acknowledging Bohr's contribution of complementarity.

Pais [4] quotes extensively from Heisenberg's recollection of events of these critical days (he was not able to find any such by Bohr): '··· *Bohr was trying to allow for the simultaneous existence of both particle and wave concepts* ··· *mutually exclusive* ··· *both needed for a complete description of atomic processes.* I disliked this approach. I wanted to start from the fact that matrix mechanics already imposed a unique physical interpretation ··· we no longer had any freedom ··· derive the correct general interpretation by strict logic ···. I definitely wanted to keep away from the wave mechanical side ··· not make any concession ··· psychologically, I came from matrix mechanics ··· a mathematical scheme which is consistent, it can be either wrong or right, but if it is right then anything added to it must be wrong because it is closed in itself. ··· I wanted to start entirely from the mathematical scheme of matrix mechanics ··· Bohr wanted to ··· *play with both schemes.*' And '··· we sometimes got a bit impatient with each other ··· we were a bit angry about it ··· utterly exhausted and rather tense.'

Pais [4] gives a detailed chronology of events surrounding Heisenberg's discovery of the Uncertainty Principle:

∼ 5Feb27 – Bohr leaves for Norway. 'Within a few days' Heisenberg had the uncertainty relations.

23Feb27 – Heisenberg sends a 14 page letter to Pauli describing the uncertainty relations.

10Mar27 – Heisenberg writes Bohr in Norway: ' ··· I have written a letter about these problems which yesterday I sent to Pauli.'

~18Mar27 – Bohr returns and criticizes Heisenberg's paper as incorrect (in details of the microscope resolving power). Heisenberg recalls ' ⋯ breaking out in tears because I just couldn't stand this pressure from Bohr.'

23Mar27 – Heisenberg's manuscript received by *Zeitschrift fur Physik*.

Apr27 – Bohr to Einstein, on Heisenberg's paper: ' ⋯ most significant ⋯ exceptionally brilliant ⋯ contribution to the discussion of the general problems of quantum theory.'

Pais quotes letters about a year later from Heisenberg to Bohr: 'I am very ashamed to have given the impression of being quite ungrateful.' and 'I have been so unhappy ⋯ and ashamed ⋯ '. But after a 1928 visit to Copenhagen '⋯ once again we understood each other so well ⋯ as in the "old days".' And from Bohr to Heisenberg: '⋯ Rarely have I felt more sincere harmony with any other human being.'

§XI-3. Concluding Remarks.

These are the premises of the orthodox Copenhagen interpretation of quantum mechanics. Should they be believed? Apparently, yes. Should they be accepted without further question? Absolutely not. They should be questioned, assaulted, explored, torn apart, laid bare; and they have been and are, with ever increasing vigor!

Have they ever been found in error? We believe that a fair answer to this question is: Not really. Many conceptual aspects of quantum mechanics, most particularly the transition between the quantum world and the classical world had been glossed over until the recent challenges by a whole new generation of ingenious and intrepid skeptics to what could be suspected of being the unquestioning sycophancy of the rest of us. Nonetheless, it can be said fairly that for every challenge, the explicit response is always implicit in the general philosophical prescriptions of the orthodox interpretation WITHOUT recourse to any extra dynamical machinery. Among these devices we include Bohmian orbits [8], spontaneous decohering events [9] , Everett parallel universes [10], and so on and on [11].

Most people will grant that the orthodox theory is *sufficient* to understand any phenomena (perhaps not including the whole universe). But is it *necessary*? Must we *really* postulate an interpretation of Schrödinger's amplitude which is somehow surrealistic and consequently unphysical and fundamentally different from classical fields like sound waves, etc.; related as it seems to be to our state of knowledge; and changing discontinuously – unconstrained by Schrödinger's equation – during some ill-defined *measurement* process?

So far as we know, there is and *can be* no proof of necessity. What evolves is a case by case demonstration that the orthodox interpretation is the most simple, economical, general, and thereby beautiful, and by inference therefore the one nearest the truth; and that not in spite of but because of the abstractions dictated in the first instance: starting with Heisenberg's first postulate that the dynamics should involve only "observables"; leading to the requirement for the dynamical variables of matrix multiplication in the space of states; and Born's introduction of transition *amplitudes* as the fundamental quantities evolving by the quantum analog of Hamiltonian mechanics; and the interpretation of these probabilities as measurable only on an *ensemble* of identically prepared systems; which by measurement are relegated to a different ensemble, closing the logical system. Well, not completely $\cdots$ For a remarkably clear, cogent, and contemporaneous view, the words of the mature Heisenberg [12] cannot be excelled.

Footnotes and References:

1) B.L. van der Waerden, *Sources of Quantum Mechanics* (North Holland, Amsterdam, 1967), Pp.21-57.

2) W. Heisenberg, Zeits. f. Phys. **43**, 172 (1927). There certainly is another completely singular and solitary achievement not yet mentioned here which distinguishes Heisenberg unique and pre-eminent role in the creation of quantum mechanics, and that is his Symmetrization Postulate. In this discovery – characteristically, from a rather rudimentary and intuitive argument – he devised the elegant and pristine modification of quantum mechanics required for the treatment of identical particles. In this, he anticipated a more sophisticated argument soon put forward independently by Dirac, and furthermore he established the foundation for the immediate construction of Quantum Field Theory, in which he played a further fundamental role. For a full discussion of these events see: I. Duck and E.C.G. Sudarshan, *Pauli and the Spin-Statistics Theorem* (World Scientific, Singapore 1997), Chs 4,5,6,7; Pp.108-203.

3) N. Bohr, W. Heisenberg and P. Jordan, Zeits. f. Phys. **35**, 557 (1926) (see here as **Paper VII·2**).

4) A. Pais, *Niels Bohr's Times* (Clarendon Press, Oxford, 1991), Pp. 308-309 for an account of Heisenberg's emotional reaction to Bohr's persistent criticism and his quick remorseful apologies.

5) B.L. van der Waerden, *Exclusion Principle and Spin* in Theoretical Physics in the Twentieth Century, A Memorial Volume to Wolfgang Pauli (Interscience, New York, 1960) M. Fierz and V.F. Weisskopf, Eds., p.199. (See also here in Footnote 2, ChX.)

6) Heisenberg credits Burkhard Drude, a fellow student at Göttingen, for the idea of a γ-ray microscope. See *Niels Bohr* (Harvard Press, Cambridge, Mass. 1985) A.P. French and P.J. Kennedy, Eds., p.169.

7) E. MacKinnon, 'Bohr on the Foundations of Quantum Theory' in *Niels Bohr – A Centenary Volume* (Harvard, Cambridge, MA, 1985) A.P. French & P.J. Kennedy, Eds., Pp.101-120.

8) D. Bohm, Phys. Rev. **85**, 166, 180 (1951).

9) G.C. Ghirardi, A. Rimini, T. Weber, Phys. Rev. D **34**, 470 (1986).

10) See many contemporary references in, for example, *Complexity, Entropy and the Physics of Information* (Addison-Wesley, Reading, Mass., 1990) W.H. Zurek, Ed., Pp.33, 405.

11) For a readable up-to-date account including modern references and views, see Sheldon Goldstein, Physics Today **51**-3, 42 (1998); and **51**-4, 38 (1998).

12) W. Heisenberg, 'The Development of the Interpretation of the Quantum Theory' in *Niels Bohr and the Development of Physics* (Pergamon, London, 1955) W. Pauli, Ed., Pp.12-29 (see here as **Paper XI·2**).

Paper XI·1: Excerpt from Zeitschrift für Physik **43**, 172 (1927).

On the Essential Content of Quantum Theoretic Kinematics and Mechanics[†]

von **W. Heisenberg** in Kopenhagen.

(Eingegangen am 23. März 1927.)

In the following, exact definitions of the words: position, velocity, energy, etc. (e.g. of electrons) are given, which are valid in quantum mechanics. It is shown that canonically conjugate dynamical variables can be determined simultaneously only with a characteristic uncertainty (§1). This uncertainty is the fundamental reason for the statistical correlations in quantum mechanics. Their mathematical formulation is given by the Dirac-Jordan theory (§2). On this basis, it is shown that the macroscopic behavior of quantum mechanics can be understood (§3). To explain the theory we discuss a particular Gedanken-experiment (§4).

We feel that we understand a physical theory only when we can qualitatively explain the experimental consequences of the theory in all simple cases, and can show that the theory involves no internal contradictions. · · · · · · The existing interpretation of quantum mechanics is still full of contradictions, such as the conflict of ideas on discontinuous and continuous theories, and on particles and waves. Already though, one can conclude that an Interpretation of Quantum Mechanics is not possible without understanding the kinematical and mechanical ideas. *Quantum mechanics was in fact derived directly from the attempt to eliminate each customary kinematical variable and to replace it by relations between specific experimentally known numbers.* [Note:Italics added.] Once this was done, no further revision in the mathematical formulation of quantum mechanics was required. · · · · · · By choosing sufficiently massive particles we can approximate the quantum mechanical laws by the classical, except when confined to small space-time intervals. That a change in kinematic and mechanical concepts is necessary follows directly from the fundamental equations of quantum mechanics.

* * * * * * * * * * * * * * * * * * **

The *italics* above are necessary to emphasize the fundamental distinction between orthodox quantum mechanics and its principal variant due to Bohm who reintroduces the concept of particle trajectories by way of the locally *defined* (but unmeasurable) velocity

$$\vec{v} \equiv \frac{\hbar}{2mi} \frac{(\psi^* \cdot \vec{\nabla}\psi - \vec{\nabla}\psi^* \cdot \psi)}{\psi^*\psi}.$$

* * * * * * * * * * * * * * * * * * **

Previously, when a particle of mass m was considered, it had been possible to speak in a simple understandable sense of the position and velocity of the particle. In quantum mechanics, however, there is a basic commutation relation

$$pq - qp = \hbar/i$$

between the particle's position q and momentum p. We have therefore good reason to be suspicious of the uncritical use of these words "position" and "velocity". When one remembers that for processes in very small space-time intervals discontinuities are typical anyway, then the traditional concepts of "position" and "velocity" fail to be directly plausible: if one thinks of the one-dimensional motion of a particle, then in a continuum theory one can draw the orbit of a particle as a smooth curve $x(t)$, whose tangent gives the velocity in the usual way. In a discontinuous theory on the other hand, in place of this curve there will be a series of points at finite intervals. *In this case it is essentially meaningless to speak of a velocity at a particular position because the velocity can be defined only by two points.* Conversely, to each point [Note added: on the discrete trajectory of individual points] there belong two different velocities. The question is whether by an exact analysis of the kinematic and mechanical ideas, it is possible to clarify the understanding of the basic quantum mechanical relations [1]. [Note the remarkable similarity to the thought processes of Dirac and the early Einstein. Only in these authors do we see thought processes so clear yet so elementary that they would be completely sterile for most others.]

§ 1. The Concepts: Position, Orbit, Velocity, Energy.

In order to follow the quantum mechanical behavior of some particle, one must know the mass of the particle and the interaction forces with other particles and any fields. Only then can the Hamiltonian of the quantum system be constructed. (The following discussion is confined to nonrelativistic quantum mechanics, since the quantum theory of electrodynamics is only very incompletely known [2].) It is unnecessary to say anything further about the "properties" of the particles, if one suitably describes all its interactions.

To be clear about the meaning of the words "position of the particle", for example of an electron relative to a given coordinate system, one must specify a definite experiment to measure the "position of the electron"; otherwise this word has no meaning.

* * * * * * * * * * * * * * * * * * **

This is the essential foundation of the Copenhagen interpretation. 'Objective reality' is attained only by an act of observation. This principle apparently originated in a remark by Jordan that *'The orbit exists only because we observe it.'* It has been carried to paradoxical extremes with Schrödinger's cat which 'was not dead until we looked' or Mermin's moon or the tree that fell or not depending on whether anyone was looking. In spite of these provocative counterexamples, it remains a cornerstone of the Copenhagen interpretation.

* * * * * * * * * * * * * * * * * * **

There is no mystery about experiments which in principle allow us to determine the "position of the electron" with arbitrary accuracy. For example: Simply illuminate the electron and observe it under a microscope. The maximum possible precision of the position determination is given by the wavelength of the light. At least in principle, one can build a γ-ray microscope and determine the position as accurately as one wishes. There is however, an essentially quantum feature: the Compton effect. Each observation of light coming from the electron involves the photoelectric effect (in the eye, in the photographic plate, in the photocell) which can be explained as a photon striking the electron, and being reflected or re-emitted through the lens of the microscope where it is again absorbed in the photo-effect. At the instant of the position determination, that is at the instant the photon is scattered from the electron, the electron momentum *changes discontinuously*.

This change increases if the wavelength of the light is decreased, that is, if the position measurement is made more accurately. At the instant the position of the electron is known, its momentum can be known only within a range corresponding to the discontinuous change; *the more accurately the position is known, the less accurately the momentum is known, and conversely.* $\cdots$ If Δq is the accuracy with which q is known (Δq is some average uncertainty in q), which here is of the order of the wavelength of the light λ; and Δp the accuracy with which the value of p is determined, which here is of the order of the change in p during the Compton scattering of a light-quantum of momentum $2\pi\hbar/\lambda$, then $\Delta p \sim \hbar/\lambda$ and $\Delta q \sim \lambda$ satisfy

$$\Delta p \Delta q \sim \hbar. \tag{1}$$

That this relation is a direct mathematical consequence of the above commutation relation will be shown later. Eqn1 is the expression of the fact that one should describe phase space as partitioned into cells of volume $\hbar$.

For the determination of the electron's position one could also use other experiments, for example a simple collision. An exact measurement of the position requires a collision with a very fast particle, since for slow particles the diffraction pattern, which according to Einstein follows from the de Broglie waves (e.g., the Ramsauer effect), prevents an exact determination of the position. During a precise position measurement, the momentum of the electron again changes discontinuously and a simple estimate for the uncertainties using the formulas for the de Broglie waves leads again to the relation Eqn1. From this discussion the concept "position of the electron" is clearly defined. $\cdots\cdots$

We go now to the concept of the "orbit of the electron". By orbit we mean a series of space points which the electron occupies one after another as "position". Since we already know what we should understand by "position at a particular time", no new difficulties are involved here. Nevertheless, it is easy to see that the frequently used expression: the "1S-orbit of the electron in the hydrogen atom" has no meaning from our point of view. In order to measure this 1S-"orbit", the atom must be illuminated with light whose wavelength is substantially less than 10^{-8} cm ($E_\gamma \sim 10$ keV). For such light, however, a single quantum scattering would completely eject the electron from its "orbit" (consequently *only one such point can be defined for any such orbit*), so the word "orbit" here has no sensible meaning. [Note added: Clearly the possible position measurements resolve themselves into a probability distribution over an ensemble, as Heisenberg immediately explains.] This can be deduced very simply from the experimental possibilities without any details of new theories.

On the other hand, such a position measurement can be carried out on many atoms in the 1S-state. The repeated position measurements just give a probability distribution function for the position of the electron, which corresponds to the average value of the classical orbit over all phases and which can be determined with arbitrary accuracy [Note added: But the details of any Bohmian orbits are completely inaccessible]. According to Born [3], this function is $\bar{\psi}_{1S}(x)\psi_{1S}(x)$, where $\psi_{1S}(x)$ is the Schrödinger wave function for the 1S-state.

* * * * * * * * * * * * * * * * * * **

Here at last is Heisenberg's unstinted acknowledgement of Born's interpretation of the wave function (absolute square) as the probability distribution for repeated experiments on an ensemble of independent identical quantum systems. And here Born is the progenitor of the Copenhagen interpretation which Heisenberg now enunciates for the first time and in very intuitive terms.

* * * * * * * * * * * * * * * * * * **

Dirac and Jordan in their formulation might say for greater generality: The probability is given by $\bar{\psi}(1S, x)\psi(1S, x)$ where $\psi(1S, x)$ is a column of the transformation matrix $\psi(E, x)$ from E to x, corresponding to $E = E_{1S}$. From this fact, that in the quantum theory for a particular state only the probability distribution function of the electron can be given, one can see – with Born and Jordan – a characteristic statistical connection of quantum theory to the laws of classical theory. One could also say – with Dirac – that the statistics is brought in by our experiments. Then even in the classical theory only the probability of a particular electron position is given, if we do not know the phase. *But the difference between classical and quantum mechanics is much more profound*: Classically, we can consider a definite phase throughout the whole experiment. In reality however this is impossible, because experiment disrupts (that is, changes) the definition of the phase of the atom. In a particular stationary "state" of the atom, the phases are indeterminable in principle, as one can see as a direct expression of the fundamental canonical commutation relations

$$Et - tE = i\hbar \quad \text{or} \quad J\phi - \phi J = \hbar/i$$

with $J=$ action variable, $\phi=$ angle variable.

Next, the "velocity" of a particle can be easily defined by measurements when it is in force-free motion. One can for example, illuminate the particle with red light and from the Doppler shift of the scattered light determine the velocity of the particle. The determination of the velocity is the more precise, the longer the wavelength of the light, since the change in the velocity of the particle due to recoil in the Compton scattering of the light-quantum is less. The position determination is correspondingly imprecise, as expressed in Eqn1. If the velocity of the electron in the atom is to be measured at a particular instant (one ignores at this instant the nuclear charge and the force of the other electrons, so that the motion is assumed to be force-free) the above determination follows. One can easily show as above, that a function $p(t)$ for a given state of the atom (e.g. 1S) cannot be defined. One gets only a probability distribution for p in this state, which has the value $\bar{\psi}(1S,p)\psi(1S,p)$. As before, $\psi(1S,p)$ is a column of the transformation matrix $\psi(E,p)$ from E to p, for $E = E_{1S}$.

Finally, depending on which experiment is done, the energy or the value of the action variable J can be measured; such experiments are particularly important because they define what we mean by the discontinuous change of the energy and the action. The Franck-Hertz impact experiment allows the measurement of the energy of the atom, establishing directly the equivalence of the energy jumps in the quantum-atom to the kinetic energy lost by the moving electrons. This measurement can be done with arbitrary accuracy only if one gives up the simultaneous determination of the electron position, that is, the phase (compare the above measurement of p) corresponding to the relation $Et - tE = i\hbar$.

For example, the Stern-Gerlach experiment measures the magnetic moment of the atom, that is, quantities which solely depend on the action variable J. The phase remains undetermined in principle. Just as it is meaningless to speak of the frequency of a light-wave at a particular instant, the energy of an atom cannot be specified at a particular moment. This corresponds in the Stern-Gerlach experiment to the fact that the precision of the energy measurement is the less, the shorter the time which the atom stays in the deflecting field [1]. An upper limit for the deflecting force is given when its potential energy difference across the particle beam to be separated, is as large as the energy difference of the stationary states whose energies are to be determined. Let ΔE be such an energy difference (ΔE is the possible precision of an energy measurement), so $\Delta E/d$ is the maximum value of the deflecting force, with d the width of the particle beam (measured by the width of the collimating slit). The angular deflection of the particles in the beam is $\Delta E \Delta t/pd$, where Δt is the time spent by the atom in the deflecting field and p the momentum of the atom in the direction of the beam. This deflection must at least equal the spread of the beam caused by diffraction at the slit, for a measurement to be possible. The angular deflection by diffraction is λ/d with λ the de Broglie wavelength. Therefore

$$\frac{\Delta E \Delta t}{pd} > \frac{\lambda}{d} \quad \text{and since} \quad \lambda = 2\pi \frac{\hbar}{p},$$

then

$$\Delta E \Delta t > \hbar. \tag{2}$$

This corresponds to the result Eqn1, and shows that an exact energy determination can only be made with a corresponding uncertainty in the time. [Note added: The analysis of the Stern-Gerlach experiment is very Bohrish, but a careful parsing of Heisenberg's acknowledgements to Bohr indicate that Bohr's contributions and critical comments did not include this development.]

§ 2. The Dirac-Jordan Theory.

* * * * * * * * * * * * * * * * * * * **

And here Heisenberg describes the full formal structure of the Copenhagen interpretation of quantum mechanics, paraphrasing and extending the work of Dirac and Jordan. Heisenberg also defines in full generality the impact of the Uncertainty Principle announced in the first section.

* * * * * * * * * * * * * * * * * * * **

The result of the preceding section can be summarized and generalized in this statement: each dynamical variable required in classical theory for the description of a mechanical system, has an exactly defined quantum analog for atomic processes. The experiment which such a definition requires, contains an intrinsic purely empirical indeterminacy when we demand from it the simultaneous determination of two canonically conjugate variables. The magnitude of this indeterminacy is given by Eqn1 (for any canonically conjugate variables whatever). It suggests that quantum theory is consistent with the special theory of relativity. [Note added: Here Heisenberg lapses into some hazy remarks on relativity, which we skip.] $\cdots\cdots$ (Special relativity) is not in conflict with the logical use of the words "position, velocity, time". Similarly, it is consistent with the concepts "electron position, velocity" in quantum theory. All experiments designed for the definition of these words, necessarily imply the indeterminacy given by Eqn1, whenever the canonically conjugate dynamical variables p, q can be exactly defined. Any experiment which required simultaneous "sharp" determinations of p and q would – according to Eqn1 – be impossible in quantum mechanics. The indeterminacy guaranteed by Eqn1 is the foremost consequence of the quantum mechanical canonical (i.e., standard) commutation relation $pq - qp = \hbar/i$; this equation requires that the classical physical interpretation of the canonically conjugate dynamical variables p and q *must be changed.*

For those physical phenomena whose quantum theory is not yet known (e.g., electrodynamics), Eqn1 is a requirement which might be useful. For quantum mechanics Eqn1 can be derived by a simple generalization of the Dirac-Jordan formulation. When we assign a (central, measured) value q' with an uncertainty Δq to the position q of an electron, we express this fact in a wave function $\psi(q)$ which is non-vanishing only in a range of magnitude Δq of q'. For example,

$$\psi(p', q) \sim e^{-(q-q')^2/2(\Delta q)^2} e^{-ip'(q-q')/\hbar}, \tag{3}$$

(for a central measured value of the momentum equal to p') for which

$$\bar{\psi}\psi \sim e^{-(q-q')^2/(\Delta q)^2}.$$

Then the probability amplitude corresponding to p is

$$\psi(p',p) = \int \psi(p',q)\psi(q,p)dq, \tag{4}$$

where according to Jordan

$$\psi(q,p) = e^{iqp/\hbar}. \tag{5}$$

Then from (4) $\psi(p',p)$ is essentially different from zero only for values of p for which $(p-p')\Delta q/\hbar \sim 1$. From Eqn3:

$$\psi(p',p) \sim \int e^{i(p-p')q/\hbar}e^{-(q-q')^2/2(\Delta q)^2} dq, \quad \text{that is}$$

$$\psi(p',p) \sim e^{-(p-p')^2/2(\Delta p)^2}e^{iq'(p-p')/\hbar}, \quad \text{and} \quad \bar{\psi}\psi \sim e^{-(p-p')^2/(\Delta p)^2} \quad \text{where}$$

$$\Delta p \Delta q \sim \hbar. \tag{6}$$

The assumption Eqn3 for $\psi(p',q)$ corresponds to the experimental fact that the measured values are p' for p, and q' for q (with precisions $\Delta p, \Delta q$ limited by Eqn6).

Purely mathematically, in the Dirac-Jordan formulation of quantum mechanics, the relations between p, q, E, etc. are written as equations between very general matrices, in which some of the quantum variables may appear as diagonal matrices. One can interpret the matrices as tensors (like moments of inertia in many dimensional spaces) between which mathematical relations exist. Ultimately one can always characterize the mathematical relation between two tensors A and B by the transformation formalism which transforms from a coordinate system oriented along the principal axes of A to another oriented along the principal axes of B. This formulation corresponds to Schrödinger's theory. Except for certain "invariants", each coordinate choice gives an apparently different formulation of quantum mechanics, whereas in the Dirac representation only the q-numbers appear.

To derive physical results in one particular mathematical scheme, we must assign numbers to the quantum variables, that is to their matrices (or "tensors" in the many dimensional space). This means that in each multidimensional space a definite direction is chosen (by the kind of experiment done) and the question asked, what is the "value" of the matrix (e.g., the moment of inertia) in this particular direction. This question only has a definite meaning when the selected direction coincides with a principal axis of that matrix; in this case there is an exact answer to the question asked. However, if the chosen direction deviates a little from one of the principal axes, then one can speak only with some uncertainty related to the degree of deviation, about the possible error of the "value" of the matrix in the given direction. One can therefore say: each quantum variable or matrix can be assigned a number which gives its "value", but only within some possible error; the possible error depends on the coordinate system; for each quantum variable there is one coordinate system for which the possible error in this variable vanishes. A specific experiment can never give exact information about all quantum variables. Rather – in a way characteristic of the experiment – physical dynamical variables are separated into "known" and "unknown" (or: more or less exactly known). The results of two experiments can only be exactly correlated when the two experiments separate dynamical variables *in the same way* into "known" and "unknown". If the separations are different, the results of the experiments can only be correlated statistically.

* * * * * * * * * * * * * * * * * * * *

This is the heart and soul of Heisenberg's physical interpretation of the matrix formulation of quantum mechanics which is fundamentally the Copenhagen interpretation. Although Heisenberg acknowledges Dirac and Jordan for their Hilbert space formalism, the results were already evident in Born's interpretation of his scattering amplitude $\sum |f'\rangle\langle f'|V|i\rangle$. The fact that Heisenberg insisted on publishing quickly (hastily?) and alone began the alienation of Bohr.

* * * * * * * * * * * * * * * * * * * *

For a more precise discussion of these statistical connections, one can perform a Gedanken-experiment. Let a Stern-Gerlach atomic beam be sent first through a field F_1, which is so strongly inhomogeneous in the beam direction that it causes many transitions by the "crossing effect". Then let the beam run free of F_1 for a certain distance, before entering a second field F_2, also inhomogeneous. Between

F_1 and F_2 and after F_2, it is possible to determine the number of atoms in different stationary states by auxiliary magnetic fields. The beam intensity should be close to zero. If the beam was in the state of energy E_n before it passed through F_1, we assign the atom a wave function – e.g., in momentum space – with energy E_n and the undetermined phase β_n

$$\Psi(E_n, p) = \psi(E_n, p) e^{-i(E_n t + \beta_n)/\hbar}.$$

After passing through the field F_1 this function evolves to [4]

$$\Psi(E_n, p) \to_{F_1} \sum_m c_{nm} \psi(E_m, p) e^{-i(E_m t + \beta_m)/\hbar}. \tag{7}$$

Here the β_m are somewhat arbitrarily defined so that the c_{nm} are uniquely determined by F_1. The matrix c_{nm} transforms the energy value from before to after passage through F_1. If after F_1 we make a measurement of the stationary state, say by using an inhomogeneous magnetic field, then we will find a probability $\bar{c}_{nm} c_{nm}$ that the atom has made a transition from state n to state m. If we experimentally determine that the atom has actually gone into the state m, then we should assign to it thereafter – not the function $\sum_m c_{nm} \psi_m$ – but in fact just the function ψ_m with undetermined phase.

* * * * * * * * * * * * * * * * * **

Here with unrivaled clarity is the endlessly criticized "collapse" of the wave function.

* * * * * * * * * * * * * * * * * **

The experimental statement is: we select a particular "state m" from all the different possibilities (c_{nm}); after the passage of the atomic beam through F_2 the situation repeats itself, just as for F_1. There is a transformation matrix d_{ml} which transforms the energy states from before to after F_2. If no measurement to determine the state was made between F_1 and F_2, then the wave function develops as

$$\Psi(E_n, p) \to_{F_1} \sum_m c_{nm} \psi(E_m, p) \to_{F_2} \sum_m \sum_l c_{nm} d_{ml} \psi(E_l, p). \tag{8}$$

Define $e_{nl} = \sum_m c_{nm} d_{ml}$. If the state of the atom is measured only after F_2, then one finds state l with the probability $\bar{e}_{nl} e_{nl}$. If on the other hand, the state m was measured between F_1 and F_2, then the probability for state l after F_2 is $\bar{d}_{ml}d_{ml}$. For many repetitions of the whole experiment (where the state is determined to be m between F_1 and F_2), one will find the state l after F_2 with the relative frequency

$$Z_{nl} = \sum_m \bar{c}_{nm} c_{nm} \bar{d}_{nl} d_{nl}.$$

This is not the same as $\bar{e}_{nl} e_{nl}$. $\cdots$ In every case which allows a determination of the stationary states m, the atom undergoes a disturbance by the apparatus between F_1 and F_2. The result is that the "phase" of the atom changes uncontrollably, just as the momentum changes during the determination of the electron position (see §1). Even for the magnetic field separation of the states between F_1 and F_2, which has the energy as an eigenvalue, *during the observation of the trajectory* [Italics emphasize that Heisenberg clearly required a classical act of 'observation' to destroy the coherence of the wave function of the observed system.] of the atomic beam (e.g., by a Wilson cloud chamber track) the atom accelerates statistically and uncontrollably, and so on. This has the result that the ultimate transformation matrix e_{nl} (from entering F_1 to leaving F_2) is no longer given by $\sum_m c_{nm} d_{ml}$, but each term has the sum over an unknown phase factor. We can only expect that the average of $\bar{e}_{nl} e_{nl}$ over all these phase changes is equal to Z_{nl}. A simple calculation shows that this is the case. Now by a specific statistical rule, we can relate the possible results of one experiment to those of others. These other experiments choose one definite case from all possibilities, and thereby limit the possibilities for all later experiments. Such a representation for the transformation matrix of the Schrödinger wave function is only possible if the sum of solutions is again a solution. In this way, we see a deep significance of the linearity of the Schrödinger equation: for this reason alone, it can be understood as the equation for waves in phase space. Also for this reason, we might consider as hopeless any attempt to replace these equations with non-linear ones, for example in the relativistic case (for many electrons).

* * * * * * * * * * * * * * * * * **

This linearity remains valid in the unitary evolution of the state vector in the Hilbert space.

* * * * * * * * * * * * * * * * * * **

§ 3. The Transition from Micro- to Macro-mechanics.

The above analysis of the words "electron position, velocity, energy, etc." clarifies the concepts of quantum kinematics and mechanics sufficiently that a complete understanding even of macroscopic processes must be [Note added: And is!? The point of endless dispute is right here.] possible from the point of view of quantum mechanics. The transition from micro- to macro-mechanics has already been discussed by Schrödinger [5], but I believe that *the Schrödinger treatment does not address the essence of the problem*: According to Schrödinger, in the highly excited states a sum of energy eigenstates can be arranged into a small wave-packet, whose periodic changes of magnitude follow the periodic motion of the classical "electron". I have the following objection: If the wave-packet has such properties, and if the radiation of the atom is expressed in a Fourier series, the frequencies of the observed vibrations should be integral multiples of a fundamental frequency. But the frequencies of the spectral lines radiated by the atom are known from quantum mechanics to be non-integral multiples of a fundamental – except in the special case of the harmonic oscillator. *Schrödinger's explanation is only possible in the harmonic oscillator case which he discussed.* In all other cases the wave-packet spreads over the whole atomic volume within a transit time. The higher the excitation of the atom, the slower the dispersion of the wave-packet. But if one waits long enough, it will occur. The above argument about the radiation from the atom can be applied to all such attempts to find a direct transition of quantum mechanics into classical mechanics for high quantum numbers. One must look further, and even should avoid arguments referring to the natural radiative stability of stationary states; they are surely incorrect, firstly because they are contradicted already by the case of low energy radiation from the highly excited states of hydrogen, and secondly because the transition of quantum mechanics into classical mechanics must be understood without the help of electrodynamics. [Note added: This is an interesting point not usually made, but surely obvious when the ultimate quantum nature of the electrodynamic interaction (not the coulomb part) is taken into account.] These well-known difficulties standing in the way of a direct connection of quantum theory with classical have been repeatedly described by Bohr [3]. We can only explain the situation once again, since it seems already to have been forgotten.

I believe that one can understand the concept of the classical "orbit" as follows: *The orbit exists only because we observe it.* For example, let an atom be in the 1000th excited state. The size of the orbit is so large that it is possible to determine the electron position with relatively long wavelength light. For the determination of the position to be not too inaccurate, the Compton recoil has the result that the atom – depending on the impulse – finds itself in some other state between, say, the 950th and the 1050th; at the same time, the momentum of the electron can be measured from the Doppler effect with a precision known from (1). These experimental facts can be summarized in a wave-packet – better a probability amplitude packet – of a size in q-space given by the wavelength of the light used, composed essentially of excited state wave functions between the 950th and the 1050th, and characterized by a corresponding packet in p-space. After some time, a new position determination with the same limitations is carried out. Its result contains the effect of all the wave-packets now already spread out with appreciable probability. This was also true in classical theory in a similar way, because even in classical theory the result of the second position measurement was only statistically related to the uncertainty of the first; even the orbits of classical theory are spread out analogously to wave-packets. Of course the statistical laws themselves differ between quantum and classical theories. The second position determination selects from all possibilities a specific "q"and uses it for all the following predictions of possibilities. After the second position determination the results of further measurements can only be calculated if one again assigns a "smaller" wave-packet of size λ (the wavelength of the light used). Each such position determination reduces the wave-packet again to its original size λ.

$* * * * * * * * * * * * * * * * * * **$

So by a selection of central events, both in classical and in quantum physics, the spatial spread of tracked events is being reduced after every measurement to keep it constant at $\Delta q \sim \lambda$. The classical selection to reduce initial condition errors is analogous to the quantum selection to reduce the wave packet to the original spread.

$* * * * * * * * * * * * * * * * * * **$

The value of "p" and "q" are known during all measurements with corresponding precisions. That the values of p and q satisfy the classical equations of motion within this range of precision can be seen from the quantum laws

$$\dot{p} = -\frac{\partial H}{\partial q}; \qquad \dot{q} = \frac{\partial H}{\partial p}. \tag{9}$$

But the orbit can only be calculated statistically from the initial conditions, which one can consider as the Uncertainty Principle for the initial conditions. The statistical laws are different for quantum and classical theories; this can lead under certain conditions to large macroscopic differences between the predictions of classical and of quantum theory.

Before discussing such an example, consider a simple mechanical system: the force free motion of a point mass allows the above discussed transition to classical theory to be formulated mathematically. The equations of motion (for one dimension) are

$$H = \frac{p^2}{2m}; \quad \dot{q} = \frac{p}{m}, \quad \dot{p} = 0. \tag{10}$$

The solution of these equations is

$$q = \frac{p_0 t}{m} + q_0; \qquad p = p_0, \tag{11}$$

where p_0 and q_0 are the initial values at $t = 0$. At $t = 0$, from Eqn3-6 of §2, we set the value of $q_0 = q'$ with uncertainty Δq, $p_0 = p'$ with uncertainty Δp. In order to find the "value" of q at time t from the "values" of p_0 and q_0, according to Dirac and Jordan the transformation function must be found which transforms all matrices with q_0 diagonal to those with q diagonal. p_0 with q_0 diagonal is the operator $p_0 = -i\hbar\partial/\partial q_0$. According to Dirac, the transformation function $\chi(q_0, q)$ satisfies the differential equation

$$\left(\frac{t}{m} \frac{\hbar}{i} \frac{\partial}{\partial q_0} + q_0 \right) \chi(q_0, q) = q\chi(q_0, q) \tag{12}$$

with the solution

$$\chi(q_0, q) = \text{const}.e^{im \int (q-q_0)dq_0/\hbar t}. \tag{13}$$

$\bar{\chi}\chi$ is therefore independent of q_0. That is, when q_0 is known exactly at $t = 0$, then for any later time $t > 0$ all values of q are equally probable; the probability that q lies in some finite range is in general zero. This is necessary since the exact determination of q_0 requires infinitely large Compton recoil. The same conclusion would hold for any other mechanical system. However if q_0 is known at $t = 0$ only with a precision Δq, and p_0 with a precision Δp (see Eqn3, §2)

$$\psi(q', q_0) = \text{const}.e^{-(q_0-q')^2/2(\Delta q)^2+ip'(q'-q_0)/\hbar},$$

then the probability amplitude for q should be calculated from

$$\chi(q', q) = \int \psi(q', q_0)\chi(q_0, q)dq_0.$$

This gives

$$\chi(q', q) = \text{const}. \int e^{im[q_0(q-tp'/m)-q_0^2/2]/\hbar t} \times e^{-(q'-q_0)^2/2(\Delta q)^2} dq_0. \tag{14}$$

If one abbreviates

$$\beta = \frac{\hbar t}{m(\Delta q)^2}, \tag{15}$$

and introduces a new "const." (independent of q), then the integration gives

$$\chi(q', q) = \text{const}.e^{-(q-tp'/m-i\beta q')^2(1-i/\beta)/[2(\Delta q)^2(1+\beta^2)]}. \tag{16}$$

From this, the probability density is

$$\bar{\chi}(q', q)\chi(q', q) = \text{const}.e^{-(q-tp'/m-q')^2/[(\Delta q)^2(1+\beta^2)]}. \tag{17}$$

The electron at time t is at $q' + tp'/m$ with the uncertainty $(\Delta q)\sqrt{1 + \beta^2}$. The "wave-packet", or better the "probability distribution" grows in width by the factor $\sqrt{1 + \beta^2}$. From Eqn8, β is proportional to t, inversely proportional to m – this is obvious – and inversely proportional to $(\Delta q)^2$. A too great precision in q_0 leads to a large uncertainty in p_0 and leads to a large uncertainty in q.

As an illustration that the difference of the statistical laws of classical and quantum theories can lead to large macroscopic effects, consider the reflection of an electron from a crystal. If the lattice constant is comparable to the de Broglie wavelength of the electron, the reflection occurs in definite discrete directions, just as in the reflection of X-rays. Classical theory predicts something grossly different. From classical theory we cannot even determine the trajectory of the electron without further arguments. If we could direct the electron to a particular point on the crystal surface, only then could we say that the scattering there takes place classically. But if we want to specify the position of the electron so exactly that we can say from what point on the surface it scattered, then by this position determination the electron must have a large momentum with such a small de Broglie wavelength that now the reflection can be described effectively in the classical approximation, without the quantum laws.

* * * * * * * * * * * * * * * * * * * **

Note here the implicit criticism of the still-to-come de Broglie and Bohm interpretation of quantum mechanics where just this precise specification of the electron's trajectory is assumed with impunity, and the spurious claims of a deterministic but equivalent Bohmian quantum mechanics put forward. Of course the proponents of Bohm quantum theory never actually *use* these physically undefinable orbits but manage to cobble together the result prescribed by orthodox quantum mechanics in any (particularly simple, usually artificial) situation, and even then only from the knowledge of the orthodox Schrödinger wave function.

* * * * * * * * * * * * * * * * * * * **

§ 4. Two Especially Interesting Gedanken-Experiments.

According to the experimental interpretation of quantum theory given here, the time of a transition by a quantum-jump must be measurable, just like the energies for stationary states. The precision with which the time can be determined is $\Delta t = \hbar/\Delta E$ [1], where ΔE is the change of energy during the quantum-jump. Consider the following experiment: An atom in state 2 at $t = 0$ makes a radiative transition to the groundstate 1. The atom's wave function is

$$\chi(p,t) = e^{-\alpha t}\psi(E_2,p)e^{-iE_2t/\hbar} + \left(1 - e^{-2\alpha t}\right)^{1/2}\psi(E_1,p)e^{-iE_1t/\hbar}, \qquad (18)$$

where we represent the radiation damping by a factor $e^{-\alpha t}$ (the actual dependence is certainly not so simple). The atom is to have its energy measured by a Stern-Gerlach experiment. The atom will follow a long broken path in the magnetic field. One measures the acceleration along the whole atomic trajectory, traced out by the small particle tracks which mark the passage of the atom. Knowing the velocity of the beam, the distance between particle tracks of the atom corresponds to a separation into small time intervals Δt which correspond to an uncertainty in the energy $\Delta E = \hbar/\Delta t$. The probability to measure a definite energy E can be calculated directly from $\chi(p, E)$ over the interval from $n\Delta t$ to $(n+1)\Delta t$ to be

$$\chi(p, E) = \int_{n\Delta t}^{(n+1)\Delta t} \chi(p, t) e^{iEt/\hbar} dt.$$

If at the time $(n+1)\Delta t$ the determination "state 2" is made, then for all later times the wave function is no longer the function in Eqn18, but one obtained from it by replacing t by $t - (n+1)\Delta t$. If on the other hand the determination is "state 1", then thereafter the atom must be assigned the groundstate wave function

$$\psi(E_1, p) e^{-iE_1 t/\hbar}.$$

In a series of intervals one observes: "state 2", $\cdots$ "state 2", "state 1", and thereafter "state 1". In order for a distinction between the two states to be possible, Δt must not be less than $\hbar/\Delta E$. The actual time of the transition can be determined only with this precision. By such an experiment as this, we interpret the meaning of the old original concept of Planck, Einstein and Bohr, where we speak of discontinuous change of the energy. Since such an experiment is doable in principle, an explanation of its outcome must be possible.

In Bohr's fundamental postulates of quantum theory, the energy of an atom – just like the value of the action-variable J for other observables (position of the electron, etc.) – has the property that it can be assigned a definite numerical value. These preferred values which the energy – in contrast to other quantum mechanical variables – can assume, result from the fact that for a closed system there is an integral of the equation of motion (for which the energy matrix is $E = \text{const}$); for non-closed systems, there is no such distinction between the energy and the other variables. In particular, if an experiment can be done to measure exactly the phase Θ of the atom, then the energy is, in principle, undetermined, corresponding to a relation $J\phi - \phi J = \hbar/i$ or $\Delta J \Delta \phi \sim \hbar$. Resonance fluorescence is such an experiment. If one irradiates an atom with light of the resonant frequency

$f_{12} = (E_2 - E_1)/2\pi\hbar$, then the atom oscillates in phase with the external radiation and it is in principle meaningless to ask which state – E_1 or E_2 – the atom is in. To do better experiments with radiation, one could measure the phase relation for an exact position measurement in the sense of §1 for the electron at different times, relative to the phase of the light (at each atom). The individual atom can be assigned the wave function

$$\chi(q, t) = c_2\psi_2(E_2, q)e^{-i(E_2 t + \beta)/\hbar} + \sqrt{1 - c_2^2}\psi_1(E_1, q)e^{-iE_1 t/\hbar}. \qquad (19)$$

The probability of a position q is

$$
\begin{aligned}
|\chi(q, t)|^2 &= c_2^2|\psi_2|^2 + (1 - c_2^2)|\psi_1|^2 + \\
&\quad c_2\sqrt{1 - c_2^2} \times 2\mathrm{Re}\left(\bar{\psi}_2\psi_1 e^{i[(E_2 - E_1)t + \beta]}\right).
\end{aligned} \qquad (20)
$$

The periodic term in Eqn20 is observable even for non-periodic experiments, since the position determination can be made for different phases of the incident light.

* * * * * * * * * * * * * * * * * * **

A note here on notation: Heisenberg of course used matrix notation, and translated matrix elements freely into Schrödinger wave
functions which he chose intuitively to illustrate his arguments. The
difficulty in following his original arguments is made greater by his
habit of using the same symbol – frequently S or ψ – in multiple
roles. For better or for worse, we have translated these multiple us-
ages into different symbols – for example χ, ϕ, Ψ – which hopefully
are more suggestive for the modern reader.

* * * * * * * * * * * * * * * * * * **

In a Gedanken-experiment devised by Bohr, the atoms of a Stern-Gerlach atomic beam are first excited at a particular place by resonant light. After an interval, they enter an inhomogeneous magnetic field; the radiation emitted by the atoms can be observed the whole way, before and after the magnetic field. Before the atoms enter the magnetic field, we have the usual resonance fluorescence and all atoms emit spherical waves in phase with the incoming light. This conclusion is opposite to that from a superficial application of light-quanta ideas, or of quantum axioms. From these, one might conclude that only a few atoms would be in the "upper state" through absorption of a light-quantum, and that the entire resonance radiation comes from these few excitation centers. It can be said: the light-quantum concept is directly useful here only for the energy-, and momentum-balance; "in reality", all atoms radiate weak coherent spherical waves into the lower state. After the atoms have passed through the magnetic field, the atomic beam has been split into two beams, one of atoms in the upper-state, the other in the lower-state. For the atoms in the lower-state beam to radiate, there would have to be a large violation of energy conservation, because the total excitation energy is in the upper-state beam. After the magnetic field, only the upper-state atoms emit light – in fact, incoherent light – and de-excite to the lower-state. As Bohr shows in this Gedanken-experiment, it is necessary to be very careful in applying the concept of "stationary state".

From the understanding of quantum theory developed here, a discussion of Bohr's experiment can be carried through without difficulty:
1) In the external radiation field, the phases of the atom are determined; therefore it makes no sense to speak of the energy of the atoms.
2) After the atom has left the radiation field, one cannot say that it is in a definite stationary state, and cannot ask about the coherence properties of the radiation.
3) One can determine the state of the atom, but the result is statistical. Such an experiment is actually done by separation of the states in the inhomogeneous magnetic field.
4) After the separation, the *energies* of the atoms are *defined*, but the *phases* are completely *undefined*. The radiation here is incoherent and only from the upper-state. The magnetic field determines the energies but destroys the phase relations.

The Bohr Gedanken-experiment is a beautiful illustration of the fact that even the energy of the atom is "in reality" not a number but a matrix. The conservation law holds for the energy matrix and therefore also for the value of the energy, only to the extent that these are measured at the time. Mathematically the preservation of the phase relationships leads to the following: Let "Q" be the coordinates of the center of mass of the atom, then according to Eqn19 the atom has the wave function

$$\phi(Q,t)\chi(q,t) = \Psi(Q,q,t), \tag{21}$$

where $\phi(Q,t)$ is a function (like $\chi(q',q)$ in Eqn16) which is different from zero only near a point in Q-space which propagates with the velocity of the atom in the beam direction. The relative probability for any value q is given by the integral

$$\int \bar{\Psi}(Q,q,t)\Psi(Q,q,t)dQ,$$

that is, by Eqn20. The function Eqn21 is different in the presence of a magnetic field. Because of the different deflection of atoms in the upper and lower states, after the magnetic field it becomes

$$\Psi(Q,q,t) = c_2\phi_2(Q,t)\psi_2(E_2,q)e^{i(E_2t+\beta)/\hbar} + \sqrt{1-c_2^2}\phi_1(Q,t)\psi_1(E_1,q)e^{iE_1t/\hbar}. \tag{22}$$

$\phi_1(Q,t)$ and $\phi_2(Q,t)$ are non-zero only near one point for ϕ_1 and another for ϕ_2, so $\bar{\phi}_1\phi_2$ vanishes everywhere. The probability for displacement q relative to Q is

$$\bar{\Psi}(Q,q,t)\Psi(q,q,t) = c_2^2|\phi_2(Q,t)\psi_2(E_2,q)|^2 + (1-c_2^2)|\phi_1(Q,t)\psi_1(E_1,q)|^2. \tag{23}$$

The cross term vanishes and, with it, the possibility to measure a relative phase. The result of the position determination will be the same regardless of the phase of the incident light.

** * * * * * * * * * * * * * * * * * * **

The ultimate result of Bohr's gedanken-experiment is that the Stern-Gerlach measurement of the energy by separation of the beam into two eigen-energy components has explicitly exacted the toll of destroying all phase information.

** * * * * * * * * * * * * * * * * * * **

In conclusion, we study the connection of $\Delta E \Delta t \sim \hbar$ with a complex problem discussed by Ehrenfest in two important papers on Bohr's Correspondence Principle [6]. Ehrenfest and Tolman speak of "weak quantization", when a quantized periodic motion is interrupted by quantum-jumps or other disturbances in time intervals not very long compared to the period of the system. In this case, there should be not only the exact quantum-dependent energy values, but also (with a small probability) energy values which are close to the undisturbed values. In quantum mechanics this behavior corresponds to: the energy is actually affected by the external forces or by quantum-jumps, because each energy measurement must take place between two disturbances. In this way an upper limit Δt is given, and a corresponding uncertainty in any energy measurement done within this interval, $\Delta E = \hbar/\Delta t$. We measure the energy value E_0 of a quantum state only with this precision ΔE. Therefore the question, whether the system "really" assumes such energy values E different from E_0, or whether its experimental value only lies within that uncertainty of the measurement, has – in principle – no meaning. If Δt is smaller than the period of the system, then it is no longer sensible to speak of discrete stationary states or of discrete energy values.

Ehrenfest and Breit posed this paradox: A rotor, here a gear-wheel, has an apparatus to measure the frequency of rotation. The rotor is meshed with a toothed-rod which can slide back and forth between two blocks; the blocks stop the rod after a definite number of rotations of the rotor, and reverse its rotation. The total period T of the system is long compared to the rotation period t of the rotor alone; the discrete energy levels of the whole system are correspondingly dense. Since all the stationary states have the same statistical weight, for large enough T practically all energy values should occur with equal probability – in contrast to what would be expected for the rotor. This paradox is sharpened by the following consideration. In order to determine whether the system is in a pure discrete energy state of the rotor alone as is frequently assumed, or whether it occupies with equal probability all possible states (corresponding to the small energy values $\hbar/T$), takes a time ΔT which is small compared to T (but $\gg t$); although the time ΔT for such measurements seems inefficient, only in this way can all possible energy eigenvalues be separated. We maintain that such experiments to determine the total energy of the system will indeed show all possible energy eigenvalues with equal probability; for this result not only the long period T, but also the motion of the rod must be considered. Even if the system is found with an energy corresponding to a particular rotor state, it can actually be changed by the forces connecting it to the rod, so that it does not correspond to just that state [7]. The coupled system – rotor and rod – have stationary states completely different from the rotor alone. The solution to the paradox lies in the following: To measure the energy of the rotor alone, we would have to eliminate the coupling between the rotor and the rod. In classical theory, for sufficiently small masses of the rod, the coupling does not change the energy. The energy of the whole system is just that of the rotor. In quantum mechanics the interaction energy between the rod and the rotor is of the same order of magnitude as the energy splitting of the rotor (even for vanishing rod mass, the interaction has a high zero-point energy!); the coupling of the rod and rotor displays this quantum-dependent energy value. Because we can only measure the energy of the system, we always find the quantum-dependent coupled energy value. Even for vanishing mass of the rod, the energy of the coupled system is different from the energy of the rotor alone; the energy of the coupled system can assume all possible values – allowed by the T-quanta – with equal probability.

§ 5. Conclusion.

Quantum kinematics and mechanics is very different from the familiar classical results. The applicability of classical ideas can be established neither from our assumptions nor from our experience; for this conclusion the statement of the Uncertainty Principle – $\Delta p \Delta q \sim \hbar$ – is critical. The momentum, position, energy, etc., of an electron are exactly defined concepts, provided one remembers that the fundamental restriction of the Uncertainty Principle allows only their limited specification. We can understand qualitatively the experimental consequences of the theory in all simple cases, so that one need no longer view quantum theory as obscure and abstract [8]. One can even see directly quantitative consequences in quantum mechanical results of the essential fundamental relation of Eqn1. Jordan has tried in this way to interpret the equation

$$\psi(q, q'') = \int \psi(q, q') \psi(q', q'') dq'$$

as a relation between probabilities. We do not agree with this interpretation (§2). So far, quantitative results have been explained on elementary grounds only by a principle of greatest possible simplicity. If, for example, the X-coordinate of the electron is no longer a "number" which can be determined experimentally, then the simplest imaginable assumption (which does not disagree with Eqn1) is that it should be a diagonal element of a matrix, whose off-diagonal terms somehow describe an unknowability manifested in other ways, such as in transformations (see, e.g., §4). The statement that "in reality" there is not just a number for the X-component of momentum, but a diagonal term of a matrix, is perhaps no more abstract and unclear than the statement that: the electric field "in reality" is the time part of an antisymmetric tensor in the space-time of the world. The word "in reality" here is justified just as much, and just as little, as for the mathematical description of any other natural phenomenon. So long as one states that all quantum quantities are matrices, then the quantum laws follow without contradiction.

If one grants that this interpretation of quantum mechanics is correct, its principal consequences can be summarized in a few words: We have *not* assumed that quantum theory – in contrast to classical theory – is an essentially statistical theory in the sense that from exact data only statistical conclusions can be deduced. Rather, in all cases for which – in classical theory – exact relations exist between dynamical variables (which are all ideally exactly measurable), the corresponding exact relations hold also in quantum theory (e.g., between momentum and energy). But in the precise formulation of the Predictability Postulate: "If we know exactly the present state, we can calculate exactly the future state," it is not the calculation that fails, but the initial postulate that is impossible. We can in principle never know the starting point in all its details. Every observation is a choice from an array of possibilities and a restriction on possibilities to come. Since the statistical character of quantum theory is dictated by the imprecision of all observations, one might suppose that behind the statistically perceived world, there hides a "real" world in which the Predictability Postulate holds. But such speculation seems to me both fruitless and senseless. Physics should only describe the formal relationship of observations. One can characterize the state of affairs even better: Since all experiments are subject to the laws of quantum mechanics, and therefore to the Uncertainty Principle, then the failure of the Predictability Postulate is a definitive result of quantum mechanics.

Added in Proof: After submission of this work, new considerations by Bohr have led to an essential deepening and refinement of the analysis of quantum mechanics attempted here. Also, Bohr has pointed out that I have overlooked essential features in some discussions here. Firstly, the uncertainty in an observation depends not only on the occurrence of discontinuities, but also directly on the requirement that the different measurements are to be made either simultaneously as in particle theory, or successively as in wave theory. For example, in the analysis of the γ-ray microscope, it is necessary to include the divergence of the rays; this has the result that in the observation of the electron position, the direction of the Compton recoil is uncertain, leading to relation (1). Secondly, it is not usually emphasized that the simple theory of the Compton effect is strictly valid only for free electrons. The resulting effect on the application of the Uncertainty Principle, as Professor Bohr has explained, is essential for a general discussion of the transition from micro- to macro-mechanics. Finally, the discussion of resonance fluorescence is not completely correct because the connection between the phase of the light and the electron's motion is not as simple as was assumed.

For permission to mention his new research on the conceptual structure of quantum mechanics, which I was privileged to learn and discuss at its genesis, I owe my heartfelt thanks to Professor Bohr.

Kopenhagen, Institut für theoret. Physik der Universität.

Footnotes and References:

1) This work arose from contributions and suggestions of other researchers. I should mention in particular Bohr's *Fundamental Postulates of Quantum Theory* (e.g., Zeits. f. Phys. **13**, 117 (1923)) and Einstein's *Discussion of the Connection between Wave Fields and Light Quanta*. The problems described here are discussed most clearly and most recently, and the questions involved partially answered, by W. Pauli (*Quantentheorie,* Handb. d. Phys., Bd. XXIII, cited as loc. cit.); quantum theory itself has changed only a little since the formulation of these problems by Pauli. It is a particular pleasure to thank Herrn W. Pauli for the great stimulation which I have received from our verbal and written discussions, which has contributed essentially to the following work.

2) Recently great progress has been made in this area by P.A.M. Dirac, Proc. Roy. Soc. **A114**, 243 (1927); and results still to be published.

3) The statistical meaning of de Broglie waves was first formulated by A. Einstein (Sitzungsber. d. preuss. Akad. d. Wiss. 1925, §3). The statistical aspect of quantum mechanics plays an essential role in M. Born, W. Heisenberg, and P. Jordan, Quantum Mechanics II (Zeits. f. Phys. **35**, 557 (1926)) esp. Ch4, §3; and P. Jordan (Zeits. f. Phys. **37**, 376 (1926)); it is mathematically analyzed in the fundamental work of M. Born (Zeits. f. Phys. **38**, 803 (1926)) and its interpretation used for scattering phenomena. The foundation of the probability interpretation from the transformation theory of matrix mechanics is found in : W. Heisenberg (Zeits. f. Phys. **40**, 501 (1926)); P. Jordan (Zeits. f. Phys. **40**, 661 (1926)); W. Pauli (Zeits. f. Phys. **41**, 81 (1927)); P.A.M. Dirac (Proc. Roy. Soc. **A113**, 621 (1926)); and P. Jordan (Zeits. f. Phys. **40**, 809 (1926)). The statistical side of quantum mechanics is discussed more generally by P. Jordan (Naturwiss. **15**, 105 (1927)); and by M. Born (Naturwiss. **15**, 238 (1927)).

4) P.A.M. Dirac, Proc. Roy. Soc. **A112**, 661 (1926); M. Born, Zeits. f. Phys. **40**, 167 (1926).

5) E. Schrödinger, Naturwiss. **40**, 664 (1926).

6) P. Ehrenfest and G. Breit, Zeits. f. Phys. **9**, 207 (1922); and P. Ehrenfest and R.C. Tolman, Phys. Rev. **24**, 287 (1924); see also the discussion by N. Bohr, *Grundpostulate der Quantentheorie,* loc. cit.

7) According to Ehrenfest and Breit, this can never or only very seldom be the result of a force which acts directly on the rotor.

8) Schrödinger characterizes quantum mechanics as deterring, indeed as repellent, for its obscurity and abstraction. Certainly for the value of the mathematical clarification of quantum mechanical laws which the Schrödinger theory has provided, there can be no praise too high. On the principal physical questions, however, in my opinion, the popular understandability of wave-mechanics has lead away from the true path defined by the work of Einstein and de Broglie on the one hand, and by the work of Bohr and quantum mechanics on the other.

Paper XI·2: Excerpt from *Niels Bohr and the Development of Physics*, (Pergamon Press, London, 1955) W. Pauli, Ed., Pp.12-29.

The Development of the Interpretation of the Quantum Theory

W. Heisenberg

Göttingen, 1955

I. That Planck's theory [1] would cause changes in the foundations of physics must have been realized immediately after Einstein's [2] work on light quanta in 1905. Nevertheless, it was another 20 years before its principles were seriously addressed by the work of Bohr, Kramers and Slater [3] in an attempt to resolve its paradoxes into rational physics. We shall briefly outline the history of this resolution from 1924 to 1927, and then discuss the criticisms which have been made against the Copenhagen interpretation of quantum theory.

Bohr, Kramers and Slater asserted that the propagation of light as waves, and its absorption and emission as quanta, are experimental facts which must be the basis of any clarification, and not explained away. They introduced the hypothesis that the waves are probability waves: they represent not reality in the classical sense, but the "possibility" of such reality. The waves defined the probability, at every point, that an atom present there is emitting or absorbing radiation in the form of quanta $h\nu$. It seemed in their work that the conservation of energy could be maintained only on a statistical average. The attempt by Bohr, Kramers and Slater contained important features of the correct interpretation, including the introduction of probability as a new kind of "objective" physical reality; and the single quantum jump as "factual" in the same nature as an event which "happens" in everyday life.

······ The mathematical equipment of the new quantum theory was complete ··· by 1926, but the physical significance was still extremely unclear.

··· Born [5] interpreted the wave in configuration space as a probability wave, in order to explain the collision processes in Schrödinger's theory. Born was able to recognize that the "probability waves" were in an abstract configuration space, not simply ordinary three-dimensional space; and that the probability wave refers to an individual process. The probability wave describes the behavior, not of a large number of electrons, but of only one system of particles whose number determines the dimension of the configuration space. The wave can be conceived as representing a statistical assembly only in so far as the experiment can be repeated as often as we please. The probability wave in $3n$ dimensions contains statistical statements about only *one* system of n electrons, which can be imagined, as in Gibbs' thermodynamics, as a sample selected arbitrarily from an infinite statistical assembly of identically constructed systems.

Born's hypothesis was extended [6] to the conclusion that the squared magnitude $|S_{ab}|^2$ of the elements of the quantum mechanical transformation matrix must be interpreted as the probability that the system will be found in state b if it is in state a. Since Schrödinger wave functions are the the transformation matrices for the transition from energy states to position states, Born's hypothesis is a particular case of this more general conclusion.

Meanwhile, Schrödinger [7] attempted to deny the existence of discrete energy values and quantum jumps, and to replace quantum theory by a simple classical wave theory. The motivation for this was the discovery that the electric charge densities, when represented as products of waves, gave the correct radiation amplitudes.

At the invitation of Bohr, Schrödinger visited Copenhagen to lecture on his wave mechanics. Long discussions took place in which Schrödinger gave a convincing picture of the new simple ideas of wave mechanics, while Bohr explained to him that not even Planck's Law could be understood without quantum jumps. "If we are going to stick to this damned quantum-jumping, then I regret that I ever had anything to do with quantum theory," Schrödinger finally exclaimed in despair, to which Bohr replied: "But the rest of us are thankful that you did, because you have contributed so much to the clarification of quantum theory." Wave mechanics had brought a new element of simplicity into quantum theory, which had to be incorporated into its interpretation.

There finally emerged the "Copenhagen interpretation of quantum theory," and I remember exhaustive discussions with Bohr which examined every new attempt at interpretation in the closest detail. Bohr worked the new simple pictures from wave mechanics into the interpretation of the theory, while I attempted to extend the physical significance of the transformation matrices into a complete interpretation of all possible experiments. Bohr developed his idea of "complementarity," while I investigated the problem of the mathematical representation of a given experimental situation, by the hypothesis that only those states which can be represented as vectors in a Hilbert space can occur in nature or be realized experimentally. It was now to be assumed in quantum mechanics that only those states which can be represented as vectors in Hilbert space can occur in nature. The uncertainty principle [8] was the simple expression of this assumption. Bohr's concept of complementarity [9] imposed the same restrictions to the applicability of classical concepts, where quite different simple pictures could co-exist without contradiction only if their range of applicability was restricted.

Some time later, Jordan, Klein and Wigner [10] showed that, starting from Schrödinger's simple three-dimensional theory of matter waves, one could quantize this theory and come back to the Hilbert space of quantum mechanics. The complete equivalence of the particle and wave pictures in the quantum theory was thus demonstrated for the first time, and Schrödinger's viewpoint of a three-dimensional wave theory of matter found its rigorous basis.

At the 1927 Solvay Conference the new interpretation of quantum theory was subject to the most ingenious criticism by Einstein. It became apparent over and over again that the new interpretation contained no internal contradictions, and led to the correct experimental results. The conference is documented in Bohr's article [11] for Einstein's 70th birthday. Ever since the 1927 Solvay Conference, the "Copenhagen interpretation" has been generally accepted as the basis of all applications of quantum theory. It has, however, been continually criticized as the "orthodox" theory.

The criticism of the theory, which we discuss in detail below, was partly concerned with another problem. The Copenhagen interpretation was not only an unambiguous prescription for the interpretation of experiments, but also a language by which one spoke of Nature on the atomic scale, and thereby a part of philosophy. Indeed, Bohr had formulated the interpretation of quantum theory in the philosophical language to which he was accustomed, and which seemed to him best suited to the problems involved. This was not, however, one of the traditional philosophies of positivism, materialism, or idealism; although it included elements from all these systems of thought.

II. The critics of the Copenhagen interpretation can be divided into three groups. The first group [12] are those who accept the conclusions of the Copenhagen interpretation in every situation, but are dissatisfied with the language used, i.e. the underlying philosophy, and would replace it by another. The second group [13,14] attempts to alter the theory so that it gives the same result in established cases, but not in all. The third [15] group expresses a general dissatisfaction with quantum theory, without making specific counter-proposals, either physical or philosophical in nature.

However, all the critics of the Copenhagen interpretation do agree on one point. All desire to return to the reality concept of classical physics or, more generally expressed, to the ontology of materialism; that is, to the idea of an objective real world, whose smallest parts exist objectively in the same way as stones and trees, independently of whether or not we observe them.

We shall see again in section **III** of this essay, that this is not completely possible. For the moment we shall subject the various counter-proposals against the Copenhagen interpretation to a short criticism.

(1a) Bohm [12] tries to connect particle orbits with the waves in configuration space. For Bohm, the particles are "objectively real" structures, like the point masses of classical mechanics; the waves in configuration space also are objective real fields, like electric fields. Only our uncertainty of the previous history of the system, and the properties of the measuring apparatus, are responsible for the statistical nature of our predictions. Bohm's results for any experiment are the same as in the Copenhagen interpretation. The first consequence is that Bohm's interpretation cannot be refuted by experiment. From the "purely physical" standpoint, we are concerned with an exact repetition of the Copenhagen interpretation, just in a different language. Bohm considers himself able to assert: "We do not need to abandon the precise, rational and objective description of individual systems in quantum theory." This objective "description" involves a kind of "ideological superstructure" which has nothing to do with immediate physical reality: the "hidden parameters" of Bohm's interpretation are designed so that they *never* occur in the description of real processes if the quantum theory remains unchanged. Bohm does express the hope that in future experiments the hidden parameters may yet play a physical part, and that the quantum theory may be proved false. In actual fact, the failure of quantum theory would take Bohm's interpretation with it. I have never understood how Bohm's proposals could be used for the description of physical phenomena. Bohm's language says nothing about physics that is different from what the Copenhagen language says. Besides introducing the superfluous "ideological superstructure" of particle orbits, Bohm's language destroys the p, q symmetry implicit in quantum theory: $|\psi(q)|^2$ is still the probability distribution in position space, but $|\psi(p)|^2$ is not that in momentum space. Since this symmetry is still present in the underlying theory, it is difficult to see what is gained by omitting it from the corresponding language. The same objection applies to de Broglie's attempts [12] to introduce pilot waves.

(1b) A similar objection can be raised against the interpretations of Bopp [12] and Fenyes [12]. These again follow entirely the Copenhagen interpretation; they are isomorphic with it, as is Bohm's. However, the language they use violates the symmetry between wave and particle, which has been regarded as an essential feature of quantum theory ever since Bohr's work in 1927 and the invention of quantized fields by Jordan, Klein and Wigner. There is no reason in quantum theory to prefer particles to waves or *vice versa*. Bopp considers the appearance or disappearance of a particle as the real fundamental process of quantum theory, and he interprets the laws of quantum mechanics as a special case of correlation statistics. The symmetry between particle and wave requires the development of correlation statistics for three-dimensional waves as well. This extension has not yet been attempted.

Whereas Bopp accepted the standpoint of the ordinary quantum theory, Fenyes supposes that large deviations are "basically" possible. He says that "the existence of the uncertainty principle by no means renders impossible the simultaneous measurement, with arbitrary accuracy, of position and velocity." Fenyes does not state what nature such measurements would have in practice.

Weizel [12] relates "hidden parameters" to a new kind of particle, the "zeron," which is not otherwise observable. The danger is that the interaction between real particles and the zerons dissipates energy into the unobservable zerons. Weizel has not explained how to avoid this danger.

(1c) Alexandrov [12] and Blochinzev [12] expressly restrict their objections to the philosophical side of the problem. At the physical level they accept the Copenhagen interpretation without alteration. Although the hypotheses in their works originate outside science, discussion of their arguments is very instructive.

Their aim is to rescue materialistic ontology, so their attack is made against the introduction of the observer into the interpretation of the quantum theory. Alexandrov writes: "We must understand by 'result of measurement,' only the objective effect of the interaction of the electron with a suitable object. Mention of the observer must be avoided, and we must treat objective conditions and objective effects. A physical quantity is an objective characteristic of the phenomenon, but not the result of an observation." According to Alexandrov, the wave function ψ characterizes the "objective" state of the electron.

Alexandrov overlooks the fact that the interaction of a system with a measuring apparatus, if the apparatus and the system are cut off from the rest of the world and treated as a whole according to quantum mechanics, does not as a rule lead to a definite result (e.g. the blackening of a photographic plate at a given point). If the defense is that "in reality" the plate is blackened after the interaction, the rebuttal is that the quantum mechanical description of the closed system electron plus plate is no longer being applied. The "factual" character of an event describable in terms of the concepts of daily life is not automatically contained in the formalism of quantum theory, but is supplied in the Copenhagen interpretation by the introduction of the observer. The introduction of the observer does not imply that some kind of subjective features are brought into the description of nature. The only function of the observer is to register decisions, i.e. processes in space time, and it does not matter whether the observer is an apparatus or a human being; *but the registration, i.e. the transition from the possible to the actual, is absolutely necessary here, and cannot be omitted from the interpretation of the quantum theory* [12]. It must be pointed out that the Copenhagen interpretation is in no way positivistic. Positivism regards the sensual perceptions of the observer as establishing the reality, whereas the Copenhagen interpretation regards things and processes which are describable in terms of classical concepts, i.e. the actual, as the foundation of any physical interpretation.

Blochinzev maintains: "In quantum mechanics we describe not the state of a particle 'itself' but the fact that the particle belongs to an ensemble. This belonging is completely objective, and does not depend on the observer." Such formulations take us very far (probably too far) from materialistic ontology. In thermodynamics for example, things are different. The temperature of a system implies that the system is just one out of a canonical ensemble, and may be considered as possibly having different energies. "In reality" it has a definite energy, and none of the others is realized; the observer would be deceived if he were to consider different energies as possible. There are indeed difficulties at this point in quantum theory, with the words "in reality." If a "completely objective" character is ascribed to a particle's belonging to a quantum-mechanical ensemble (especially for a mixture of states), the word "objective" is used in a different sense from classical physics; in classical physics "belonging to an ensemble" always means a statement about the observer's knowledge of a system based on a past observation of that system. Thus one sees: such concepts as "objective reality" have no immediately evident meaning when applied to situations arising in atomic physics.

(2) Unlike the above critics, Janossy [13] attacks the "orthodox" quantum theory entirely on the firm ground of physics. His point of attack is the "reduction of wave-packets," i.e. the fact that the wave function changes discontinuously when the observer takes cognizance of a result of measurement. Janossy asserts that this reduction cannot be deduced from Schrödinger's equation, and from this he concludes that there is an inconsistency in the "orthodox" interpretation. The reduction of the wave-packets occurs in the Copenhagen interpretation when the

measurement is completed from the possible to the actual, i.e. the actual is selected from the possible by the "observer." The assumption is that any interference terms are removed in the actual experiment by undefined interactions of the apparatus with the system and with the rest of the world. Janossy now tries to alter quantum theory by introducing damping terms so that the interference terms disappear by themselves after a finite time. Even if this corresponded to reality, there are still a number of alarming consequences, as Janossy himself points out (e.g. superluminal velocities), so we are not ready to sacrifice the simplicity of quantum theory until we are compelled by experiments to do so.

(3) Among the critics of the "orthodox" quantum theory, Schrödinger [15] takes an exceptional position. He would ascribe the "objective reality" not to the particles, but to the waves, and is not prepared to interpret the waves as "probability waves only." In his work "Are there quantum jumps?" he attempts to deny the existence of quantum jumps altogether. Schrödinger's work, first of all, contains some misunderstandings of the usual interpretation. Only the waves in configuration space, that is the transformation matrices, are probability waves in the usual interpretation, while the three-dimensional material or radiation waves are not. The latter, according to Bohr and to Klein, Jordan and Wigner, have just as much (and just as little) "objective reality" as particles; they have no direct connection with probability waves, but have a continuous density of energy and of momentum, like a Maxwell field. Schrödinger therefore rightly emphasizes that at this point the processes can be conceived of as being more continuous than they usually are. But he cannot remove the element of discontinuity which is found everywhere in atomic physics. In the usual interpretation of quantum theory, it is contained in the transition from the possible to the actual. Schrödinger makes no counter-proposal as to how he intends to introduce the element of discontinuity, everywhere observable, in a manner different from the usual interpretation.

The criticism of quantum theory expressed by Einstein [14] and others start from the fear that the quantum theory might deny the existence of an objectively real world, and so might cause the world to appear in some way (by a misunderstanding of idealistic philosophy) as an illusion. The physicist must, however, postulate in his science that he is studying a world which he himself has not made, and which would be present, essentially unchanged, if he were not there. We shall give here a brief review of the extent to which this basis of all physics is maintained in the Copenhagen interpretation of quantum theory.

III. ······ An individual atomic system can be represented by a wave function or by a statistical mixture of such functions, i.e. by an ensemble (mathematically, by a density matrix). If the system interacts with the external world, only the second representation is possible, since we do not know the details of the "external world." If the system is closed, we may have a "pure case," and the system represented by a vector in Hilbert space. In this closed case, the representation is completely "objective," i.e. it contains no features dependent on the observer's knowledge; it is also completely abstract and incomprehensible, since the wave function does not refer to real space or to a real property. The representation becomes a part of a description of Nature only by being linked to the outcome of experiment. From this point we must take account of the interaction of the system with the measuring apparatus and use a statistical mixture in the mathematical representation of the larger system of the original system plus the apparatus. Bohr has constantly emphasized that the connection with the external world is a necessary condition for the result of the measuring apparatus to be registered as something actual. The compound system of quantum system plus measuring apparatus must therefore be described mathematically by a mixture. The full description now contains, in addition to the objective features describing the quantum system, also non-objective statements about the observer's knowledge. If the observer registers a certain behavior of the measuring apparatus as actual, then the mathematical representation changes discontinuously because a certain one among the various possibilities has proved to be the real one. The discontinuous "reduction of the wave packet," which cannot be derived from Schrödinger's equation, is a consequence of the transition from the possible to the actual. It is exactly analogous in Gibbs' thermodynamics to a measurement restricting a system from a large ensemble to a smaller one.

We see that a system cut off from the external world cannot be described in terms of classical concepts. We may say that the state of the closed system represented by a Hilbert vector is indeed objective, but not real, and that the classical idea of "objectively real things" must to this extent be abandoned. The characterization of a system by its Hilbert vector is complementary to its description in classical terms, similar to the way that a microscopic state is complementary in Gibbs' thermodynamics to the statement of the temperature. The measuring apparatus must be characterized as a statistical mixture, and the individual states in the mixture are altered by interaction with the observer. *Knowledge of the "actual" is thus,* from the point of view of quantum theory, *by its nature always an incomplete knowledge.* For the same reason, the statistical nature of the laws of microscopic physics cannot be avoided.

The criticism of the Copenhagen interpretation of quantum theory reflects an anxiety that, with this interpretation, the concept of "objective reality" which forms the basis of classical physics might be driven out of physics. As we have shown here, this anxiety is groundless, since the "actual" plays the same decisive part in quantum theory as it does in classical physics. The Copenhagen interpretation is in fact based on the existence of processes which can be simply described in terms of space and time, i.e. in terms of classical concepts which compose our "reality." If we attempt to penetrate behind this reality into the details of atomic events, this "objectively real" world dissolves – not in the mist of a new but unclear idea of reality, but in the transparent clarity of a mathematics whose laws govern the possible and not the actual. It is not by chance that "objective reality" is limited to the realm of what Man can describe simply in terms of space and time. We must realize that natural science is not Nature itself but a part of the relation between Man and Nature, and therefore dependent on Man. The idealistic argument that certain ideas are *a priori* ideas, i.e. come before all natural science, is here correct. The ontology of materialism rests upon the illusion that the kind of existence, the direct "actuality" of the world around us, can be extrapolated into the atomic realm. This extrapolation, however, is impossible.

Since all counter-proposals made against the Copenhagen interpretation sacrifice essential properties of quantum theory, we may suppose that the Copenhagen interpretation is unavoidable if these properties, like Lorentz invariance, are to be genuine features of Nature; and every experiment supports this view.

References and Footnotes:

1) M. Planck, Verhandl. Deutsch. Phys. Ges. **2**, 237 (1900).

2) A. Einstein, Ann. Phys. **17**, 132 (1905).

3) N. Bohr, H. Kramers and J.C. Slater, Zeits. f. Phys. **24**, 69 (1924).

4) W. Bothe and H. Geiger, Zeits. f. Phys. **33**, 639 (1925).

5) M. Born, Zeits. f. Phys. **37**, 863 (1926); and **38**, 803 (1926).

6) W. Heisenberg, Zeits. f. Phys. **40**, 501 (1926).

7) E. Schrödinger, Ann. Phys. **79**, 361, 489, 734 (1926).

8) W. Heisenberg, Zeits. f. Phys. **43**, 172 (1927).

9) N. Bohr, Naturwiss. **16**, 245 (1928).

10) P. Jordan and O. Klein, Zeits. f. Phys. **45**,751 (1927); and P. Jordan and E. Wigner, Zeits. f. Phys. **47**, 631 (1928).

11) N. Bohr, in *Albert Einstein, Philosopher-Scientist.* The Library of Living Philosophers, Inc., Vol. 7, p.199. Evanston, 1949.

12) A. Alexandrov, Dokl. Akad. Nauk **84**, 2 (1952); D. Blochinzev, Sowjetwiss. **6**, 4 (1953); D. Bohm, Phys. Rev. **84**, 166 (1951); and **85**, 180 (1952); F. Bopp, Z. Naturforsch. **2a**, 202 (1947); **7a**,82 (1952); and **8a**, 6 (1953); L. de Broglie, *La physique quantique resterat elle indeterministe?* Gauthier et Villars, Paris 1953; I. Fenyes, Zeits. f. Phys. **132**, 81 (1952); and W. Weizel, Zeits. f. Phys. **134**, 264 (1953), **135**, 270 (1953).

13) L. Janossy, Ann. Phys. **11**, 324 1952).

14) A. Einstein, in *Albert Einstein, Philosopher-Scientist.* The Library of Living Philosophers, Inc., Vol. 7, pp.665 ff. Evanston, 1949.

15) M. von Laue, Naturwissenschaft. **38**, 60 (1951); E. Schrödinger, Brit. J. Phil. Sci. **3**, 109, 233 (1952); M. Renninger, Zeits. f. Phys. **136**, 251 (1953); C.F. Weizäcker, Ann. Phys. **36**, 275 (1939); and G. Ludwig, Zeits. f. Phys. **135**, 483 (1953).

Chapter XII

Niels Bohr:

The situation is peculiar $\cdots$ The quantum postulate requires giving up the causal space-time description of atomic processes. $\cdots$ the space-time description and the claim of predictability of classical theory are complementary but exclusive $\cdots$

"$\cdots$ the quantum postulate requires us to develop a 'complementarity' theory – whose consistency depends on the possibility of definition and observation."
Niels Bohr, Sept 16, 1927, at the Volta Celebration in Como.

§ XII-1. Introduction.

Bohr was to remain fixated for the rest of his life – and with him, most of the physics community, then and now – on his principle of complementarity; which became repetitive and tedious and was never fully satisfying because it never fully addressed problems beyond the inadequacy of classical mechanics to describe wave-particle duality, i.e., beyond the Heisenberg Uncertainty Principle. He never entered into the question of the *necessity* of quantum mechanics, or of the *uniqueness* or even the *correctness* of quantum mechanics – all of that seems to have been taken for granted: it was the only game in town. Nor did he really bridge the gap between the micro-world of the fundamental quantum events and the macro-world where necessarily only classical measurements are available. He never commented on the nature of the wave function ψ as something completely new and different from our preconceived notions based on familiar classical fields, in spite of his defeat of Schrödinger on this very point. He really never emphasized ψ as a probability-cum-information theoretic device; nor did he ever really explore the question of why the deeper level? Why a probability *amplitude* underlying the probability itself. Nor did he explicitly discuss the quantum impact on our

374

conceptions of reality, until it was forced into the open seven years later by the perceptive questioning of Einstein, Podolsky, and Rosen.

Bohr was to repeat his mantra for twenty more years and dominated the first and even second generation of critics who – either worn out by war and old age, or simply more caught up in exploiting quantum mechanics as a tool than in exploring it *per se* – left the subject to a third generation. Perhaps the greatest impact was that of David Bohm who set the stage for even a fourth generation, notably John Bell. But now at last experimental advances made possible wonderful *real* experiments – most notably by Alain Aspect – rather than theoreticians' *gedanken* experiments, to actually demonstrate the *peculiar situation* that Bohr had described – however opaquely – some fifty years before.

§ XII-2. On the 'Copenhagen' Interpretation of Quantum Mechanics.

§ 2a. Preliminary Remarks.

Even the name 'Copenhagen Interpretation' is hard to understand, because the only clear statements of it are due to Dirac [1] and to Jordan [2], and based on the original observations of Born [3] – and even earlier of Born and Jordan [4] – of the indeterministic statistical predictions available from quantum mechanics. These authors recognized Heisenberg's original prescription [5] as matrix multiplication and immediately generalized it to the properties of a Hilbert space spanned by a complete orthonormal set of state vectors $\{|\Psi_n\rangle; n = 1, 2, \cdots\}$ rotating in time under a unitary transformation generated by the Hamiltonian. Ultimately, we are instructed by Dirac and Jordan that the transition probability from the state $|\Psi_n\rangle$ to the state $|\Psi_m\rangle$ is the projection of the state vector $|\Psi_n(t)\rangle$ onto the state vector $|\Psi_m(0)\rangle$, and the transition probability after a time t:

$$P_{n\to m}(t) = |\langle\Psi_m(0)|\Psi_n(t)\rangle|^2$$

is the basic probabilistic information content of quantum mechanics.

Much has been written on the Copenhagen Interpretation of quantum mechanics, but there is no better statement of the situation than that by Langevin [6]: "$\cdots$ Heisenberg $\cdots$ has shown how the very existence of quanta excludes the possibility of knowing precisely – and at the same time – all the quantities which might be the object of measurement. $\cdots$ Quantum mechanics $\cdots$ is an admirable translation of this new situation. The wave function which describes the object *no longer depends solely on the object* [Note: All Italics added.], as was the case in the classical representation, but instead *states what the observer knows* and what are his possibilities for prediction about the evolution of the object. For a given object, the wave function consequently *is modified in accordance with the information possessed by the observer.* The wave function $\cdots$ recognizes that certain noncommutable quantities cannot be exactly known simultaneously. It characterizes the system by a certain number of observable quantities, different forms of 'maximum knowledge' corresponding to different 'pure states'.

$\cdots$ the formalism of quantum theory expresses this by the choice of wave function representing the information possessed by the observer, and by the manner in which each new measurement intervenes to modify this choice.

$\cdots$ The essential character of the new physics emerges with complete clarity [!] in the two stages of change of the wave function: by coupling the system observed with the measuring device; and by the intervention of the observer $\cdots$ using the new data to reconstitute his information [and update the wave function]."

This non-objective observer-dependent split of the world into a quantum-part which evolves according to Schrödinger's equation and a classical-part – the observer (and his mental processes?) and the apparatus – which does not, is the major source of the pervasive dissatisfaction with the interpretation of quantum mechanics on the part of people who want a return to a classically objective world.

§ **2b.** Review of Relevant Classical Mechanics.

Quantum mechanics is deduced from classical mechanics [7] in the Action-Lagrangian-Hamiltonian formulation [8]. To specify completely the instantaneous configuration of an N-dimensional dynamical system we require knowledge of the N-coordinates $q_n(t)$ of the system point in a suitable configuration space $Q(N)$. The time evolution of the $q_n(t)$ is determined by the Principle of Least Action:

$$S = \int L(q_n(t), \dot{q}_n(t), t)dt \quad \text{and} \quad \delta S = 0,$$

which leads to the Euler-Lagrange equations:

$$\frac{d}{dt}\frac{\partial L}{\partial \dot{q}_j} = \frac{\partial L}{\partial q_j}, \qquad j = 1 \cdots N.$$

The dynamical system is characterized by the Lagrangian L (which ultimately has to be postulated); and the initial conditions $q_j(0), \dot{q}_j(0)$ which must be pre-determined by a series of measurements, however idealized, but which are in no way constrained by the underlying dynamics. Any choice of coordinates which is a $1 \leftrightarrow 1$ map will serve.

The Hamiltonian formulation – derivable from the Lagrangian formulation – replaces:
i) the Lagrangian with $N - q$'s, N second-order E-L equations, an N-dimensional configuration space, and $2N$-initial conditions on $q_j(0), \dot{q}_j(0)$ by
ii)the Hamiltonian with $N - q$'s, $N - p$'s, $2N$ first-order Hamilton equations, a $2N$-dimensional phase space of the q's and p's, and $2N$-initial conditions on the $q_j(0), p_j(0)$.
The generalized momentum p_j canonically conjugate to the generalized coordinate q_j is

$$p_j = \frac{\partial L}{\partial \dot{q}_j};$$

and the Hamiltonian

$$H(q, p, t) = \sum_j p_j \dot{q}_j - L(q, \dot{q}, t)$$

with the Hamilton equations of motion

$$\dot{q}_j = \frac{\partial H}{\partial p_j} \quad \text{and} \quad \dot{p}_j = -\frac{\partial H}{\partial q_j}.$$

Any function F(q(t),p(t);t) satisfies

$$\frac{dF}{dt} = [F, H]_{PB} + \frac{\partial F}{\partial t}$$

where the Poisson bracket

$$[F, H]_{PB} \equiv \sum_j \frac{\partial F}{\partial q_j}\frac{\partial H}{\partial p_j} - \frac{\partial F}{\partial p_j}\frac{\partial H}{\partial q_j}.$$

Of particular interest are the fundamental 'canonical' Poisson brackets:

$$[q_j, p_k]_{PB} = \delta_{jk} \quad \text{and} \quad [q_j, q_k]_{PB} = [p_j, p_k]_{PB} = 0.$$

Using the Hamilton equations we recover

$$[q_j, H]_{PB} = \dot{q}_j \quad \text{and} \quad [p_j, H]_{PB} = \dot{p}_j.$$

The Hamilton formulation of classical mechanics has many general properties: for example, it permits a much wider choice of variables (q, p) than those suggested by the 'contact' transformations $(q \to q')$ of the Lagrangian configuration space formulation. These are the 'canonical' transformations $(q, p \to q', p')$ (including e.g., $p \to p' = -q$ and $q \to q' = p$) which preserve the Hamilton equations of motion and the canonical form of the fundamental Poisson bracket.

§ 2c. Quantum Preliminaries.

Born and Jordan [4], and Dirac [9], immediately recognized that the Heisenberg rule [5] for differentiation of quantum dipole oscillator strengths implied a matrix generalization of classical mechanics: replace the canonical Poisson bracket by the canonical commutator bracket

$$[q, p]_{PB} \to \frac{1}{i\hbar}[q, p]_- \equiv \frac{1}{i\hbar} \sum_k (q_{jk}p_{km} - p_{jk}q_{km}) = \delta_{jm}.$$

The number-valued (q, p) of classical mechanics are replaced by the operator-valued – here matrix-valued – (q, p) of quantum mechanics, where the matrix indices refer to the quantum states. The Hamilton equations of motion for these Heisenberg operators

$$i\hbar\dot{q} = [q, H]_- \quad \text{and} \quad i\hbar\dot{p} = [p, H]_-$$

can be solved formally:

$$q_H(t) = e^{i \int H dt'/\hbar} q_H(0) e^{-i \int H dt'/\hbar}, \quad \text{etc.,}$$

and their matrix elements evaluated in a Hilbert space of states $|\Psi_n\rangle_H$, returning us to Heisenberg's original transition amplitudes

$$q_{mn}(t) = \langle \Psi_m | q_H(t) | \Psi_n \rangle_H = e^{i(E_m - E_n)t} q_{mn}(0).$$

These are conveniently reinterpreted from the Heisenberg representation (fixed states, time dependent operators) to the Schrödinger representation (time dependent states, fixed operators) by identifying

$$q_S = q_H(0), \quad |\Psi(t)\rangle_S = e^{-i\int H dt'/\hbar}|\Psi\rangle_H;$$

with the familiar Schrödinger equation for the Schrödinger state vector:

$$i\hbar\frac{\partial}{\partial t}|\Psi\rangle_S = H|\Psi\rangle_S.$$

The Schrödinger state vectors $|\Psi(t)\rangle_S$ evolve in time by a unitary transformation generated by the Hermitian Hamiltonian, retaining the original orthonormality and completeness properties of the Heisenberg state vectors (dropping the subscripts):

$$\langle\Psi_m|\Psi_n\rangle = \delta_{mn} \quad \text{and} \quad \sum_m |\Psi_m\rangle\langle\Psi_m| = 1.$$

The unitary transformation can be visualized as a continuous rigid rotation around the origin of an orthogonal coordinate system with axes defined by the orthogonal unit vectors $\{|\Psi_n(t)\rangle; n = 1\cdots\infty\}$, into a different one with axes defined by the new orthogonal unit vectors $\{|\Psi_n(t')\rangle; n = 1\cdots\infty\}$.

The concept of 'probability amplitude' made its first appearance as Heisenberg's dipole matrix element q_{nm}. The basic dynamical quantity in quantum mechanics is not the probability itself as naively might have been anticipated from a classical stochastic description of emission of quanta, but the 'field amplitude q_{nm}' whose *absolute square* determines the $m \to n$ transition probability between states m and n.

The Hilbert space can be as simple as the two dimensional space of spin-1/2 states $\alpha_z = |S_z = +1/2\rangle$ and $\beta_z = |S_z = -1/2\rangle$ or it can be as imposing as the infinite dimensional Hilbert space $|x\rangle$ necessary to characterize the position x of a particle on a line; and overlaying this the related Hilbert space of the momentum eigenstates $|p\rangle$. These have the familiar properties

$$\langle x'|x\rangle = \delta(x - x'), \quad \langle p'|p\rangle = \delta(p - p'), \quad \langle x|p\rangle = e^{ipx},$$

and the completeness properties

$$\sum_x |x\rangle\langle x| = 1, \quad \sum_p |p\rangle\langle p| = 1.$$

There is of course even more mind-boggling structure for any realistic situation, but we can often illustrate the general problem with explicit reference only to the spin-1/2 case, and quantum measurement by the spin projection onto various axes.

§ 2d. An Elementary Quantum Measurement.

Dirac instructs us that the only observable values of a dynamical variable q are eigenvalues of the corresponding Hermitian quantum operator. This is illustrated by a beam of unpolarized spin-1/2 particles incident on a Stern-Gerlach apparatus oriented to separate magnetic moments along the $\hat{z}$-axis: $S - G(\hat{z})$. Two separated beams will be produced, Beam$_1$ with $S_z = +1/2$ and Beam$_2$ with $S_z = -1/2$. If Beam$_1$ is put through another $S - G(\hat{z})$, we get one beam further deflected and no opposing beam, corresponding to the absence of $S_z = -1/2$ in the fully filtered (or analyzed or selected) Beam$_1$ of $S_z = +1/2$ particles. Each particle emerging in Beam$_1$ is assigned a wave function α_z (with uncorrelated phases).

Next suppose that Beam$_1$ is passed through another Stern-Gerlach apparatus $S - G(\hat{z}')$ which selects the spin along a different axis $\hat{z}'$ at an angle θ to $\hat{z}$. As with the original $S - G(\hat{z})$, Beam$_1$ is split into two beams, Beam$_{11'}$ with $S_{z'} = +1/2$ and Beam$_{12'}$ with $S_{z'} = -1/2$, as required by Dirac and by the quantization of angular momentum along the $\hat{z}'$-axis. From the basic rotation property of spin-1/2:

$$\langle \alpha_{z'}|\alpha_z\rangle = \cos\theta/2, \quad \langle \beta_{z'}|\alpha_z\rangle = \sin\theta/2.$$

If we somehow count coincident events between Beam$_1$ and the secondary beams, we must find a fraction $\cos^2\theta/2$ in Beam$_{11'}$ and $\sin^2\theta/2$ in Beam$_{12'}$. Note that we have no definite knowledge in which secondary beam the coincidence will occur, in accord with Born's indeterminacy.

Beam$_{11'}$ consists of particles with $S_{z'} = +1/2$, which are assigned wave function $\alpha_{z'}$ (with arbitrary phases), and an amplitude renormalized to unity, to be compared to the wave function before the measurement (i.e., before $S - G(\hat{z}')$):

$$\alpha_z = \alpha_z' \cos\theta/2 + \beta_z' \sin\theta/2 \Rightarrow \alpha_z'.$$

The 'wave function has collapsed' due to the measurement.

The Copenhagen Interpretation of this result is the following: The initial Stern-Gerlach apparatus $S - G(\hat{z})$ prepares an ensemble of identical systems whose wave function is α_z. The second $S - G(\hat{z}')$ separates members of this ensemble into two new ensembles,

1) one with $S_{z'} = +1/2$, wave function $\alpha_{z'}$, and probability $\cos^2 \theta/2$;
2) the other with $S_{z'} = -1/2$, wave function $\beta_{z'}$, and probability $\sin^2 \theta/2$.

The wave function assigned to each ensemble describes our best knowledge of the configuration of the elements of the ensemble as objectively defined by the measurement(s), i.e., the selection procedure, by which the members are assigned to the ensemble. The wave function makes no specific prediction for the result of any particular measurement on any individual member of the ensemble [10], but only predicts the probability of various results, e.g. $\cos^2 \theta/2$ for $S_{z'} = +1/2$ in the ensemble of particles Beam$_1$ defined by $S - G(\hat{z})$.

§ **2e.** Density Matrices, Mixtures, and Pure States.

The above illustration presumes a perfect selection of $S_z = +1/2$ in Beam$_1$ from an initial (unpolarized) beam. If the Stern-Gerlach apparatus does not completely resolve the $S_z = \pm 1/2$ components, then we assign a 'density matrix' $\mathcal{P}_1$ to the Beam$_1$ ensemble [6]

$$\mathcal{P}_1 = |\alpha_z\rangle P_+ \langle\alpha_z| + |\beta_z\rangle P_- \langle\beta_z| = \begin{pmatrix} P_+ & 0 \\ 0 & P_- \end{pmatrix},$$

with $P_\pm$ the probabilities of $S_z = \pm 1/2$. The probability of measuring $\alpha_{z'}$ in Beam$_{11'}$ is now

$$
\begin{aligned}
P_{11'} &= |\langle\alpha_{z'}|\alpha_z\rangle|^2 P_+ + |\langle\alpha_{z'}|\beta_z\rangle|^2 P_- \\
&= \text{Trace}|\alpha_{z'}\rangle\langle\alpha_{z'}| \{|\alpha_z\rangle P_+ \langle\alpha_z| + |\beta_z\rangle P_- \langle\beta_z|\} \\
&= \text{Trace}\mathcal{F}_\alpha \mathcal{P}_1 = P_+ \cos^2 \theta/2 + P_- \sin^2 \theta/2.
\end{aligned}
$$

Here we have introduced a 'filter' matrix

$$\mathcal{F}_\alpha = |\alpha_z'\rangle\langle\alpha_z'| = \begin{pmatrix} \cos^2 \theta/2 & 0 \\ 0 & \sin^2 \theta/2 \end{pmatrix}$$

to characterize the (assumed perfect) selection of $S_{z'} = +1/2$ by $S - G(\hat{z}')$.

The density matrix $\mathcal{P}$ generalizes to any number of states Ψ_n with probabilities $1 \geq P_n \geq 0$,

$$\mathcal{P} = \begin{pmatrix} P_1 & 0 & 0 \\ 0 & \cdots & 0 \\ 0 & 0 & P_N \end{pmatrix},$$

with Trace$\mathcal{P} = 1$. In a different representation $\Phi \neq \Psi$ the density matrix is no longer necessarily diagonal except in the special case where all probabilities are equal and $\mathcal{P} \sim 1$ the unit matrix. In the special case of a pure state Ψ_a with probability $P_a = 1$, the density matrix satisfies $\mathcal{P}^2 = \mathcal{P}$ independent of the representation. The ensemble average of any operator G follows as

$$\langle G \rangle = \sum_{jkm} P_j \langle \Psi_j | \psi_k \rangle \langle \psi_k | G | \psi_m \rangle \psi_m | \Psi_j \rangle = \sum_{km} \mathcal{P}_{mk} G_{km} = \text{Trace}\mathcal{P}G.$$

The probability of any particular eigenvalue g_a of G is $\mathcal{P}_{aa}$ in the representation ϕ which diagonalizes G, and for a pure state Ψ_n is $|\langle \Psi_n | \phi_a \rangle|^2$ as expected.

§ 2f. Schrödinger's Cat.

The paradox of Schrödinger's cat [11] is conveniently summarized by Keller [12]. The cat is part of a Rube Goldberg device involving a radioactive nucleus whose decay precipitates a chain of events breaking a bottle of poison gas and killing the cat. Described quantum mechanically the cat's wave function is facetiously and heuristically supposed to be

$$|\text{Cat}\rangle = e^{-\Gamma t/2}|\text{Live Cat}\rangle + (1 - e^{-\Gamma t})^{1/2}|\text{Dead Cat}\rangle,$$

with both live and dead components. Quantum mechanically the cat is both alive AND dead! But surely any cat is either alive OR dead. In fact, when we look we see one or the other and the cat's wave function after looking is immediately

$$|C'\rangle = |L\rangle \quad \text{OR} \quad |D\rangle.$$

The wave function has collapsed and we run the risk by looking of ending whatever chance the cat had.

One great accomplishment of the Copenhagen ensemble interpretation of quantum mechanics is that *nothing* about the Schrödinger cat paradox needs to worry us. Quantum mechanics says nothing about any individual event, except of course to include all the possibilities. It does *not* say that any particular cat is half dead and half alive. When the observer looks to find out the state of the cat, there is no implication in quantum mechanics that the cat's demise is in any way caused by the looking. *The wave function is not a physical attribute of the cat, but rather an information theoretic attribute of the ensemble of identically prepared cats.* At any given time quantum mechanics tells us that they are dying at some rate Γ. At time T, if we looked at all the cats we would find a fraction $(1 - e^{-\Gamma T})$ dead and $e^{-\Gamma T}$ alive. The dead cats are relegated to a new (stationary?) ensemble. The live cats are relegated to a new ensemble with a new wave function and a new survival probability $e^{-\Gamma(t-T)}$ for $t \geq T$, but the same risk after the observation as before. Nothing could be simpler or make more common sense.

§ XII-3. Contemporary Remarks.

There are a multitude of fundamental quantum systems – both microscopic and macroscopic – that pose problems reminiscent of Schrödinger's cat, and their resolution by the Copenhagen ensemble interpretation is not unequivocal. What arises here is most sensibly and economically understood by an extension of the concept of an ensemble to include a single system subjected to repeated observations [13]. We return to these recent developments in the work of Griffiths [14], Omnès [10], and Gell-Mann and Hartle [15] in a later chapter.

In spite of these modern improvements, the traditional Copenhagen interpretation has strong support within broad limits. Gell-Mann and Hartle [16], in response to an article advocating the Bohm alternative (a deterministic non-local hidden variable theory of classical particle trajectories propagating on a Schrödinger pilot wave [17]) say:

"There is no question that the 'orthodox' Copenhagen interpretation works in measurement situations and accurately predicts the outcome of laboratory experiments. It is not wrong. Rather, it is a special case of the [Read: our] more general interpretation in terms of decoherent histories of the universe. The Copenhagen picture is too special to be fundamental, and it is clearly inadequate for quantum cosmology."

And in the same context, Griffiths states [16]:
"··· the fundamental ideas have not changed ···. Thus, wave functions are the building blocks out of which histories are constructed ···"

In the same discussion, Zeilinger [16] describes the standard understanding of the Schrödinger cat paradox:
"··· the Copenhagen interpretation remains one of the most significant intellectual achievements of our century. ··· [that] Schrödinger's cat is paradoxical ··· reflects a serious misunderstanding. All the quantum state is meant to be is a ··· catalog of our knowledge of the system. ··· there is never a paradox if we realize that quantum mechanics is about information. ···"

At the same time, in another exchange in the same journal, Beller defends the communication techniques of post-modern deconstructionist literary critics in comparison to those of the quantum physics pioneers. Beller states [18]:
"··· Bohr was notorious for the obscurity of his writing. Yet physicists ··· [attribute Bohr's obscurity] to a 'depth and subtlety' that mere mortals are not equipped to comprehend. ··· they blamed themselves, not Bohr (i). ··· Bohr intended ··· complementarity ··· [to be] applicable to physics, biology, psychology and anthropology ··· a substitute for the lost religion (ii). Heisenberg ··· [hoped] the results of quantum physics will ··· influence the world of ideas ··· [like a new renaissance] (iii). Max Born ··· begged Bohr ··· [to include him in his] 'Institute of Complementarity' (iv)." Others have responded to Beller's comments, and she to them [18]; but let us try also:
i) It was given to fewer than perhaps ten people to be prime movers in the new quantum revolution: Heisenberg, Born, Jordan, Dirac, Bohr (from a previous generation), de Broglie, Schrödinger, Pauli, and perhaps Einstein (as a critic from an even earlier generation) ···. So, yes, when physicists try to find the ultimate origin of the ideas of our subject, we do study their every word.
ii) Is it a 'religion'? If so many of us end up defending the faith, it is only because
– despite our own best efforts and those of outstanding physicists of the last sev-

enty years – there has been no substantial change or even fundamental advance in the accepted view of the subject, with the possible exception of the Feynman Path Integral formulation to go along with the Heisenberg and the Schrödinger formulations. Technical refinements, computational techniques, difficult problems solved, subtle and ingenious applications, $\cdots$ but nothing that was not substantially outlined before 1928 by the founders of the subject. If we have all become sycophants, it has not been for lack of trying or lack of motivation to make fundamental changes.

iii) As far as Bohr or Heisenberg over-reaching their competence with the complementarity principle, what they proposed was a world without absolutes or dogma, a world of compromise and tolerance, an 'I'm OK, you're OK' world. Not a bad idea for a world in the shadow of WWII, the Cold War, and the H-bomb. Not a bad idea still.

iv) If Dirac's and Wigner's somewhat unctuous courtesy to us on occasion is any indication, there were a lot of extreme (and not always sincere) expressions of courtesy among people of this generation, so Born's remarks are not necessarily a declaration of philosophical congruence.

We tend to agree with Beller that hyperbolic remarks such as 'the moon is not there when nobody looks' or Schrödinger's 'cat' and many more, are antithetical to serious understanding. But has she ever sat through a lecture on decoherent histories? Not much fun. A little levity is sometimes a needed stimulus.

Beller quotes Wilshire [19] that "[in quantum mechanics], emotion, passion, and wild speculation become essential to science." This statement seems difficult to relate to the actuality of Heisenberg's revelation to replace differentiation by finite differences to get the sum rules; or to Dirac's solitary walk during which he recognized the similarity of Heisenberg's work to dimly remembered Poisson brackets; or to Born and Jordan's recognition of matrix multiplication. All this was no doubt generated by a deep passion, and accompanied by intense emotion since certainly they realized that they had done something very important, but wild speculation? Their work more closely resembles that of a master craftsman who has succeeded in making a clock that ran on the first try and kept perfect time thereafter. And surely everything since has been derivative, and made necessary by

Heisenberg's initial premise, and Born and Jordan's recognition of its mathematical structure.

Finally, we offer some advice (surely gratuitous) to Beller when dealing with the great and their writings:

i) Even the great are great only about 10% of the time;

ii) Do not judge the great by unreferreed papers, conference proceedings, public lectures, books, statements made in their dotage, $\cdots$. All that matters is that at least *once* they did something *original* that *really, really* mattered. That is why in 1925 people cared what Bohr and Einstein thought. That is why what matters to us is not so much whether or not Heisenberg was a perfect person in his later years, but that he invented quantum mechanics.

§ XII-4. Seeing Schrödinger's Cat!

Physics is ultimately an experimental science and it takes new experiments to drive us to new understanding. Haroche [20] has done the Schrödinger Cat experiment without hurting the cat! Speaking metaphorically, Haroche sends a 'quantum mouse' through a small hole in the box containing the 'quantum cat'. The probability amplitude for the q-mouse to reappear is expected to be one thing (large, classically) if the q-cat is dead, another (small, classically) if the q-cat is alive, and some non-classical quantum combination of the two for coherent combinations of alive and dead. In Haroche's experiment, the time for the coherent combination $|$live$\rangle$ plus $|$dead$\rangle$ to decohere (because of unavoidable interactions with the environment) can be measured and is found to be shorter for more complicated q-cats. Of course, with a real c-cat and a c-mouse, the decoherence times would be unobservably small, and no coherent state of the c-cat could be observed.

Haroche's q-cat is observed in an indirect measurement by the q-mouse in a coherent quantum non-demolition (QND) experiment in which the phase of the q-mouse wavefunction is sensitive to the state of the q-cat, but does not decohere it. This extends the sense of quantum measurement in a way not directly anticipated in the original Copenhagen interpretation. Before the miracles of modern quantum optics, measurements invariably involved decoherence, usually by coarse graining of multiple quantum processes into a single classical observation. Such decoherence does occur eventually in Haroche's experiment when the phase of the q-mouse is ultimately determined, but that is well clear of the q-cat and does not decohere the q-cat's quantum state.

In brief, Haroche's experiment unfolds in four steps:
1) A resonant high-Q microwave cavity is loaded with a few (1 to 6) quanta in an undetermined coherent state $|\phi_0\rangle$; then
2) a Rydberg atom in a coherent state $|R_0\rangle_1$ of $n = 50$ and 51 traverses the cavity at time $t = 0$, leaving the cavity and the atom in an entangled coherent two component state $|\phi^*\rangle \times |R^*\rangle_1$ – this is the q-cat in a coherent ($|$live$\rangle + |$dead$\rangle$). The resonant cavity is coupled to a resistor which eventually dissipates the coherence of the q-cat. Next
3) a second identical Rydberg atom – the q-mouse – traverses the cavity and undoes the phase entanglement imposed by the first one and acquires a characteristic phase opposite that of the first Rydberg atom if the q-cat is still in its $t = 0$ coherent state.
4) The detailed relative phase of the first and second Rydberg atoms determines the degree of decoherence of the q-cat state at time t.

The paradox of the Schrödinger cat is resolved. A q-cat in a coherent state ($|$live$\rangle + |$dead$\rangle$), with sufficiently large decoherence time T, *does* permit an indirect non-destructive observation of its state by a coherent quantum non-demolition measurement. For a c-cat, of course, the decoherence time T is far too short to allow the successive operations of the QND experiment.

Collins [21] quotes Zurek on Haroche's experiment "the sheer ability to manipulate three quantum systems $\cdots$ with this precision is stunning" and should lead to "dramatic new demonstrations of the essential quantum nature of reality" by exploring the continuous interface between the q-world and the c-world.

Footnotes and References:
1) P.A.M. Dirac, *The Principles of Quantum Mechanics* (Oxford, London, UK 1930, 1934, 1947), ChII; also Proc. Roy. Soc. **A113**, 621 (1926).
2) P. Jordan, Zeits. f. Phys. **40**, 809 (1926); also Naturwiss. **15**, 105 (1927).

3) M. Born, Zeits. f. Phys. **37**, 863 (1926),**38**, 803 (1926), here as **Papers X·1,2**; also Naturwiss. **15**, 238 (1927).

4) M. Born and P. Jordan, Zeits. f. Phys. **34**, 858 (1925). Here as **Paper VII·1**.

5) W. Heisenberg, Zeits. f. Phys. **33**, 879 (1925). Here as **Paper VI·1**.

6) P. Langevin, in: F. London and E. Bauer, *La théorie de l'Observation en mécanique quantique* (Hermann, Paris 1939); English translation in *Quantum Theory and Measurement* (Princeton, Princeton, NJ 1983) J.A. Wheeler and H.W. Zurek, Eds., Pp.217-259.

7) M. Born, W. Heisenberg, and P. Jordan, Zeits. f. Phys. **35**, 557 (1926). Here as **Paper VII·2**.

8) E.C.G. Sudarshan and N. Mukunda, *Classical Dynamics: A Modern Perspective* (Wiley, New York, NY 1974), Ch9; H. Goldstein, *Classical Mechanics* (Addison-Wesley, Reading, Ma 1981), Ch9.

9) P.A.M. Dirac, Proc. Roy. Soc. **A109**, 642 (1925). Here as **Paper VIII·1**.

10) R. Omnès, *The Interpretation of Quantum Mechanics* (Princeton, Princeton, NJ 1994), Pp. 356-358. Omnès usefully distinguishes 'true' statements from 'reliable' statements. For example: 'The result of the measurement of S_z at time $t = 0$ is the datum $S_z(t = 0) = +1/2$' is an unalterably 'true' statement, provided of course that the measurement is trustworthy. A corresponding 'reliable' statement would be: 'The result of the measurement of S_z at time $t = 0$ is $S_z = +1/2$.' This statement is 'reliable' in the sense that it can be verified with certainty – repeated measurements of S_z will return the value $+1/2$ with probability unity, as required by Dirac. But the 'reliable' statement is not an exclusive truth because other results (e.g., $S_{z' \neq z} = \pm 1/2$) are possible at later times. It was the classical but quantum mechanically superficial tendency to equate 'true' and 'reliable' which created much of the early confusion in the interpretation of quantum measurements.

11) E. Schrödinger, Naturwiss. **23**, 807, 823, 844 (1935); English translation in *Quantum Theory and Measurement* (Princeton, Princeton NJ 1983) J.A. Wheeler and W.H. Zurek, Eds., p. 157: "One can even set up quite ridiculous cases. A cat $\cdots$ living and $\cdots$ dead $\cdots$ in equal parts." For a modern perspective, see: P. Yam, Scientific American **276**-6, 124 (June 1997); and: W.H. Zurek, Physics Today **44**-10, 36 (1991).

12) Evelyn Fox Keller, Am. J. Phys **47**-8, 718 (1979). We recommend Keller's psychoanalytic discussion of the situation with only a few reservations, but high praise for such liberating remarks as: "$\cdots$ the belief that nature is 'knowable' $\cdots$ we call magical $\cdots$ omniscience $\cdots$ satisfies a primitive need. $\cdots$ acceptance of a more realistic, more mature, and more humble relation to the world in which the boundaries between the subject and object are acknowledged to be never quite rigid, and in which knowledge of any sort is never quite total. $\cdots$ debate reflects the difficulties even quantum physicists have in completely

relinquishing some adherence to at least one of the two basic premises of classical physics – the objectifiability and knowability of nature." Keller objects to the use of an ensemble to describe macroscopic objects (including Schrödinger's cat) on the grounds that this view " ··· permits the retention of the classical view of the particle as having a well defined position and momentum (and hence a classical trajectory), albeit unknowable." We find this – if true – to be irrelevant in the face of a quantum definition of reality. It really *does* depend on what the meaning of 'is' is. We can think what we please, and believe what we will, but that which we must all accept as real ultimately is only that which is measurable and in the quantum sense objectifiably knowable.

13) E.P. Wigner, Am. J. Phys. **31**, 6 (1963); J.B. Hartle, Am. J. Phys. **36**, 704 (1968); L. Cooper and D. Van Vechten, Am. J. Phys. **44**, 99 (1969).

14) R. B. Griffiths, Phys. Rev. A**54**, 2759 (1996); A**57**, 1604 (1998).

15) M. Gell-Mann and J.B. Hartle, in *Complexity, Entropy and the Physics of Information* (Addison-Wesley, Reading Ma 1990), W.H. Zurek, Ed., Pp.425-459; they also have an extensive bibliography.

16) M. Gell-Mann and J.B. Hartle, Physics Today **52**-2, 11 (1999); also, R.B. Griffiths, ibid.; and, A. Zeilinger, ibid.

17) S. Goldstein, Physics Today **51**-3, 42 (1998); **51**-4, 38 (1998).

18) M. Beller, Physics Today **51**-9, 29 (1998); also responses by L. Lerner, T. Lawry, C.P. Enz, N. Byers, P. Roman, M.C. Gutzwiller, and S.W. Keyes; and a reply by M. Beller, Physics Today **52**-1, 15 (1999).

19) D. Wilshire, *Gender/Body/Knowledge: Feminist Deconstructions of Being and Knowing* (Rutgers, New Brunswick, NJ 1989) A. Jaggar, S. Bordo, Eds., p.105.

20) M. Brune, E. Hagley, J. Dreyer, X. Maître, A. Maali, C. Wunderlich, J.M. Raimond, and S. Haroche, Phys. Rev. Lett. **77**, 4887 (1996); summarized by G. Taubes, SCIENCE **274**, 1615 (1996). For more on QND experiments, see ChXIV.

21) G.P. Collins, Scientific American **281**(4), 16 (OCT 1999).

Paper XII·1: Excerpt from Nature **121**, 580 (1928).

THE QUANTUM POSTULATE AND THE
RECENT DEVELOPMENT OF ATOMIC THEORY

Prof. N. BOHR, For. Mem. R. S.

Regarding the physical interpretation of quantum theory, I make the following general remarks [1] on the principles underlying the description of atomic phenomena. I hope these remarks will help to rationalize the many different and seemingly divergent views on the subject.

§ 1. QUANTUM POSTULATE AND CAUSALITY.

(1) Quantum theory imposes a fundamental limitation on classical physical ideas as applied to atomic phenomena. The situation is peculiar, because the interpretation of experiments does still depend on classical concepts. In contrast, in spite of any difficulties in its formulation, the essence of quantum theory is the quantum postulate: every atomic process has an essential discreteness – completely foreign to classical theories – characterized by Planck's quantum of action $\hbar$.

(2) This postulate requires giving up the causal space-time description of atomic processes. [Note added: The word 'causal' should be replaced by 'deterministic'. 'Causal' has acquired the separate meaning of 'cause preceding effect', which is not in question in these discussions. What is in question, and in fact *has to be given up*, is the requirement on the theory that it determine a unique outcome from any physically possible initial situation.] *Our usual description of physics is based on the idea that objects can be observed without disturbing them.* [Note: All italics added, as are all comments in square brackets.] For example, in the theory of relativity, every observation ultimately rests on the coincidence of two events at the same space-time point. These coincidences are not affected by the differences of various observations. In direct contrast, the quantum postulate states that any quantum observation involves an essential interaction with the apparatus [in order to bring the quantum micro-event into the observers classical macro-world. This is Bohr's crucial distinction between a 'classical [c-]reality' (pre-existing and independent of the means of observation)

and a 'quantum [q-]reality (depending upon the choice of of quantum observation and therefore *not* pre-existing.] No independent reality can be ascribed either to the atom or to the apparatus. The concept of observation is arbitrary to the extent that it depends on which objects are included in the system to be observed. But ultimately every observation must be reduced to our [classical, macroscopic] sense perceptions. In interpreting observations, we can use theoretical notions, and in any particular case *it is a question of convenience* at what stage of the observation the quantum postulate with its inherent 'discreteness' is brought in ['judgement' would be a better choice of word than 'convenience'].

* * * * * * * * * * * * * * * * * * **

(3) The detection of a discrete photon by a photomultiplier tube is an example where the actual detection could conceivably be described – with horrendous effort – quantum mechanically, at least for a few stages. The photo-emission of an electron in the first stage could be treated quantum mechanically with some degree of credibility. The electron multiplication by a factor ~ 30 at the second stage presents increasing difficulty, and a many component wave function with proliferating variables such as the continuous energy distributions and model dependent phases. Even at this early stage, one would expect that an average over initial resolving power would result in a random phase average over the result and a rapid convergence to the classical expectation. To show this explicitly in a real situation is never attempted. Succeeding stages with electron multiplication eventually of millions obviously compels one to give up any quantum description and accept the resulting current pulse as a *classical* indicator of the incident *quantum* event.

* * * * * * * * * * * * * * * * * * **

(4) This situation has far-reaching consequences:
i) the usual [i.e., *precise, unique, classical*] definition of the [pre-existing] state of a physical system requires the elimination of all external disturbances. But then, by the *quantum* postulate, any observation would be impossible. Therefore,

ii) in order to make an observation we must have interactions with measuring instruments not belonging to the system, with the result that the *unambiguous definition of the* [pre-existing] *state of the system is lost.*

There can be no predictability [Again, as in the case of the word 'causal', 'predictable' needs some qualification. What Bohr means is 'uniquely predictable', i.e. 'deterministic'.] in the ordinary sense that precise, observed values can be assigned for each of the space-time and momentum-energy variables present in the system. The very nature of quantum theory forces us to regard the *space-time description* and the *claim of predictability* [again, 'determinism'], which characterize classical theories, as *complementary* but *exclusive* features of idealized observations and definition. Quantum theory teaches us that the validity of our usual space-time description depends on ignoring the small value of Planck's quantum of action $\hbar$ compared to the actions involved in ordinary sense perceptions. In the description of atoms, the quantum postulate requires us to develop a 'complementarity' theory [i.e., 'complete' *including* eventual interpretation] – whose consistency depends on the possibility of definition *and* observation. [Here Bohr already precludes the possibility of a physics based on the de Broglie–Bohm classical particle trajectories following rays of the 'pilot-wave'. Such classical trajectories are ruled out as unobservable *in principle* and therefore undefinable as quantum dynamical concepts. Not for seven more years, and only then in response to criticism from Einstein, Podolsky, and Rosen, will Bohr explicitly address the full impact of his ideas on the concept of reality. Only classically can a unique [c-]reality be assumed to precede the classical measurements which determine it. In quantum mechanics, where the result of a measurement depends on the *choice* of experiment, the [q-]reality is necessarily only *realized* – i.e., made real – *after* the measurement.]

(5) This view is dictated by the nature of light and the ultimate constituents of matter. For light, its propagation in space-time is determined by electromagnetic theory. Interference phenomena *in vacuo* and the optical properties of matter are governed by the wave superposition principle. However, the conservation of energy and momentum during the interaction of radiation and matter is described *only* by the quantum concept. The validity of the superposition principle on the one hand and the conservation laws on the other, have been demonstrated by direct experiments. [Bohr finally adopts Einstein-Compton kinematics which he had at first

disputed along with coauthors Kramers and (a reluctant) Slater.] This clearly shows the impossibility of a predictable [again, 'deterministic'] space-time description of light because:

i) in attempting to construct the laws of space-time propagation of light from the quantum postulate, we are restricted to statistical considerations. And,

ii) predictability for individual light processes, characterized by $\hbar$, requires abandoning the space-time description.

There can be no question of a simultaneous validity of the [classical] ideas of [a detailed] space-time [description] and of predictability. The two views of light are different interpretations of experiments in which the limitation of the classical concepts enter in complementary [but exclusive] ways.

* * * * * * * * * * * * * * * * **

Bohr was not without humor, however unintended. "This *clearly shows* · · ·" should then be followed by: "the impossibility of a deterministic wave description of light because:

i) the classical wave description of light is only statistically related to the quantum description. And,

ii) determinism for individual light processes, characterized by $\hbar$, requires abandoning the classical wave description.

There is no possibility of the simultaneous validity of the classical wave description and of determinism. · · ·"

* * * * * * * * * * * * * * * * **

(6) Particles with mass present an analogous situation. The discreteness of the elementary electrical charge is forced upon us by the evidence. But other experience (diffraction of electrons from crystals) requires the use of the de Broglie wave theory. Just as for light, we have a dilemma which is the very essence of the experimental evidence. Here again we are dealing not with contradictory but with complementary pictures, *which together generalize the classical description.* It must be kept in mind that radiation in free space as well as isolated particles with mass (and charge) are abstractions. *Their properties in quantum theory are definable only through their interactions with other systems.* Nonetheless, these ab-

stractions are indispensable for a description of experiments [even] in our ordinary space-time view.

(7) The difficulties with a predictable space-time [Read 'a deterministic, unique, classical wave-'] description in the quantum theory have been discussed at length, and are now the subject of a recent important development. Heisenberg [2] has deduced a fundamental uncertainty which affects all measurements of quantum dynamical variables. Before we discuss Heisenberg's results we show how *this uncertainty is unavoidable*, by an analysis of the most basic concepts used in interpreting experiments.

§ 2. QUANTUM OF ACTION AND KINEMATICS.

(8) The fundamental contrast between the quantum and the classical concepts is apparent in the basic formulas which are the foundation both of the theory of light quanta and of the wave theory of matter:

$$E\tau = P\lambda = 2\pi\hbar, \tag{1}$$

where E and P are energy and momentum, and τ and λ are the period and wavelength of the associated wave. In these formulas the two notions of light, and also of particles with mass, enter in sharp contrast. Energy and momentum are associated with the concept of particles which classically can be assigned definite space-time coordinates. [In contrast,] The period and wavelength refer to a wave train of infinite extent in space and time. Only with the superposition principle is it possible to obtain the localized particle mode of description. A limited extent of the wave packet in space and time requires the interference of a group of elementary harmonic waves. de Broglie showed that the translational velocity of the particle associated with the wave packet is just the group-velocity of the waves. An elementary wave is $A\cos(\omega t - kx + \delta)$, with A and δ the amplitude and phase, $\omega = 2\pi/\tau = E/\hbar$ in terms of the period τ and energy E, and $k = 2\pi/\lambda = P/\hbar$ in terms of the wavelength λ and the momentum P. The phase velocity is ω/k; the group velocity relevant here is

$$v_g = \frac{d\omega}{dk} \Rightarrow \frac{dE}{dP} = \frac{P}{m},$$

just what is required for a non-relativistic particle with energy $E = P^2/2m$ and velocity $v = P/m$. Here the use of wave packets causes a lack of sharpness in the definition of the period and the wavelength, and consequently in the definition of the corresponding energy and momentum given by Eqn1.

(9) A limited wave packet [of finite space-time extent] can only be obtained by a superposition of elementary plane waves with all values of k and ω. The range of values of these variables $(\Delta k, \Delta \omega)$ required for space-time localization of the wave packet in the interval $(\Delta x, \Delta t)$ is $\Delta x \Delta k \sim \Delta t \Delta \omega \sim 1$. These relations are the condition that the wave trains cancel each other by interference at the space-time boundary of the wave packet. They also mean that the group as a whole *has no phase* in the same sense as elementary waves. The greatest accuracy possible in defining the energy and momentum of the particles described by such a wave packet is therefore

$$\Delta t \Delta E \sim \Delta x \Delta p \sim \hbar. \tag{2}$$

In general, the situation is much less favorable because of a spreading of the wave packet. The limitation on classical concepts is connected with the limited validity of classical mechanics, which corresponds to geometrical optics where the propagation of waves is described by 'rays'. Only in this limit can momentum and energy be unambiguously defined on the basis of space-time pictures.

(10) Eqn2 can be stated in the language of relativity: In quantum theory there is a general reciprocal relation between
i) the precision with which we can specify the momentum-energy vector of a particle, and
ii) the precision with which we can specify its space-time vector.
This expresses the complementary nature of the space-time (particle-)description and the momentum-energy (wave-)description.

(11) In order to reconcile the momentum-energy conservation laws with space-time observations, we must replace pointlike space-time events with events defined within finite space-time regions. This avoids paradoxes in describing the scattering of radiation by free particles and in the collision of two such particles. Classically, the scattering requires a finite extent of the radiation in space-time, whereas the

change in motion of the electron according to the quantum postulate is an instantaneous effect taking place at a definite point in space-time. As in the case of radiation, *it is impossible to define the momentum and energy for an electron without considering a finite space-time region.* The conservation laws require the same accuracy of definition of the momentum-energy vector for the electron as for the radiation. Therefore, the space-time regions must be the same size for both quanta in the interaction. [This spells an end to any absolute objective [measurable] [c-]reality such as required by the precise localization of the system-point in the phase space of classical Hamiltonian mechanics. Attempts to evade this requirement by the introduction of an underlying but so far invisible 'hidden variable' [c-]reality are becoming more constrained (see Ch14).]

(12) A similar conclusion applies to the collision of two particles, although here the significance of quantum mechanics was disregarded before the wave concept was introduced. Here the quantum postulate transcends the space-time description of individual particles. For a detailed description of the collision between charged particles, we must take into account the electrical interaction of the particles. Such a procedure requires a further departure from the usual visualization.

§ 3. MEASUREMENTS IN THE QUANTUM THEORY.

(13) Heisenberg gives Eqn2 as the maximum precision with which the space-time coordinates and the momentum-energy components can be measured simultaneously. His view is based on:

i) the coordinates of a particle can be measured with arbitrary accuracy only by using, for example, a microscope with light of sufficiently short wavelength. However, the scattering of the light from the charged particle is always accompanied by a finite change of momentum which increases as the wavelength of the light decreases; and,

ii) the momentum of a particle can be determined with arbitrary accuracy only by measuring, for example, the Doppler shift of the scattered light, provided that the wavelength of the light is so large that the effect of recoil can be neglected, but then the determination of the space coordinates of the particle is correspondingly less accurate.

(14) *The essence of the quantum postulate is this inevitable limitation on the precision of measurement.* A closer investigation brings out the complementary nature of the measurements. A discontinuous change of energy and momentum during observation does not prevent us having accurate values for the space-time coordinates, or for the momentum-energy components before and after the process. *The uncertainty relation between canonically conjugate quantum dynamical variables is an essential result of the limited accuracy with which changes in energy and momentum can be defined, when the wave packet defining the space-time coordinates of the particle is sufficiently small.*

(15) A microscope useful for position determination must have a convergent beam of light in order to form an image. With λ the wavelength of the light and θ the aperture (the maximum angle of convergence [i.e., of collection]), the resolving power of the microscope is $\Delta x \geq \lambda/\theta$. Even if the object is illuminated by parallel incident light of momentum $P = 2\pi\hbar/\lambda$ known in magnitude and direction, the finite size of the aperture precludes an exact knowledge of the recoil momentum accompanying the scattering. Even if the momentum of the particle were accurately known before the scattering, our knowledge of the momentum parallel to the focal plane after the observation would be uncertain by $\Delta P \simeq \hbar \times \theta/\lambda$. The product of the uncertainty Δx of the position coordinate and the uncertainty ΔP of the parallel component of momentum is given by Eqn2 $\Delta x \Delta P \geq \hbar$. One might have expected that the accuracy of determining the position would depend not only on the convergence but also on the length of the wave train, because the particle could change its place during the measurement. However, the exact knowledge of the wavelength is irrelevant for the above estimate. For any value of the aperture, the wave train can be taken so short that any change of position during the measurement can be neglected compared to the uncertainty of position due to the finite resolving power of the microscope.

(16) In measuring the momentum with the Doppler shift in the Compton effect, one assumes a parallel wave train. The accuracy with which the change in wavelength of the scattered radiation can be measured, depends in an essential way on the length of the wave train in the direction of propagation. If we assume that the incident and scattered radiations are parallel and antiparallel to the posi-

tion coordinate and momentum to be measured, then $\Delta v = c\lambda/2l$ is the accuracy in the determination of v, where l is the length of the wave train and we assume $c \gg v$.

* * * * * * * * * * * * * * * * * * **

For a finite wave train, $\lambda \sim l/(n \pm 1) \sim \lambda_0 \pm \lambda_0^2/l$ with $\lambda_0 = l/n$. From the Doppler shift, $v/c \sim \delta\lambda/\lambda$. The uncertainty in v is $\Delta v/c \sim \Delta(\delta\lambda)/\lambda$. Adding the possible errors in the difference $\delta\lambda$ gives

$$\frac{\Delta v}{c} \sim \lambda/l.$$

* * * * * * * * * * * * * * * * * **

The uncertainty of the momentum after observation is $\Delta P \sim mc\lambda/2l$. The recoil momentum, $\sim \hbar \times 2/\lambda$, does not give rise to an appreciable uncertainty. The Compton effect gives the momentum in the direction of the radiation before and after the recoil from the wavelengths of the incident and scattered radiation. However, even if the position of the particle was accurately known in the beginning, our knowledge of the position after observation will be uncertain. Because of the impossibility of defining the precise instant of recoil, we know the mean velocity in the direction of observation during the scattering process only with an accuracy $\hbar \times 2/m\lambda$. The uncertainty in the position after observation is $\Delta x \sim \hbar \times 2l/mc\lambda$. Again, the product of the uncertainties in the position and in the momentum is the general result Eqn2.

(17) Collisions using particles with mass lead to the same conclusions about the possible accuracy of measurements of position and momentum of a particle as does the scattering of light. In both cases the uncertainty involves the description of the 'apparatus' as well as the 'object'. *The uncertainty cannot be avoided* in any description of an individual quantum particle with respect to an ordinary fixed external coordinate system of solid bodies and unperturbable clocks. The experiments – opening and closing shutters, etc. – only permit conclusions about the space-time extension of the associated wave packets.

(18) In tracing our observations back to our senses, at every stage one must take into account the effect of the quantum postulate on the observation. The

resulting statistical elements introduce inevitable uncertainties into the description of the object. It might be supposed that the arbitrariness in the division between object and apparatus would allow the elimination of the uncertainty altogether. In the position measurement, one might ask if the momentum transfer could not be determined precisely by the conservation laws and a measurement of the change of momentum of the microscope – including light source and photographic plate – during observation. *Such a measurement is impossible* if at the same time one wants to know the precise position of the microscope. It follows from the wave nature of matter that the position and the momentum of the microscope itself can be defined only within the limits of Eqn2.

(19) The idea of observation is essential to the predictable space-time description. [Read 'The concept of *a perfectly precise* observation *with no other affect* is essential to the *classical deterministic* space-time description.'] Due to the limitation of Eqn2, however, this description is possible in quantum theory only if the restrictions of the Heisenberg Uncertainty Principle are taken into account. [Read '··· this description is *impossible* in quantum theory because the restrictions of the Heisenberg Uncertainty Principle *must* be taken into account.']

(20) It is instructive to compare the fundamental uncertainties of the quantum description of microscopic phenomena with the ordinary imprecision – due to imperfect measurements – inherent in any classical observation. *Macroscopic phenomena are defined by repeated observations.* [Bohr's recognition of this fact is basic to modern interpretations of quantum mechanics (see Ch15).] In classical theory, each succeeding observation improves our knowledge of the initial state of the system. But in quantum theory, the impossibility of neglecting the effect of the observation on the object means that every observation introduces a new uncontrollable uncertainty. It follows from the Heisenberg Uncertainty Principle that any measurement of the position of the particle inevitably results in a corresponding finite change in its momentum, and a sacrifice in the description of its dynamical behavior. Conversely, the determination of its momentum always implies a loss in the knowledge of its position. This limitation on the description of atomic phenomena is an inevitable consequence of the quantum postulate, and contrasts with the classical distinction between object and instrument which is inherent in our very idea of

observation. [Read '⋯ contrasts with our classical concept of a pre-existing [c-]reality for the object, independent of the apparatus, which is inherent in our *pre-Heisenberg classical concept* of observation.']

§ 4. CORRESPONDENCE PRINCIPLE AND MATRIX THEORY.

(21) So far we have only considered general features of quantum mechanics. The emphasis has been on the interaction between isolated particles and radiation. It has been possible – using classical concepts and the quantum postulate – to understand some essential aspects of experience. For example, the excitation of spectra by electron impact and by radiation are understood in terms of discrete stationary states and individual transition processes. No more detailed description of the space-time behavior of the process is required.

(22) The contrast with the classical description is striking. Spectral lines – which classically would be ascribed to one state of the atom – in the quantum description correspond to individual transition processes in which the excited atom has a choice. A formal connection with the classical ideas can be obtained only in the limit where the differences in the involved states are small and the discontinuities can be disregarded. Only in this approximate way, is it possible to interpret the spectra on the basis of classical ideas. In this limiting description, the individual transition processes are each associated with a harmonic in the expected classical atomic motions.

＊＊＊＊＊＊＊＊＊＊＊＊＊＊＊＊＊＊＊＊

Here Bohr presents a totally atypical – for him – classic and clear summary of Heisenberg matrix mechanics. He emphasizes the foundation of quantum mechanics in the Action Principle-Lagrangian-Hamiltonian formulation of classical mechanics, and the fundamental role of the Poisson bracket-commutator bracket canonical prescription for quantization. In this way, Heisenberg's original quest for a theory of *observables* is realized in the matrix representation of the canonically conjugate quantum dynamical variables dictated by the choice of the classical Action.

＊＊＊＊＊＊＊＊＊＊＊＊＊＊＊＊＊＊＊＊

(23) Heisenberg broke completely with the classical concepts by replacing the ordinary kinematical and mechanical variables by ones which refer to the individual quantum processes. He replaced the Fourier coefficients of the classical motion by a matrix whose elements are associated with the purely harmonic transitions between stationary states. By requiring the frequencies ascribed to the matrix-elements to obey the combination principle for spectral lines, Heisenberg was able to introduce simple rules of matrix multiplication, which led to a direct quantum generalization of the fundamental laws of classical mechanics. Through the work of Born and Jordan and of Dirac, the theory was formulated in a way as general and consistent as classical mechanics. Planck's constant $\hbar$, characteristic of quantum theory, appears explicitly only in the fundamental relation between matrices representing canonically conjugate variables of the Hamiltonian mechanics. These matrices do not obey the commutative law of multiplication, but must satisfy the commutation relation

$$qp - pq = i\hbar. \tag{3}$$

This commutation relation is the foundation of the matrix formulation of quantum theory. The matrix theory has often been called a calculus of directly observable quantities. It must be kept in mind, however, that the remark is limited to those problems in which the space-time description is not directly applicable, and 'observation' is not to be understood in the usual sense. [Bohr lost his nerve here, and faltered in his advocacy of canonical quantum theory. It is the quantum theory – through the full Action-Lagrangian-Hamiltonian formulation – *including* the coupling to the measuring instrument, which *defines* the canonical variable to be measured (observed). The matrix elements between stationary atomic states of the atomic electron's dipole moment x_{nm}, their Heisenberg equations of motion, and their linear coupling to the radiation field – all present in Heisenberg's original paper – provide the universal metaphor for 'directly observable quantities'.]

(24) In the correspondence of quantum mechanics with classical mechanics, the statistical character of the quantum description is fundamental. Heisenberg has developed an analysis of the physical content of quantum theory resulting from the paradoxical character of the commutation relation Eqn3. He has found the

Heisenberg Uncertainty Principle

$$\Delta x \Delta p \sim \hbar \tag{4}$$

as a general restriction on the maximum accuracy with which two canonically conjugate variables can simultaneously be observed. Heisenberg has been able to clarify many paradoxes in the application of the quantum postulate, and to demonstrate the consistency of the quantum theory. In connection with the complementary nature [Here 'complementary' meaning 'necessary to complete' has also the sense of a 'reciprocal', 'either-or' coexistence relationship of mutually exclusive alternatives.] of the quantum description implicit in the Heisenberg Uncertainty Principle, we must constantly keep in mind the fundamental equivalence between *'the possibility of definition'* and *'the possibility of observation'*. [This fundamental requirement – which led directly to Heisenberg's discovery of quantum mechanics – is the *first* thing to go in *every* attempt to make quantum theory more classically palatable by returning to the Bohr-Sommerfeld-de Broglie-Bohm semi-classical orbital mechanics, which is precisely what Heisenberg wanted to leave behind. Anyone who actually *does* a calculation of a transition rate, or of an inelastic cross-section and angular distribution, or any other quantum process, knows what a completely retrograde move that would be.] For this question, Schrödinger wave mechanics and the principle of superposition have proven of great help. Next, we will review the relation of wave mechanics to the general formulation of the quantum theory using matrix transformation theory.

§ 5. WAVE MECHANICS AND THE QUANTUM POSTULATE.

(25) In his wave theory of particles, de Broglie showed that the stationary states of an atom can be visualized as an interference effect of the phase wave associated with a bound electron. This point of view at first did not lead beyond the early quantum theory of Sommerfeld. Schrödinger, however, developed a wave equation which has been of decisive importance for great progress in atomic physics. The proper vibrations of the Schrödinger wave equation represent the stationary states of an atom, meeting all the requirements. Schrödinger associated the solutions of the wave equation with a continuous distribution of charge and current which gives the electric and magnetic properties of the corresponding atomic state. The superposition of two characteristic solutions corresponds to a vibrating distribution

of electric charge, which in classical electrodynamics would give rise to radiation, illustrating the quantum postulate and the correspondence principle for the transition between two stationary states. Another application of Schrödinger wave mechanics has been made by Born to the collision of atoms and free electrons. He obtained a *statistical interpretation* of the wave function, allowing a calculation of the transition processes required by the quantum postulate.

* * * * * * * * * * * * * * * * * * **

As for matrix mechanics, Bohr's overview of wave mechanics is without peer. His final conclusion is that the Schrödinger wave function is no more an observable physical reality (and no less) than the Heisenberg matrix elements which characterize it by radiation or (impulsive) inelastic scatterings. In particular, he points out that Schrödinger's resurrection of the Bohr-Kramers-Slater resonant transfer of energy between two atoms – in order to avoid the quantum jump must fail – because it makes the two atoms a single quantum system isolated from observation, and evades the issue of quantum measurement.

* * * * * * * * * * * * * * * * * * **

(26) Schrödinger hoped that the wave theory would remove the discontinuous character required by the quantum postulate and make possible a complete description of atomic processes along the lines of the classical theories. Schrödinger [3] proposed that the discontinuous exchange of energy between atoms required by the quantum postulate be replaced in the wave theory by a simple resonance phenomenon. The idea of stationary states would be an illusion. It must be kept in mind however, that in the resonance problem we are concerned with a closed system which – according to the view presented here – is not accessible to observation. In fact, wave mechanics – just as matrix theory – must be interpreted by an explicit use of the quantum postulate. The two formulations for bound states and interactions are complementary in the same sense as the wave and particle description of free particles.

(27) It is not possible to demand an understanding of atomic processes through

ordinary space-time pictures. All our knowledge of the internal properties of atoms is derived from experiments on their radiation or collision reactions. The interpretation of experimental facts ultimately depends on the abstractions of radiation in free space, and free particles. Our whole space-time view of physical phenomena, as well as the definition of energy and momentum, depends on these abstractions. In the application of auxiliary ideas we should only demand internal consistency in the possibilities of definition and observation. [What Bohr appears to be saying is that the Schrödinger wave function can only be observed indirectly and piece-meal by a multitude of measurements of Heisenberg matrix elements; and judged by its consistent success in their prediction; and is thereby an 'auxiliary' device (but very useful and in a practical sense, essentially essential).]

(28) In the Schrödinger wave equation we have a representation of the stationary states of an atom with an unambiguous definition of the energy of the system. However, in the interpretation of observations, a fundamental loss of the space-time description is unavoidable. The concept of stationary states precludes any specification of the behavior of 'separate' particles in the atom. In problems where such a description is essential, we must use the 'general' solution of the wave equation. Here we run into a complementarity quite analogous to that met earlier in the case of light and free particles. While the precise definition of energy and momentum requires a purely harmonic elementary wave, every space-time description requires interferences inside a group of such elementary waves. Again in this case, *we have the same limitations of observation and definition that are required by the Heisenberg Uncertainty Principle.*

(29) According to quantum theory any observation on the behavior of the electron in the atom is accompanied by a change in the state of the atom, in general the ejection of the electron from the atom. A description of the 'orbit' of the electron by subsequent observations is impossible, due to the fact that an infinite number of wave functions must be superposed to represent the 'motion' of the particle. The complementary nature of the description of the electrons in the atom depends on the neglect of their interactions. This means that the duration of the ejection process must be short compared to the electron periods in the atom, which in turn means that the uncertainty of the energy transfer in the ejection is

large compared to the energy differences between atomic states.

§ 6. REALITY OF STATIONARY STATES.

(30) Stationary states are a characteristic application of the quantum postulate. This concept involves completely giving up any time description. This loss of the time description is the price for an unambiguous definition of the energy. Moreover, a stationary state means the exclusion of all interactions with particles outside the system. That such a closed system has a well defined energy is an immediate consequence of conservation of energy. This justifies the assumption of stability for stationary states: the atom, *before* and *after* an external influence, will be in a state of well defined energy. This is the basis for the quantum postulate in atomic processes. [Bohr cannot help being confusing and vague here, because – with the single exception of the truly stationary ground state – the characterization of the system by a sequence of 'stationary' excited states is an idealization. In atomic physics it is frequently an excellent approximation to neglect the coupling of the electrons to the radiation field; and then include it perturbatively to give decay widths Γ for the excited states, which are much less than their energy differences s ΔE. If this had not been a good approximation, or if one had tried to solve the problem exactly from the beginning, progress would have been much more difficult.]

(31) To understand the well-known paradoxes to which this assumption leads in the description of collision and radiation processes, it is essential to consider the limitations on the definition of the reacting free particles as expressed by the Heisenberg Uncertainty Principle. If the definition of the energy of the reacting particles is to be precise enough to speak of conservation of energy during the reaction, the reaction time interval Δt must be long compared to a natural period of the transition process $\hbar/\Delta E$, for the energy difference between the stationary states ΔE. This is important in the passage of fast particles through an atom. The transit time is very small compared with the natural periods of the atom, so it seems possible to understand the conservation of energy with the assumed stability of stationary states. In the wave representation, however, the reaction time is connected with the accuracy of the knowledge of the energy of the colliding particle, and there can never be the possibility of a contradiction of the law of

energy conservation.

(32) The concept of stationary states demands that in any observation permitting a distinction between different stationary states, we can disregard the history of the atom. Quantum theory ascribes a phase to each stationary state which depends on the previous history of the atom, and would seem to contradict the very idea of stationary states. In any time-dependent problem, however, a strictly closed system is insufficient. The use of purely harmonic proper vibrations in the interpretation of observations is only an idealization. In a more rigorous description it must be replaced by a wave packet over a finite frequency interval, which has no definable phase.

(33) The unobservability of the phase is made clear in the Stern-Gerlach experiment. Atoms with different orientation in the field can be separated only if the deflection of the beam is larger than the diffraction by the slit of the corresponding de Broglie waves. This means that the product of the time of passage of the beam and its energy uncertainty due to its finite width, is $\geq \hbar$. The condition for separation of the beam can be expressed as

$$d = P_\perp \Delta T / m \text{ and } \Delta d = \hbar / \Delta P_\perp << d,$$

which, with $P_\perp \Delta P_\perp / m = \Delta E$, requires $\Delta E \Delta T >> \hbar$. This result was used by Heisenberg to illustrate the uncertainty principle for energy and time. Here we are not simply dealing with a measurement of the energy at a given time. But since the period of the normal vibrations of the atom in the field is $\tau = 2\pi\hbar / E$, the condition for separation of different states just means the loss of phase.

(34) To consider an atom as a closed system means to neglect the spontaneous emission of radiation which puts an upper limit on the lifetime of stationary states. This neglect is justified in many applications because the coupling between the atom and the radiation field is very small compared to that between the particle and the atom. It is possible to neglect the reaction of the radiation on the atom, and thus to disregard the finiteness of the lifetime and the corresponding energy uncertainty of the stationary states.

* * * * * * * * * * * * * * * * * * * **

Compare the $2P - 1S$ hydrogen transition with a decay probability

$$\frac{1}{\tau_{rad}} = \frac{\Gamma}{\Delta\omega} \sim \frac{e^2 a^2}{\Delta\omega} \sim e^2 (\Delta\omega^2 R)^2$$

with $a \sim \Delta\omega^2 R$ the acceleration of the electron, and an oscillation period $\tau_{osc} \sim \frac{1}{\Delta\omega}$. The number of atomic oscillations in a radiative lifetime is

$$N = \frac{\tau_{rad}}{\tau_{osc}} = \frac{(2\pi)^2 R^2}{137\lambda^2} \sim 4 \times 10^6,$$

justifying the assumption of a stationary state.

* * * * * * * * * * * * * * * * * * * **

(35) The treatment of the radiation problem in quantum theory began with Heisenberg's reformulation of the correspondence principle. In the rigorous form of the theory developed by Dirac [4], the radiation field itself is included in the closed quantum system. It is possible to take account of the quantum character of the radiation and to construct a theory in which the width of the spectral lines is taken into consideration. The complete absence of space-time pictures in this theory is a striking illustration of the complementary character of quantum theory. This is a particularly radical departure from the classical predictable [Read: 'deterministic'] description of Nature. [In fact, all these conclusions are contained in Born's initial calculation of inelastic scattering; and of course Feynman diagrams are a visualization if not a classical space-time picture of these processes.]

(36) Because of the understanding of atomic properties by classical electrodynamics through the correspondence principle, the apparent incompatibility of
i) stationary states and
ii) individual particles in the atom
might seem a difficulty. This means that at large quantum numbers, classical pictures of electron motion can be constructed. However, this is not a gradual transition to classical theory. The quantum postulate does not lose its significance for large quantum numbers. In fact, the conclusions obtained from the correspondence principle with the aid of classical pictures still depend on the concept of stationary states and discrete transition processes, even in this limit.

(37) This question is a particularly instructive example of the new methods. As shown by Schrödinger [5], it is possible by superposition of proper vibrations to construct wave packets small compared to the 'size' of the atom, which propagate like classical particles when the quantum numbers are large enough. In the special case of the simple harmonic oscillator, the wave packet will keep together for any length of time, and will oscillate like the classical motion. This result led Schrödinger to the hope of constructing a classical wave theory without the quantum postulate. Heisenberg showed that the simplicity of the oscillator case is exceptional and due to the simple harmonic nature of the classical motion. Even in this example there is no possibility for an asymptotic description of free particles. In general, the wave packet will spread out over the whole atom. The 'motion' of a bound electron can be followed only for a number of periods comparable to the quantum numbers involved [6]. Here again we see the contrast between the wave theory superposition principle and the assumption of individual localized particles. No predictable connection can be made between observations leading to the definition of a stationary state and other observations on the behavior of localized particles in the atom.

(38) Summarizing: The concepts of stationary states and discrete transition processes have just as much or as little 'reality' as the very idea of individual particles. In both cases we are limited by the demands of predictability [Read:'limited by the *inevitability of indeterminacy*'] complementary to the space-time [Read: *'classical'*] description, due to the restricted possibilities of definition and observation [Read: 'due to the *quantum impossibility* of *perfect* definition and observation'] as stated in the Heisenberg Uncertainty Principle.

§ 7. THE PROBLEM OF ELEMENTARY PARTICLES.

(39) Using the complementarity feature [Read: 'Using the Heisenberg Uncertainty Principle which is an inevitable result of ⋯'] of the quantum postulate, it is possible to construct a consistent theory of atomic phenomena which is a rational generalization of the predictable [Read: 'predictive, uniquely determined'] space-time description of classical physics. This does not mean, however, that classical theory is simply the limiting case of $\hbar \to 0$. The connection of that theory with experi-

ence cannot be separated from the problems of quantum theory. [Note: The rest of this paragraph and in fact of the lecture is without substance.] An indication of this is the well-known difficulty of accounting for elementary particles using classical mechanical and electrodynamic principles. Even general relativity does not fulfill such requirements. A solution to such problems is possible only by a quantum-theoretical field theory, in which the ultimate quantum is an intrinsic feature of the theory. Klein [7] has connected this problem with the five-dimensional unified theory of electromagnetism and gravitation proposed by Kaluza. Here the conservation of charge is an analog of the conservation of energy and momentum. Just as these dynamical variables are complementary to the space-time description, the ordinary four-dimensional space-time would seem to result from the fact that electric charge always occurs in well-defined units, with the consequence that the conjugate fifth dimension is unobservable.

(40) Aside from these deep unsolved problems, classical electron theory has been the guide for a further development first advanced by Compton, that the electron, besides its mass and charge, has a magnetic moment due to an intrinsic spin angular momentum $\hbar/2$. This assumption, introduced independently by Goudsmit and Uhlenbeck to understand the anomalous Zeeman effect, has proved most fruitful. Indeed, the hypothesis of the spinning electron, together with the antisymmetrization postulate of Heisenberg [8], which is essential for the quantum description of atoms with several electrons, have brought the interpretation of the spectral laws and the periodic system to completion. The underlying principles have even made it possible to understand some properties of atomic nuclei. Dennison [9] has recently shown that the specific heat of hydrogen can be understood on the assumption that the proton has a spin angular momentum the same magnitude as that of the electron. Due to its larger mass, however, the proton magnetic moment is much smaller than that of the electron.

(41) The problem of the elementary particles remains unsolved. There is no unambiguous explanation for the different behavior of the electron and the light quantum under the Pauli exclusion principle. This principle is important for the problem of atomic structure as well as for the recent development of statistical theories, selecting one among several possibilities each of which fulfills the corre-

spondence requirement. The difficulty of satisfying the requirements of relativity in quantum theory seemed impossible to overcome after the promising attempts of Darwin and Pauli failed to agree with the kinematical considerations of Thomas, which are so essential for the interpretation of experimental results. Recently, however, Dirac [10] has solved the problem of the relativistic spinning electron by an ingenious new method. His fundamental equation contains quantities of a still higher degree of complexity, four component relativistic generalizations of Pauli's two component spinors.

(42) Relativity implies a predictive space-time description of phenomena. The rationalization of relativity with the quantum postulate requires that we give up visualization in the usual sense in the formulation of quantum laws. The difficulty is that every word in the language refers to our ordinary perception. In quantum theory this problem is an inevitable consequence of the discreteness characterizing the quantum postulate.

Footnotes and References:

1) This paper is a lecture on the present state of quantum theory delivered on Sept 16, 1927, at the Volta celebration in Como.

2) W. Heisenberg, Zeits. f. Phys. **43**, 172 (1927).

3) E. Schrödinger, Ann. d. Phys. **83**, 956 (1927).

4) P.A.M. Dirac, Proc. Roy. Soc. **A114**, 243 (1927).

5) E. Schrödinger, Naturwiss. **44**, 664 (1926).

6) C. Darwin, Proc. Roy. Soc. **A117**, 258 (1927); E. Kennard, Zeits. f. Phys. **47**, 326 (1927).

7) O. Klein, Zeits. f. Phys. **46**, 188 (1927).

8) W. Heisenberg, Zeits. f. Phys. **41**, 239 (1927).

9) D. Dennison, Proc. Roy. Soc. **A115**, 483 (1927).

10) P.A.M. Dirac, Proc. Roy. Soc. **A117**, 610 (1928).

Chapter XIII

EINSTEIN, PODOLSKY, and ROSEN:

This makes the reality of P and Q depend on the process of measurement carried out on the first system, which does not disturb the second system in any way. No reasonable definition of reality could be expected to permit this.

Albert Einstein, Boris Podolsky, and Nathan Rosen, Phys. Rev. **47**, 477 (1935).

§ XIII-1. Introduction.

From Rosenfeld [1], we learn that the Einstein-Podolsky-Rosen "onslaught came down upon us as a bolt from the blue $\cdots$ as soon as Bohr heard my report of Einstein's argument, everything else was abandoned: we had to clear up such a misunderstanding at once. $\cdots$ Bohr immediately started dictating a reply. Soon, he became hesitant: 'No, this won't do, we must try all over again $\cdots$ we must make it quite clear $\cdots$ What *can* they mean? Do *you* understand it?'. $\cdots$ Eventually, he broke off with the familiar remark that he 'must sleep on it.' The next morning he took up the dictation again $\cdots$ there was no trace of the previous day's sharp dissent. He seemed to take a milder view of the case: 'That's a sign that we are beginning to understand the problem.' Now the real work began in earnest: day after day, week after week, the argument was patiently scrutinized with simpler and more transparent examples. $\cdots$ the weakness in the critics' reasoning became evident, and their whole argument – for all its false brilliance – fell to pieces. 'They do it smartly,' Bohr commented, 'but what counts is to do it right.' "

EPR base their rejection of quantum mechanics on a perceived inconsistency in the possible results of the wave function collapse. Consider measurements on an 'entangled' quantum system of two initially interacting and correlated particles, which have become widely separated. They are no longer interacting but have retained their initial correlation. In the simplest example, the particles are assumed to fly apart at time T with equal and opposite momenta $p_1 = -p_2$ and wave function

$$\Psi(x_1, x_2) = \int_{-\infty}^{+\infty} dp e^{ip(x_1 - x_2)/\hbar}.$$

A measurement of the momentum of particle-1 giving $p_1 = p_0$ collapses the wave function into

$$\Psi(x_1, x_2) \Rightarrow \left(e^{ip_0 x_1/\hbar}\right) \times \left(e^{-ip_0 x_2/\hbar}\right),$$

corresponding to the momentum $p_2 = p_0$ for the far-removed particle-2.

Alternatively, a measurement of the position of particle- giving $x_1 = x_0$ would collapse the wave function into

$$\Psi(x_1, x_2) \Rightarrow \delta(x_1 - x_0) \times \delta(x_2 + x_0),$$

corresponding to the position $x_2 = -x_0$ for the separated particle-2.

Now EPR would like to conclude – because they 'could have' indirectly determined the momentum of particle-2 'without in any way' disturbing it; and likewise they 'could have' indirectly determined its position – that they were free to consider both as properties of particle-2 and as 'elements of [c-]reality' which could not reasonably be supposed to depend on some far-distant measurements that could not disturb it. But this is in violation of quantum mechanics as expressed equivalently by the wave functions and by the Heisenberg Uncertainty Principle based on the commutation relations and made intuitive in specific gedanken experiments by the Bohr complementarity principle.

Paraphrasing Rosenfeld: The error that EPR made was to fall into the habits of thought of classical physics where 'elements of reality' and 'physical concepts' are, in Einstein's words, 'free creations of the mind.' His 'criterion of reality' thus created 'has an essential ambiguity' [Rosenfeld's words, what he should have said was *is fundamentally impossible*]. It is necessary to give up our preconceived notions of what 'elements of reality' ought to be.

Rosenfeld characterizes EPR as the last clash between Bohr and Einstein, but Einstein – although silenced – was never to alter his opinions.

Rosenfeld's private account to Pais [2] was not so politically correct and included the statement that Bohr was at first infuriated, and the next day when he saw his way clear to a rebuttal of EPR he sang a gibberish song lampooning Podolsky's name in his elation.

Pais himself views EPR as much ado about nothing. He quotes Uhlenbeck as saying that at the time, no workaday physicists paid any attention to one more Bohr-Einstein debate; and says that to characterize EPR as a paradox is inappropriate. Pais dismisses EPR as simply and correctly concluding that their definition of objective [c-]reality is incompatible with the assumption that quantum mechanics is complete. This failed attempt to define an objective [q-]reality will be the only surviving feature of EPR, standing as it does in contrast to quantum mechanics. The eventual dramatization of this last fact by Bell's Theorem and Aspect's experiment, however, assure EPR's lasting importance.

Pais suggests that the rejection of quantum theory by Einstein can be summarized as a disagreement with Bohr over the concept of *phenomenon*: Bohr ultimately defined *phenomenon* "to refer exclusively to observations under specified circumstances, including an account of the whole experiment." In contrast, Einstein believed [see also 3] in a "deeper-lying theoretical framework which describes 'phenomena' independently of these conditions. This is what Einstein meant by the term *objective reality*. It was his conviction that quantum mechanics is logically consistent, but is an incomplete manifestation of an underlying theory in which an objectively real description is possible, a position he maintained until his death."

There were at the time other responses to EPR in addition to Bohr's. Kemble [3] gives the 'orthodox' and still accepted interpretation: "··· When we say that the 'state' of an electron in motion is $\psi(x, t)$ we mean that an ensemble of a very large number of similarly prepared electrons would have statistical properties described by this function, and that we cannot know more about an individual electron than the fact that it belongs to a suitably chosen potential ensemble of this character. Every analysis of a wave function into a linear combination of orthogonal functions can be interpreted as a resolution of the complete ensemble into a collection of sub-ensembles. The expansion by EPR of the complete wave function $\Psi(x_1, x_2)$ for the combined system-$(1, 2)$ into eigenfunctions of a dynamical variable A of the system-1 [$\sum_a \phi_A^a(1) \psi^a(2)$] means that in each sub-ensemble we have a correlation between a particular pure state [$\phi_A^a(1)$] of the variable A for system-1 and a corresponding state [$\psi^a(2)$] for system-2. The act of measuring the variable A [to be a'] is the physical counterpart of the mathematical expansion and shows in which sub-ensemble a particular pair of system-1 and system-2 belongs [i.e., $\phi_A^{a'}(1) \psi^{a'}(2)$]. The 'reduction of the wave packet' is the process of substituting the wave function of the sub-ensemble for that of the original complete ensemble which must accompany the transfer of one's attention from an arbitrary member or members of the original ensemble to systems known to belong to the particular sub-ensemble.

There seems to be no reason to doubt the completeness of the quantum-mechanical description of atomic systems within the frame of our present experimental knowledge."

Everything that Kemble said was and is [conventionally believed to be] true. So, yes, [q-]reality does differ from EPR's objective [c-]reality.

Schrödinger [4] gave an amazingly perceptive and complete analysis of the situation. In the same series of 1935 papers in which he introduced his 'cat paradox', he introduced the concept of 'entanglement' and completely analyzed its consequences for the EPR paradox – even to the point of a very modern and amusing dialog of hypothetical objections and responses. In this context, Schrödinger expressed a most sophisticated attitude which has become part of the modern orthodoxy – not at all what we might have expected from the caricature of him that has come down to us of the defeated and retrograde opponent of Copenhagen abstractionism. Here he elucidates the role of the wave function ψ as an 'expectation-catalog' which changes discontinuously in a measurement to incorporate the new knowledge obtained. "Things are not at all simple. *It is the most difficult and most interesting point of the theory.*" Then, even for two completely separated bodies,

"Maximal knowledge of a total system does not necessarily include total knowledge of all its parts, not even when these are fully separated from each other and not influencing each other at all." and "if a particular measurement on the first system yields *this* result, then for a particular measurement on the second system the valid expectation statistics are [Which we read as: 'the valid wave function is'] such and such [Read: one thing]; but if the measurement in question on the first system should have *that* result, then some other expectation [Add: statistics] [And as above, replace: 'expectation statistics' by 'wave function'] holds for the second system". And "if one knows the ψ-function and makes a particular measurement with a particular result, then one again knows the [new] ψ-function, that's all. It's just because the combined system consists of two *separate* [Our italics.] parts that the situation seems particularly strange." This can be stated as: "Best possible knowledge of the whole does not necessarily include the same for its parts." or "The whole is in a definite state, the parts taken individually are not."

Then Schrödinger employs the following amusing dialog [in tones more associated now with Wheeler]:
"How so? Surely a system must be in some sort of state."
"No. State is ψ-function, is maximal sum of knowledge. I didn't necessarily provide myself with this, I may have been lazy. Then the system is in no [one] state."
"Fine, but then too the agnostic prohibition of questions is not yet in force [whatever that means] and I can tell myself: the subsystem is really in *some* state, I just don't know which."
"Wait. Unfortunately not. There is no 'I just don't know.' $\cdots$"

"$\cdots$ Any 'entanglement of predictions' can obviously only go back to the fact that the two bodies at some earlier time formed *one* system $\cdots$ and have left behind *traces* on each other."

Schrödinger eventually over-reaches himself: "We dare hope to clear up, if not altogether avoid, the singular jump of the ψ-function." – and much, much more [5]. Nonetheless, we believe that for all his misstatements, Schrödinger understood the orthodox view perfectly clearly but simply would not stop talking.

§ XIII-2. Concluding Remarks.

EPR included the rebuttal of their paradox (see their paragraph 19) – "··· one would not arrive at our conclusion if one insisted that two]*cdots* physical quantities can be regarded as simultaneous elements of [q-]reality *only when they they can be simultaneously measured* ···." – but they rejected it because "··· No reasonable definition of [c-]reality could be expected could be expected to permit this." From our present vantage point it seems bizarre that EPR would not at least question their postulate of [c-]reality in the face of this incompatibility with quantum behavior. According to Pais [2] and Sachs [6], this failure was deeply and permanently rooted in Einstein's conception of classical phenomena as objective space-time events with a reality independent of any history, as distinct from Bohr's view limited by the restrictions of complementarity. According to Pais and Sachs, Einstein continued to believe that there was a deeper reality – based not on classical Bohmian orbits or hidden variables, but on some undiscovered non-linear field relationship whose self consistency would be the ultimate source of the quantum enigmas.

It was the great step by Heisenberg to discover the dynamical theory of the quantum probability *amplitudes*. These amplitudes underly the observed probabilities themselves and change physical laws – so far, apparently forever – to describe a [q-]reality whose abstract depth and subtlety is beyond anything conceivable on the basis of a classical understanding.

The result is Bohr's 'peculiar situation'; and Schrödinger's 'most difficult and interesting point of the theory'. Strangely enough, it was Schrödinger's initial attempt to understand quantum transitions on the basis of semi-classical radiation theory and wave mechanics which was – and still is – seductively intuitive but *ultimately misleading*. It has left the false impression that the Schrödinger wave function is somehow – in some respect not fully defined – a classical material field with an objective [c-]reality similar perhaps to the classical electrostatic field, or to (the square root of) a hydrodynamic density. This initial understanding is planted in every beginning student's mind for the perfectly defensible purpose of 'getting on with the physics'; and in fact it serves a useful – even essential – purpose in providing intuitive guidance to everyone interested in phenomenological applications of quantum mechanics. The vast majority of us at least start each problem with a hand-waving semi-classical wave-mechanical model replete with bells and whistles. And that informal approach to physics has proven most productive in many applications. The formal theoretical mathematically rigorous approach is appropriately usually left for 'mopping up exercises'.

EPR explicitly challenged this semi-classical understanding of the Schrödinger wave function and *demanded* – not for the first time [7] – that the true abstraction of the [q-]reality of the wave function be fully acknowledged. Born [8] and Dirac [9] had implicitly pointed out the same necessity. Jordan [10] had stated the same conclusion very explicitly. After EPR, Schrödinger [5] explained the situation explicitly and completely, although with some obvious reluctance to completely accept his own conclusions.

After EPR – or rather after the eventual Bohm-Bell-Aspect demonstration of it [see our next chapter] – it was no longer acceptable to deny

i) the necessity of the 'collapse of the wave function', and

ii) the abstract information-theoretic nature of the wave function in quantum mechanics; and with it,

iii) the ensemble interpretation of quantum measurements.

The orthodox Copenhagen interpretation of quantum mechanics had been demonstrated to be not wrong and at least sufficient in a limited class of experiments. No one would be fool enough to say that the game is over. The game is never over. But orthodox quantum mechanics has survived one more experimental and intellectual challenge in its long unblemished history.

Just exactly what was Bohr's contribution to all this? Bohr used his complementarity principle to give intuitive expression to the Heisenberg Uncertainty Principle. With it, he was able to show that EPR-states possessing a precisely defined [c-]reality are an unattainable idealization precluded by fundamental principle intrinsic to the wave-particle duality of the quantum world. These classical [c-]real EPR-states must be replaced by quantum states which satisfy a less restrictive [q-]reality. The operative principle dictating true facts which are the possible subjects of sensible discussion is that each must be *defined* – not just *definable*) – in an experiment. The net result of such restrictions on sensible discussion is a [q-]reality free from the incompleteness claimed by EPR.

Bohr shows that a classical description of wave-particle duality is not possible because of the restrictions of quantum mechanics made explicit in the Heisenberg Uncertainty Principle. Unfortunately, what he does not do is show explicitly what the consequences must be for our understanding of the wave function.

Footnotes and References:

1) L. Rosenfeld, 'Niels Bohr in the Thirties: Consolidation and Extension of the Conception of Complementarity' in *Niels Bohr: His Life and Work as Seen by His Friends and Colleagues* (North-Holland, Amsterdam, 1967) S. Rozental, Ed., Pp. 114-136; alternatively see *Quantum Theory and Measurement* (Princeton, Princeton, NJ, 1983) J.A. Wheeler and W.H. Zurek, Eds., Pp. 142-143.

2) A. Pais, *Niels Bohr's Times* (Clarendon, Oxford, UK, 1991), Pp. 430-435.

3) E.C. Kemble, Phys. Rev. **47**, 973 (1935). See also A.E. Ruark, Phys. Rev. **48**, 466 (1935) for a discussion along the same lines as Bohr; and H. Margenau, Phys. Rev. **49**, 240 (1936), for a proposal to alter quantum mechanics.

4) E. Schrödinger, Die Naturwissenschaften, **23**, 807, 823, 844 (1935); translated by J.D. Trimmer in Proc. Am. Phil. Soc. **124**, 323 (1980), available in *Quantum Theory and Measurement* (Princeton, Princeton NJ, 1983) J.A. Wheeler and W.H. Zurek, Eds., Pp. 153-167, especially § 10,11.

5) W.H. Furry, Phys. Rev. **49**, 393, 476 (1936). Furry's analysis is "essentially the same" as Schrödinger's, and his conclusions are the same as Schrödinger's first statements about the results of a measurement on system-1 determining the wave function of system-2. In his second paper Furry generalizes Schrödinger's discussion in a now obvious way. It seems to us that Furry errs in his statement that their interpretations are "exact opposites".

6) M. Sachs, Am. J. Phys. **36**, 463 (1968); **37**, 229 (1969).

7) N. Mott, Proc. Roy. Soc. **A126**, 79 (1929); available in *Quantum Theory and Measurement* (Princeton, Princeton NJ, 1983) J.A. Wheeler and W.H. Zurek, Eds., Pp. 128-134.

8) M. Born, Zeits. f. Phys. **37**, 863 (1926); ibid, **37**, 803 (1926). Here as **Paper X·1,2**.

9) P.A.M. Dirac, *The Principles of Quantum Mechanics* (Oxford, London, UK 1930, 1937, 1947), ChII; also Proc. Roy. Soc. **A113**, 621 (1926).

10) P. Jordan, Zeits. f. Phys. **40**, 809 (1926); also Naturwiss. **15**, 621 (1926).

Paper XIII·1: Excerpt from Phys. Rev. **47**, 777 (1935).

CAN QUANTUM-MECHANICAL DESCRIPTION OF PHYSICAL REALITY BE CONSIDERED COMPLETE?

A. EINSTEIN, B. PODOLSKY, AND N. ROSEN

Institute for Advanced Study, Princeton, New Jersey
(Received March 25, 1935)

In a [c-]complete theory there is an element [a dynamical variable] corresponding to each element of [c-]reality. A sufficient condition for the [c-]reality of a physical quantity is the possibility of predicting it with certainty, without disturbing the system. In quantum mechanics in the case of two physical quantities described by noncommuting operators, the knowledge of one precludes the knowledge of the other. Then either (i) the description of [c-]reality given by the wave function in quantum mechanics is not [c-]complete; or (ii) these two quantities cannot have simultaneous [c-]reality [for 'either – or' read 'both – and' for a correct statement of the actual situation]. Consideration of the problem of making predictions concerning a system on the basis of measurements made on another system that had previously interacted with it leads to the result that if (i) is false then (ii) is also false [this *is* the case]. One is thus led to conclude that the description of [c-]reality as given by a wave function is not [c-]complete [and so is this].

I: (1) Any serious consideration of a physical theory must take into account the distinction between the objective [c-]reality, which is independent of any theory [not so, it turns out], and the physical concepts with which the theory operates. These concepts are intended to correspond with the objective [c-]reality, and by means of these concepts we picture this [c-]reality to ourselves. [Note added: Throughout, we distinguish e.g., [c-]reality (classical-) from [q-]reality (quantum-). Also we separate our own comments between square brackets.]

(2) In judging the success of a physical theory, we must ask two questions:
i) 'Is the theory correct?' and
ii) 'Is the description given by the theory complete?'
Only if the answers to both of these questions is yes, may the theory be said to be satisfactory. The correctness of the theory is judged by the degree of agreement

between the conclusions of the theory and human experience. This experience, which alone enables us to make inferences about [c/q-]reality, in physics takes the form of experiment and measurement. It is the question of completeness that we consider here, as applied to quantum mechanics.

(3) Whatever the meaning of *complete*, the following requirement seems to be necessary for a [c/q-]complete theory: *every element of the physical* [c/q-]*reality must have a counterpart in the physical theory.* We shall call this the condition of [c/q-]completeness. The question of completeness is thus answered when we determine what are the elements of the physical [c/q-]reality.

(4) The elements of physical [c-]reality cannot be determined by *a priori* philosophical considerations, but must be found by an appeal to results of experiments and measurements. A comprehensive definition of [c-]reality is, however, unnecessary for our purpose. [In fact, such a comprehensive definition would seem to be available in the canonical phase space variables of classical Hamiltonian mechanics [1] – which satisfy EPR's definition of [c-]reality – and are taken directly into quantum mechanics [2] – where they do not.] We shall be satisfied with the following criterion, which we regard as reasonable. *If, without in any way disturbing a system, we can predict with certainty (i.e., with probability equal to unity) the value of a physical quantity, then there exists an element of physical* [c-]*reality corresponding to this quantity.* It seems to us that this criterion, while far from exhausting all possible ways of recognizing a physical [c-]reality, at least provides us with one such way, whenever the conditions set down in it occur. Regarded not as a necessary, but merely as a sufficient, condition of [c-]reality, this criterion is in agreement with classical as well as quantum-mechanical ideas of reality. [Not so, thanks largely of course to EPR. We are now confronted with the seemingly inescapable conclusion that quantum mechanics precludes [c-]completeness; and therefore requires a change from the concept of [c-]reality to a new [q-]reality.]

(5) To illustrate the ideas involved consider the quantum-mechanical description of the behavior of a particle having a single degree of freedom. The fundamental concept of the theory is the concept of *state*, which is supposed to be [q-]completely characterized by the wave function ψ, which is a function of the

variables chosen to describe the particle's behavior [in fact, the same canonical variables as enter the classical Hamiltonian [3] (with the notable extension to include spin)]. Corresponding to each physically observable quantity A [classical dynamical variable] there is an operator which will be designated by the same letter.

(6) If ψ is an eigenfunction of the operator A, that is, if

$$A\psi = a\psi, \qquad (1)$$

where a is a number, then the physical quantity A has with certainty the value a whenever the particle is in the state ψ [Following Omnès [4], this is a 'reliable' statement as distinct from a 'true' statement]. In accordance with our criterion of [c-]reality, for a particle in the state ψ for which Eqn1 holds, there is an element of physical [q-]reality corresponding to the physical quantity A. For example, let

$$\psi = e^{ip_0 x/\hbar}, \qquad (2)$$

where $\hbar$ is Planck's constant, p_0 is some constant number, and x the independent variable. Using the momentum operator of the particle we obtain

$$p\psi = \frac{\hbar}{i}\frac{\partial \psi}{\partial x} = p_0\psi. \qquad (4)$$

Thus in the state ψ of Eqn2, the momentum has with certainty the value p_0. It thus has meaning to say that the momentum of the particle in the state given by Eqn2 is [q-]real [a 'reliable' property, a result of a 'true' datum].

(7) On the other hand if Eqn1 does not hold, we can no longer speak of the physical quantity A having a particular value. This is the case, for example, with the coordinate of the particle. The operator corresponding to it, say q, is the operator of multiplication by the independent variable. Thus,

$$q\psi = x\psi \neq a\psi. \qquad (5)$$

In accordance with quantum mechanics we can only say that the relative probability that a measurement of the coordinate will give a result lying between a and b is

$$P(a,b) = \int_a^b \bar{\psi}\psi dx = \int_a^b dx = b - a. \qquad (6)$$

Since this probability depends only on the difference $b - a$, we see that all values of the coordinate are equally probable.

(8) A definite value of the coordinate for a particle in the state ψ of Eqn2 is thus not predictable, but may be obtained only by a direct measurement. Such a measurement however disturbs the particle and thus alters its state. After the coordinate is determined, the particle will no longer be in the state given by Eqn2. The usual conclusion from this in quantum mechanics is that *when the momentum of a particle is known, its coordinate has no physical* [q-]*reality.* [No dispute here.]

(9) Also, it is known in quantum mechanics that if the operators A and B corresponding to two physical quantities do not commute, that is if $AB \neq BA$, then the precise knowledge of one of them precludes such knowledge of the other. Furthermore, any attempt to determine the latter experimentally will alter the state of the system in such a way as to destroy the knowledge of the first. [Agreed.]

(10) From this follows that either
i) *the quantum-mechanical description of* [c-]*reality given by the wave function is not* [c-]*complete* or
ii) *when the operators corresponding to two physical quantities do not commute the two quantities cannot have simultaneous* [c-]*reality.*
[Again, for 'either – or' read 'both – and' to recover the [q-]reality.] For if both of them had simultaneous [c-]reality – and thus definite values – these values would enter into the [c-]complete description, according to the condition of [c-]completeness. If the wave function provided such a [c-]complete description of [c-]reality, it would contain these values; these would then be predictable. This not being the case, we are left with the alternatives stated. [Either alternative (i) above, or (ii). It is astonishing that EPR did not follow there own criterion in Para(2) – "The correctness of the theory is judged by the degree of agreement between the conclusions of the theory and human experience." – and alter their criteria for [c-]reality to conform to the evident [q-]reality. Obviously the evidence was then not so clear, but still $\cdots$.]

(11) In quantum mechanics it is usually assumed that the wave function *does* contain a [q-]complete description of the physical [q-]reality of the system in the state to which it corresponds. At first this assumption is entirely reasonable, for

the information obtainable from a wave function seems to correspond exactly to what can be measured without altering the state of the system. We shall show, however, that this assumption, together with the criterion of [c-]reality given above, leads to a contradiction.

II: (12) For this purpose let us suppose that we have two systems I and II, which we permit to interact from the time $t = 0$ to $t = T$; we suppose that there is no longer any interaction between the two parts after time T. We suppose further that the states of the two systems before $t = 0$ were known. We can then calculate with the help of Schrödinger's equation the state of the combined system $I + II$ at any subsequent time $t > T$. Let us designate the corresponding wave function by Ψ. We cannot, however, calculate the state in which either one of the two systems is left after the interaction. This, according to quantum mechanics, can be done only with the help of further measurements, by a process known as the *reduction of the wave packet*. Let us consider the essentials of this process.

(13) Let $a_1, a_2, a_3, \cdots$ be the eigenvalues of some physical quantity A of system I and $u_1(x_1), u_2(x_1), u_3(x_1), \cdots$ the corresponding eigenfunctions, where x_1 are the variables of the first system. Then Ψ, considered as a function of x_1, can be expressed as

$$\Psi(x_1, x_2) = \sum_{n=1}^{\infty} \psi_n(x_2) u_n(x_1), \tag{7}$$

where x_2 are the variables of the second system. Here $\psi_n(x_2)$ are to be regarded merely as the coefficients of the expansion of Ψ into a series of orthogonal functions $u_n(x_1)$. Suppose now that the quantity A of I is measured and it has the value a_k. It is then concluded that after the measurement the first system is left in the state $u_k(x_1)$, and that the second system is left in the state $\psi_k(x_2)$. This is the reduction of the wave packet: the wave packet given by the infinite series of Eqn7 is reduced to a single term $\psi_k(x_2) u_k(x_1)$.

(14) The set of functions $u_n(x_1)$ is determined by the choice of the physical quantity A. If we had instead chosen another quantity B having the eigenvalues $b_1, b_2, b_3, \cdots$ and eigenfunctions $v_1(x_1), v_2(x_1), v_3(x_1), \cdots$ we would have obtained

instead of Eqn7, the expansion

$$\Psi(x_2, x_1) = \sum_{m=1}^{\infty} \phi_m(x_2) v_m(x_1), \tag{8}$$

where ϕ_m's are the new coefficients. If now the quantity B is measured and is found to have the value b_r, we conclude that after the measurement the first state is $v_r(x_1)$ and the second is left in the state $\phi_r(x_2)$.

(15) We see therefore that as a consequence of two different measurements performed on the first system, the second system may be left in states with two different wave functions. On the other hand, since at the time of measurement the two systems no longer interact, no [c-]real change can take place in the second system as a consequence of anything that may be done to the first system. This is, of course, merely a statement of what is meant by the absence of an interaction between the two systems. [This is one more [c-]reality that will have to be modified to reach accord with the new [q-]reality dictated by these results. It is not "of course, merely a statement ⋯". It is a whole new [q-]reality which is made [q-]real by 'doing' not by 'talking'.] Thus, *it is possible to assign two different wave functions* (in our case ψ_k and ϕ_r) *to the same* [c-]*reality* (the second system after the interaction with the first). [This result – 45 years later in Aspect's experiments [5] – forced the way of thinking dramatically put forward by Wheeler [6] and Mermin [7]: The second system is in some strange quantum limbo until the actual observation makes the suspended [q-]reality of the wave function an actualized [q-]reality. What was it before? It is not entirely meaningless to ask but the answer – $\Psi(x_2, x_1)$ – is a 'reliable' result which reproduces with certainty *only* the measurement result which was used to select it in the first place, a 'true' datum for time T. Thereafter, a different measurement (of A or B) produces a *new* 'true' datum with *new* 'reliable' results.]

(16) It may happen that the two wave functions – ψ_k and ϕ_r – are eigenfunctions of two noncommuting operators corresponding to some physical quantities P and Q, respectively. That this may actually be the case can best be shown by an example. Let us suppose that the two systems are two particles, and that

$$\Psi(x_1, x_2) = \int_{-\infty}^{+\infty} e^{i(x_1 - x_2 + x_0)p/\hbar} dp, \tag{9}$$

where x_0 is some constant. Let A be the momentum of the first particle; then, as we have seen in Eqn4, its eigenfunctions will be

$$u_p(x_1) = e^{ipx_1/\hbar} \tag{10}$$

corresponding to the eigenvalue p. Since we have here the case of a continuous spectrum, Eqn7 will now be written

$$\Psi(x_1, x_2) = \int_{-\infty}^{+\infty} \psi_p(x_2) u_p(x_1) dp, \text{ where } \psi_p(x_2) = e^{-i(x_2-x_0)p/\hbar}. \tag{11}$$

This ψ_p is also an eigenfunction of the momentum operator P_2 corresponding to the eigenvalue $-p$ of the momentum of the second particle. On the other hand, if B is the coordinate of the first particle, it has for eigenfunctions

$$v_x(x_1) = \delta(x_1 - x), \tag{14}$$

corresponding to the eigenvalue x, where $\delta(x_1 - x)$ is the Dirac delta-function. Eqn8 in this case becomes

$$\Psi(x_1, x_2) = \int_{-\infty}^{+\infty} \phi_x(x_2) v_x(x_1) dx, \tag{15}$$

where

$$\phi_x(x_2) = \int_{-\infty}^{+\infty} e^{i(x-x_2+x_0)p/\hbar} dp = 2\pi\hbar\delta(x - x_2 + x_0). \tag{16}$$

This ϕ_x, therefore, is an eigenfunction of the operator Q_2 with eigenvalue $x_2 = x + x_0$ for the coordinate of the second particle. Since

$$PQ - QP = \hbar/i, \tag{18}$$

we have shown that it is possible for ψ_k and ϕ_r to be eigenfunctions of two non-commuting operators, corresponding to physical quantities.

(17) Returning now to the general case contemplated in Eqns7 and 8, we assume that ψ_k and ϕ_r are indeed eigenfunctions of some noncommuting operators P and Q, corresponding to the eigenvalues p_k and q_r, respectively. Thus by measuring either A or B we are in a position to predict with certainty, and without in any way disturbing the second system, either the value of the quantity P (that is

p_k) or the value Q (that is q_r). In accordance with our criterion of [c-]reality, in the first case we must consider the quantity P as being an element of [c-]reality, in the second case the quantity Q is an element of [c-]reality. But, as we have seen, both wave functions ψ_k and ϕ_r belong to the same [c-]reality [but *not* the same [q-]reality, which must be made real by a *measurement* to make the dynamical variable a *reliable* property of the state of the system].

(18) Previously we proved that either
i) the quantum-mechanical description of [c-]reality given by the wave function is not complete, or
ii) when the operators corresponding to the two physical quantities do not commute the two quantities cannot have simultaneous [c-]reality.
Starting then with the assumption that the wave function *does* give a [c-]complete description of the physical [c-]reality, we arrived at the conclusion that two physical quantities with noncommuting operators, *can* have simultaneous [c-]reality. Thus the negation of (i) leads to the negation of the only alternative (ii). [Of course, there *is* another alternative, and we are amazed that EPR did not conclude their paper on a positive note: an expanded quantum conception of [q-]reality.] We are thus forced to conclude that the quantum-mechanical description of physical [c-]reality given by wave functions is not [c-]complete [which we now accept as fact, necessitating a change in our concept of reality from [c-]reality to [q-]reality].

(19) One could object that our criterion for [c-]reality is not sufficiently restrictive [rather that it is *too* restrictive; requiring as it does a perfect localization of the classical Hamiltonian mechanical system-point in phase space – a requirement inconsistent with the wave-particle duality of quantum mechanics, best expressed in the Heisenberg Uncertainty Principle [8] as the formal statement of Bohr's complementarity principle [9]]. Indeed, one would not arrive at our conclusion if one insisted that two or more physical quantities can be regarded as simultaneous elements of [q-]reality *only when they can be simultaneously measured or predicted.* On this point of view, since either P or Q can be predicted [measured] – but not both simultaneously – they are not simultaneously [c-]real. This makes the [q-]reality of P and Q depend upon the process of measurement carried out on the first system, which does not disturb the second system in any way. [No longer a tenable conclusion.] No reason-

able definition of [c-]reality could be expected to permit this. [But in fact this is just what is needed for an acceptable [q-]reality.]

(20) While we have thus shown that the wave function does not provide a [c-]complete description of the physical [c-]reality, we left open the question of whether or not such a description exists. We believe, however, that such a theory is possible. [After 65 years we are still waiting [10].]

Footnotes and References:
(There were none in the original. The ones below refer only to our own added comments.)
1) E.C.G. Sudarshan and N. Mukunda, *Classical Dynamics: A Modern Perspective* (Wiley, New York, 1974), Chapter 9; also H. Goldstein, *Classical Mechanics, Second Edition* (Addison-Wesley, Mass., 1981), Chapter 8.
2) P.A.M. Dirac, *The Principles of Quantum Mechanics* (Oxford, London, 1956), Chapter IV & V. See also P.A.M. Dirac, Proc. Roy. Soc. **A109**, 642 (1925) (here as **Paper VIII·1**), §4.
3) M. Born and P. Jordan, Zeits. f. Phys. **34**, 858 (1925) (here as **Paper VII·1**), Chapter II.; M. Born, W. Heisenberg, and P. Jordan, Zeits. f. Phys. **35**, 557 (1926) (here as **Paper VII·2**), Chapters 1 & 2.
4) R. Omnès, *The Interpretation of Quantum Mechanics* (Princeton, Princeton NJ, 1994), p.356.
5) A. Aspect, P. Grangier, and G. Roger, Phys. Rev. Lett. **49**, 91 (1982); A. Aspect, J. Dalibard, and G. Roger, Phys. Rev. Lett. **49**, 1804 (1982).
6) J.A. Wheeler, *Quantum Theory and Measurement* (Princeton, Princeton NJ, 1983) J.A. Wheeler and W.H. Zurek, Eds., p.182.
7) N.D. Mermin, Physics Today **38**-4, 38 (1985); also in *Niels Bohr: A Centenary Volume* (Harvard, Cambridge MA, 1985) A.P. French and P.J. Kennedy, Eds., Pp. 141-147.
8) W. Heisenberg, Zeits. f. Phys. **43**, 172 (1927) (here as **Paper XI·1**).
9) N. Bohr, Nature **121**, 580 (1928) (here as **Paper XII·1**).
10) J.S. Bell, Physics **1**, 195 (1964); also in *Quantum Theory and Measurement* (Princeton, Princeton NJ, 1983) J.A. Wheeler and W.H. Zurek, Eds., p.403; also in J.S. Bell, *Speakable and Unspeakable in Quantum Mechanics* (Cambridge, Cambridge UK, 1987), p.14.

Paper XIII·2: Excerpt from Phys. Rev. **48**, 696 (1935).

CAN QUANTUM-MECHANICAL DESCRIPTION OF PHYSICAL REALITY BE CONSIDERED COMPLETE?

N. BOHR

Institute for Theoretical Physics, University, Copenhagen
(Received July 13, 1935)

The 'criterion of physical [c-]reality' proposed by Einstein, Podolsky, and Rosen contains an essential ambiguity [Read: 'an error' or 'a fundamentally impossible requirement' or 'a requirement incompatible with the Heisenberg Uncertainty Principle and the complementarity principle'] when applied to quantum phenomena. A viewpoint called 'complementarity' is explained by which the quantum-mechanical description of physical phenomena is [q-]complete.

Einstein, Podolsky, and Rosen [1] have presented an argument which leads them to conclude that the quantum-mechanical description of physical [c-]reality is not [c-]complete. Their argument, however, does not describe the actual situation in atomic physics. I will explain in greater detail the general viewpoint called 'complementarity' which I have presented before [2], in which quantum mechanics emerges as a [q-]complete rational description of atomic processes. [Note that we include our own comments within square brackets. Also italics are added.]

The meaning of the expression 'physical reality' cannot be deduced from *a priori* philosophical concepts [alone], but must be based directly on experiments and measurements. EPR propose a 'criterion of [c-]reality' as follows: "If, without in any way disturbing a system, we can predict with certainty the value of a physical quantity, then there exists an element of physical [c-]reality corresponding to this physical quantity." As an example, they [claim to] show that in quantum mechanics, just as in classical mechanics, it is possible to predict the value of any given variable of a mechanical system from measurements on another system which has been in interaction with the first. They ascribe an element of [c-]reality to each such variable. Since it is a fundamental feature of quantum mechanics that it is *not* possible to give definite values to both of two canonically conjugate variables, they conclude [at first correctly] that quantum mechanics is not [c-]complete, and [then erroneously] that a more satisfactory theory remains to be developed.

* * * * * * * * * * * * * * * * * * * **

The meaning of 'elements of physical reality' *includes* a priori philosophical concepts. Those familiar from classical Hamiltonian mechanics – the canonically conjugate p's and q's – have *failed* the tests of experience, that is to say actual *measurement*, as expressed in the quantum wave-particle dualism and formalized in the Heisenberg Uncertainty Principle. The HUP *precludes in principle* the ideal of *perfect simultaneous localization* of the system point in (p, q)-phase space, essential for the [c-]reality of classical Hamiltonian mechanics. Quantum Hamiltonian mechanics dictates and incorporates *that knowability* of the canonically conjugate Hamiltonian dynamical variables which is *consistent* with the HUP. The HUP *itself* is a consequence and a convenient expression of the effect of the canonical commutation relations which are the fundamental feature distinguishing quantum from classical Hamiltonian dynamics. The ultimate result is a surprisingly profound, abstract, and subtle [q-]reality, whose enigmatic and paradoxical manifestations were first dramatized by Einstein, Podolsky, and Rosen; and – it seems – only indirectly and inadequately appreciated until clarified by Bohm (1952), Bell (1964), and eventually Aspect (1982) (see Ch14).

* * * * * * * * * * * * * * * * * * * **

Such an argument is unsuited to judge the soundness of quantum mechanics which is based on a coherent mathematical formalism that covers automatically any measurement like that suggested [3]. The apparent contradiction is a result of the essential inadequacy of *classical* physics as the basis for a rational account of quantum phenomena. It is the *finite interaction between the object measured and the measuring instrument*, dictated by the very existence of the Planck quantum of action $\hbar$, which – because of the impossibility of controlling the reaction of the object on the instrument – requires that we abandon finally the classical ideal of predictability [Read: 'determinacy'.] and necessitates a radical revision of our concept of physical [c-]reality.

The criterion of reality proposed by EPR – however cautious its formulation might appear – contains an essential ambiguity [Read: $\cdots$ an essential impossibility] when applied to actual problems. To make my argument clear, I describe in detail a few simple examples of basic measurements and their fundamental limitations.

I begin with the simplest example of a particle passing through a slit in a diaphragm, which is part of a larger apparatus. Even if the incoming momentum of the particle is precisely known [i.e., is 'reliable' but not 'true', in the sense of Omnès.] before it reaches the diaphragm, the diffraction by the slit of the associated plane wave implies an uncertainty in the outgoing momentum after the particle passes through the slit. The width of the slit is the uncertainty Δq of the position of the particle in the plane of the diaphragm, perpendicular to the slit. From de Broglie's relation between momentum and wavelength, the uncertainty Δp of the momentum of the particle in this direction is given by the Heisenberg Uncertainty Principle as $\Delta p \Delta q \sim \hbar$, which is a direct consequence of the quantum-mechanical canonical commutation relation for any pair of conjugate variables. The uncertainty Δp requires an exchange of momentum between the particle and the diaphragm. The question is – what consequence does this unknown momentum exchange have on the further description of the phenomena being studied by the whole apparatus?

As in the usual experiments of electron diffraction, we assume that the diaphragm and other parts of the apparatus (including perhaps a multi-slit diaphragm and a photographic plate suitably arranged) are rigidly fixed to a support which defines the space frame of reference. The momentum exchanged between the particle and the diaphragm passes into the support. We have no possibility of taking account of these reactions in predictions of the final result of the experiment, for example the position the particle hits on the photographic plate. The impossibility of a closer analysis of the reaction between the particle and the apparatus is a general feature of any study of this type, involving the feature of *individuality* [Read: $\cdots$ of *inseparability* $\cdots$] completely foreign to classical physics.

Any possibility of taking into account [exactly] the momentum exchanged between the particle and the apparatus would permit a conclusion about the 'course' of such phenomena – say, through what particular slit of a second diaphragm the particle passes on its way to the photographic plate. This would be incompatible with the fact that the diffraction pattern on the photographic plate is determined not by the presence of any one slit, but by the positions of *all* the slits of the second diaphragm. [This is precisely the point where the de Broglie-Bohm hidden variable theory – involving classical particle trajectories following Schrödinger 'pilot'-waves – departs from Heisenberg's quantum mechanics of 'observables'.]

A different experiment, with the first diaphragm not rigidly connected to the apparatus, would allow the measurement of its momentum with any accuracy [4] before and after the passage of the particle and would predict the momentum of the particle after passing through the slit. Such measurements would require only an unambiguous [precise] application of the classical law of conservation of momentum, applied to a collision between the diaphragm and some test body whose momentum can be precisely known before and after the collision. If all space and time intervals are sufficiently large, there is no limit on the the possible accuracy of a measurement of the momentum of the test bodies, but the price is a loss of knowledge of their space-time coordinates. This is the analog of the loss of knowledge of the momentum in experiments designed to determine the position. Both experiments depend ultimately on a purely classical account of the apparatus, but include the corresponding quantum uncertainties of the *conjugate quantities.*

The difference between the two experiments is that the 'momentum detector' has a free diaphragm whose position becomes an 'object' to be measured. The quantum uncertainty relations for its position and momentum *must be taken into account.* Even if we knew its position before the first measurement of its momentum, and again knew its position after the last measurement, because of the uncontrollable displacement of the diaphragm during each collision with the test bodies, we lose the knowledge of its position when the particle passed through the slit. The 'momentum detector' is unsuitable as a 'position detector'. It can be shown that if the momentum of the diaphragm is measured with an accuracy sufficient to specify the particle [as] passing through a particular slit in the second diaphragm, then the minimum uncertainty of the position of the first diaphragm consistent with such knowledge, will totally wipe out any interference effect (the diffraction pattern on the photographic plate) which would have resulted if all parts of the apparatus had been rigidly fixed together.

After the initial measurement of the momentum of the first diaphragm, and after the passage of the particle through the slit, we are still left with the *free choice* whether to know the momentum of the particle or its initial position relative to the rest of the apparatus. In the first case, we make the second measurement of the momentum of the diaphragm, losing forever its position when the particle passed. In the second case, we determine the position of the diaphragm, with the inevitable loss of all knowledge of the momentum exchanged between it and the particle.

> * * * * * * * * * * * * * * * * * **
> This is as explicit as Bohr ever got in distinguishing EPR's [c-]reality
> – which can only be achieved in the idealization that the measure-
> ments do not affect the apparatus and do not preclude a 'true' state-
> ment about its *pre-existing* [c-]reality – from [q-]reality which is de-
> pendent on and *realized by* (i.e., made [q-]reality by) the 'free-choice'
> of which measurement is made.
> * * * * * * * * * * * * * * * * * **

My purpose in repeating these simple, well-known facts is to emphasize that we are not dealing with an incomplete description characterized by some arbitrary selection of different elements of physical reality at the expense of others, but with a rational distinction between essentially different experimental procedures which are designed on the one hand for a precise measurement of position, and on the other for a precise measurement of momentum. The loss in each experimental arrangement of one or the other of the two specifications of the physical situa-tion – which together characterize classical physics, and which may be said to be complementary to (in the sense of completing) one another – is characteristic of quantum theory. It is the result of the impossibility at the quantum level of ac-curately controlling the reaction of the object being measured on the measuring apparatus: the transfer of momentum in the case of a position measurement, and the displacement in the case of a momentum measurement. In any experimental study of quantum phenomena, we are confronted not merely with an *ignorance* of the value of *certain physical quantities*, but with the *impossibility* of *precisely defining* these quantities in an unambiguous way. [Read: ⋯ *ignorance* of both of any pair of canonically conjugate variables, but with the *impossibility of knowing* (in the sense of *measuring*) – precisely and simultaneously – both of any pair of canonically conjugate variables in *any* way.]

---

* * * * * * * * * * * * * * * * * * **

EPR also pointed out but rejected the possibility (which turns out to be the [q-]reality situation) that we can measure x_1 or p_1, but not both simultaneously. Now Bohr makes explicit that it is also impossible to determine both for the same event, and therefore no [c-]completeness or [c-]reality is possible.

* * * * * * * * * * * * * * * * * * **

These remarks apply to the problem raised by EPR, which does not involve any greater complications than the simple examples discussed above. Their quantum state of two free particles can be produced by a simple experimental arrangement consisting of a rigid diaphragm with two parallel slits, narrow compared to their separation, through each of which one particle with given initial momentum passes independent of the other. If the momentum of the diaphragm is measured before and after the emission of the particles, we know the sum of the particles' momenta, $p_1 + p_2$ (perpendicular to the slits), as well as the difference of their positions, $x_1 - x_2$, in the same direction. The conjugate quantities, $x_1 + x_2$ and $p_1 - p_2$, are entirely unknown [5]. It is clear that a subsequent single measurement either of the position x_1 or of the momentum p_1 will automatically determine either x_2 or p_2, respectively, with any desired accuracy (at least if the wavelength of the each particle's motion is sufficiently small compared with the width of the slits). As pointed out by EPR, we have at this stage a completely free choice whether we want to determine one or the other of these conjugate quantities by a process which does not directly involve the particle concerned.

Like the above simple case of the choice between experiments suitable for either a position determination or a momentum determination of a particle which has passed through a slit, we have, in this 'freedom of choice', just *a distinction between different experimental procedures which allow the unambiguous use of complementary classical concepts.* To measure the position x_1 of one particle means to correlate it with some instrument rigidly fixed to a support which defines the space frame of reference. Such a measurement also provides the knowledge of the position of the diaphragm when the particles passed through the slits. Only in this way do we obtain the initial position x_2 of the other particle. By allowing an essential but uncontrollable momentum to pass from the first particle into the support, we have lost any future possibility of applying the law of conservation of momentum to the system of the diaphragm and the two particles. Therefore we lose our ability to predict [Read: $\cdots$ to determine from a subsequent measurement of p_1 $\cdots$]the momentum p_2 of the second particle. Conversely, if we choose [first] to measure the momentum p_1 of the first particle, because of the uncontrollable displacement inevitable in such a measurement, we lose any possibility of deducing from the behavior of this particle the position of the diaphragm and have no basis whatever for a prediction [Read: $\cdots$ for a determination from a subsequent measurement of x_1 $\cdots$] of the position x_2 of the other particle.

We now see that the criterion of physical [c-]reality proposed by EPR contains an ambiguity in the meaning of the expression 'without in any way disturbing the system.' Of course there is no question of a mechanical disturbance of the system during the last critical stage of the measurement. But there is the essential question of *an influence on the very conditions which define the possible types of predictions for the future behavior of the system.* [Just how Bohr's statement relates to the spin correlations in various realizations of Bohm's experiment will be discussed in later chapters.] Since these conditions constitute an inherent element of the description of any 'physical [c-]reality', we see that the EPR argument does not justify their conclusion that the quantum mechanical description is essentially not [c-]complete. On the contrary, the quantum mechanical description may be characterized as the rational [[q-]complete] expression of all possibilities of unambiguous interpretation of measurements, compatible with the finite and uncontrollable interaction in quantum theory between the object being measured and the measuring instruments.

* * * * * * * * * * * * * * * * * * * *

Bell in his essay *'Bertlmann's socks and the nature of reality'*† admits
to having 'very little idea' what Bohr meant in the above paragraph.
The expression 'mechanical disturbance' and the italicized comment
beginning *'an influence* · · ·', and especially the expression 'finite and
uncontrollable interaction · · ·', Bell finds do not address the reservations about quantum mechanics expressed by EPR. On the contrary,
he feels that Bohr simply 'ignored the essential point of EPR that
in the absence of action at a distance, only the first system could be
supposed disturbed by the first measurement and yet definite predictions become possible for the second system. Is Bohr just rejecting
the premise – no action at a distance – rather than refuting the
argument?'
† – J.S. Bell, *Speakable and Unspeakable in Quantum Mechanics*
(Cambridge, Cambridge UK, 1987), Appendix 1, Pp. 155-156.

* * * * * * * * * * * * * * * * * * * **

It is the impossibility – manifested in the new physical laws – of any two
experiments which would unambiguously [simultaneously and precisely] define two
complementary physical quantities [canonically conjugate dynamical variables] which
at first sight [make these new laws] appear inconsistent with the basic principles of
science. It is this entirely new situation in the description of physical [quantum-]
phenomena, that is described by the notion of *complementarity* [best expressed by
the HUP, and requiring the new concepts of [q-]reality and [q-]completeness].

[Bohr next analyzes the effect of the energy-time uncertainty relation.] These experimental arrangements are especially simple because of the minor role played by
the idea of time. We have used the notion of 'before' and 'after'; but in each case
the time intervals are supposed to be large compared to the characteristic periods
of the motions involved. If we want a more accurate time description of quantum
phenomena, we meet new paradoxes, again involving the interaction of the objects
and the instruments. We deal in this case with apparatus having movable parts,

like shutters blocking the slits, controlled by clock-mechanisms. Besides the transfer of momentum, we now have to consider an exchange of energy between the object and the clock.

The decisive point for time measurements in quantum theory is completely analogous to that for position measurements. Just as the transfer of momentum to the apparatus – whose position is to be defined – is entirely uncontrollable, so the transfer of energy between the object and the clock – which defines the time of an event – will defy any closer analysis. It is *impossible in principle to control the energy which goes into the clocks without interfering with their precision as time indicators.* We must allow for a possible discrepancy in the energy balance corresponding to the Heisenberg Uncertainty Principle for the conjugate time and energy variables. The *complementary* relation between a detailed time account of atomic processes and a precise definition of energy transfers in atomic reactions is analogous in quantum theory to the *mutually exclusive* character of precise position and momentum determinations. [Without further detail on the energy-time uncertainty relation, Bohr returns to the general discussion.]

The necessity of distinguishing between the object to be investigated and the measuring instruments is a [Bohr might better have said: The separability of the object ··· *from* the measuring instruments is the ···] *principal distinction between classical and quantum descriptions of physical phenomena.* It is true that the place in each measurement *where* this distinction [separation] is made is largely a matter of convenience [judgement]. In classical physics the distinction [separation] between the object and the measuring apparatus does not require any difference in the character of their descriptions. In contrast, quantum theory is based on the indispensable use of classical concepts for the interpretation of all measurements, even though the classical theories do not account for the new types of [quantum-] phenomena which occur in atomic physics. There is no question possibility] of any unambiguous interpretation of quantum mechanics other than the well-known rules which predict the results of a given experiment in a totally classical way, based on conservation laws. A proper correspondence with the classical theory through these conservation laws excludes any inconsistency in the quantum description, due to a change in the place where the distinction [separation] is made between object and apparatus.

Obviously, in any experiment we have a free choice of this place *only* where the quantum and classical descriptions coincide.

I must emphasize the great lesson learned from general relativity theory on the nature of physical [c-]reality in quantum theory. The situations in these two generalizations of classical theory are quite analogous. The unique role of measuring instruments in quantum mechanics is comparable to the necessity in general relativity of an ordinary [non-relativistic, even static] description of all measuring processes, including a classical distinction between space and time coordinates, even though the very essence of the theory is the establishment of new laws with no such distinction [6]. The dependence on the reference system in relativity theory, of all readings of scales and clocks can be compared with the essentially uncontrollable exchange of momentum or energy between the objects of measurements and all instruments defining the space-time system of reference. In quantum theory this confronts us with the situation described by complementarity. This new feature of quantum physics requires a radical revision of our concept of 'physical reality': [from the naive, precise and exhaustively defined, unique and *pre-existing* [c-]reality of EPR, to a more subtle, less restrictive [q-]reality which is made [q-]real only by the choice of (and performance of) a measurement; and is only a pre-existing [q-]*potentiality* waiting to be [q-]realized by a measurement, but not a pre-existing [c-]*reality*]. There is a close parallel with the fundamental modification of ideas about the absolute character of physical phenomena in the general theory of relativity. [Bohr was obviously pandering to Einstein here.]

Footnotes and References:

1) A. Einstein, B. Podolsky, and N. Rosen, Phys. Rev. **47**, 777 (1935).

2) Cf. N. Bohr, *Atomic Theory and The Description of Nature* (Cambridge 1934).

3) The conclusions in this article are a direct consequence of the transformation theorems of quantum mechanics, which more than any other feature of the formalism establish its mathematical completeness and its logical correspondence with classical mechanics. It is always possible when discussing a mechanical system consisting of two parts (1) and (2), whether interacting or not, to replace any two pairs of conjugate variables $(q_1 p_1)$, $(q_2 p_2)$ of systems (1) and (2) which satisfy canonical commutation relations

$$[q_1, p_1] = [q_2, p_2] = i\hbar; \quad [q_1, q_2] = [p_1, p_2] = [q_1, p_2] = [q_2, p_1] = 0,$$

by two pairs of new conjugate variables $(Q_1 P_1)$, $(Q_2 P_2)$ related to the first variables by a simple orthogonal transformation, corresponding to a rotation of angle θ in the planes

(q_1, q_2), (p_1, p_2)

$$q_1 = Q_1 \cos\theta - Q_2 \sin\theta \qquad p_1 = P_1 \cos\theta - P_2 \sin\theta$$
$$q_2 = Q_1 \sin\theta + Q_2 \cos\theta \qquad p_2 = P_1 \sin\theta + P_2 \cos\theta.$$

These variables satisfy $[Q_1, P_1] = i\hbar$, $[Q_1, P_2] = 0$. It follows that for the state of the combined system definite exact numerical values cannot be assigned to both Q_1 and P_1, but can to both Q_1 and P_2. From the expression of these variables in terms of (q_1, p_1) and (q_2, p_2),

$$Q_1 = q_1 \cos\theta + q_2 \sin\theta, \qquad P_2 = -p_1 \sin\theta + p_2 \cos\theta,$$

it follows that a subsequent measurement of either q_2 or p_2 will allow us to predict the value of q_1 or p_1 respectively.

4) The obvious impossibility of actually carrying out such measuring procedures as are discussed here and in the following does not affect the theoretical argument, since the procedures in question are essentially equivalent to atomic processes like the Compton effect, where a corresponding application of conservation of momentum is well established.

5) This description corresponds to the transformation of footnote [3] with $\theta = -\pi/4$. The wave function Eqn9 of EPR corresponds to $P_2 = 0$ and the limiting case of two infinitely narrow slits.

6) This fact, together with the relativistic invariance of the uncertainty relations of quantum mechanics, ensures the compatibility of the arguments of this article with relativity theory. The question will be explored further in another paper, where I will discuss a very interesting paradox due to Einstein on the use of gravitation theory for energy measurements. The solution of the paradox is an instructive illustration of the generality of the concept of complementarity. A more thorough discussion of space-time measurements in quantum theory will also be given with mathematical and experimental details which had to be left out of this article, where the emphasis is on the dialectic aspects of the question. [In this regard, see Ch17 for DeWitt's remarkable self-consistency requirements in application of the Wheeler-DeWitt Schrödinger equation for canonically quantized space-time.]

Paper XIII·3: Excerpt from Niels Bohr, *Atomic Physics and Human Knowledge,* (Wiley, NY, 1958) Pp.32-66.

DISCUSSION WITH EINSTEIN ON EPISTEMOLOGICAL PROBLEMS IN ATOMIC PHYSICS†
– 1949 –

NIELS BOHR

To honor Albert Einstein and acknowledge our indebtedness for the guidance his genius has given us, I thought the best way would be to explain how much I owe him. The many occasions discussing atomic physics come vividly to mind and I felt that I should give an account of these discussions, which have been of greatest value to me. I hope the account might convey how essential the open-minded exchange of ideas has been for progress in a field where new experience has time after time demanded a reconsideration of our views.

(1) The main debate has been the departure from classical principles of natural philosophy made necessary by the novel development of physics begun by Planck's discovery of the universal quantum of action $\hbar$. This discovery of the fundamental feature of atomicity of nature, has taught us that classical physics is an idealization which applies only in the limit where all actions are large compared to the quantum $\hbar$. The question has been whether the loss of predictability [Read: unique determinacy] in the description of atomic processes is reparable, or whether we are faced with an irrevocable change in the fundamental analysis of [quantum-]physical phenomena. To describe our discussions and the arguments for the contrasting viewpoints, I have reviewed developments to which Einstein himself contributed so decisively.

(2) Planck's analysis of thermal radiation led him to his fundamental discovery. Planck's considerations were statistical and he made no conclusions about the impact of the quantum on mechanics and electrodynamics. Einstein's original contribution to quantum theory was the recognition of how physical phenomena depend directly on individual quantum effects [1]. By exploring the novel features of atomicity which point beyond classical physics, he was led to the conclusion that each radiation process involves the emission of individual light-quanta or 'photons'

with energy and momentum $E = \hbar\omega$ and $P = \hbar k$, where $\omega = 2\pi f$ and $k = 2\pi/\lambda$ in terms of the frequency f and wavelength λ of the corresponding wave. The idea of the photon led to a dilemma, since any *particle picture* of radiation would conflict with interference effects which require a *wave picture*. The dilemma is made worse by the fact that interference effects are our only way of defining frequency and wavelength in the very expressions for the energy and momentum of the photon.

(3) Furthermore, there could be no predictive analysis [Read: $\cdots$ no unique, precise, continuous, classical, space-time description] of radiative processes, but only an estimate of the probability for the occurrence of the individual [discontinuous quantum-]process. Here the use of probability is essentially different from the familiar statistical analysis of systems of great complexity. We face not complexity, but the inability of classical concepts to describe the peculiar feature of indivisibility, or 'individuality', characterizing the elementary [and fundamental primitive quantum emission or absorption] processes.

(4) $\cdots$ Rutherford's atomic nucleus revealed the inadequacy of classical concepts to explain the inherent stability of the atom. Here the quantum theory offered a clue $\cdots$ any reaction of the atom resulting in a change of its energy involved a complete transition between two stationary quantum states $\cdots$ by a step-like process $\cdots$ each transition is accompanied by the emission of a photon of the energy given by Einstein. $\cdots$ an atom in an excited state will have the possibility of transitions with photon emission to one or another of its lower states. $\cdots$ the choice between such transitions leaves only the notion of the relative probabilities of the individual transition processes. $\cdots\cdots$ [2].

(5) Einstein himself showed that Planck's law for thermal radiation followed from assumptions on the basic ideas of the quantum theory of the atom [3]. He formulated statistical laws for the radiative transitions, assuming that (i) when the atom is exposed to an external radiation field, absorption and emission of photons will occur with a probability proportional to the intensity of the radiation, and (ii) – even in the absence of the external field - *spontaneous* emission processes take place with a certain *a priori* probability. Einstein emphasized the fundamental statistical character of the description by the close analogy between the spontaneous radiative

transitions and the transformations of radioactive substances.

(6) Einstein further pointed out that the radiation was 'directed' in the sense that not only is a momentum corresponding to a photon with the direction of propagation transferred to the atom in the absorption process, but also the emitting atom will get an equivalent momentum in the opposite direction, even though there can be no question in the wave picture of any preference for a single direction of emission. Einstein's own attitude was: "These features of the elementary processes ··· make the development of a proper treatment of radiation ··· unavoidable. ······ no closer connection with the wave concepts is obtainable and ··· it leaves to chance the time and direction of the elementary processes ···."

(7) Similar paradoxes were raised by Compton's discovery of the change in wave-length accompanying the scattering of X-rays by electrons. This provided a direct proof of Einstein's view regarding the transfer of energy and momentum in radiative processes [4,5,6].

(8) ··· The clarification of the situation was begun by the development of a more comprehensive quantum theory; first by de Broglie's recognition that the wave-particle duality was not restricted to radiation, but was equally unavoidable for material particles. ······ The paradoxical aspects of quantum theory were even more emphasized by the apparent contradiction between the wave description and the discreteness of particles with mass and charge.

(9) ··· Quantum mechanics provided an essentially statistical description of atomic processes combining features of discreteness and of superposition, equally characteristic of quantum theory. ··· There could be no doubt of the adequacy of the quantum-mechanical formalism, but its abstract character gave rise to a widespread feeling of uneasiness. An elucidation of the situation demanded a thorough examination of the very observational problem in atomic physics.

(10) This was begun by Heisenberg [7], who pointed out that the knowledge of the state of an atom [Read: any quantum object] will always involve a fundamental 'indeterminacy'. ··· Any measurement of the position of an electron must use some device, like a microscope, scattering short wavelength radiation with a

corresponding momentum transfer which is larger the more accurate the position measurement. Heisenberg pointed out that the fundamental quantum-mechanical commutation relation $qp - pq = i\hbar$ imposes a limit on the on the precision with which conjugate variables q and p can be specified, $\Delta q \Delta p \sim \hbar$, where Δq and Δp are inevitable quantum uncertainties in these variables. This 'Heisenberg Uncertainty Principle' is fundamental for understanding the paradoxes that occur in analyzing quantum effects with classical physical pictures.

(11) At the same time, I advocated a point of view [8] called 'complementarity', to describe quantum phenomena and to clarify the peculiar observational problems in this field. It is decisive to recognize that *however far the phenomena transcend the scope of classical physical explanation, the ultimate account of all evidence must be expressed in classical terms.* The crucial point is *the impossibility of any sharp separation between the behavior of the quantum object and the interaction with the measuring instruments which define the phenomena.*

(12) An essential ambiguity is involved in ascribing classical physical attributes to quantum objects, as in the dilemma between the particle and wave properties of electrons and photons, where we have contrasting [and mutually exclusive] pictures referring to essentially different aspects of empirical evidence [Read: $\cdots$ of experimental measurements]. An example where complementary aspects appear is the Compton effect. Any experiment designed to study the exchange of momentum and energy between electron and photon must involve an uncertainty in the space-time coordinates of the collision sufficient for the determination of the wave-number and frequency. Conversely, any experiment to locate the collision more accurately, would [according to the HUP] preclude any precise account of momentum and energy. In quantum theory the uncontrollable interaction between the quantum object and the measuring instrument makes impossible a predictive [precise, classically complete] description [of a pre-existing [c-]reality *independent* of the *choice* of which measurement is actually made]. This is not *simply* a limitation of the scope of the quantum-mechanical description, but the whole argument of complementarity must be regarded as [an essential part of] the logical quantum generalization of the classical ideal of predictability [This word again! Read: $\cdots$ of the classical ideal of the naive [c-]reality to a new and subtle [q-]reality.]

(13) $\cdots$ Einstein raised the simple example of an electron passing through a narrow slit in a diaphragm placed in front of a photographic plate. Because of the diffraction of the incident wave, it is not possible to predict the point where the electron will hit the plate, but only to calculate the probability that it will hit in any given area. The difficulty, which Einstein felt so acutely, is that – if the electron is recorded at point A, then there is no effect at point B – although the laws of ordinary wave propagation suggest a correlation between two such events. Einstein's attitude gave rise to intense discussions. The situation is not [completely] analogous to the application of probability to complicated systems, but rather it recalls Einstein's own early conclusions about the unidirection of individual photons which contrast so strongly with the simple wave picture [of, say, the $\sin^2 \theta$ angular distribution of classical dipole radiation patterns which *do* have a simple interpretation in terms of a statistical ensemble]. The discussions centered on whether quantum mechanics [c-]completely described the possibilities or, as Einstein maintained, whether a fuller description could be obtained from the balance of energy and momentum for individual processes. [Einstein did not say where a fuller description was to come from, only that the quantum description could not be [c-]complete.]

(14) To understand Einstein's arguments, consider the momentum and energy balance together with the location of the quantum particle in space and time. The simple case of a particle passing through a hole in a diaphragm which can be opened or closed with a shutter, illustrates the effect of the Heisenberg Uncertainty Principle. The state of the particle after the diaphragm is a spherical wave of angular opening θ and, in case 2b, of limited radial extent. This state involves a range Δp in the particle momentum parallel to the diaphragm and, in case 2b, a range ΔE in the kinetic energy.

(15) The uncertainty Δq in the location of the particle (in the plane of the diaphragm) is the radius of the hole a; the angle of diffraction $\theta \sim \lambda/a$ so the uncertainty of the momentum conjugate to q is $\Delta p \sim \theta P \sim \hbar/\Delta q$, in agreement with the Heisenberg Uncertainty Principle. This result can also be obtained from the limited extent of the wave $\Delta q \sim a$, which requires the superposition of wave-numbers with range $\Delta k \sim 1/\Delta q$. Similarly, the spread of frequencies in the limited

wave-train of $2b$ must be $\Delta f \sim 1/\Delta t$, where Δt is the time the shutter was open. Therefore the energy uncertainty of the particle localized in the time interval Δt is $\Delta E \Delta t \sim \hbar$, again in accord with the Heisenberg Uncertainty Principle for the conjugate variables E and t.

(16) The uncertainties in the state of the quantum particle can be traced to the momentum and energy exchange with the diaphragm or the shutter. The shutter must move with a speed $v \sim a/\Delta t$, and in a collision with a particle with parallel momentum Δp will exchange an energy with the particle

$$\Delta E \sim v\Delta p \sim \frac{\Delta q \Delta p}{\Delta t} \sim \frac{\hbar}{\Delta t},$$

allowing momentum and energy conservation between the quantum particle and the apparatus. If we want to know the momentum and energy of the measuring apparatus with sufficient accuracy to know the momentum and energy exchanged with the quantum particle, then the apparatus too must be subject to the Heisenberg Uncertainty Principle, and we lose *its* precise location in space and time. This crucial point clearly brings out the complementary (i.e., mutually exclusive) character of the simultaneous determination of conjugate variables.

(17) It is the assumption that the apparatus has well defined position that makes possible, within the limitations of the Heisenberg Uncertainty Principle, the prediction of where the particle will be recorded. If we allowed a large uncertainty in the position of the diaphragm, it would be possible know the momentum transfer to the diaphragm and to predict the direction of the particle from the hole to the plate. We now deal with the two-body system of diaphragm and particle, analogous to the Compton effect where observation of the recoiling electron allows us to know the direction of the scattered photon. These important considerations were clarified by an apparatus involving another diaphragm with two slits. A beam of quantum particles from the left leads finally to a two-slit interference pattern on the photographic plate. This pattern is the accumulation of a large number of individual particle tracks on the plate, and the distribution of these tracks follows a simple law derivable from the wave theory. The momentum transferred to the first diaphragm must be different if the particle passed through the upper slit in the second diaphragm. Einstein suggested that knowing the momentum transfer would

make it possible to know through which of the two slits the electron had passed before it arrived at the plate, making the two-slit interference pattern paradoxical.

(18) However, a closer examination shows that the measurement of the momentum transfer requires an uncertainty in the position of the first diaphragm which would destroy the interference pattern. The momentum difference between the two paths is $\Delta p \sim P\theta$ so the uncertainty in position of the first slit must be $\Delta q \sim \hbar/\Delta p \sim 1/k\theta$, with k the de Broglie wave-number of the particle. The phase uncertainty between paths 1 and 2 due to the displacement Δq is

$$\Delta \phi \sim k\Delta L \sim 2k\theta\Delta q \sim 2,$$

which obliterates the two-slit interference pattern.

(19) This point is of great logical significance: we have a choice of *either* observing the path of the quantum particle *or* observing the interference effects. This resolves the paradox that the behavior of the quantum particle might appear to depend on the presence of a slit through which it could be proved not to pass. We have here a typical example of the *complementarity* of *mutually exclusive* experiments and the *impossibility* of *separating* the behavior of *quantum objects* from their interaction with the *measuring instruments* which <u>define</u> that behavior.

(20) The attitudes taken in this novel situation touched naturally on many aspects of philosophical thinking, but, in spite of all divergences of approach and opinion, a most humorous spirit animated the discussions. Einstein mockingly asked us if we could really believe that God played dice, to which I replied by pointing out the great caution called for in ascribing attributes to providence in every day language. Ehrenfest, in his affectionate manner of teasing his friends, jokingly hinted at the similarity between Einstein's attitude and that of the opponents of relativity theory; but instantly Ehrenfest added that he would not be able to find relief in his own mind before concord with Einstein was reached.

(21) Einstein's criticism provided incentive for us all to reexamine various aspects of the description of quantum phenomena. The main point is the distinction between [Read: inseparability of] *quantum objects* under investigation and from] the *measuring instruments* which define the phenomena in classical terms. It is deci-

sive that, in contrast to the measuring instruments proper, macroscopic objects (e.g., slits, shutters) together with the quantum particles constitute the system to which the quantum-mechanical formalism must be applied. It is essential that the *whole experimental arrangement* be taken into account.

(22) The extent to which the visualization of quantum phenomena must be abandoned because of the impossibility of their subdivision [separation], is illustrated by an example due to Einstein: A semi-reflecting mirror is placed in the path of a photon, leaving the possibilities of transmission or reflection. Two separate types of experiment can be performed:

i) The photon can be recorded on one, and only one, of two photographic plates in the reflected and transmitted paths; or

ii) with mirrors instead of plates, interference effects between the two wave-trains can be observed.

Any pictorial representation of the photon meets with the difficulty: on the one hand

i') the photon always chooses *one* of the two ways; but on the other hand

ii') it behaves as if it had passed *both* ways.

It is arguments of this kind which emphasize the impossibility of subdividing separating quantum phenomena and reveal the ambiguity in ascribing classical physical attributes to quantum objects. It must be kept in mind that the only unambiguous use of space-time concepts in the description of quantum phenomena is confined to the recording of observations on a photographic plate or some similar irreversible amplification effect. Of course the quantum of action is ultimately responsible ······ but this is not relevant here.

(23) A discussion arose about how to speak of phenomena for which only statistical predictions can be made. The question was whether, for individual events, we should follow Dirac and speak of a choice on the part of 'nature', or Heisenberg, and speak of a choice on the part of the 'observer'. Such terminology seems dubious since it is not reasonable to endow nature with volition in any ordinary sense, nor is it possible for the observer to influence the events. To my mind, there is no alternative except to admit that in this realm of experience, the possible measurements on individual phenomena only allow us to study complementary

aspects.

(24) At the 1930 Solvay Conference, our discussions with Einstein took a dramatic turn. He objected to the view that control of momentum and energy exchange between objects and measuring instruments was invariably limited by the Heisenberg Uncertainty Principle. Einstein argued that control was possible when the requirements of relativity were considered. In particular, the relation $E = mc^2$, should allow us to measure the total energy of any system simply by weighing it, and in principle to know the energy transferred when it interacts with a quantum object.

(25) Einstein proposed a box with a hole in it, which could be opened or closed by a shutter moved by a clock inside the box. In the beginning, the box contains a certain amount of radiation. The clock is set to open the shutter for a very short interval at a precisely known time, and a single photon is released. It is also possible to weigh the whole box before and after, and thereby determine the energy of the photon with arbitrary precision, in apparent contradiction to the energy-time Heisenberg Uncertainty Principle of quantum mechanics. This argument was a serious challenge and required a thorough re-examination of the whole measurement problem. The outcome of the discussion, however, made it clear that the argument failed to circumvent the requirements of quantum theory. To resolve the paradox, it was necessary to include the identity of inertial and gravitational mass, and also the dependence of the rate of a clock on its position in a gravitational field – following from Einstein's Principle of Equivalence between gravity and accelerated reference frames.

(26) Consider an apparatus with a box suspended on a spring-balance. The box can be weighed with any given accuracy Δm by adjusting its zero position. The essential point is that an accuracy Δq in this position involves a corresponding uncertainty $\Delta p \sim \hbar/\Delta q$ in the control of the momentum of the box. This uncertainty Δp must be small compared to the total momentum imparted by the gravitational field g to a body of mass Δm during the time interval T of the balancing procedure, $\Delta p \sim \hbar/\Delta q < T \cdot g \cdot \Delta m$. The greater the accuracy of the reading q, the longer the measuring time T for a given accuracy Δm.

(27) According to general relativity, a clock displaced in a gravitational field g by an amount Δq, will change its reading in a time interval T by an amount ΔT

$$\frac{\Delta T}{T} = \frac{1}{c^2} g \Delta q.$$

After the weighing there will be a time uncertainty $\Delta T > \hbar/(c^2 \Delta m)$, consistent with the energy-time Heisenberg Uncertainty Principle of quantum theory. Consequently, use of the apparatus to accurately measure the energy of the photon prevents us from accurately controlling the time of its escape.

(28) $\cdots$ Einstein nevertheless expressed a feeling of unease over the apparent lack of specific principles for the explanation of nature, on which all could agree. I could only answer that in dealing with entirely new experiences, we could hardly trust any accustomed principles, however broad, apart from the demand of avoiding logical inconsistencies and in this respect, the mathematical formalism of quantum mechanics should surely meet all requirements.

(29) $\cdots$ Einstein's critical attitude towards quantum theory was soon brought to public attention in the paper 'Can Quantum-Mechanical Description of Physical Reality be Considered Complete?' [11] by Einstein, Podolsky and Rosen. Their argument is based on the criterion: "If, without in any way disturbing a system, we can predict with certainty the value of a physical quantity, then there exists an element of physical reality corresponding to this physical quantity." Their example considers the quantum-mechanical formalism for the state of a system consisting of two separated parts which had been in interaction. It is shown that quantities for one system – not previously specified in its representation – can be determined by a measurement performed on the other system. The authors conclude that quantum mechanics does not "provide a complete description of the physical reality," and they maintain that a more complete account of the phenomena should be possible.

(30) The EPR paper created a stir among physicists. Certainly the issue is a very subtle one, well suited to emphasize how far quantum theory is beyond the reach of pictorial visualization. I tried to show [12] from the point of view of complementarity, that the apparent inconsistencies were completely removed. The conclusion of this article was: "$\cdots$ the argument of EPR does not justify their

conclusion that the quantum-mechanical description is essentially incomplete. On the contrary, the quantum mechanical description – as expressed in the Heisenberg Uncertainty Principle – may be characterized as the logical utilization of all possibilities of unambiguous interpretation of measurements, compatible with the finite and uncontrollable interaction between the quantum objects and the measurement apparatus····."

* * * * * * * * * * * * * * * * * * **

Here we are left still uncomfortable. There is no doubt that the HUP must certainly render EPR's [c-]reality and [c-]completeness invalid, as argued from Bohr's complementary principle. But the conclusion – '··· *may* be characterized ···'??? Surely we were hoping for a definitive '···*must* be····.' But even more, we were hoping for an authoritative statement on the interpretation of quantum mechanics, perhaps the essence of ψ as an information theoretic device, or the distinction between quantum probabilities and ordinary (?) probabilities, and much more. Bohr's endless and repetitive complementarity arguments – no matter how clever or polished they might become – are not enough. The 'need of a radical change' was not in dispute. What is missing is any argument – besides its success so far – that quantum mechanics is the *end* of such radical change.

* * * * * * * * * * * * * * * * * * **

(31) Rereading these passages, I am deeply aware of the problems of communication which made it very difficult to explain the essential ambiguity involved in describing the physics of quantum objects where no sharp distinction can be made between the behavior of the objects themselves and their interaction with the measuring apparatus. I hope this account makes clear the necessity of a radical change in the basic principles of physical explanation and restores logical order in this field of experience.

(32) Einstein's own views were described at that time [13]. He argued there that the quantum mechanical description is to be considered merely a means of accounting for the average behavior of a large number of atomic systems. His

attitude to the contrary belief that it is an exhaustive description of the individual phenomena was expressed in the following words: "To believe this is logically possible without contradiction; but it is so very contrary to my scientific instinct that I cannot forgo the search for a more complete description."

(33) This attitude is a rejection of the whole argument that I have developed above, showing that in quantum mechanics we are dealing not with an arbitrary selection from a more detailed analysis of atomic phenomena, but with the recognition that such a 'deeper' analysis is *in principle* impossible. The discreteness of quantum effects is a novel situation unforeseen in classical physics and inconsistent with conventional ideas based on ordinary experience. Quantum theory calls for changes in elementary concepts characteristic of modern science.

(34) In the following years, the more philosophical aspects of the situation in atomic physics aroused the interest of ever larger circles ⋯⋯. I tried to point out the analogy between the limitation imposed on the description in atomic physics and situations met with in other fields of knowledge [14]. ⋯⋯ I had little success in convincing my listeners, for whom the debate among physicists themselves was naturally a cause of skepticism about the necessity of going so far in abandoning customary demands on the explanation of natural phenomena. ⋯⋯ In recent years I have on several occasions met with Einstein. Our continued discussions have not led to a common view about the epistemological problems in atomic physics. Our opposing views have been summarized recently [17]. ⋯

(35) The discussions with Einstein left a deep and lasting impression, and when writing this I have, so-to-speak, been arguing with Einstein. I am of course, relying on my own memory. Many of the developments of quantum theory, in which Einstein played so large a part, may appear to him in a different light. I trust, however, that I have conveyed how much it has meant to me to benefit from the inspiration which we all derived from Einstein.

* * * * * * * * * * * * * * * * * * **

These were Bohr's final opinions in the Bohr-Einstein exchange. They represent an advance in clarity over his original 1935 response; but do not fully satisfy one's hopes and still leave us without an authoritative 'last word'. Nevertheless, the essence of our present understanding seems to be here if we look hard enough:

i) the *free-choice* that makes [q-]reality depend upon the measurements actually performed and thereby precludes the pre-existing [c-]reality invoked by EPR;

ii) the half-silvered mirror 'gedanken experiments' to be to be realized 30 years later by Aspect (and to be met again in Ch14);

iii) and generally, the HUP as a quantitative expression of the essential complementarity and mutual mutual exclusiveness of the wave-particle duality; which

iv) precludes any hope of [c-]completeness of the EPR sort, and makes impossible the purely classical description of quantum events, depending as it does on perfect localization of the system-point in the phase space of classical Hamiltonian mechanics.

Disappointingly, Bohr's arguments do not go beyond complementarity to justify any specific quantum dynamics or even any meaning to be attached to the probabilities produced by it. The result is a nagging discontent that is only now being satisfied – not so much by new ideas, but mainly by the development of a more suitable vocabulary. The ideas sketched by Bohr always seemed to lack – not conviction or credibility, perhaps, but – a certain intimacy and persuasiveness as distinct from being an edict from on high. Perhaps the only solution to such feelings is what has happened – a gradual growth of familiarity leading to comfort and confidence in a quantum way of thinking which is no longer new, but which is to be embarked upon with wonder and awe.

* * * * * * * * * * * * * * * * * * **

Footnotes and References:

† – Also in *Albert Einstein: Philosopher-Scientist* (The Library of Living Philosophers, Evanston,Il, 1949) Vol7, p.199.

1)A. Einstein, Ann. Phys. **17**, 132 (1905) (here as **Paper II·1**)

2) N. Bohr, *The Theory of Spectra and Atomic Constitution*, (Cambridge, 1922).

3) A. Einstein, Zeits. f. Phys. **18**, 121 (1917) (here as **Paper II·3,3a**).

4) A. Einstein and P. Ehrenfest, Zeits. f. Phys. **11**, 31 (1922).

5) N. Bohr, H.A. Kramers and J.C. Slater, Phil. Mag. **47**, 785 (1924) (here as **Paper III·2**).

6) A. Einstein, Berl. Ber., 221 (1924); 3,8 (1925).

7) W. Heisenberg, Zeits. f. Phys. **43**, 172 (1927) (here as **Paper XI·1**).

8) N. Bohr, Nature **121**, 580 (1928) (here as **Paper XII·1**).

11) A. Einstein, B. Podolsky and N. Rosen, Phys. Rev. **47**, 777 (1935) (here as **Paper XIII·1**).

12) N. Bohr, Phys. Rev. **48**, 696 (1935) (here as **Paper XIII·2**).

13) A. Einstein, J. Franklin Inst. **221**, 349 (1936).

14) N. Bohr, Philosophy of Science **4**, 289 (1937).

17) N. Bohr, Dialectica **1**, 312 (1948).

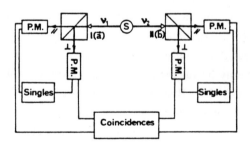

Schematic of Aspect's experimental setup to test Bell's inequality (See our ChXIV, p.459). Polarimeters $I(\vec{a})$ and $II(\vec{b})$ select linear polarizations $\vec{a}$ and $\vec{b}$ on photons ν_1 and ν_2 from EPR-source S. [From A. Aspect, P. Grangier, and G. Roger, Phys. Rev. Lett. **49**, 91 (1981).]

PART FOUR. Chapter XIV

BOHM and BELL,

CLAUSER and ASPECT:

"EPR argued that quantum mechanics $\cdots$ should be supplemented by additional variables to restore causality and locality. $\cdots$ That idea is formulated mathematically, and is shown to be incompatible with quantum mechanics."

J.S. Bell, *On the Einstein-Podolsky-Rosen Paradox*, Physics 1-3, 195 (1964).

§ XIV-1. Introduction.

EPR [1] knew the resolution of their paradox: "$\cdots$ one would not arrive at our conclusion if one insisted that two $\cdots$ physical quantities can be regarded as simultaneous elements of [q-]reality *only when they can be measured simultaneously* $\cdots$ No sensible definition of [c-]reality could be expected to permit this." And Bohr [2] – in his own peculiar way – along with others [3], tried to set dismiss it. But Bohm [4] kept the paradox alive in the form of a theoretically *and experimentally* accessible spin-correlation measurement, and with Aharanov suggested the first experiment [5,6] to demonstrate the presence of the mysterious anti-intuitive long-range quantum correlations. Bell [7] proved the purely quantum nature of the correlations – as distinct from those of any underlying deterministic local hidden variable theory – introducing the criterion of 'Bell's Inequality'. The whole was made real by Aspect's observation [8] of the violation of the Bell's inequality, verifying the presence of the mysterious and paradoxical quantum correlations which are quite literally wonderful, that is to say full-of-wonder, to be wondered at, and about.

Bohm introduced the now familiar correlations of two spin-1/2 particles in a spin-0 singlet state with the spin wave function

$$\psi(1,2) = \{\alpha(1)\beta(2) - \beta(1)\alpha(2)\}/\sqrt{2}.$$

α and β are Pauli spinors with $S_z = \pm 1/2$; the particles are assumed to fly apart leaving their spin wave function unchanged and possessing the full long-range quantum correlations expressed in $\psi(1,2)$. This is the prototype *entangled-* or *EPR-state* fundamental to Quantum Non-Demolition or QND-measurements. This ability to perform 'reversible quantum measurements' (really a *comparison* of quantum states) has made possible teleportation of quantum information and is a fundamental ingredient in proposed quantum computation scenarios.

The orientation of the $\hat{z}$-axis can be chosen at will. For a $\hat{z}'$-axis rotated at an angle θ with respect to $\hat{z}$, the rotated spinors are

$$\alpha' = \alpha\cos(\theta/2) - \beta\sin(\theta/2) \text{ and } \beta' = \alpha\sin(\theta/2) + \beta\cos(\theta/2),$$

leaving the spin-0 wave function invariant in form

$$\psi(1,2) = \{\alpha'(1)\beta'(2) - \beta'(1)\alpha'(2)\}/\sqrt{2}.$$

The EPR-paradox appears in the various spin measurements that can be made on particle-1 and the now far-separated particle-2:

i) A measurement of $S_z(1) = +1/2$ will coincide identically with a subsequent measurement of $S_z(2) = -1/2$. The $\hat{z}$-components of the two spins are perfectly anti-correlated, as they would be classically for overall spin-0. Furthermore

ii) the reduced wave function $\alpha(1)\beta(2)$ after the measurement corresponds to no correlations of the other two components of the spins. But because of the form invariance of the wave function under rotations,

iii) the same perfect anti-correlation must be found for a measurement of the spin components along any axis, in particular along the $\hat{x}$- and $\hat{y}$-axes. Now EPR argued that the spin measurement on particle-1 was done "without in any way disturbing" the spin of particle-2; so particle-2 must have had these spin components (whether $\pm 1/2$ we don't know) already – as EPR "elements of reality". EPR are forced to conclude that the wave function – which contains no hint of such elements of reality – must be an incomplete description of reality.

Bohm's experiment makes clear that EPR's concept of [q-]reality was over-stated. EPR's resolution of their own paradox is the correct one. The physical origins of the limitations imposed by the quantum theory can be expressed in alternative ways:

i) in the Heisenberg matrix mechanics, through the non-commutation of certain dynamical variables, leading to the Heisenberg Uncertainty Principle; and equivalently

ii) in the Schrödinger wave mechanics through the realization of the algebra of the dynamical variables in the wave function, which puts the appropriate consistent limitations on the actual evaluation of these variables. These two perspectives are united in an intuitive and powerful way – essential for understanding –

iii) in Bohr's complementarity principle. In a final analysis though,

iv) we must give up any naive understanding of the wave function based on semi-classical models of quantum events and ultimately simply

v) accept the [q-]reality of quantum mechanics as an information-theoretic device which propagates in-states (input information) – in a way consistent with the requirements of the non-commutative nature of quantum dynamics – to out-states (output information).

Bohm and Aharanov point out that quantum correlations similar to those for spin-1/2 particles occur between the polarization vectors of the two opposite-going photons from the annihilation of singlet positronium: $e^+e^- (J = 0^-) \rightarrow \gamma_1\gamma_2$. The same holds true for the two photons from the decay of the pseudoscalar neutral pi-meson: $\pi^0(J = 0^-) \rightarrow \gamma_1\gamma_2$. The symmetric pseudoscalar wave function for the two photons with momenta $k^z_{1,2}$ along the $\hat{z}$-axis is

$$\psi(1,2) = (k^z_1 - k^z_2) \{\epsilon^x_1 \epsilon^y_2 - \epsilon^y_1 \epsilon^x_2\}/\sqrt{2}$$

corresponding to mutually perpendicular polarizations $\hat{\epsilon}_1, \hat{\epsilon}_2$. The familiar EPR-paradox arises when photon-1 is found to be polarized in the $\hat{x}$-direction, say, and then the far-removed photon-2 is known to be polarized in the $\hat{y}$-direction. Bohm and Aharanov's analysis of the angular correlations of the rescattering planes of such decay photon pairs (already measured by Wu [6]) supported the essential integrity of the quantum correlations even at macroscopic distances where they might have been thought somehow to have lost their validity.

§ XIV-2. Bell's Inequality for Bohm's Experiment.

Bell's profound accomplishment [7,9] was to construct a test which *distinguished* in a rigorous and quantitative way the quantum correlations present in a Bohm-type experiment from those predicted by deterministic classical local hidden variable (CLHV-)theories. Bell went beyond the consideration of spins $S_z(1) = +1 \Rightarrow S_z(2) = -1$ in opposite directions and analyzed the correlations for arbitrary orientations in which the spin of particle-2 is measured. He found a restriction on the possible CLHV-correlations [11] which is not obeyed by the correlations of quantum mechanics, and which is clearly violated in Aspect's subsequent experimental results, ruling out this whole class of CLHV-alternatives to quantum theory. Bohm, Bell, and Aspect combined to *experimentally reject* EPR's criterion of [q-]reality. The quantum correlations *do exist* above and beyond any correlations possible with CLHV-theories.

Bell imitated the predictions of quantum mechanics using an underlying classical deterministic theory with hidden variables λ of a quite general sort. The hidden variables are distributed over a range of λ with a probability $\rho(\lambda) \geq 0$ normalized so $\int \rho(\lambda)d\lambda = 1$ in some general sense of integration. The observables of quantum mechanics are assumed to be the sum over this probability distribution of all the classical unique deterministic results for each value λ of the unobserved hidden variables.

Bell considers observables modeled after the possible values of the spin component along the direction $\hat{a}$, $S_a = \pm 1$, so

$$A(\hat{a}, \lambda) = \pm 1 = -A(-\hat{a}, \lambda) \text{ and } B(\hat{b}, \lambda) = \pm 1 = -B(-\hat{b}, \lambda)$$

are the possible values of A along $\hat{a}$ and B along $\hat{b}$ for any λ. The expected value of the product will be a classical weighted average

$$P(\hat{a}, \hat{b}) \equiv \int A(\hat{a}, \lambda)B(\hat{b}, \lambda)\rho(\lambda)d\lambda.$$

Bohm's experiment measures spins from the breakup of a spin-0 state, so with certainty

$$P(\hat{a}, -\hat{a}) \equiv 1 = \int A(\hat{a}, \lambda) B(-\hat{a}, \lambda) \rho(\lambda) d\lambda.$$

To be consistent with the normalization condition – because A and B can only be ± 1 – this requires that

$$B(-\hat{a}, \lambda) = A(\hat{a}, \lambda) = -A(-\hat{a}, \lambda).$$

So the hidden variable spin correlation – normalized at the particular configuration $\hat{b} = -\hat{a}$ considered by Bohm – is

$$P(\hat{a}, \hat{b}) = -\int A(\hat{a}, \lambda) A(\hat{b}, \lambda) \rho(\lambda) d\lambda.$$

Bell constructs a triangle inequality from

$$P_{ac} - P_{cb} = -\int \rho d\lambda \left\{ A_a A_c - A_c A_b \right\}$$

(in an obvious abbreviated notation). Taking the absolute magnitudes

$$|P_{ac} - P_{cb}| = |\int \rho d\lambda A_a A_c \left\{ 1 - \frac{A_c A_b}{A_a A_c} \right\}| = |\int \rho d\lambda A_a A_c \left\{ 1 - A_a A_b \right\}|,$$

which follows because $A = \pm 1$ only, so $A_c A_b / A_a A_c = A_a A_b$. Further, since $\rho d\lambda$ and $1 - A_a A_b$ are both ≥ 0, and $|A_a A_c| = 1$, taking the absolute value signs inside the integral gives a Schwartz inequality for the hidden variable correlation function

$$|P_{ac} - P_{cb}| \leq \int \rho d\lambda \left\{ 1 - A_a A_b \right\} \leq 1 + P_{ab}.$$

Bell shows that the quantum correlation function [10] $P_{ab}^{QM} = -\hat{a} \cdot \hat{b}$ can violate the inequality by taking θ_{ab} small but for arbitrary direction of $\hat{c}$. Then the left side of the inequality $|\hat{a} \cdot \hat{c} - \hat{c} \cdot \hat{b}| \simeq |\sin \theta_{ac}||\theta_{ab}|$ should be less than the right side $1 - \cos \theta_{ab} \simeq \theta_{ab}^2 / 2$. For sufficiently small θ_{ab} the inequality cannot hold. So the Bell inequality – true for any local [11,12] probability based theory – is violated by quantum mechanics. Bell further shows that the difference between the quantum and the hidden variable correlations cannot be made arbitrarily small even on average over some small finite resolution.

§ XIV-3. Aspect's Experiment.

Clauser [13] generalized Bell's theorem to describe a realistic version of Bohm's experiment with finite resolving power and detector efficiencies. The primary experiment – after a number of pioneering precursors, all of which were considered somewhat equivocal – is that of Aspect [8,14]. This experiment has been repeated many times in many variants, using photon correlations over ever greater distances (10 km; with 9σ precision), with ever brighter sources (with 242σ precision in ≤ 3-minutes), and with rapid switching (precluding any communication between the detectors at signal velocities $v \leq c$) [15]. It is still interesting to push the limits for spin-1/2 particles.

The Bell inequality

$$|P(a,b) - P(a,b')| \leq \int_\Gamma \rho d\lambda |A_a B_b - A_a B_{b'}| = \int_\Gamma \rho d\lambda \left(1 - B_b B_{b'}\right),$$

is relaxed by Clauser to require only a less-than-perfect correlation

$$P(a',b) = 1 - \delta, \quad 0 \leq \delta \leq 1 \quad \text{for some} \quad (a',b).$$

The parameter space can be split into $\Gamma_\pm$ where $A(a',\lambda)B(b,\lambda) = \pm 1$ and

$$\int_{\Gamma_-} \rho d\lambda = \frac{1}{2}\delta.$$

Then (in a further abbreviated notation)

$$\int_\Gamma B_b B_{b'} = \int_\Gamma A_{a'} B_{b'} - 2 \int_{\Gamma_-} A_{a'} B_{b'} \geq P(a',b') - \delta = P(a',b') + P(a',b) - 1,$$

and Bell's inequality for data nowhere perfectly correlated becomes

$$|P(a,b) - P(a,b')| \leq 2 - P(a',b) - P(a',b').$$

For the experiment considered, the data depends only on the difference of the orientations – i.e., a, b – of the two linear polarizers.

Clauser takes as his binary data the probabilities $A(a), B(b) = +1$ or -1 to be *emergence* or *non-emergence* probabilities from the respective filters. With the assumption that subsequent joint detection is independent of a, b, the coincidence rate $R(a, b)$ will be proportional to the probability that $A(a) = B(b) = +1$ defined to be w_{++}. With R_0 the detection rate without the polarizers, we specify the various rates

$$\frac{R(a, b)}{R_0} = w_{++}; \quad \text{and} \quad \frac{R_1}{R_0} = w_{++} + w_{+-}, \quad \frac{R_2}{R_0} = w_{++} + w_{-+}$$

where $w_{++} + w_{+-} + w_{-+} + w_{--} = 1$. The quantity satisfying the generalized Bell inequality is

$$\begin{aligned} P(a, b) &= w_{++} - w_{-+} - w_{+-} + w_{--} = \int A_a B_b \rho d\lambda \\ &= 4\frac{R(a, b)}{R_0} - 2\frac{R_1}{R_0} - 2\frac{R_2}{R_0} + 1. \end{aligned}$$

Here the counting rates R_1 and R_2 will be assumed to be constant independent of the orientation of the single polarizer not summed over in their definition.

Finally, the generalized Bell inequality is expressed as a linear, symmetric, and homogeneous function of the experimental counting rates which remains valid in the presence of experimental asymmetries and inefficiencies

$$\left| \frac{R(a, b)}{R_0} - \frac{R(a, b')}{R_0} \right| + \frac{R(a', b)}{R_0} + \frac{R(a', b')}{R_0} - \frac{R_1}{R_0} - \frac{R_2}{R_0} \leq 0.$$

Clauser analyzes the $6^1 S_0 \to 4^1 P_1 \to 4^1 S_0$ cascade in calcium [16] for which nominal values are [17]

$$\frac{R(\phi)}{R_0} = \frac{1}{4}(1 + \cos 2\phi); \quad \text{and} \quad \frac{R_1}{R} = \frac{R_2}{R} = \frac{1}{2},$$

with ϕ the angle between polarizers. The maximum violation of the inequality occurs for $2\phi_{ba} = \pi/4$, $2\phi_{b'a} = 3\pi/4$, $2\phi_{b'a'} = \pi/4$, and $2\phi_{a'b} = \pi/4$ and equivalents [18]. In this case the inequality reads

$$\frac{1}{4}\left| \left(1 + \frac{\sqrt{2}}{2}\right) - \left(1 - \frac{\sqrt{2}}{2}\right) \right| + \frac{1}{4}\left(1 + \frac{\sqrt{2}}{2}\right) + \frac{1}{4}\left(1 + \frac{\sqrt{2}}{2}\right) - \frac{1}{2} - \frac{1}{2} \leq 0.$$

which would require $\sqrt{2} \leq 1$ and is a distinct violation of Bell's inequality by the quantum correlation function.

Aspect's first experiment also used the Ca-cascade but had the added capability of actually detecting events $R_{\pm\pm}$ of all types by separating orthogonal polarizations in an analyzer (which transmits 'parallel' polarization and reflects (at $90°$) the orthogonal component), followed by photomultiplier detection. The high degree of experimental symmetry in the four channels gives confidence in the suppression of systematic errors by all possible checks (e.g., $\sum R_{\pm\pm}(a,b) =$ independent of a,b). The analyzer efficiencies are $T^{\parallel}, R^{\perp} \simeq .95$ and $T^{\perp}, R^{\parallel} \simeq .01$. Photomultiplier efficiencies were $\simeq 10^{-3}$. Aspect's result renormalized to Clauser's above was $1.348 \pm .008 > 1$, and also reproduced the quantum correlation function within 1% for all analyzer angles.

In the second experiment, Aspect *et al* closed one final loophole in the logic. Instead of static analyzers, they randomly fast-switch each individual beam between two analyzers of different orientations. The orientations a,b can be changed while the photons are in flight to foreclose any conceivable causal ($v \leq c$) effect of the analyzers on the emission process in the source. The switching time is $\simeq 10$ ns, the flight time is $\simeq 40$ ns, and the cascade delay is $\simeq 5$ns, so the source and the analyzers have a spacelike separation. Count rates were reduced about a factor of ten by the more restrictive geometry required by long flight-times, and the reduced efficiency of the optical switching. Aspect's result (again renormalized) $\sqrt{2} - 1 = .414 \rightarrow (.404 \pm .08) > 0$ violates the Bell inequality by 5σ.

§ XIV-4. Modern Developments. 'Alice to Bob' Teleportation.

Bennett [19] described the use of perfect EPR-correlations to 'teleport' a quantum state $|\phi\rangle$ from Alice to Bob with perfect quantum 'fidelity' $\mathcal{F}$. Kimble [20] has achieved the first experimental teleportation with fidelity clearly in the quantum domain, using Vaidman's [21] extension of Bennett's prescription from Bohm-Bell discrete valued EPR-states to continuum EPR-states made from Glauber squeezed states of two oscillators.

Teleportation is achieved in the following steps sketched by Vaidman:
i) Initially, Alice and Bob prearrange to share an EPR-state of particles 2 and 3. Each gets an 'e-bit' of information $q_2 + q_3 = 0$ and $p_2 - p_3 = 0$ (note that this information on a 'Bell-state' [22] of *two* quantum systems is not precluded by the

Heisenberg Uncertainty Principle).

ii) Then Alice makes similar measurements on particles 1 and 2 obtaining $q_1 + q_2 = a$ and $p_1 - p_2 = b$ and sends these two c-bits of classical information to Bob.

iii) With this classical information, Bob has the correlation between particles 1 and 3: $q_3 = q_1 - a$ and $p_3 = p_1 - b$.

iv) If the initial state of particle 1 is $\phi(q_1)$, then after the measurements the state of the perfectly correlated particle 3 is

$$e^{ibq_3}\phi(q_3 + a),$$

and the exact state ϕ can be obtained by Bob using the offsets a, b.

Alice measures her offset quantum state by a non-unitary process, obtaining two c-bits of information a, b at the expense of destroying her e-bit. Alice sends her two c-bits of classical information over the airways where Bob and everyone else can detect it in its offset form. But by using his e-bit *only* Bob can remove the offset to reconstitute the state and obtain the original q-bit, the state $|\phi\rangle$ (modulo a constant phase). The net result is that $|\phi\rangle$ has been securely teleported from Alice to Bob.

The process illustrates a number of general properties of quantum information:

i) There is no way for Alice to send classical information about $|\phi\rangle$ without destroying the quantum state: q-bits cannot be 'cloned'. And even then,

ii) if $\phi(x)$ were a single magnitude and phase, they would be an Action-Angle conjugate pair of variables satisfying the Heisenberg Uncertainty Principle. They could not be simultaneously measured precisely, and would defy reconstitution. So,

iii) it takes at least a two component state to be teleported. Finally,

iv) it is worth emphasizing the limited sense in which teleportation is here conceived: it is teleportation not of 'matter', and not even of 'information' in any free-standing or absolute sense, but of *correlations* as existing in EPR-states.

Bennett describes the prototype teleportation of the spin-1/2 state of particle-(1) (α, β are the usual Pauli-spinors)

$$\phi_1 = a\alpha_1 + b\beta_1, \text{ with } |a|^2 + |b|^2 = 1$$

using the Alice-Bob entangled $(2,3)$ EPR-state

$$\Psi_{23}^- = \sqrt{\frac{1}{2}} \left(\alpha_2 \beta_3 - \beta_2 \alpha_3 \right).$$

The product state $\Psi_{123} = \phi_1 \times \Psi_{23}$ involves no correlation between particle-(1) and the $(2,3)$ EPR-pair *until* Alice makes a measurement to determine the state of the pair $(1,2)$. In the complete, orthonormal Bell-basis for Alice's particles-$(1,2)$ [22]

$$\Psi_{12}^\pm = \sqrt{\frac{1}{2}} \left(\alpha_1 \beta_2 \pm \beta_1 \alpha_2 \right) \text{ and } \Phi_{12}^\pm = \sqrt{\frac{1}{2}} \left(\alpha_1 \alpha_2 \pm \beta_1 \beta_2 \right),$$

the product state becomes

$$\Psi_{123} = \frac{1}{2} \left(\Psi_{12}^- \times (-1)_3 + \Psi_{12}^+ \times (-\sigma_z)_3 + \Phi_{12}^- \times (\sigma_x)_3 + \Phi_{12}^+ \times (-i\sigma_y)_3 \right) \phi_3.$$

Alice's measurement onto the Bell-basis (with uniform probability $1/4$) determines Bob's e-bit (for particle-3) to be the teleported q-bit ϕ (originally of particle-1) within a specified π-rotation and an undetermined overall phase. The fidelity of the perfect teleportation

$$\mathcal{F} = |\phi_1^\dagger \phi_3|^2 = \left\{ |a|^2 + |b|^2 \right\}^2 \equiv 1,$$

is to be contrasted to a random classical result for Bob of

$$\mathcal{F}_{cl} = \left\{ |a \cos\theta + b \sin\theta|^2 \right\}_{avg} = \frac{1}{2}$$

when averaged over Bob's spinor amplitude $(\cos\theta, \sin\theta)$.

§ XIV-5. Concluding Remarks.

How much – in the final analysis – should we make of EPR and Bohm's experiment and Bell's inequality? According to Pais – speaking, of course, before the most modern experimental developments making use of EPR-pairs for quantum non-demolition experiments (quantum 'measurements' of a new but limited sense) with application to teleportation and potential for quantum computation and cryptography – not much [23]. The *cognoscenti* always knew the right answer, of course, certainly ever since Mott's analysis of cloud chamber tracks [24]. But for many , unless pressed, it was a rote response to a familiar stimulus and not the considered response of any independent thought. Everyone could calculate the right answer but very few paused to really think about it. The EPR-paradox attained its true importance only after it was sharpened by Bohm's *gedanken* experiment, and made quantitatively testable by Bell's inequality. Then Aspect's experimental dramatization finally revealed Einstein's quantum "spookiness" for everyone to marvel at.

Mermin [9] characterizes the dismissal of EPR prior to Bell as being "entirely metaphysical", and cites Pauli's advice to Born that Einstein's arguments about quantum mechanics were like debating how many angels can sit on the head of a pin. After Bell and Aspect, we are finally *forced* by *one* experimental fact to reject EPR's [c-]reality in favor of a less restrictive [q-]reality. The full implications of [q-]reality as encoded in quantum mechanics are difficult to comprehend and accept even with the compelling motivation of Aspect's experiments and all their successors.

We have suggested a culprit for this state of affairs – it is our perfectly natural attraction to the intellectual comfort of our first intuitive introduction to quantum mechanics through wave mechanics. Even Born found Schrödinger's wave mechanics and his semi-classical theory of radiation and scattering to be essential to progress. This remains true today for many applications. One great exception is the interpretation of Bohm's experiment and the results of Clauser and especially those of Aspect. These experiments *demand* that we adopt the Heisenberg-Born-Jordan-Dirac interpretation of quantum mechanics as an *information theoretic* device.

Quantum mechanics propagates *information* in a way consistent with the limitations of *knowability* inherent in the quantization of the Action in units of the Planck h quantum. The key step was Heisenberg's discovery of a dynamical theory of 'observables'. Heisenberg's theory was immediately recognized by Born and Jordan, and by Dirac, as a matrix version of classical Lagrangian-Hamiltonian mechanics. The fundamental difference is the replacement of the Poisson brackets of the classical number-valued dynamical variables (p, q) by commutator brackets of quantum matrix-valued dynamical variables (P, Q). The matrix indices refer to a Hilbert space of vectors identified with (as) the quantum states. The limitations on knowability are made explicit in Heisenberg's Uncertainty Principle, and made intuitive by Bohr's principle of complementarity based on the wave-particle dualism.

The Schrödinger wave function is a preliminary stage usually essential to actual computation and to intuitive and phenomenological progress. It too must be interpreted in an *information theoretic* way and *not* – fundamentally – as a classical wave or the square root of a density. Such elementary wave mechanical ways of thinking are deeply rooted in our intuitions and are unarguably valuable in many circumstances. The exceptional circumstances of the greatest epistemological interest are just those brought to everyone's attention by EPR, Bohm, Bell, Clauser, and Aspect. These require us to *give up* the semi-classical notions of the wave function and *accept* the full abstractions of a less definite [q-]reality.

As Bohr labored so hard to explain, the [q-]reality only becomes a [c-]reality (one of many) when – when what? And here is another sticking point. When we make a 'measurement' and change our 'information' (and 'collapse' the wave function) and propagate 'it' anew.

This state of affairs leaves many (most? all?) unhappy that the [q-]world should progress in fits and starts when everything *really* must be one large system obeying one grand Schrödinger equation. Why should the world care what we know? or what we do? What about the quantum mechanics of the whole universe? These are questions of the ultimate fundamental [q-]interest. Our understanding of quantum mechanics cannot be considered complete until the quantum microworld is joined smoothly and seamlessly to the classical macro-world. We examine recent great progress on this front in the next chapters.

And now, at the start of the new millennium – with the great experimental impetus of Aspect, Heroche, Kimble, and many others – we have a new excitement to re-examine, re-focus, and revolutionize our understanding of quantum mechanics as *quantum*-information theory.

Footnotes and References:

1) A. Einstein, B. Podolsky, and N. Rosen, Phys. Rev. **47**, 777 (1935); Para. 19. Here as

Paper XIII·1.

2) N. Bohr, Phys. Rev. **48**, 696 (1935). Here as **Paper XIII·2**.

3) E.C. Kemble, Phys. Rev. **47**, 973 (1935); W.H. Furry, Phys. Rev. **49**, 393, 476 (1936); and E. Schrödinger, reprinted in translation in *Quantum Theory and Measurement* (Princeton, Princeton NJ, 1983) J.A. Wheeler and W.H. Zurek, Eds., Pp. 153-167.

4) D. Bohm, *Quantum Theory* (Prentice-Hall, New York, 1955), Pp. 611-623.

5) D. Bohm and Y. Aharanov, Phys. Rev. **108**, 1070 (1957).

6) C.S. Wu and I. Shaknov, Phys. Rev. **77**, 136 (1950).

7) J.S. Bell, Physics **1**, 195 (1964); in 'On the Einstein-Podolsky-Rosen Paradox' reprinted in *Speakable and Unspeakable in Quantum Mechanics* (Cambridge, Cambridge UK, 1987), Pp. 14-22.

8) A. Aspect, P. Grangier, and G. Roger, Phys. Rev. Lett. **49**, 91 (1982).

9) N.D. Mermin, Physics Today **38**-4, 38 (1985); also in *Niels Bohr: A Centenary Volume* (Harvard, Cambridge MA, 1985) A.P. French and P.J. Kennedy, Eds., Pp. 141-147.

10) To calculate P_{ab}^{QM} choose $\hat{a} = \hat{z}$ and $\hat{b} = \hat{z}\cos\theta + \hat{x}\sin\theta$.

$$
\begin{aligned}
P_{ab}^{QM} &= \{\alpha_1\beta_2 - \beta_1\alpha_2\}^\dagger \, \sigma_1 \cdot \hat{a}\sigma_2 \cdot \hat{b} \{\alpha_1\beta_2 - \beta_1\alpha_2\}/2 \\
&= \left\{\beta^\dagger \sigma \cdot \hat{b}\beta - \alpha^\dagger \sigma \cdot \hat{b}\alpha\right\}/2 = -\cos\theta = -\hat{a} \cdot \hat{b}.
\end{aligned}
$$

11) *Local* because the distribution $B(\hat{b}, \lambda)$ for one measurement does not depend on the direction $\hat{a}$ of the other. Bell points out that only a non-local hidden variable distribution with A depending also on $\hat{b}$ and B on $\hat{a}$ can reproduce P_{ab}^{QM}. Bell also presents a simple example of a local hidden variable theory which can reproduce elementary results. He chooses $\hat{\lambda}$ as a unit vector uniformly distributed over the upper hemisphere $\hat{\lambda} \cdot \hat{z} \geq 0$. The HV value of $\langle \sigma \cdot \hat{a} \rangle - A(\hat{a}, \hat{\lambda})$ – is defined to be the sign of $\hat{a}' \cdot \hat{\lambda}$ where $\hat{a}'$ depends on $\hat{a}$ in a way chosen to reproduce the quantum expectation $P_a^{QM} = \cos\theta_a \equiv P_a^{HV}$. A simple calculation produces $P_a^{HV} = 1 - 2\theta_{a'}/\pi$, which can be solved for $\theta_{a'}$ as function of θ_a. The problem is to reproduce the correlation $\langle \sigma_1 \cdot \hat{a}\sigma_2 \cdot \hat{b} \rangle$, which requires the direction $\hat{a}'$ to depend on both $\hat{a}$ and $\hat{b}$, i.e., a non-local hidden variable theory. (Ironically, Bohm's original hidden variable theory *is* non-local [12], and its proponents are undeterred by all this.)

12) D. Bohm, Phys. Rev. **85**, 166, 180 (1957). For a contemporary advocate see S. Goldstein, Physics Today **51**-3, 42 (1998); **51**-4, 38 (1998).

13) J.F. Clauser, M.A. Horne, A. Shimony, and R.A. Holt, Phys. Rev. Lett. **23**, 880 (1969). For an exhaustive review of work leading up to Aspect's, see J.F. Clauser and A. Shimony, Rep. Prog. Phys. **41**, 1981 (1978). Earlier experiments are pointed out *not* to test Bell's inequality for a variety of reasons. For example, Wu's experiment [6] uses the weak polarization dependence of Compton scattering which cannot yield a violation

of Bell's inequality (M.A. Horne, PhD thesis, Boston University, 1969).

14) A. Aspect, J. Dalibard, and G. Roger, Phys. Rev. Lett. **49**, 1804 (1982).

15) Physics Today **51**-12, 9 (1998); W. Tittel, J. Brendel, H. Zbinden, and N. Gisin, Phys. Rev. Lett. **81**, 3563 (1998); W.T. Buttler, R.J. Hughes, P.G. Kwiat, S.K. Lamoreaux, G.G. Luther, G.L. Morgan, J.E. Nordholt, C.G. Peterson, and C.M. Simmons, Phys. Rev. Lett. **81**, 3283 (1998); G. Weihs, T. Jennewein, C. Simon, H. Weinfurter, and A. Zeilinger, Phys. Rev. Lett. **81**, 5039 (1998).

16) C.A. Kocher and E.D. Commins, Phys. Rev. Lett. **18**, 575 (1967).

17) See [13,14] for the effect of finite detector solid angles ($\simeq 1.5\%$) and polarizer efficiencies ($\geq 95\%$).

18) N.D. Mermin, Phys. Rev. Lett. **65**, 1838 (1990).

19) C.H. Bennett, G. Brassard, C. Crépeau, R. Jozsa, A. Peres, and W.K. Wooters, Phys. Rev. Lett. **70**, 1895 (1993). Bennett defines teleportation as "$\cdots$ a term from science fiction meaning to make a person or thing disappear while an exact replica appears somewhere else" and goes on to say " The net result of [our example of] teleportation is completely prosaic: the removal of ϕ from Alice's hands and its appearance in Bob's hands a suitable time later. The only remarkable feature is that $\cdots$ the information in ϕ has been cleanly separated into classical and nonclassical parts." And it is only the *quantum information* which has been 'teleported', and that in a prosaic, classical, causal way by the prearranged *sharing* of an EPR-pair between Alice and Bob. It is an unbreakable, un-copiable, and un-reusable code. Alice destroys her copy in encrypting the message, Bob his as the price of de-encrypting it and resurrecting the quantum state ϕ from its limbo.

20) A. Furusawa, J.L. Sorensen, S.L. Braunstein, C.A. Fuchs, H.J. Kimble, and E.S. Polzik, *Science* **282**, 706 (Oct 23, 1998). Kimble defines fidelity $\mathcal{F}$ to measure the quality of the teleported state. Perfect teleportation corresponds to $\mathcal{F} \equiv |\langle\phi_A|\phi_B\rangle|^2 = 1$. For a two component state, the threshold for quantum teleportation is $\mathcal{F}_{cl} = 1/2$. Kimble's experiment achieved $\mathcal{F} = .58 \pm .02$.

21) L. Vaidman, Phys. Rev. **A49**, 1473 (1994).

22) S. Braunstein, A. Mann, and M. Revzen, Phys. Rev. Lett. **68**, 3259 (1992).

23) A. Pais, *Niels Bohr's Times* (Oxford, Oxford UK, 1991) Pp. 429-431.

24) N. Mott, Proc. Roy. Soc. **A126**, 79 (1929); reprinted in *Quantum Theory and Measurement* (Princeton, Princeton NJ, 1983) J.A. Wheeler and W.H. Zurek, Eds., Pp. 129-134.

Biographical Notes:

David Bohm (1917-1992) – From Goldstein[†] we learn of Bohm's profound and unique contributions – most of them unorthodox and some of them highly controversial [1] – in pursuit of a deeper understanding of quantum theory. His early contributions were in nucleon scattering (after his PhD in 1943 with J.R. Oppenheimer), on plasma oscillations (at Princeton, with Eugene Gross), and on collective excitations in the electron gas (with David Pines). He then wrote his definitive exposition of the Copenhagen interpretation in *Quantum Theory* (Prentice-Hall, NY, 1951) where he gave his version of the EPR-paradox involving particle spins, which produced results (with Yakir Aharanov) leading directly to Bell's inequalities and the Clauser-Aspect experiment. Concurrently, Bohm refined the de Broglie-Schrödinger 'pilot-wave' interpretation of wave mechanics to his *'non-local* hidden variable' version of nonrelativistic quantum mechanics which transcends not only von Neumann's original no-go theorem for hidden variables but also – ironically – Bell's inequalities derived for *local* hidden variables [1]. In addition to these highly original contributions, he contributed the Bohm-Aharonov effect demonstrating the non-local quantum role of the potentials of electrodynamics. Bohm's then [1951] radical political stance led to his indictment for contempt of Congress by the House Un-American Activities Committee resulting in his black-balling from American universities and his subsequent international career at the University of Sao Paulo, Brazil (1951-55), Technion in Haifa, Isreal (55-57), Bristol (57-61), and Birkbeck College, London (61-83).

† – S. Goldstein, Physics Today **47**-8, 72 (1994).

1) S. Goldstein, Physics Today **51**-3, 42 (1998); **51**-4, 38 (1998).

    John Bell (1928-1990) – Romer[††], in evaluating Bell's contributions to the interpretation of quantum mechanics, quotes the status of the EPR-paradox *before* Bell, summarized in Pauli's put-down of Einstein: "One should no more rack one's brains about the problem of whether something one cannot know anything about exists all the same, than about the ancient question of how many angels can sit on the point of a needle. · · · Einstein's questions are ultimately always of this kind." What Bell's work did, was to make Einstein's questions *subject to experiments* of the most beautiful kind which – although not totally conclusive (Bell assumed *local* hidden variables) – all result in complete agreement with the predictions of quantum mechanics according to the Copenhagen interpretation. Romer's editorial responded to an article titled "The Man Who Proved Einstein Was Wrong" [1]. Romer concludes that it was Pauli who was wrong, and Einstein who was right to continue questioning the orthodox quantum theory and to demand that the ultimate arbiter in physics must be experiment.

†† – R.H. Romer, Am. J. Phys. **59**-4, 299 (1991).

1) John Gribben, New Scientist **128**, 43 (1990).

Chapter XV

FEYNMAN'S PATH INTEGRAL

FORMULATION OF QUANTUM MECHANICS:

"The contribution from a single path is postulated to be an exponential whose phase is the classical action $\cdots$ The total $\cdots$ from all paths reaching x, t from the past is the wave function $\psi(\mathbf{x}, \mathbf{t})$."

R.P. Feynman, Reviews of Modern Physics **20**, 367 (1948).

§ XV-1. Introduction.

The Path Integral formulation of quantum mechanics [1] stands as perhaps the greatest of the two great monuments – Feynman diagrams being the other – which will forever memorialize Richard Feynman.

Feynman postulated the amplitude for a quantum system to evolve from the initial configuration $q(t_i)$ at time t_i to the final configuration $q(t_f)$ at time t_f in the remarkable form:

$$A\left(q(t_f), t_f; q(t_i), t_i\right) = \int \exp\left[\frac{iS(q(t))}{\hbar}\right] \mathcal{D}q(t).$$

The expression is deceptively simple but its interpretation is profound, as explained at length in simple examples by Feynman and Hibbs [2]. The phase S is the classical action – the time integral of the Lagrangian – on a particular classical path connecting the initial and final configurations $q(t_i)$ and $q(t_f)$. The path is labeled at each time t by a configuration $q(t)$. *Each* path contributes an amplitude of unit magnitude with its particular phase. The integration indicated by $\int \mathcal{D}q(t)$ is over *all* possible values of the coordinate $q(t)$ at each time t. The actual definition of these operations can be made explicit in simple cases, and the integrals actually done in the elementary cases of the free particle in one dimension and the simple harmonic oscillator. The equivalence of the Feynman Path Integral formulation to the Heisenberg and Schrödinger formulations also follows.

What is remarkable is the broad generality of the formulation. It is as general as the Lagrangian *cum* Action Principle formulation of dynamics itself. It can be extended to many coordinates q; to fields, where $q(t)$ is replaced by $q(x, y, z, t)$ and the action by a space-time integration over a Lagrangian density; to relativistic field dynamics with explicit space-time symmetry; and much more [3].

The particular use we emphasize for Feynman path integrals is application to the evolution, and especially the decoherence and classical limit, of quantum systems interacting with an environment. The advantage of Feynman's formulation here is the ease with which the coordinates of the environment can be 'integrated out' in favor of an 'influence function' acting on the quantum system [4].

The Feynman Path Integral formulation of quantum mechanics was viewed by many for a long time as interesting but inessential, an embellishment on a quantum mechanics already adequately in place with the Heisenberg and Schrödinger formulations. Feynman himself set in motion the reversal of this attitude by noting that the path integral formulation of quantum field theory is a *sine qua non* for a graviton field theory. In this case, many auxiliary restrictions are required to reduce the apparent degrees of freedom of the (traceless, symmetric, hermitian) tensor field $h_{\mu\nu}$ to the actual ones of the graviton. The auxiliary conditions prevent assigning canonical commutation relations to the physical degrees of freedom in a manifestly covariant and gauge invariant way and make old-fashioned canonical techniques unmanageable. The same holds true for the non-Abelian gauge fields of the Standard Model [3] and, apparently, *all* of its conjectured progenitor structures.

No obvious *essential* role for Feynman path integrals can be claimed in the restricted realm of non-relativistic quantum mechanics. Here it is a matter of the *intuitive descriptive power* and *convenience* that a facility with path integrals gives us to *visualize* quantum processes in a global, qualitative, and semi-classical way combining the de Broglie ray-optic limit in one extreme – continuously and formally – with the full quantum mechanics in the other. Our immediate motivation to discuss path integrals – in addition to the intrinsic interest in a third formulation of quantum mechanics – is their importance in modern discussions of the interpretation of quantum mechanics; especially for the quantum-classical connection which has finally been made more formally complete [4,5,6,7] with their help.

Feynman's Path Integral formulation of quantum mechanics had its origin in Dirac's profound and far-reaching recognition [8] of the role in quantum mechanics of the Lagrangian and the Action Principle. Both are viewed in classical mechanics as more fundamental than the Hamiltonian, but quantum mechanics had until then been based exclusively on the Hamiltonian form of mechanics. Dirac's immediate conclusion was that *no meaning* can be given in quantum mechanics to the differential operations – of the Lagrangian with respect to the coordinates and velocities – which occur in classical Lagrangian mechanics. Dirac had shown – in the very first steps of his original pursuit [9] of Heisenberg's new quantum mechanics [10] – that "the only differentiation process that can be carried out with respect to the [matrix-]dynamical variables of quantum mechanics is that of forming Poisson brackets and this leads to the Hamiltonian theory." Dirac concluded "we must take over the *ideas* of the classical Lagrangian theory, not the *equations* · · ·."

Early praise for Feynman's achievement was lavish but – it turns out – never lavish enough. Yourgrau and Mandelstam [11] wrote in 1952 "One cannot fail to observe that Feynman's principle in particular – and this is no hyperbole – expresses the laws of quantum mechanics in an exemplary neat and elegant manner, notwithstanding the fact that it employs somewhat unconventional mathematics · · · · · · possesses a definite advantage over · · · the Hamiltonian version · · · [which] is not manifestly covariant · · · [but then they lose their prescience] · · · this seldom presents great difficulty · · ·".

The Feynman Path Integral formulation of quantum field theory has turned out to be critical to the progress of the last thirty-five years in the Standard Model and beyond, which got a major impetus from Feynman's famous long-unpublished 1961 lecture tape-recorded late at night in a Warsaw beer hall. DeWitt [12] first formulated the theory of radiative corrections for non-abelian gauge fields using the FPI formulation with 'ghost' fields to impose the gauge restrictions to all orders in perturbation theory. This theoretical structure was used by 't Hooft and Veltman to prove the renormalizability of the Standard Model, work recognized in their 1999 Nobel Prize.

§ XV-2. Feynman Path Integrals.

a) Elementary Remarks.

Feynman [1] characterizes as *classical* events $a, b, c \cdots$ those developing with a probability P_{ca} which satisfies

$$P_{ca} = \sum_b P_{cb} P_{ba} \tag{1}$$

for event a developing to event b with probability P_{ba} and then event b developing to event c with probability P_{bc}, summed over all mutually exclusive alternative events b. *Quantum* probabilities are different in an essential way and evolve according to an underlying complex quantum probability *amplitude* ϕ as

$$\phi_{ca} = \sum_b \phi_{cb} \phi_{ba} \tag{2}$$

and only then the quantum probability

$$P_{ca} = |\phi_{ca}|^2 = \left| \sum \phi_{cb} \phi_{ba} \right|^2 \neq \sum P_{cb} P_{ba} \quad \text{in general.} \tag{3}$$

The crucial logical distinction between the classical and the quantum evolution is the classical statement that the event b actually happened. This requires a careful definition of what constitutes an *event* and what we mean by *happened*. It is an old story, often repeated:

i) event b corresponds to observable B having the value b [13]; and

ii) in order for the event to have 'happened', an actual observation of B with result b *must* have been done.

The accompanying *inevitable* phase disruption of the quantum amplitude cause the cross terms to cancel so

$$P_{ca} = \sum |\phi_{cb} \phi_{ba}|^2 \Rightarrow \sum |\phi_{cb}|^2 |\phi_{ba}|^2$$

and replaces the quantum evolution with the classical.

The generalization to many such (quantum-) events is immediate

$$\phi_{fi} = \sum \phi_{fa}\phi_{ab}\cdots\phi_{wi} \tag{4}$$

and again $P_{fi} = |\phi_{fi}|^2$. The probability is the absolute square of the total amplitude summed over all possible event sequences $a, b, c \cdots w$, i.e., *over all paths through the space of the events.* Any measurement to determine that an event c has actually happened will break the quantum chain of coherent events at c into a succession of incoherent (or decohered) classical events, as above.

Feynman discusses in detail the case that events $a, b, c \cdots$ are space-time events $(x_1, t_1), (x_2, t_2), \cdots$ where the times $t_{i+1} = t_i + \Delta$ are a succession of times separated by a small interval Δ (eventually taken infinitesimal), and the x_i are possible values of the system coordinate. Then the classical rule for compounding probabilities

$$P_{fi} = \int \cdots \int P(\cdots x_{a+1}, x_a \cdots) \cdots dx_{a+1} dx_a \cdots$$

is replaced by the quantum rule for compounding amplitudes

$$\phi_{fi} = \int \cdots \Phi(\cdots x_{a+1}, x_a \cdots) \cdots dx_{a+1} dx_a \cdots,$$

and $P_{fi} = |\phi_{fi}|^2$ as usual. The two most prominent problems faced by Feynman were:

i) to define the sum or integral measure over all possible paths; and

ii) to settle upon a suitable choice for the amplitude Φ for a given path.

Feynman solved the first problem by application of Occam's razor: simply integrate over cartesian coordinates. And for the second problem he borrowed from Dirac the classical action, and made a remarkable leap in his basic postulate: *"The paths contribute equally in magnitude but the phase of their contribution is the classical action (in units of $\hbar$); i.e., the time integral of the Lagrangian taken along the path."* So

$$\Phi[x(t)] = e^{iS[x(t)]/\hbar}, \tag{5}$$

where

$$S[x(t)] = \int dt L[\dot{x}(t), x(t)]$$

along the path $x(t)$. For classical Lagrangians quadratic in the velocities $\dot{x}$ it is possible to write the action as a sum over time-slices

$$S[x(t)] = \sum_a S_{cl}(x_{a+1}, x_a), \tag{6}$$

with

$$S_{cl}(x_{a+1}, x_a) = \text{Min} \int_{t_a}^{t_{a+1}} L_{cl}(\dot{x}, x)dt.$$

Finally,

$$\phi_{fi} = \lim_{\Delta \to 0} \int \prod_a \left(\frac{dx_a}{A} e^{iS_{cl}(x_{a+1}, x_a)/\hbar} \right). \tag{7}$$

The normalization factor A has to be chosen appropriately. The integrations are over N time-slices $a = 1, N$, equally spaced by $\Delta = (t_f - t_i)/(N + 1)$. The paths connect the initial coordinate $x_i \equiv x(a = 0)$ at t_i to the final value $x_f \equiv x(a = N + 1)$ at t_f. The actual paths are a jagged sequence of connected line segments with velocity discontinuities at the endpoints. The difficulties of understanding the innermost workings of these integrations will have to be left to someone with more strength than we can muster. Suffice it to say, following Feynman [1], that the naive treatment considered there is sufficient for elementary Lagrangians.

The simplest example of a free particle in one dimension is reproduced in footnote [14]. The important case of the the quantum harmonic oscillator linearly coupled to an external classical force is also easily done [1,4].

b) Equivalence to Ordinary Quantum Mechanics.

The Feynman path integral amplitude ϕ_{fi} can be +expressed as an integral over all values of the coordinate at some particular time-slice k corresponding to time t. The integration variable x_k can be labeled simply x. Then

$$\phi_{fi} = \int dx_k \phi_{fk} \phi_{ki}$$

can be written – in a suggestive and familiar form which we now justify – as

$$\phi_{f,i} = \int dx \chi_f^*(x, t) \psi_i(x, t). \tag{8}$$

The function $\psi_i(x, t)$ is the amplitude ϕ_{ki} at x, t of all paths which originated at x_i at time t_i. These paths propagate the *initial* data x_i at t_i always forward in time to $x = x_k, t = t_k$. Therefore the function $\psi_i(x,t)$ can depend only on x and t and on the initial value data, and not at all on anything that happens *after* time t. We conclude that $\psi_i(x,t)$ can be characterized as the *initial state* wave function.

Conversely, $\chi_f^*(x,t)$ depends only on the *final* values x_f propagated always backward in time from t_f to t and can be characterized as the *final state* wave function.

To see that $\psi(x,t)$ satisfies the Schrödinger equation, write the last time-slice integration explicitly as

$$
\begin{aligned}
\psi(x,t) &= \int \frac{dx'}{A} e^{iS(x,x')} \psi(x', t - \Delta) \\
&= \int \frac{dx'}{A} e^{i[(x-x')^2/2\Delta - V(x)\Delta]} \psi(x', t - \Delta) \\
&\simeq \int \frac{dz}{A} e^{iz^2/2\Delta} \left(1 - iV\Delta\right) \psi(x + z, t - \Delta).
\end{aligned}
$$

Expand $\psi(x + z, t - \Delta)$ as

$$
\simeq \psi(x,t) - iV\Delta\psi - \frac{\partial \psi}{\partial t}\Delta + \frac{\partial \psi}{\partial x}z + \frac{1}{2}\frac{\partial^2 \psi}{\partial x^2}z^2.
$$

The integral over ψ

$$
\int \frac{dz}{A} e^{iz^2/2\Delta} \psi = \frac{\sqrt{2i\pi\Delta}}{A}\psi \equiv \psi
$$

fixes $A = \sqrt{2i\pi\Delta}$ [14]. The integral over z vanishes, and the integral over z^2 is just $i\Delta$. The net result (restoring the units) is Schrödinger's equation as required.

Feynman makes further contact with familiar results of ordinary quantum mechanics. The matrix element of an operator $F(x)$ can be expressed as usual with wave functions χ_f^* and ψ_i obtained (at least in principle) from the path integrations. The Green's function formulation [14] can be recovered from the path integral result. With

$$
\psi(x_f, t_f) = \int G_+(x_f, t_f; x_i, t_i)\psi(x_i, t_i)dx_i \tag{9}
$$

the Green's function can be identified from Eqn9 above as

$$
G_+(x_f, t_f; x_i, t_i) = \int \prod_a \left(\frac{dx_a}{A} e^{iS_{cl}(x_{a+1}, x_a)/\hbar}\right) \equiv \int \exp\left[\frac{iS}{\hbar}\right] \mathcal{D}x(t), \tag{10}
$$

in terms of the abstract path integration $\int \mathcal{D}x(t)$. In the usual quantum calculation, this is written

$$G_+(x_f, t_f; x_i, t_i) = \langle x_f | \mathcal{G}_+(t_f, t_i) | x_i \rangle \tag{11}$$

with

$$\mathcal{G}_+(t_f, t_i) = \mathcal{T} \exp\left(-i \int_{t_i}^{t_f} H_{qu}(t) dt\right). \tag{12}$$

Here $\mathcal{T}$ is the time-ordering operation on the power series expansion of the exponential containing the quantum Hamiltonian, which in general is time dependent.

It is worth emphasizing the profoundly different formal structure of the two approaches:

i) the path integral involves infinitely many functional integrations over the *classical action* as a function{*al*} of {*all possible values of all*} coordinate variables $x(t)$ integrated over {*summed over discrete*} t. Compare this to

ii) the standard quantum formulation, which involves time integrations over *non-commuting quantum operators*, and eventually the evaluation of Hilbert space matrix elements using orthogonality and completeness. This is certainly a different look, one very reminiscent of de Broglie's initial vision [16] as made formal by Schrödinger [17]. From Feynman [1]: "We may state Huygen's principle $\cdots$ in an analogous manner starting with Hamilton's first principle of *least action* for *classical* or 'geometrical' mechanics. If the amplitude of the wave ψ is known on a given surface of all x at time t, its value at a nearby point at time $t + \Delta$ is the sum of contributions from all points of the surface at t. Each contribution is delayed in phase by an amount proportional to the *action* required to get from the surface to the point along the path of *least action* of classical mechanics."

§ XV-3. Influence Functional and Decoherence.

Feynman and Vernon [3] study a quantum oscillator Q linearly coupled to an array of oscillators X intended to represent an environment – perhaps measuring instruments, perhaps external fields coupled to lossy resonators at some temperature. The Path Integral formulation of quantum mechanics is ideally suited for this purpose because the coordinates X of the environment oscillators can be integrated out and their full effect on the quantum oscillator represented by an influence function $\mathcal{F}(Q, Q')$. Feynman and Vernon express the amplitude for the combined system with coordinates $q = Q, X$ as the path integral from q_τ at time τ to q_t at time T as

$$K(q_T, T; q_\tau, \tau) = \int \exp\left[\frac{iS[q(t)]}{\hbar}\right] \mathcal{D}q(t). \tag{13}$$

The amplitude to go from a state ϕ_n at τ to a state ϕ_m at T is

$$
\begin{aligned}
A_{mn} &= \int \phi_m^*(q_T) K(q_T, q_\tau) \phi_n(q_\tau) dq_T dq_\tau \\
&= \int \phi_m^*(q_T) \exp\left[\frac{iS[q(t)]}{\hbar}\right] \phi_n(q_\tau) \mathcal{D}q(t) dq_T dq_\tau. \tag{14}
\end{aligned}
$$

The general form is now separated into quantum system variables Q and environment system variables X. The states $\phi(q)$ are taken to be product states $\phi_m(q) = \psi_m(Q)\chi_f(X)$, and the action is taken to be

$$S(q) = S_0(Q) + S(X) + S_I(Q, X).$$

With these substitutions the transition *probability* $|A_{mn}|^2$ can be written:

$$
\begin{aligned}
P_{mf,ni} &= \int \psi_m^*(Q_T)\psi_m(Q_T') \exp\left[\frac{i}{\hbar}(S_0(Q) - S_0(Q'))\right] \times \\
&\quad \times \mathcal{F}_{fi}(Q, Q')\psi_n^*(Q_\tau')\psi_n(Q_\tau) \times \\
&\quad \times \mathcal{D}Q(t)\mathcal{D}Q'(t) dQ_\tau dQ_\tau' dQ_T dQ_T'. \tag{15}
\end{aligned}
$$

The *influence function* containing *all* the effects of the environment is:

$$
\begin{aligned}
\mathcal{F}_{fi}(Q, Q') &= \int \chi_f^*(X_T)\chi_f(X_T') \times \\
&\quad \times \exp\left[\frac{i}{\hbar}(S(X) - S(X') + S_I(Q, X) - S_I(Q', X',))\right] \times \\
&\quad \times \chi_i^*(X_\tau')\chi_i(X_\tau)\mathcal{D}X(t)\mathcal{D}X'(t) dX_\tau dX_\tau' dX_T dX_T'. \tag{16}
\end{aligned}
$$

This can also be expressed in the more familiar Green's function formalism as

$$
\begin{aligned}
\mathcal{F}(Q, Q') &= \left(\int \chi_f^*(X_T)G_Q^+(X_T, X_\tau)\chi_i(X_\tau) dX_T dX_\tau\right) \times \\
&\quad \times \left(\int \chi_f^*(X_T')G_{Q'}^+(X_T', X_\tau')\chi_i(X_\tau') dX_T' dX_\tau'\right)^*. \tag{17}
\end{aligned}
$$

Here the environment Green's function G_Q^+ depends on the quantum system coordinate Q through the interaction $S_I(Q, X)$.

Feynman and Vernon prove as general properties of the influence function $\mathcal{F}_{fi}(Q, Q')$:

i) For indefinite environment states, $\mathcal{F}_{fi}$ should be replaced by an average $\langle \mathcal{F} \rangle = \sum_{f,i} w_{f,i} \mathcal{F}_{f,i}$ where w_{fi} is the probability of the various environmental configurations. This can include the direct sum over final states f, and the sum over a thermal distribution of initial states i to describe temperature effects.

ii) For a number of independent environmental systems k, the total influence function is the product $\mathcal{F} = \prod_k \mathcal{F}_k$. Feynman and Vernon introduce the corresponding influence phases Φ_k ($\mathcal{F}_k \equiv \exp[i\Phi_k]$) which are additive.

iii) $\mathcal{F}^*(Q, Q') = \mathcal{F}(Q', Q)$ follows by inspection.

iv) If the final state of the environment is averaged over, then $\mathcal{F}(Q, Q) \equiv 1$. This is most easily seen from the Green's function form above. If $Q = Q'$, $\mathcal{F}_{fi}(Q, Q)$ is the transition probability to state χ_f of χ_i propagated through the 'potential' Q forward to the time T, summed over all states f, and so equal to unity.

v) In the presence of noise and dissipation, $\mathcal{F}(Q, Q') \to 0$ for $Q \neq Q'$, rapidly decohering the quantum system.

Feynman and Vernon also develop applications including:

i) The effect of the environment on the density matrix of the quantum system follows from the assumption of statistically independent systems as

$$\rho(Q_T, Q_T') = \int \mathcal{F}(Q, Q') \exp\left[\frac{i}{\hbar}(S_0(Q) - S_0(Q'))\right] \times \\ \times \rho(Q_\tau, Q_\tau') \mathcal{D}Q(t) \mathcal{D}Q'(t) dQ_\tau dQ_\tau'. \tag{18}$$

ii) The effect on $\mathcal{F}$ of finite temperature dissipative environments represented by arrays of linearly coupled oscillators is expressed by a complex impedance function Z coupling the quantum system Q to a classical potential with a specified noise power spectrum.

We must refer to the original paper for their many applications. Our primary interest has been to gain some facility in the path integral formalism to prepare ourselves for the discussion in the following chapter of Gell-Mann and Hartle's theory of decoherence based on these techniques.

§ XV-4. Concluding Remarks.

The Feynman Path Integral formulation of quantum mechanics – based on the Dirac quantum action principle – opens into a vast theoretical world we can only hint at [3]. This is the world of relativistic quantum field theory of Feynman-Schwinger-Tomonaga; leading to the Yang-Mills non-Abelian gauge theories; then to the Standard Model of Weinberg-Glashow-Salam; and further to contemporary theories still developing of strings and higher structures incorporating gravity. At the present state of development this theoretical world is seemingly – at least to bewildered spectators – infinite in all directions and it is the great challenge to find *the* path through this maze. Hopefully, Dirac's principle of truth in simplicity and beauty will eventually prevail. Hopefully, too, the great energy, enthusiasm, and – above all – *genius* which has brought humankind this far will be equaled and even exceeded by future generations who confront daunting problems in this subject.

Footnotes and References:

1) R.P. Feynman, Rev. Mod. Phys. **20**, 367 (1948) (here as **Paper XV·1**).

2) R.P. Feynman and A. R. Hibbs, *Quantum Mechanics and Path Integrals* (McGraw-Hill, New York, NY, 1965).

3) E.g., M. Kaku, *Quantum Field Theory* (Oxford, New York, NY, 1983), Ch.8. See Kaku for the functional integration over anti-commuting fields, and over fields subject to constraints.

4) R.P. Feynman and F.L. Vernon, Annals of Physics **24**, 118 (1963) (here as **Paper XV·2**).

5) J. B. Hartle, Phys. Rev. D **44**, 3173 (1991); and Am. J. Phys. **36**, 704 (1968).

6) M. Gell-Mann and J.B. Hartle, Phys. Rev. D **47**, 3345 (1993) (here as **Paper XVI·1**).

7) J.A. Wheeler, 'Information, Physics, Quantum: The Search for Links' in *Complexity, Entropy, and the Physics of Information* (Addison-Wesley, Reading, MA, 1990) W.H. Zurek, Ed., p.3.

8) P.A.M. Dirac, Phys. Zeits. Sowjetunion **3**, 64 (1933), available in *Selected Papers on Quantum Electrodynamics* (Dover, New York, NY, 1958) J. Schwinger, Ed., p. 312; here as **Paper XV·1**. Also, P.A.M. Dirac, Rev. Mod. Phys. **17**, 195 (1945); and P.A.M. Dirac, *The Principles of Quantum Mechanics* (Oxford, London, UK, 1947), p. 125.

9) P.A.M. Dirac, Proc. Roy. Soc. **A109**, 642 (1925) (here as **Paper VIII·1**).

10) W. Heisenberg, Zeits. f. Phys. **33**, 879 (1925) (here as **Paper VI·1**).

11) W. Yourgrau and S. Mandelstam, *Variation Principles in Dynamics and Quantum Theory* (Pitman, London, 1960), p.128. This source gives also an exemplary exposition of the Feynman-Schwinger, -Schrödinger, -Heisenberg, and -classical equivalences.

12) B. DeWitt, Phys. Rev. Lett. **12**, 742 (1964) (here as **Paper XVII·1D**). DeWitt cites Feynman's unpublished researches on Yang-Mills field theory and quantum gravity in his *Theory of Radiative Corrections for Non-Abelian Gauge Fields.* See also our ChXVII for DeWitt's canonical quantum theory of gravity, and his immediate extension to the quantum field theory of gravity based on the FPI formulation of quantum mechanics. See also R.P. Feynman, W. Wagner, and F. Morinigo, *Feynman Lectures on Gravitation* (Addison Wesley, Cambridge MA, 1994) B. Hatfield, Ed. See also *Search and Discovery,* Physics Today **52**(12), 17 (1999) for other references.

13) Here we quote Feynman's footnote 8 in reference [1]: "If **A** and **B** are the operators corresponding to measurements A and B, and if ψ_a and χ_b are solutions of

$$\mathbf{A}\psi_a = a\psi_a$$

and

$$\mathbf{B}\chi_b = b\chi_b,$$

then

$$\phi_{ba} = \int \chi_b^* \psi_a dx = (\chi_b^*, \psi_a).$$

Thus ϕ_{ba} is an element $\langle b|a \rangle$ of the transformation matrix for the transformation from a representation in which **A** is diagonal to one in which **B** is diagonal."

14) The Feynman path integral over paths connecting the point x_1 on the time-slice t_1 to the point x_3 on the time slice $t_3 = t_1 + 2\Delta$ can be calculated explicitly for an infinitesimal time interval Δ in the elementary case of a free particle in one dimension. In this simplest case, a single gaussian integral over intermediate positions x_2 at time $t_2 = t_1 + \Delta$ is required and the result can be generalized after the first step to positions x_0, x_f at times $t_0, t_f = t_0 + T$ separated by a finite time T. We have the Feynman amplitude for propagation from $1 \to 3$ compounded from the propagation from $1 \to 2$ and then from $2 \to 3$ integrated over all x_2 (all at fixed times $t_1, t_2 = t_1 + \Delta, t_3 = t_2 + \Delta$

$$A(31) \sim \int A(32)dx_2 A(21).$$

Feynman uses Dirac's result

$$A(21) \sim e^{i \int_1^2 dt L(\dot{x}, x)/\hbar}$$

with the crucial postulate that all $\sim$'s become strict equalities, and that all individual quantum paths should contribute with *equal a priori probabilities*. Writing the most naive finite difference approximation for the classical action (not always adequate [see 1]) for a free particle of mass m (in units where $\hbar = c = m = 1$)

$$S = \int_1^2 dt\, L = \int_1^2 dt\, \dot{x}^2/2,$$

we get

$$A(21) = N_{21} e^{i(x_2 - x_1)^2/(2\Delta)},$$

and

$$
\begin{aligned}
A(31) &= N_{32} N_{21} \int_{-\infty}^{+\infty} dx_2 e^{i(x_3 - x_2)^2/(2\cdot\Delta)} e^{i(x_2 - x_1)^2/(2\cdot\Delta)} \\
&= \sqrt{\Delta}\sqrt{i\pi} N_{32} N_{21} e^{i(x_3 - x_1)^2/(4\cdot\Delta)} \equiv N_{31} e^{i(x_3 - x_1)^2/(2\cdot[2\Delta])}.
\end{aligned}
$$

Here we have used the Gaussian integral

$$\int_{-\infty}^{+\infty} dz\, e^{iz^2} = \sqrt{i\pi}.$$

The normalization

$$N_{21} = N_{32} = 1/\sqrt{2i\pi\Delta}$$

gives

$$N_{31} = 1/\sqrt{2i\pi[2\Delta]}$$

as required for replication. Repeated integrations over time-slices separated by $2\Delta \cdots$ and so on, and finally by $T/2$ produce the result

$$A\left(\{x_f, t_f = t_0 + T\} \leftarrow \{x_0, t_0\}\right) = \sqrt{\frac{m}{2i\pi\hbar T}} e^{im(x_f - x_i)^2/(2\hbar T)},$$

(resupplying the units). We recognize the Feynman path integral amplitude – the sum over all classical paths $\{\alpha\}$

$$\sum_\alpha e^{i\Phi_\alpha},$$

with each phase Φ_α equal to the classical action (in units of $\hbar$) along the path α, all paths taken with the same weight – as the free particle Green's function

$$G(x_f, t_f \leftarrow x_0, t_0) = \langle x_f | e^{-iHT} | x_0 \rangle = \int \frac{dk}{2\pi} e^{(+ik(x_f - x_0) - ik^2 T/2)}$$

of the usual formulation of non-relativistic quantum mechanics [15].

15) L.I. Schiff, *Quantum Mechanics* (McGraw-Hill, New York, NY, 1985), Ch. 9.

16) L. de Broglie, Phil. Mag. **47**, 446 (1924) (here as **Paper IV·1**).

17) E. Schrödinger, Ann. d. Physik **79**, 489 (1926) (here as **Paper IX·1b**).

Reminiscences about Feynman.

Much has been written about Feynman, usually by spectators of his performances as a comedian. Very few people knew him in his primary role of research theoretical physicist. In his whole career he might have had no more than ten students and a comparable number of senior collaborators. Few people were closer to him than **Frank Vernon**, his PhD student around 1960, whose reminiscences follow after a brief introduction.

Frank Lee Vernon, Jr., had a super-abundance of the *sine qua non* for working with Feynman. It was understood that you had to be smart, but everyone was smart (or still thought so). To make it past the protective barriers surrounding Feynman and actually get repeat action, a person had to bring something extra. Vernon brought charisma. He was a quarterback on a championship Texas high school football team, a champion tennis player, a pianist, a USNavy veteran, an EE graduate of SMU and Berkeley, and an AEROSPACE CORP scientist and executive. And wise enough to plan his approach to Feynman with great care. The campus scuttlebutt was that Feynman demanded independent students with imagination, initiative, and their own idea for a thesis problem. Vernon presented four problems of interest to himself as an electrical engineer, and Feynman – somewhat taken by storm – chose masers. Feynman's criterion for a thesis problem was that it not be too interesting or he would rush to solve it himself; but if it was not interesting enough, he wouldn't want to think about it at all.

Their joint effort exceeded all expectations. Vernon recalls several results of his which surprised Feynman at their weekly conference but didn't necessarily please him. When Vernon independently introduced temperature into the formalism, Feynman's competitive instincts made him almost angry that some interloper should steal his thunder with his own formalism.

Paper XV·1: Abstract of Rev. Mod. Phys. **20**, 267 (1948).

Space-Time Approach to Non-Relativistic

Quantum Mechanics

R.P. FEYNMAN

Cornell University, Ithaca, New York

Abstract: Non-relativistic quantum mechanics is formulated here in a different way. It is, however, mathematically equivalent to the familiar formulation. In quantum mechanics the probability of an event which can happen in several different ways is the absolute square of a sum of complex contributions, one from each alternative way. The probability that a particle will be found to have the path $x(t)$ lying somewhere within a region of space-time is the square of a sum of contributions, one from each path in the region. The contribution from a single path is postulated to be an exponential whose (imaginary) phase is the classical Action (in units of $\hbar$) for the path in question. The total contribution from all paths reaching x, t from the past is the wave function $\psi(x, t)$. This is shown to satisfy Schrödinger's equation. The relation to matrix and operator algebra is discussed. Applications are indicated, in particular to eliminate the coordinates of the field oscillators from the equations of quantum electrodynamics.

··· there is a pleasure in recognizing old things from a new point of view. ··· the general concept of superposition of probability amplitudes in quantum mechanics ··· can be extended to define a probability amplitude for any motion or path (position *vs.* time) in space-time. ··· This is true when the Action is the time integral of a quadratic function of the velocity ··· but the formulae are very suggestive ··· [for] ··· a wider class of Action functionals. ···

We may summarize these ideas in our first postulate:

I. If an ideal measurement is performed to determine whether a particle has a path lying in a region of space-time, then the probability that the result will be affirmative is the absolute square of a sum of complex contributions, one for each path in the region.

$\cdots$ The second postulate gives a particular content to this framework by pre-scribing how to compute the important quantity [amplitude] Φ for each path:

II. The paths contribute equally in magnitude but the phase of their contribution is the classical Action (in units of $\hbar$); i.e., the time integral of the Lagrangian taken along the path.

$\cdots\cdots$ the influence of a perturbing measuring instrument can be integrated out $\cdots$ as we did for the oscillator. The statistical density matrix $\cdots$ results from $\cdots$ integrations over dx_i, dx'_i $\cdots$ of $\exp i(S - S')/\hbar$ $\cdots$ required to describe the result of elimination of the field oscillators where the final state of the oscillators is unspecified $\cdots$

* * * * * * * * * * * * * * * * * **

This ability to synthesize original and powerful theoretical structures from the most informal, nebulous, and heuristic of beginnings dis-tinguishes the earliest work of Einstein, of Heisenberg and here of Feynman. Their abilities to turn such simple questions into such profound conclusions by the application of such pure logic are truly miraculous.

* * * * * * * * * * * * * * * * * **

Paper XV·2: Abstract from Annals of Physics **24**, 118 (1963).

The Theory of a General Quantum System Interacting

with a Linear Dissipative System

R.P. FEYNMAN[1] and F.L. VERNON[2]

*California Institute of Technology[1], Pasadena, California
and The Aerospace Corporation[2], El Segundo, California*

Abstract: $\cdots$ using Feynman's space-time formulation of nonrelativistic quan-tum mechanics $\cdots$ the behavior of a system $\cdots$ coupled to other external quantum systems, may be calculated in terms of its own variables only. $\cdots$ the effect of the external systems $\cdots$ can always be included in a general class of 'influence functionals' of the coordinates of the system only. $\cdots$ specific forms of the influence functionals representing the effect of definite and random classical forces, linear dissipative systems at finite temperatures, and combinations of these are analyzed in detail. $\cdots$ Influence functionals for all linear systems are shown to have the same form in terms of their classical response functions. In addition, a fluctuation-dissipation theorem is derived relating temperature and dissipation of the linear system to a fluctuating classical potential acting on the system of interest which reduces to the Nyquist-Johnson relation for noise in the case of electrical circuits. Sample calculations of transition probabilities for the spontaneous emission of an atom in free space and in a cavity are made. $\cdots$

Paper XV·3: Excerpt from Phys. Zeits. Sowjetunion **3**, 64 (1933).

THE LAGRANGIAN IN QUANTUM MECHANICS

By P.A.M. Dirac.
(Received November 19, 1932.)

1) Quantum mechanics was built on analogy with the Hamiltonian theory of classical mechanics. The classical notion of canonical coordinates and momenta has a very simple quantum analog; so the whole classical Hamiltonian theory, which is based on this notion, could be taken over into quantum mechanics.

2) There is an alternative formulation of classical mechanics based on the Lagrangian, in terms of coordinates and velocities instead of coordinates and momenta. The two formulations are closely related, but the Lagrangian formulation is the more fundamental.

3) The Lagrangian method deduces all the equations of motion from the stationary property of the Action function defined as the time-integral of the Lagrangian. There is no corresponding Action Principle in terms of the coordinates and momenta of the Hamiltonian theory. Moreover, the Lagrangian method can be expressed relativistically; whereas the Hamiltonian theory is essentially non-

relativistic in form, since it uses a particular time variable as the canonical conjugate of the Hamiltonian.

4) One should ask what corresponds in the quantum theory to the Lagrangian method of the classical theory. However, one cannot expect to take over the classical Lagrangian equations in any direct way. These equations involve partial derivatives of the Lagrangian with respect to the coordinates and velocities and *no meaning can be given to such derivatives in quantum mechanics*. The only differentiation process that can be carried out with respect to the dynamical variables of quantum mechanics is that of forming Poisson brackets and this process leads to the Hamiltonian theory [1, 2].

5) We must therefore seek the quantum Lagrangian theory in an indirect way. We must take over the *ideas* of the classical Lagrangian theory, not the *equations*.

Contact Transformations.

6) We begin with an analogy between classical and quantum contact transformations. Let the two sets of variables be p_r, q_r and P_r, Q_r, $(r = 1, 2, \cdots n)$. The q's and Q's are each complete sets of independent coordinates, and any function of the dynamical variables can be expressed in terms of them. In the classical theory the transformation equations can be written $p_r = \partial F/\partial q_r$, and $P_r = -\partial F/\partial Q_r$, where F is some function of the q's and Q's.

7) In quantum theory we can take one representation where the q's are diagonal, and a second where the Q's are diagonal. There will be a transformation function $(q'|Q')$ connecting the two representations. We now show that this transformation function is the quantum analog of $e^{iF/\hbar}$.

8) If f is any function of the dynamical variables in the quantum theory, it will have a "mixed" representation [Read: matrix-element] $(q'|f|Q')$, which can be written in terms of the usual matrix-elements $(q'|f|q'')$ or $(Q'|f|Q'')$ [using $\int |q'')dq''(q''| = \int |Q'')dQ''(Q''| \equiv 1$] as

$$(q'|f|Q') = \int (q'|f|q'')dq''(q''|Q') = \int (q'|Q'')dQ''(Q''|f|Q'). \tag{1}$$

From these we obtain

$$(q'|q_r|Q') = q'_r(q'|Q'), \text{ and } (q'|p_r|Q') = \frac{\hbar}{i}\frac{\partial}{\partial q'_r}(q'|Q'); \tag{3}$$

also $(q'|Q_r|Q') = Q'_r(q'|Q')$, and $(q'|P_r|Q') = -\frac{\hbar}{i}\frac{\partial}{\partial Q'_r}(q'|Q')$. $\tag{5}$

Note the sign difference between Eqn3 and Eqn5.

9) These equations can be generalized to any function $f(q,Q)$ which can be "well-ordered", that is expressed as $f(q,Q) = \sum_k g_k(q)h_k(Q)$ where the g_k's and h_k's are functions only of the q's and Q's. For such functions,

$$(q'|f(q,Q)|Q') = f(q',Q')(q'|Q'). \tag{6}$$

This remarkable equation gives a connection between $f(q,Q)$, a function of operators, and $f(q',Q')$, a function of numerical variables [their eigenvalues, i.e., q is the operator, and q' is the eigenvalue of the state $|q'\rangle$: $q|q'\rangle = q'|q'\rangle$]. Putting $(q'|Q') = e^{iU/\hbar}$, where U is a new function of the q's and Q''s, we get from Eqn3

$$(q'|p_r|Q') = \frac{\partial U}{\partial q'_r}(q'|Q'), \tag{7}$$

Using Eqn6, we get the operator equation $p_r = \partial U(q,Q)/\partial q_r$, which holds provided $\partial U/\partial q_r$ is well-ordered. Similarly we get $P_r = -\partial U(q,Q)/\partial Q_r$, provided $\partial U/\partial Q_r$ is well-ordered. These equations are of the same form as the classical equations and show that U in Eqn7 is the quantum analog of the classical function F which generates the classical contact transformation.

The Lagrangian and the Action Principle.

11) In the classical theory the dynamical variables evolve in such a way that their values q_t, p_t at time t are connected with their values q_T, p_T at any other time T by a contact transformation in the form Eqn1 with $q, p = q_t, p_t$ and $Q, P = q_T, p_T$ and F equal to the Action S, the time integral of the Lagrangian from T to t. In the quantum theory q_t, p_t will still be connected with q_T, p_T by a contact transformation and there will be a transformation function $(q_t|q_T)$ connecting the representation in which q_t is diagonal to that in which q_T is diagonal. The preceding

section shows that

$$(q_t|q_T) \sim \exp\left[i\int_T^t Ldt/\hbar\right], \tag{8}$$

where L is the Lagrangian. For an infinitesimal time difference Δt

$$(q_{t+\Delta t}|q_t) \sim \exp\left[iL\Delta t/\hbar\right]. \tag{9}$$

12) The transformations Eqns8 and 9 are fundamental in quantum theory and it is satisfying that they are [not just *analogous* as Dirac supposed, but *equal* as Feynman postulated] expressible simply in terms of the Lagrangian. This is the *natural extension* of the well-known result that the phase of the wave function corresponds to Hamiltons principle function [i.e., the Action S] of classical theory. Eqn9 suggests that we ought to consider the classical Lagrangian not as a function of coordinates and velocities at time t, but rather as a function of the coordinates at time t and the coordinates at time $t + \Delta t$.

13) For simplicity we take the case of a single degree of freedom, which is easily generalized. With the notation $\exp\left[i\int_T^t Ldt/\hbar\right] = A(tT)$, $A(tT)$ is the classical analog of $(q_t|q_T)$.

14) Divide up the time interval $T \to t$ into a large number of small intervals $T \to t_1 \to t_2, \cdots, t_{m-1} \to t_m \to t$. Then

$$A(tT) = A(tt_m)A(t_m t_{m-1})\cdots A(t_2 t_1)A(t_1 T). \tag{10}$$

In quantum theory we have

$$(q_t|q_T) = \int (q_t|q_m)dq_m(q_m|q_{m-1})dq_{m-1}\cdots(q_2|q_1)dq_1(q_1|q_T). \tag{11}$$

Eqn11 at first sight does not seem to correspond properly to Eqn10, since in Eqn11 we must integrate whereas in Eqn10 there is no integration.

15) To resolve this apparent discrepancy we consider Eqn11 as $\hbar$ becomes very small. From Eqns8 and 9 the integrand in Eqn11 must be of the form $\exp[iS/\hbar]$ where S remains finite as $\hbar$ tends to zero. Imagine one of the intermediate q's, say q_k, varying while all the others remain fixed. In general $S/\hbar$ will vary extremely rapidly, and $\exp[iS/\hbar]$ will oscillate with very high frequency around zero and

its integral will be practically zero. *The only important part of the domain of integration of q_k is that for which S is stationary for small variations of q_k.*

16) We can apply this argument to each of the variables of integration in Eqn11. The result is that the only important part of the domain of integration is that for which S is stationary for small variations in all the intermediate q's. From Eqn8 we get

$$S = \int_{t_m}^{t} L dt + \int_{t_{m-1}}^{t_m} L dt \cdots \int_{T}^{t_1} L dt = \int_{T}^{t} L dt,$$

which is just the Action function which classical mechanics requires to be stationary for small variations in all the intermediate q's. This shows the way Eqn11 goes over into the classical result when $\hbar$ can be considered very small.

17) For comparison with quantum theory when $\hbar$ cannot be considered small, Eqn10 must be interpreted as [a sum over all possible intermediate q's] corresponding to the integration over all these q's in Eqn11.

18) Eqn11 is the quantum analog of the Action Principle. If we take particular q_T and q_t, then the importance of any set of intermediate q's is determined by the integration of Eqn11. If we make $\hbar$ tend to zero, this statement goes over into the statement of the classical Action Principle. The importance of any set of values of the intermediate q's is zero unless they make the Action function stationary. This is another way [the most fundamental] of formulating the classical Action Principle.

Application to Field Dynamics.

19) The problem of fields in the classical Lagrangian theory is a natural generalization of that for particles. We choose as coordinates q suitable field or potential amplitudes. Each field amplitude is a function of the four space-time variables x, y, z, t. The one independent variable t of particle theory must be generalized to four independent variables x, y, z, t [3].

20) We introduce at each space-time point a Lagrangian density $\mathcal{L}$ which must be a function of the coordinates [field amplitudes] $q(x, y, z, t)$ and their first derivatives $q_x, q_y, q_z, q_t \equiv \dot{q}$, corresponding in particle theory to the coordinate and velocity. The integral $\int \mathcal{L} d^4 x$ must be stationary for small variations of the coordinates

[i.e., field amplitudes for given x, y, z, t] $\cdots$.

21) It is easy to see what the field analog of the above discussion must be. If S is the integral of the classical field Lagrangian over a particular region of space-time, then the field analog of $(q_t|q_T)$ must correspond to $\exp[iS/\hbar]$. This is a *functional* of the coordinates [field amplitudes] on the boundary of the space-time region labeled by t and T. It is a four-dimensional generalization of the $(q_t|q_T)$ in the particle case in the following sense.

22) The composition law $(q_t|q_T) = \int (q_t|q_1)dq_1(q_1|q_T)$ relating coordinates [field amplitudes] in two regions must be integrated over all values of the coordinates [field amplitudes] on the common boundary of the two regions.

23) Repeated application of Eqn12 will give Eqn11 and the quantum analog of the Action Principle for fields.

24) The absolute square $|(q_t|q_T)|^2$ of the transformation function in the *particle* case can be interpreted as the probability of observing q_t at t for a state which at $T < t$ gave the result q_T. A corresponding interpretation in the *field* case will exist only when the generalized transformation function refers to a region of space-time bounded by [two infinite time-like surfaces] $\cdots$.

St. John's College, Cambridge.

References and Footnotes:
1) M. Born and P. Jordan, Zeits. f. Phys. **34**, 858 (1925) (here as **Paper VII·1**); M. Born, W. Heisenberg, and P. Jordan, Zeits. f. Phys. **35**, 557 (1926) (here as **Paper VII·2**).
2) P.A.M. Dirac, Proc. Roy. Soc. A**109**, 642 (1925) (here as **Paper VIII·1**).
3) It is customary to regard the field values for two different values of (x, y, z) but the same t as two different coordinates, instead of as two values of the *same coordinate* for two different points in the domain of independent variables, and in this way to keep the idea of a single independent variable t. This point of view is necessary for the Hamiltonian treatment, but for the Lagrangian treatment the point of view adopted here seems preferable for its space-time symmetry.

Chapter XVI

HARTLE'S DECOHERENT SUM

OVER COARSE-GRAINED HISTORIES

"$\cdots$ is more general than the usual Hamiltonian quantum mechanics. It predicts probabilities even though the process of prediction cannot be organized in terms of states, their unitary evolution, and their reduction."

J.B. Hartle, Phys. Rev. D44, 3173 (1991).

§ XVI-1. Decoherent Histories.

In a series of pioneering studies, Hartle [1,2], Hartle and Gell-Mann [3], Griffiths [4], and Omnés [5], have constructed an interpretation of quantum mechanics for *closed systems*, suitable for application to the ultimate problem of the quantum cosmology [6] of an isolated closed universe [7]. The aim of their work is to construct an alternative to the widely perceived artificiality and arbitrariness of the Copenhagen interpretation of quantum mechanics. The Copenhagen scenario requires a split of the world into a quantum object subjected to an external classical apparatus in an idealized and formalized measurement process. The objection, of course, is that – presumably – one *grand* Schrödinger equation should govern the *whole* world at all times; so any interpretation which has to postulate a classical regime, and – compounding the felony – even has to *interrupt* the Schrödinger evolution of that part supposed to be governed by quantum mechanics, is begging some very important questions of principle. The split of the world into a classical apparatus and a quantum object is supposed to be most dramatically untenable in quantum cosmology where we seek a quantum description of the whole world with us inside [7].

Hartle generalizes conventional Hamiltonian quantum mechanics to an *encom-*

passing structure based on *decoherent histories*. The notion of decoherent histories is conveniently formulated in terms of Feynman path integrals [8]. It can be reduced in special situations to the usual Schrödinger or Heisenberg time evolution of quantum states in a Hilbert space, interrupted at particular times by the traditional measurement processes [9]. In general, however, Hartle's decoherent histories have many advantages, including those resulting from the space-time symmetry of the Lagrangian formulation of quantum mechanics [10].

A. Fundamental Definitions and Overview.

Decoherent histories are sets of Feynman paths $\{\alpha\}$ which can be assigned classical probabilities $P\{\alpha\}$ satisfying

$$0 \leq P\{\alpha\} \leq 1, \quad \sum_{\{\alpha\}} P\{\alpha\} = 1, \text{ and } P(\{\alpha\} + \{\beta\}) = P\{\alpha\} + P\{\beta\}, \quad (1)$$

which requires that the sets of Feynman paths $\{\alpha\}$ be exhaustive (i.e., complete) and mutually exclusive. Based as they are on the Lagrangian formulation, Feynman path integrals are defined by the generalized coordinates (q's) of the system, and the paths are labeled by the trajectories $q(x_k, t_k)$ at (a continuum of) space-time points $(x_k, t_k), k = 0, 1, \cdots N$. The condition that the sets $\{\alpha\}$ be mutually exclusive and satisfy the classical probability condition is the defining criterion for decoherent histories (termed 'consistent histories' by Griffiths and Omnés). This condition requires special "coarse-graining" – i.e., (quasi-)macroscopic sums over individual (discrete) quantum trajectories to construct sets $\{\alpha\}$ which decohere; i.e., that have no quantum interference as in

$$P\{\alpha + \beta\} =_{Qu} |\psi_\alpha + \psi_\beta|^2 \Rightarrow_{Cl} |\psi_\alpha|^2 + |\psi_\beta|^2 = P\{\alpha\} + P\{\beta\}. \quad (2)$$

Hartle introduces *fine-grained* histories $\{f\}$ corresponding to unique paths through the space of individual (time-ordered) quantum states of the closed system (characterized by a complete orthonormal set of wave functions ϕ), so $\{f\} = \{\alpha_1, \alpha_2 \cdots \alpha_N\}$ is the history of the system in states $\phi_{\alpha_1} \phi_{\alpha_2} \cdots$ at ordered times $t_1 < t_2 < \cdots < t_N$ and propagated via some Hamiltonian H in between. *Coarse-grained* histories are basically sums over selected intervals of fine-grained histories. These are conveniently expressed as Feynman path integrals, but are simple only when the paths are specified by the ordinary coordinates of a single particle, and only then for the simplest interactions.

Analogous to the elementary product $\psi_\alpha \psi_\beta^*$, Hartle defines a *decoherence functional* $D(f, f')$ for fine-grained histories which is similar in form to terms which occur in N^{th}-order time-dependent perturbation theory, as we will see in a moment. It is the amplitude for the system to propagate from an initial state ψ_0 at time $t = 0$ to the final state ψ_T at time $t = T$ through the N ($\to \infty$) steps $\alpha_1, \alpha_2 \cdots$ on the path f, multiplied by the complex conjugate of a similar amplitude for the (in general, independent) path f'.

Decoherence functionals for coarse-grained histories are constructed as

$$D(h, h') = \sum_{f \in h} \sum_{f' \in h'} D(f, f'). \tag{3}$$

The requirement that decoherent histories possess classical probabilities puts severe restrictions on the definition of the coarse-graining.

Hartle characterizes the Feynman path integral formulation of decoherent histories as being both *more* and *less* general than conventional Hamiltonian quantum mechanics. More general because coarse-graining allows the discussion of *events* defined over a space-*time* interval, not just at a fixed time. In general these coarse-grained events are not describable by a Hamiltonian or a *state* at a given time. Less general because we are restricted to Euclidean coordinates in configuration space for the definition of the measure in the Feynman path integrals. This is not a very severe restriction, and Feynman path integrals can be extended to the full phase space [3].

Here we summarize the concepts of decoherent histories. They are almost self-evident, especially in the Feynman path integral formulation of quantum mechanics. Hartle provides many simple, elementary, and even analytic examples. Unfortunately, realistic decoherence properties are manifested only in more complicated – especially multi-dimensional – situations which cannot be treated explicitly. The central ideas include:

1) Fine-grained histories $\{f\}$: e.g., the closed system passes through a time-ordered

succession of states $(\phi_a(t_1), \phi_b(t_2) \cdots \phi_c(t_n))$, where the ϕ's refer to traditional quantum states a, b, c of a complete, orthonormal set; at times $t_1 < t_2 < \cdots < t_n$.

2) Coarse-grained histories $\{h\}$: e.g., $\{h\} = \sum_a \sum_b \cdots \{f(\phi)\}_{(a,b,\cdots \in h)}$ are defined to be *mutually exclusive*. They can be further coarse-grained as $\{\bar{h}\} = \sum_{h \in \bar{h}} \{h\}$.

3) Hartle introduces a coarse-grained *decoherence functional*

$$D(h, h') = D^*(h', h) \sim \sum_{(f \in h)} \sum_{(f' \in h')} \phi_f \phi_{f'}^* \text{ which is}$$

(i) hermitian,

(ii) positive, i.e., $D(h, h) \geq 0$,

(iii) mutually exclusive and complete so $\sum_h = \sum_f$,

(iv) normalized so $\sum_h D(h, h) = \sum_f D(f, f) = 1 \quad \left(\sim \sum_f |\phi_f|^2 \right)$,

(v) *required* to be decohered so that

$$\sum_h \sum_{h'} D(h, h') = \sum_h D(h, h) + \{ \text{ off-diagonal elements } \Rightarrow 0 \}.$$

The *fundamental decoherence condition* – Re $D(h, h') = 0, h \neq h'$ – guarantees

$$\sum_{h,h'} D(h, h') \to \sum_h D(h, h) = 1. \tag{4}$$

This permits *classical probabilities* $P_h \equiv D(h, h)$ – which satisfy $0 \leq P_h \leq 1$ and $\sum_h P_h = 1$ – to be assigned to each *decoherent history h*.

The crucial point is that decoherent histories with classical probabilities have been defined which – in Hartle's words – provide the "minimal defining elements" for the "interpretation" – i.e., classical manifestation – of quantum mechanics. This has been achieved with *no mention* of the conventional *bête noires* of critics of the Copenhagen interpretation such as 'observer', 'measurement', 'state of the system', and 'collapse of the wave function'.

In brief, Hartle's *decoherent histories* interpretation of quantum mechanics is derived from the conventional Hamiltonian quantum mechanics starting with the *fine-grained histories* involving some complete orthonormal set ϕ_n, then *coarse-graining* over these to achieve decoherent histories. The corresponding *decoherence functional* $D(h, h')$ will next be constructed by *Feynman path integral* techniques following Feynman and Vernon [11].

B. Feynman Path Integral Decoherence Functional.

From ChXV we have the Feynman path integral amplitude required to construct the *decoherence functional* for particular *fine-grained* paths $X(t), X'(t')$. The generalized coordinates X are assumed to be a (multi-dimensional) Euclidean space, and the paths are specified by the value of $X(t)$ at a succession of times $0 \leq t \leq T$. The *fine-grained* decoherence functional is

$$D[\{X(t)\}, \{X'(t')\}] = \delta(X_T - X'_T) \exp\left[\frac{i}{\hbar}(S(\{X\}) - S(\{X'\}))\right] \rho(X_0, X'_0). \quad (5)$$

Here, $\rho(X_0, X'_0) = \sum_j \phi_j(X_0) P_j \phi_j^*(X'_0)$ is the density matrix of the initial state. The density matrix of the final state – $\delta(X_T - X'_T)$ – is characterized as 'future indifference', i.e., complete ignorance of the final state. The initial state propagates along the path $\{X\} = X(t)$ from $X(t = 0) = X_0$ to $X(t = T) = X_T$ and acquires the characteristic phase equal to the classical action $S(\{X\})$ (with $\hbar$ set to one as usual). This is the time integral of the classical Lagrangian L along the path $\{X\}$

$$S(\{X\}) = \int_{\{X\}} L dt, \text{ (and similarly for } X'(t')).$$

The *total* decoherence functional for *all* fine-grained paths involves
1) an integral over *all* paths $\{X\}, \{X'\}$. Each path – as visualized by Feynman for a single particle in ordinary x, y, z configuration space – is an infinite sequence of straight-line segments connecting $X(t) \to X(t + \Delta) \to X(t + 2\Delta) \cdots$ for infinitesimal Δ and for any possible *independent* sequence $X(t), X(t + \Delta) \cdots$. The sum over these is *taken to be* an integration over all $X(t_k), X'(t_{k'})$ at a succession of infinitesimally separated time-slices $0 < t_1 < t_2 < \cdots < t_N < T$ with $N \to \infty$. This general path integration is abbreviated $\int \mathcal{D}X\mathcal{D}X'$. A final
2) integration over *all* initial positions (configurations) $\int dX_0 dX'_0$, and then
3) over *all* final positions $\int dX_T dX'_T$ leads to the familiar results.
From ChXV we have

$$\int_{X_0}^{X_T} \mathcal{D}X e^{iS(X)} = \langle X_T | e^{-iHT} | X_0 \rangle.$$

With $\int dX_T |X_T\rangle\langle X_T| = 1$ and $\langle X_0'|X_0\rangle = \delta(X_0 - X_0')$ this gives

$$\int dX_T dX_0 dX_0' \langle X_0'|e^{iHT}|X_T\rangle\langle X_T|e^{-iHT}|X_0\rangle\rho(X_0, X_0')$$

$$= \int dX_0 \rho(X_0, X_0) = \text{Tr}\rho \equiv 1.$$

The normalization condition on the decoherence functional Eqn3 is required for its interpretation as a sum of classical probabilities for all paths.

Hartle separates the fine-grained paths into coarse-grained sets

$$\mathcal{C}_\alpha = \sum_{f \in \alpha} \{f\} \text{ which are}$$

1) exhaustive (i.e., complete), so $\bigcup \mathcal{C}_\alpha = 1$; and
2) mutually exclusive, so $\mathcal{C}_\alpha \cap \mathcal{C}_\beta = 0, \alpha \neq \beta$.
Then the coarse-grained decoherence functional is

$$D(\mathcal{C}_\alpha, \mathcal{C}_{\alpha'}) = \int_{\mathcal{C}_\alpha} \mathcal{D}X \int_{\mathcal{C}_{\alpha'}} \mathcal{D}X' \delta(X_T - X_T') e^{i(S(X) - S(X'))} \rho(X_0, X_0'). \tag{6}$$

This is the Feynman path integral foundation of Hartle's decoherent histories formulation of quantum mechanics.

The actual separation of Feynman paths into distinct classes $\{f_\alpha\}$ is possible in simple cases, as Hartle demonstrates. The demonstration that a coarse-graining $\mathcal{C}_\alpha$ is in fact achieved, is possible only in the simplest cases. Unfortunately, these do not seem to possess sufficient structure (i.e., dimensionality, dense level structure) to be realistic or relevant. To return to ordinary Hamiltonian quantum mechanics Hartle writes

$$D(\mathcal{C}_\alpha, \mathcal{C}_{\alpha'}) = \sum_{ij} P_j K_{ij}(\mathcal{C}_\alpha) K_{ij}^*(\mathcal{C}_{\alpha'}) \text{ with} \tag{7}$$

$$K_{ij}(\mathcal{C}_\alpha) = \int_{\mathcal{C}_\alpha} \mathcal{D}X \phi_i^*(X_T) e^{iS(X)} \psi_j(X_0) = \langle \phi_i|\mathbf{C}_\alpha|\psi_j\rangle. \tag{8}$$

$\mathbf{C}_\alpha$ is the "class operator" of the coarse-grained history $\{\mathcal{C}_\alpha\}$. Since $\sum_\alpha \{\mathcal{C}_\alpha\} = 1$,

and $\sum_{\mathcal{C}_\alpha} K_{ij}(\mathcal{C}_\alpha) = \int_{all\{f\}} \mathcal{D}X \cdots = \langle \phi_i|e^{-iHT}|\psi_j\rangle$, then $\sum_\alpha \mathbf{C}_\alpha = e^{-iHT}$.

The individual class operators $\mathbf{C}_\alpha$ can be written in terms of projection operators $\mathbf{P}^n_{\alpha_n}$ onto the configuration space region α_n at time t_n as

$$\mathbf{C}_\alpha = \mathbf{C}_{\alpha_N \cdots \alpha_1} = e^{-iHT} \mathbf{P}^N_{\alpha_N} \cdots \mathbf{P}^n_{\alpha_n} \cdots \mathbf{P}^1_{\alpha_1} \text{ and finally as}$$
$$\Rightarrow \mathbf{P}^N_{\alpha_N}(T) \cdots \mathbf{P}^n_{\alpha_n}(t_n) \cdots \mathbf{P}^1_{\alpha_1}(t_1). \tag{9}$$

These are projection operators in the Heisenberg representation onto configuration space regions α_n at ordered times t_n. They satisfy $\sum_\alpha \mathbf{C}^*_\alpha \mathbf{C}_\alpha = 1$, but in general are neither Hermitian nor unitary.

It is further possible to generalize the partitioning in the space-time coarse-graining into *disjoint* regions R_i satisfying

$$\bigcup R_i = 1, \text{ and } R_i \bigcap R_j = 0 \text{ for } i \neq j.$$

Each history is now labeled by an index $\alpha_i = +1$ if the path enters region R_i, or $\alpha_i = 0$ if it does not. The path is characterized by a history such as

$$C_{(\alpha_1 \alpha_2 \cdots \alpha_n)} = 101 \cdots 0010 \cdots.$$

Hartle shows that the class operators and decoherence functional for such general space-time partitions are calculable in principle. Feynman and Hibbs [8], and others [3,11], extend path integration from the configuration space of the generalized coordinates Q to the phase space of the Q's and P's and construct a coarse graining on momentum variables. The deeper significance of coarse graining over space-time regions is that it permits coarse-graining on the basis of time-intervals, the only kind possibly relevant for quantum cosmology.

Hartle's primary thesis is that "classical probabilities can be assigned only to sets of alternative coarse-grained histories that decohere". The general criterion for decoherence is the (near-)vanishing of the real part of the off-diagonal elements of the decoherence functional $\mathrm{Re}D(h \neq h') \simeq 0$. It is pedagogically unfortunate that the analytic examples are too simple to exhibit realistic mechanisms of decoherence of coarse-grained histories because they have *far* too few degrees of freedom.

The model of Feynman and Vernon [11], and Caldeira and Leggett [12], in which a test oscillator is coupled linearly with a strength γ to an array of environmental oscillators at temperature Θ can be simplified in the Fokker-Planck limit

($\Theta \gg \Lambda \gg \hbar\omega$, where Λ is a high energy cutoff on the spectrum of the environmental array and $\omega/2\pi$ is the characteristic frequency of the test oscillator). The Feynman-Vernon influence function – which contains the integrated effect of the environment in the decoherence functional – becomes *complex* and *non-local*. Caldeira and Leggett find two characteristic times:

1) a dissipation time $T_D \sim 1/\gamma$, and

2) a decoherence time $\tau \sim T_D \times \left\{\hbar^2/(2Mk\Theta d^2)\right\}$.

For sufficiently large M ($\sim$grams) and d ($\sim$ cm), the decoherence time can be small $\tau \ll T = 2\pi/\omega \ll T_D$. The decoherence mechanism is reminiscent of Heisenberg's criticism of Schrödinger's 'coherent wave-packet' [13]: Frequent random interruptions of the test oscillator can *de-phase* the quantum coherence in times τ small compared to the quantum period T without being affected by the (average-)classical dissipation over the long time T_D. To simulate this source of decoherence in a realistic way requires – in Hartle's words – "*sufficient* and *particular* coarse-graining $\cdots$. Even though a coarse-graining partitions the classical paths, it can be *too fine* to decohere quantum mechanically. $\cdots$ coarse-grainings $\cdots$ must be chosen with care."

C. Recovering Conventional Quantum Mechanics.

Decoherent coarse-grained histories define expectation probabilities for the quantum evolution of a closed system without the necessity of intervening measurements to provide the decoherence. The Feynman path integral formulation of decoherent histories includes 'measurement situations' in a way closely related to Mott's original discussion of cloud chamber tracks [14]. Here we *do* separate the system into a quantum subsystem and a classical apparatus which records events and thereby *restricts* the paths to *one* coarse-grained path ($\alpha_1\alpha_2\cdots\alpha_n$). The decoherence functional then collapses to a probability

$$P(\{\alpha\}) = D(\{\alpha\},\{\alpha\}) = \text{Tr}\left(\mathbf{P}_{\alpha_n}\cdots\mathbf{P}_{\alpha_2}\mathbf{P}_{\alpha_1}\rho\mathbf{P}_{\alpha_1}\mathbf{P}_{\alpha_2}\cdots\mathbf{P}_{\alpha_n}\right). \qquad (10)$$

There is free unitary propagation between the 'measurement situations' ($\alpha_1 \to \alpha_2 \to \cdots \to \alpha_n$) and reduction of the wave packet (by the Heisenberg projection operator which restricts the quantum system to be in the coarse-grained region

R_{α_i} in a time interval which includes t_i). Hartle emphasizes that the fundamental limitation of *coarse-graining* rules out any possible fundamental role in the interpretation of quantum mechanics for *ideal* measurements, which are necessarily defined only for *fine-grained* events.

Any *realistic* description of *real* measurements clearly requires that we go beyond the idealizations of the Copenhagen interpretation to a space-time coarse-graining as in the Feynman path integral decoherent histories.

All the idealized devices of conventional Hamiltonian quantum mechanics find their (approximate) place in the decoherent histories interpretation of quantum mechanics. For example, we still have the generalized notions of

1) the 'state' of the system at time t as a summary of data events $(\alpha_1\alpha_2\cdots\alpha_n)$ on a particular decoherent history; and

2) the 'reduction of the wave packet' , which no longer corresponds to an explicit physical intervention, but to the <u>consideration</u> of a new conditional probability on a <u>new</u> decoherent history $(\alpha_1\alpha_2\cdots\alpha_n\alpha_{n+1})$ with a <u>new</u> decoherence functional serving as a new (updated) density matrix.

§ XVI-2. The On-Going Debate.

The '*real*' meaning of it all is the subject of an interminable and unproductive debate which shows no sign of closure, nor any sign or even hint of a revelation of *any* error or short-coming of the doctrine of the founders of the subject, going back to the very first statements of Born and Jordan [15]. These appeared in the very first paper following the discovery by Heisenberg [16] of the rule for multiplying quantum variables, which they identified as matrix-multiplication. Born and Jordan identified the canonical commutation relations as the essential ingredient of the new mechanics of quanta. They immediately recognized the time development of the quantum system to be the *unitary evolution* in the '*Hilbert space*' of the '*state vector*', as generated by the '*Hamiltonian operator*' [17].

The debate about the meaning of it all sensibly begins here. The problem is to correlate this pristine mathematical structure with the world we '*see*'. The challenge is *not* to invent baroque bells and whistles to fulfill every classical requirement and uninformed expectation, but quite the opposite. The challenge is to assume truth in pristine mathematical beauty and to explore what that beautiful mathematics requires us to believe and how it permits us to think.

And here the arguments begin.

What is beautiful? Unequivocal answer: The Action Principle [18]. And the Action Principle is used most directly and beautifully in the Feynman Path Integral formulation of quantum mechanics [8,10], which (after the fact, of course) contains the quantum mechanics of Heisenberg, and Schrödinger [19], and ultimately classical physics [2,3,4,5].

The debate has been aired recently [20], and it is amusing to highlight remarks from both sides which exemplify the difference in views:

(i) From Gell-Mann and Hartle: "··· the'orthodox' Copenhagen interpretation works in measurement situations and accurately predicts the outcomes of laboratory experiments. It is not wrong. ··· [it] is too special to be fundamental, and it is clearly inadequate for quantum cosmology." Our only comment would be that quantum cosmology is a rapidly moving target and a rather abstract one.

(ii) From Griffiths: "··· the formalism is shown to be complete as a fundamental theory, without need of the additional principles [e.g., Bohmian trajectories]. ··· one aspect of consistent histories [i.e., decoherent histories] [is] perfectly clear ···: A quantum history ··· correspond[s] to subspaces of the Hilbert space. ··· a wave function is associated with a one-dimensional subspace ··· wave functions are the building blocks out of which histories are constructed ··· Bohmian mechanics, not consistent histories, needs ('hidden') variables in addition to the standard Hilbert space of wave functions."

(iii) And closest to our views, from Zeilinger: "··· the Copenhagen interpretation remains one of the most significant intellectual achievements of our century. ··· the very austerity ··· unsurpassed ··· speaks very much in its favor. ··· a major intellectual step forward over naive classical realism. ··· Schrödinger's cat [paradox] ··· reflects a serious misunderstanding. All the quantum state is meant to be is a representation of our knowledge ··· it is even claimed that something happens to the cat because it is being observed. There is no basis for any such claim. ··· there

is never a paradox if we realize that quantum mechanics is about information.···"
(iv) And in rebuttal, from Goldstein: "··· merely an appeal to prevailing preju-
dices. ··· does Zeilinger truly believe that 'quantum mechanics is about information'? ··· does Zeilinger really wish to deny that the change of the state vector
that occurs during the measurement process is 'a real physical process' even when
it leads to the destruction of the possible interference? [Yes! Precisely that.] Can
quantum interference be genuinely understood by invoking a wave function that is
nothing more than 'a representation of our knowledge'? [Yes! and *that* is as good
as it gets.] ··· not at all sensible for a theory to acknowledge that 'any statement
about the world has to make reference to observation', [so] Zeilinger's assertion is
plainly false. [There follow examples which are completely gratuitous invocations
of naive classical realism as the basis of *knowing*. These have no bearing on the
subject of quantum mechanics.] ··· very much like the austerity of solipsism ···
not merely implausible, but also deficient ····." [What is missing in Goldstein's
accusation of solipsism, is any appreciation of the non-subjective constraints of
the correlation between a non-trivial wave function and the ensemble of systems
described by it.]

And on and on.

§ XVI-3. Concluding Remarks.

It is a delicious irony – and not atall a pretty sight – to see the most powerful
theoreticians struggle to derive freshman physics from quantum mechanics. And
even this is only a baby step in the connection of the world we see to the ultimate
reality we are currently told to believe of Planck sized objects in $9 + 1$-dimension
space-time ······

Footnotes and References:
1) J.B. Hartle, Am. J. Phys. **36**, 704 (1968); see related discussion in E.P. Wigner, Am.
J. Phys. **31**, 6 (1963), especially his Eqns 11,12 (here as Eqn(11) which Wigner dates to J.
von Neumann, *Mathematical Foundations of Quantum Mechanics* (Princeton, Princeton
NJ, 1955) translated from the original German Edition of 1932 by R.T. Beyer, esp. ChVI.
Wigner also pays homage to F. London and E. Bauer, whose 1939 text on the Copenhagen
interpretation is conveniently found in translation in *Quantum Theory and Measurement*
(Princeton, Princeton NJ, 1983) J.A. Wheeler and W.H. Zurek, Eds., p.217.

2) J.B. Hartle, Phys. Rev. **D44**, 3173 (1991).

3) M. Gell-Mann and J.B. Hartle, Phys. Rev. **D47**, 3345 (1993) (here as **Paper XVI·1**). See also *Complexity, Entropy and the Physics of Information* (Addison-Wesley, Reading MA, 1990) W.H. Zurek, Ed., p.425.

4) R.B. Griffiths, Phys. Rev. **A54**, 2759 (1996); Phys. Rev. **A57**, 1604 (1998).

5) R. Omnès, Rev. Mod. Phys. **64**, 339 (1992); also R. Omnès, *The Interpretation of Quantum Mechanics* (Princeton, Princeton NJ, 1994); also R. Omnès, *Understanding Quantum Mechanics* (Princeton, Princeton NJ, 1999) for the most recent word, but certainly not the last one, on this subject.

6) B.S. DeWitt, Phys. Rev. **160**, 1113 (1967) (here as **Paper XVII·1**).

7) A. Vilenkin, Phys. Lett. **117B**, 25 (1982) (here as **Paper XVII·2**; but see A. Linde, Scientific American Quarterly – *Magnificent Cosmos*, Spring, 1998 Vol **9**(1), 98 (1998); also A. Linde, *Particle Physics and Inflationary Cosmology* (Harwood Academic Publishers, 1990) for the role of quantum fluctuations. A recent extension of these ideas consistent with string theory is developed in: G. Veneziano, Cern Courier **39**(2), 18 (1999); see also, e.g., M. Gasperini and G. Veneziano, Phys. Rev. **D59**, 043503 (1999); and http://www.to.infn.it/ gasperin/ for an updated collection of papers on the PBB (Pre-Big-Bang) QST (Quantum String Theoretic) DDI (Dilaton Driven Inflation) from APT (Asymptotic Past Triviality) cosmological scenario which "neatly explains why the Universe, at the big bang, looks so fine-tuned (without being so) and provides a natural arrow of time in the direction of higher entropy."

8) R.P. Feynman, Rev. Mod. Phys. **20**, 267 (1948) (here as **Paper XV·1**). See also R.P. Feynman and A.R. Hibbs, *Quantum Mechanics and Path Integrals* (McGraw-Hill, New York NY, 1965).

9) L. Diósi, Phys. Lett. **B280**, 71 (1992).

10) P.A.M. Dirac, Phys. Zeits. Sowjetunion **3**, 64 (1933) (here as **Paper XV·3**).

11) R.P. Feynman and F.L. Vernon, Ann. Phys. (NY) **24**, 118 (1963) (here as **Paper XV·2**).

12) A. Caldeira and A. Leggett, Physica **A121**, 587 (1983).

13) W. Heisenberg, Zeits. f. Phys. **43**, 172 (1927) (here as **Paper XI·1**).

14) N.F. Mott, Proc. Roy. Soc. (London) **126**, 79 (1929).

15) M. Born and P. Jordan, Zeits. f. Phys. **34**, 858 (1925) (here as **Paper VII·1**).

16) W. Heisenberg, Zeits. f. Phys. **33**, 879 (1925) (here as **Paper VI·1**).

17) M. Born, W. Heisenberg, and P. Jordan, Zeits. f. Phys. **35**, 557 (1926) (here as **Paper VII·2**).

18) W. Yourgrau and S. Mandelstam, *Variation Principles in Dynamics and Quantum Theory* (Pitman, London, 1960), p.127; see also C. Lanczos, *The Variation Principles of*

Mechanics (Toronto, Toronto, 1949), esp. Pp.161, 229.

19) E. Schrödinger, Annalen der Physik **79**, 361, 489, 734 (1926) (here as **Paper IX·1,2,3**).

20) Physics Today **52**(2), 11 (1999): Letters from M. Gell-Mann and J. Hartle, R.B. Griffiths, and A. Zeilinger in response to S. Goldstein's *Quantum Mechanics without Observers* in Physics Today **51**(3), 42 (1998).

Paper XVI·1: Abstract of Phys. Rev. D**47**, 267 (1993).

Classical equations for quantum systems

Murray Gell-Mann[1] and James B. Hartle[2]

Los Alamos National Laboratory (T-6)[1], Los Alamos, NM 87545
and Santa Fe Institute[1], Santa Fe, NM 87501
and Department of Physics, UCSB[2], Santa Barbara, CA 93106

Received 23 October 1992

Abstract: The origin of the phenomenological deterministic laws that approximately govern the quasi-classical domain of familiar experience is considered in the context of the quantum mechanics of closed systems such as the universe as a whole. A formulation of quantum mechanics is used that predicts probabilities for the individual members of a set of alternative coarse-grained histories that *decohere*, which means that there is negligible quantum interference between the individual histories in the set. We investigate the requirements for coarse-grainings to yield decoherent sets of histories that are quasi-classical, i.e., such that the individual histories obey, with high probability, effective classical equations of motion interrupted continually by small fluctuations and occasionally by large ones. We discuss these requirements generally but study them specifically for coarse-grainings of the type that follows a distinguished subset of a complete set of variables while ignoring the rest. More coarse-graining is needed to achieve decoherence than would be suggested by naive arguments based on the uncertainty principle. Even coarser graining is required in the distinguished variables for them to have the necessary inertia to approach classical predictability in the presence of noise consisting of the fluctuations that typical mechanisms of decoherence produce. We describe the derivation of phenomenological equations of motion for a particular class of models. Those models assume configuration space and a fundamental Lagrangian that is the difference between a kinetic energy quadratic in the velocities and a potential energy. The distinguished variables are taken to be a fixed subset of coordinates of configuration space. The initial density matrix of the closed system is assumed to factor into a product of a density matrix in the distinguished subset and another in the rest of the coordinates. With these restrictions, we improve the derivation from quantum mechanics of the phenomenological equations of motion governing a quasi-classical domain in the following respects: Probabilities of the correlations in time that define equations of motion are explicitly considered. Fully nonlinear cases are studied. Methods are exhibited for finding the form of the phenomenological equations of motion even when these are only distantly related to those of the fundamental action. The demonstration of the connection between quantum mechanical causality and causality in classical phenomenological equations of motion is generalized. The connections among decoherence, noise, dissipation. and the amount of coarse-graining necessary to achieve classical predictability are investigated quantitatively. Routes to removing the restrictions on the models in order to deal with more realistic coarse-grainings are described.

Reversible primitive elements necessary and sufficient for Deutsch's Universal Quantum Computer. See our ChXVIII, p.520. [From R.P. Feynman, *Foundations of Physics* **16**, 507 (1986).]

Chapter XVII

DeWITT'S
WAVE FUNCTION of the UNIVERSE:

"In the quantum theory $\cdots$ of finite worlds, these [constraints] alone govern the dynamics $\cdots$ 'time' must be determined intrinsically $\cdots$ the quantized Friedmann universe is studied in detail $\cdots$ it is necessary to introduce a clock $\cdots$."

B.S. DeWitt, Phys. Rev. **160**, 1113 (1967).

§ XVII-1. Introduction: The Wheeler-DeWitt Equation.

The very phrase – "Wave Function of the Universe" – does it proclaim humankind's grandest quest? Or is it 'much ado about nothing'? Is it the ultimate intellectual endeavor of Physics? Or is it a facetious game to be played, tongue-in-cheek? In fact, it is a game that *must* be played, compelled as we *must* be by the structure of fundamental physical theory which is the only basic foundation of any possible ultimate *understanding*, short of invoking theological arguments.

The canonical [i.e., standard] prescription [i.e., drill] for constructing fundamental theory proceeds in the steps defined by Born and Jordan [1], by Born, Heisenberg, and Jordan [2], and by Dirac [3]:

1) Start with the classical Action (inspired by experience); $\Rightarrow$

2) leading to a Lagrangian and the Euler-Lagrange equations of motion; $\Rightarrow$

3) then to a Hamiltonian and Hamilton equations of motion; $\Rightarrow$

4) which produce the canonical Poisson brackets for the generalized coordinates q and the conjugate momenta p. Next,

5) we infer the underlying quantum theory by elevating the classical variables (q_c, p_c) to quantum operators (q_q, p_q); $\Rightarrow$

6) whose quantum algebra is obtained by replacing their classical Poisson brackets

by their quantum commutator brackets.

7) The quantum algebra can be realized by the prescription

$$q_c \to q_q \equiv q, \quad p_c \to p_q = \frac{\hbar}{i} \frac{\partial}{\partial q},$$

8) which operate in a Schrödinger equation on the wave function $\Psi(q)$; $\Rightarrow$

9) which gives the probability $|\Psi(q)|^2$ for the occurrence of q.

The drill is rarely unobstructed, the usual difficulty being to define a minimal set of dynamically independent coordinates q. What is often encountered is an overabundance of candidate coordinates $\bar{q}$ subject to some classical restriction $\mathcal{F}(\bar{q}) = 0$ (an example is the gauge invariance restriction on the vector potentials of electrodynamics). Dirac imposed this classical restriction on the variables as a *quantum subsidiary condition* $\mathcal{F}(\bar{q})\Psi(\bar{q}) \equiv 0$, to select the physically allowable states $\Psi(q)$ from all those generated by the unrestricted variables. This restriction is most naturally done in the Feynman Path Integral formulation of quantum mechanics as DeWitt develops in the second paper of his trilogy on the subject [4], but we will continue with the more familiar Schrödinger representation.

The canonical program has been carried out by DeWitt [4] with the Wheeler-DeWitt equation for the Schrödinger quantization of Einstein's classical general relativity. DeWitt's work is powerful, elegant, remarkably prescient and complete, and anticipated by almost fifteen years any application or further development of the notion of the wave function of the universe. The Wheeler-DeWitt equation is especially noteworthy for its remarkable abstraction and completeness.

DeWitt expresses the Einstein Lagrangian $L = \int \sqrt{g}R d^3x$ in terms of the scalar curvature R, $g = det\gamma$ and the metric coefficients of the 3-geometry γ_{ij} in the Friedmann form of the interval

$$ds^2 = dt^2 - \gamma_{ij}dx^i dx^j.$$

The path to a Hamiltonian formulation is complicated by two classes of constraints:

1) *primary constraints* resulting from the prescription $g_{00} = 1, g_{0i} = g_{i0} = 0$, which are more or less trivial; and

2) *secondary or dynamical constraints* which are necessary to maintain the primary

constraints for all time.

Expressed in terms of vanishing Poisson brackets with the Hamiltonian, these require the Hamiltonian to vanish in any Ricci-flat space where the difference between the extrinsic curvature (^4R) and the intrinsic curvature (^3R) – which can be identified with the Hamiltonian – must vanish. In this purely geometric case, the 3-geometry is determined entirely by the Hamiltonian constraint and not by a dynamical time evolution. In the quantum theory we must require

$$\int d^3x \xi(x) \mathcal{H}\Psi = 0$$

for arbitrary smooth weight functions $\xi(x)$.

The immediate question arises: What Hamiltonian quantum dynamics survives? if the only state of the geometry is the zero-energy – presumably time-independent – state? DeWitt obtains the Einstein equations for the 3-metric in the WKB approximation to the Wheeler-DeWitt functional Schrödinger equation in an infinite-dimensional manifold – *superspace* – of 3-geometries. The state functional depends only on the 3-geometry at an (irrelevant) initial time. There is a problem defining the inner-product of two state functionals – a result of the failure of the Hamiltonian to have a lower bound, and consequently of any wave-functional to be at hazard of gravitational collapse. The inner-product problem can be solved only by introducing a phenomenological time with a clock whose period is commensurate with that of the (bouncing) universe.

DeWitt concludes with a muted optimism for the prospects of the canonical quantum theory of gravity. He emphasizes the compelling 'extraordinary economy' of the theory. In the analysis of any situation the theory always demands a precise *quid pro quo*. It will say nothing about time unless a clock is provided; it will say nothing about geometry unless an instrument to measure it is provided. DeWitt characterizes the combined quantum and gravitational theories as the operational theory *par excellence*, even to the extent of compelling its own interpretation. DeWitt strongly favors the Everett 'many universes' interpretation of quantum mechanics for the interpretation of his equation, a notion we prefer not to pursue.

It is impossible to fully appreciate DeWitt's profound contributions without detailed study of his original papers. We have included highlights in excerpts at the end of this chapter. We quote here: "$\cdots$ the quantized Friedmann universe is studied in detail, and its wave functions in the WKB approximation are obtained. In order to obtain non-static wave functions which resemble a dynamical universe evolving it is necessary to introduce a clock. $\cdots$ normalizability of the wave functions requires a precise commensurability between the periods of universe and clock. $\cdots$ time is only a phenomenological concept. $\cdots$ probability flows in a closed finite circuit in configuration space. $\cdots$ wave packets do *not* ultimately spread in time. $\cdots$ the problem of time-reversal invariance and entropy is briefly discussed." And further: " $\cdots$ denote by $\mathcal{M}$ the set of all possible 3-geometries $^3\mathcal{G}$ which a finite world may possess. $\cdots$ $\mathcal{M}$ is the *domain manifold* of the state functional Ψ, and the $^3\mathcal{G}$ are its "points." $\cdots$ The only way in which dynamics can enter $\cdots$ is through the *Hamiltonian constraint*:

$$\left(G_{ijkl} \frac{\delta}{\delta^3 g_{ij}} \frac{\delta}{\delta^3 g_{kl}} + \sqrt{^3 g}\, ^3 R \right) \Psi \left[^3\mathcal{G} \right] = 0 \quad \cdots$$

it is $\cdots$ useful to regard G_{ijkl} as a metric tensor and to study the properties of the manifold $\mathcal{M}$ which it defines."

DeWitt resolves the "no-time" paradox of the Wheeler-DeWitt-Schrödinger equation by introducing an internal particle dynamics. The time-dependence of R is made visible by correlation with the particle coordinates q_i. The zero-energy constraint equation

$$(H + \mathcal{H}) \Psi = 0$$

is made highly degenerate by combining the negative spectrum of the geometrical Hamiltonian $\mathcal{H}$ with the degenerate positive spectrum of the particles' Hamiltonian H. This requires commensurate energy levels (actions) for the particle and the geometrical systems.

The problem of defining a positive definite norm for the wave function Ψ satisfying the Hamiltonian constraint equation is similar to that for the relativistic Klein-Gordon equation, for the same reason that second time-derivatives are involved. No clear-cut general answer can be given but DeWitt finds a satisfactory construction for the quantized Friedmann universe. We forgo the explicit development and turn instead to the subsequent application of DeWitt's results to the problem of understanding the first instant of the Big-Bang universe as a quantum tunneling event from 'nothing'. The full Wheeler-DeWitt equation is too intractable and for 'practical applications' (ludicrous as that expression is in this context), DeWitt ultimately reduced the general equation remarkably by restricting the Friedmann space-time metric

$$ds^2 = dt^2 - \gamma_{ij} dx^i dx^j$$

to be of the one-parameter Friedmann-Robertson-Walker (FRW) form for a closed universe

$$ds^2 = dt^2 - a^2(t) \left(\frac{dr^2}{1 - r^2} + r^2 d\Omega^2 \right)$$

depending only on the single scale parameter $a(t)$ – a *mini-superspace*.

§ XVII-2. Quantum Creation of the Universe from Nothing.

Albrow [5] and Tryon [6] were first to suggest that the universe began as a spontaneous quantum fluctuation. The idea was made specific in the calculation of Atkatz and Pagels [7] who supposed the universe to be created as a quantum FRW universe by tunneling from zero scale through a potential barrier to a finite scale which continued smoothly into the Big-Bang. The principal result obtained by Atkatz and Pagels was that creation from nothing was possible only for a closed universe. Their result is analogous to the original Gamow tunneling solution for the α-decay of a radioactive nucleus and was not very informative.

Atkatz and Pagels' tunneling calculation immediately preceded Guth's invention [8] of the Inflationary cosmological theory. Guth postulated that the universe starts with a Planck-scale period of immense exponential expansion of a deSitter space-time – driven by an initial intrinsic and persistent positive pressure in a so-called 'false-vacuum'. Guth's inflation scenario provides a unified explanation for the fundamental assumptions required by the original Big-Bang:
1) the observed homogeneity and isotropy of regions of the universe that were previously apparently causally disconnected;
2) the flatness of the observed universe which required an incredible energy balance

as an initial condition; and,

3) the absence of magnetic monopoles which would be expected as defects produced between regions with different realizations of the spontaneous symmetry breaking of the underlying super-symmetric GUTS.

All the arbitrary assumptions necessary to fine-tune the original Big-Bang are eliminated by the inflation postulate which leads to the observed universe being a very small patch – causally connected, homogeneous, and flat – on a much larger space-time structure. The false-vacuum is simulated by a scalar field which starts on a broad positive-potential energy plateau whose positive pressure drives the exponential expansion. The scalar field slowly evolves to a potential minimum where oscillations convert potential energy into thermal energy. At this point the exponential inflation interval stops and the original Big-Bang starts from a baseball-sized beginning on which the observable universe is the above-mentioned patch.

Vilenkin [9] observed that the only role left for the quantum theory of gravity was to produce the Planck size initial condition for Guth's inflation scenario. Vilenkin found that the spontaneous creation of such an initial state of false-vacuum is possible and in some sense natural. In agreement with Atkatz and Pagels, Vilenkin also concluded that the only verifiable prediction of the theory is that the universe is closed. The most recent observational results indicate that this conclusion is not correct but that it is a very close-run thing, and presumably Vilenkin's result can accommodate the new results by including a suitable – but so far inexplicably small – cosmological constant required to modify general relativity at large distances. Anticipating that such a modification can be found, we continue Vilenkin's demonstration that quantum tunneling from 'nothing' "···gives a cosmological model which does not have a singularity at the Big-Bang and does not require any initial or boundary conditions. The structure and evolution of the universe are determined entirely by the laws of physics." This quotation from Vilenkin is obviously quite as hyperbolic as the equations he solves.

The notion that the world starts from *nothing* is an over statement of the theory which has, of course, considerable intellectual pre-structure and requires certain assumptions however economical.

The 'nothing' referred to is the initial length scale $a(0)$ in the FRW metric

$$ds^2 = dt^2 - a^2(t) \left(\frac{dr^2}{1 - r^2} + r^2 d\Omega^2 \right).$$

Insertion of the FRW metric into the Einstein equation of general relativity yields for the time evolution of the scale factor $a(t)$

$$\dot{a}^2 = H^2 a^2 - 1,$$

where $H = (8\pi G \rho_v / 3)^{1/2}$ is the Hubble constant, G the Newton gravitational constant, and ρ_v the mass density of a primordial false-vacuum gas. A bounce-solution suggests itself

$$a(t) = H^{-1} \cosh(Ht)$$

which collapses to minimum radius $a(0^-) = H^{-1}$ with zero velocity $\dot{a}(0^-) = 0^-$ as $t \to 0^-$; then expands at $t > 0^+$ from $a(0^+) = H^{-1}$ with $\dot{a}(0^+) = 0^+$. The classical bounce solution has a quantum interpretation as a tunneling event: For $0 < a < H^{-1}$, no classically allowed solutions exist but a classically forbidden continuation can be constructed at imaginary time $\tau = -it$

$$a(\tau) = H^{-1} \cos(H\tau)$$

which analytically connects the two branches of the bounce-solution with an excursion into Euclidean time $0 < \tau < \infty$.

The suggestion is to replace the classical Einstein equations of general relativity by the quantum Wheeler-DeWitt equation for a wave function of the universe determining a transition probability for tunneling through a potential barrier from $a = 0$ to some $a = H^{-1}$, in a classically allowed regime. We get an important fundamental insight but it is not a detailed or even very informative description of the event.

The situation is even more profoundly difficult. In the purely geometric case in which the Einstein curvature tensor $R_{\mu\nu} = 0$ and the space-time is Ricci-flat, the Hamiltonian in the Wheeler-DeWitt-Schrödinger equation is identically zero and the tunneling event would appear to be time independent. Vilenkin's conclusion is that it is possible to calculate only relative transition probabilities from $a = 0$ (nothing) to various $a \neq 0$. The only prediction possible, and the only measure of success of the theory, would be the overwhelming preference for a highly symmetric universe which – in Vilenkin's view – would connect smoothly to Guth's inflation scenario.

§ XVII-3. Concluding Remarks.

The Einstein-Wheeler-DeWitt-Schrödinger equation plays an essential role in the quantum theory of gravity [QTG]. As DeWitt explains in his great trilogy, it is necessary for a fundamental conceptual understanding of QTG. QTG itself is an essential preliminary to investigating the quantum field theory [QFT] of the graviton [QGD]. DeWitt uses the Feynman Path Integral formulation of QFT as the only feasible way to impose the gauge constraints of Einstein's metric theory on QFT in a manifestly covariant way. DeWitt's original investigation of the role of 'ghost' fields to impose the gauge constraints to all orders of perturbation theory, was instrumental in the subsequent proof by 't Hooft and Veltman [11] that the Standard Model – including quantum chromodynamics, electroweak theory and the Higg's mechanism of massive gauge bosons – is renormalizable and, in that sense, is a 'theory', and not just a lowest-order 'phenomenology'.

Guth's inflation scenario [8] has served the necessary role for which it was originally constructed, but the actual detailed mechanism of the inflation continues to be hotly debated. Recent suggestions for the origin of the universe [12] are more dynamical in the quantum field theoretic sense than portrayed by Vilenkin's tunneling process. Vilenkin's quantum tunneling assumes a deSitter Hubble constant H small compared to the Planck mass, so quantum gravity *per se* – in the sense of gravitons – is unimportant. The rapidly evolving string theory involves a Kaluza-Klein compactification of a Euclidean space-time of some 10-dimensions and a continuation to the 3+1-Minkowski space. These dynamics presumably precede Vilenkin's tunneling.

Linde [12] employs a dynamical entry to Guth's inflationary universe which involves continuous quantum field fluctuations in an underlying scalar field. All sufficiently large fluctuations put the scalar field up on the potential plateau necessary for Guth's inflation to start, leading to an infinite sea of foam. Each bubble in the sea of foam is an inflating Guth universe with a possibly different realization of the broken fundamental symmetries of physical laws. Linde's vision of infinitely chaotic inflation is perhaps the ultimately humbling prospect for humankind and an impenetrable barrier to the scientific resolution of all existential questions. Linde's idea still seems to leave the role suggested by Vilenkin of a Wheeler-DeWitt description of the birth of our bubble universe as a single quantum tunneling event.

Veneziano foregoes a Big-Bang event entirely. He starts with a finite size patch of space-time with a string-based scalar dilaton field in an asymptotically past trivial [APT] state. Fluctuations in the dilaton field drive inflation [DDI] and large fluctuation regions 'neck-off' from the original space-time patch. Our bubble universe is the interior of one such region. Veneziano's scenario converges to Linde's, which in turn follows Guth's, which for late times can be described by Vilenkin's tunneling solution of DeWitt's equation. Veneziano is in addition able to argue an ever-increasing entropy to give a cosmological arrow of time.

Footnotes and References:

1) M. Born and P. Jordan, Zeits. f. Phys. **34**, 858 (1925) (here as **Paper VII·1**).

2) M. Born, W. Heisenberg, and P. Jordan, Zeits. f. Phys. **35**, 557 (1926) (here as **Paper VII·2**).

3) P.A.M. Dirac, Proc. Roy. Soc. **A129**, 642 (1925) (here as **Paper VIII·1**). The canonical prescription is a 'bottom-up' induction of the 'more fundamental' from the 'less', useful for inference from our classical-world experience. It has, of course, eventually to be replaced by a 'top-down' prescription.

4) B.S. DeWitt, Phys. Rev. **160**, 1113 (1967) (here as **Paper XVII·1A**); and the application of Feynman Path Integrals to the *Quantum Theory of Gravity II, III* in Phys. Rev. **162**, 1195, 1239 (1967) (here as **Paper XVII·1B,C**); also Phys. Rev. Lett. **12**, 742 (1963) (here as **Paper XVII·1D**); also B.S. DeWitt, private communication (1999).

5) M.G. Albrow, Nature **241**, 56 (1973).

6) E.P. Tryon, Nature **246**, 396 (1973).

7) D. Atkatz and H. Pagels, Phys. Rev. **D25**, 2065 (1982). See also D. Atkatz, Am. J. Phys. **62**(7), 619 (1994).

8) A.H. Guth, Phys. Rev. **D23**, 347 (1981).

9) A. Vilenkin, Phys. Lett. **117B**, 25 (1982); Phys. Rev. **D27**, 2848 (1983) (here as **Paper XVII·2**).

10) J.B. Hartle and S.W. Hawking, Phys. Rev. **D27**, 2960 (1983) (here as **Paper XVII·3**).

11) See *Search and Discovery*, Physics Today **52**(12), 17 (1999) for further references.

12) A. Linde, Phys. Lett. **116B**, 340 (1982). For a most readable account, see A. Linde, Scientific American Quarterly – *Magnificent Cosmos*, Spring, 1998. Vol **9**(1), 98 (1998), expanded and updated from Scientific American, November 1994. For a full technical exposition of chaotic inflation, see A. Linde, *Particle Physics and Inflationary Cosmology* (Harwood Academic Publishers, 1990). For reference to a recent extension of these ideas consistent with string theory, see G. Veneziano, Ref(7) in our ChXVI. Veneziano replaces the $a(0) = 0$ initial singularity of the Big-Bang scenario by the Pre-Big-Bang condition of Asymptotic Past Triviality on a finite size world, which eventually evolves in near coincidence with Linde's scenario.

Paper XVII·1A: Excerpts from Phys. Rev. **160**, 1113 (1967).

Quantum Theory of Gravity I. The Canonical Theory

Bryce S. DeWitt

Institute for Advanced Study, Princeton, New Jersey
and Department of Physics, University of North Carolina, Chapel Hill, NC

(Received 25 July 1966)

Abstract: $\cdots$ the canonical formulation of general relativity theory is presented. $\cdots$ In the quantum theory the $\cdots$ constraints become conditions on the state vector, and in the case of finite worlds these conditions alone govern the dynamics. $\cdots$ the WKB approximation and Hamilton-Jacobi theory. Einstein's equations are $\cdots$ geodesic equations in the manifold of 3-geometries $\cdots$ The label x^0 itself is irrelevant and "time" must be determined intrinsically. $\cdots$ the quantized Friedmann universe is studied in detail $\cdots$ it is necessary to introduce a clock.

$\cdots$ the quantum theory $\cdots$ is an extraordinarily economical theory. $\cdots$ it will say nothing about time unless a clock to measure time is provided, and $\cdots$ nothing about geometry unless a device is introduced to tell when and where the geometry is measured. $\cdots$ the formalism determines its own interpretation. $\cdots$ forces us to abandon all use of externally imposed coordinates (in particular x^0) and to look instead for an internal description of the dynamics.

[27]The Feynman sum-over-histories $\cdots$ method has not yet been applied to any "practical" problem $\cdots$. It will be encountered $\cdots$ in the following papers of this series. $\cdots$

Paper XVII·1B: Excerpt from Phys. Rev. **162**, 1195 (1967).

Quantum Theory of Gravity II.
The Manifestly Covariant Theory

Bryce S. DeWitt

(Received 25 July 1966)

Because of the non-Abelian character of the coordinate transformation group $\cdots$ [a Feynman propagator is found which] propagates physical quanta only $\cdots$ [for] amplitudes for scattering, pair production, and pair annihilation by the background field. $\cdots$ A fully covariant generalization of the complete S-matrix is proposed $\cdots$ The big problem of radiative corrections is then confronted $\cdots$

Paper XVII·1C: Excerpt From Phys. Rev. **162**, 1239 (1967).

Quantum Theory of Gravity III.
Applications of the Covariant Theory

Bryce S. DeWitt

(25 July 1966)

The basic momentum-space propagators and vertices $\cdots$ are given for both Yang-Mills and gravitational fields. $\cdots$ Problems arising in renormalization theory and the role of the Planck length are discussed. $\cdots$ The physical significance of the renormalization terms is discussed.

$\cdots$ The canonical theory leads almost unavoidably to speculations about the meaning of "amplitudes for different 3-geometries" or "the wave function of the universe." The covariant theory, on the other hand, concerns itself with "micro-processes" such as scattering, vacuum polarization, etc.

Paper XVII·1D: Excerpt from Phys. Rev. Lett. **12**, 742 (1964).

Theory of Radiative Corrections for Non-Abelian Gauge Fields

Bryce S. DeWitt

(Received 12 May 1964)

$\cdots$ Manifest covariance is essential for correct renormalization, but covariant propagators <u>necessarily</u> propagate nonphysical as well as physical quanta around closed loops. To compensate $\cdots$ one must [add] another closed loop $\cdots$ describing $\cdots$ a set of fictitious particles [ghosts] and their interaction with real quanta. In higher order radiative corrections $\cdots$ the complexity of the difficulties has frustrated further progress. $\cdots$ this note [indicates] how the difficulties can be overcome. $\cdots$

Paper XVII·2: Excerpt from Phys. Lett. **117B**, 25 (1982).

CREATION OF UNIVERSES FROM NOTHING

Alexander VILENKIN

Physics Department, Tufts University, Medford MA 02155, USA

Received 11 June 1982

Abstract: A cosmological model is proposed in which the universe is created by quantum tunneling from literally nothing into a deSitter space. After the tunneling, the model evolves along the lines of the inflationary scenario. This model does not have a big-bang singularity and does not require any initial or boundary conditions.

$\cdots$ The only relevant question seems to be whether or not the spontaneous creation of universes is possible. The existence of the instanton [Note added: the deSitter solution continued to Euclidean imaginary time] suggests that it is. One can assume that instantons, being stationary points of the Euclidean action, give a dominant contribution to the path integral of the theory. There may be several relevant instantons ⋯ we must live in one of the rare universes which tunneled to the symmetric vacuum state.

Paper XVII·3: Excerpt from Phys. Rev. D**28**, 2960 (1983).

Wave Function of the Universe

J.B. Hartle[1] and S.W. Hawking[2]
ITP, UCSB[1], Santa Barbara, CA 93106
and DAMTP[2], Silver Street, Cambridge, England
(Received 19 July 1983)

⋯ One has a complete spectrum of excited states which show that a closed universe similar to our own and possessed of a cosmological constant can escape the big crunch and tunnel through to an eternal deSitter expansion. We are able to calculate the probability for this transition. ⋯ the Euclidean functional integral ⋯ does single out a reasonable candidate ⋯ yields a basis for constructing quantum cosmologies.

⋯ the prescription applied to simple mini-superspace models yields a semiclassical wave function which corresponds to the classical solution of Einstein's equations of highest spacetime symmetry and lowest matter excitation.

⋯ The Euclidean functional integral prescription sheds light on one of the fundamental problems of cosmology: the singularity. In the classical theory ⋯ the field equations, and hence predictability, break down. ⋯ analogous ⋯ an electron orbiting a proton. ⋯ in the quantum theory there is no singularity or breakdown. ⋯ the amplitude for a zero-volume three-sphere in our mini-superspace model is finite and nonzero.

⋯ the initial boundary conditions of the universe: ⋯ are that it has no boundary.

Chapter XVIII

DEUTSCH: "THE UNIVERSAL
QUANTUM COMPUTER

$\cdots$ [has] many remarkable properties not reproducible by any Turing machine. These do not include the computation of non-recursive functions, but they do include 'quantum parallelism', $\cdots$ by which certain probabilistic tasks can be performed faster $\cdots$ than by any classical restriction of it."

D. Deutsch, *Proc. Roy. Soc. Lond.* A **400**, 97 (1985).

§XVIII-1. Introduction: Elements of Quantum Computation.

The great excitement today in non-relativistic quantum mechanics – theoretical and especially experimental – is driven by the prospect of constructing the universal quantum computer [1,2]. The UQC will have vast numbers of quantum bits operating in parallel arrays, making computable a whole new class of problems [3,4]. These problems include

i) finding prime factors of very large integers ($\sim 10^{200}$), an ability important for safe encryption of passwords and codes [5,6];

ii) the conceptually simple but computationally complex 'traveling salesman' problem;

iii) searches of large data bases;

iv) actual simulation of quantum systems [1]; computable problems which involve a large number of elements N and a solution which requires a number of operations more than polynomial in N. Not all problems are computable, as we might guess from Gödel's theorem.

The basic entity of q-computation is the q-bit of information. The classical c-bit is the two-valued binary pair $(0, 1)$. The corresponding q-bit is derived from the superposition of the standard Pauli spinors (α, β): the superposition $|Q\rangle = a\alpha + b\beta$

defines the complex pair (a, b) constrained by $|a|^2 + |b|^2 = 1$. The q-bit (a, b) with its double continuum of possible (complex-)values has immensely more computational potential than the corresponding two valued c-bit. Physical examples of potential q-bits include the two spin states of a spin-1/2 particle, a (well-chosen) two level atomic system, and – perhaps most promising – the polarization states of a laser beam.

The process of q-computation consists of the coherent unitary (quantum mechanical) evolution according to some 'program' of the N amplitude pairs (a_i, b_i) of an N q-bit array, followed by an interrogation (classical quantum measurement) of the N pairs for conceivable answers. For some special problems, correct answers can be arranged to coincide with coherent responses of magnitude N^2, compared with incoherent responses of magnitude $\leq N$ for incorrect answers. The result is analogous to the Bragg refraction of X-rays traversing a crystal. For some directions, many paths through the crystal arrive in-phase and give intensity points characteristic of the geometry of the atomic array. The response is probabilistic, but the answer is exact. Not all computation problems can be formulated in this way. For those which cannot, q-computation holds no advantages [2,3]. It is just for those (few?) which can – like the search for prime factors mentioned above – where q-computation holds great interest [5,6].

The full advantages of q-computation are achieved only with the added subtlety of long range EPR-correlations of "entangled" pairs of q-bits, in which the phase of the superposition $(|1\rangle|0\rangle \pm |0\rangle|1\rangle)$ is an added e-bit of information. This turns out also to be crucial to error-control in the extremely fragile quantum coherence of arrays of q-bits.

Some initial pessimism has been expressed [7] over the fragile character of quantum coherence and the resulting difficulty of constructing a robust q-computer. The fear that arrays of q-bits would inevitably rapidly decohere was met by the dramatic counter-example of 'a q-computer in a cup of coffee'. The idea [8] is that nuclear spins are remarkably well protected from the external inter-atomic environment and they offer a mundane almost trivial structure for a NMR q-processor. However, without knowledge of the individual spin phases, the states of such a spin-1/2 ensemble can only serve as a quantum (trans-silicon chip) parallel c-processor. A second concern exists over error detection, correction, and stabilization of individual q-bits. Solutions to this problem are suggested by running three or more q-bits in tandem and comparing their content frequently [9]. This can be done with the non-destructive quantum non-demolition experiments now available [9].

Bennett [4] summarizes the fundamental quantum mechanical processes employed in q-computation as:

1) Superposition: The state Ψ of a q-computer is a complex superposition (entanglement) of the complete set of quantum states of N q-bits,

$$\Psi = \sum_x c_x |x\rangle \text{ where } x \sim 10111001011100\cdots,$$

for states with spins 'uduuudd$\cdots$', etc. The complex expansion coefficients are normalized $\sum |c_x|^2 = 1$. The computation consists of the parallel reversible unitary evolution of these states generated by the 2×2 Pauli matrices $1, \sigma_x, \sigma_y, \sigma_z$ acting on each q-bit (spin). Feynman describes the fundamental unitary operation required for quantum computation [1] as the 'NOT' operation. Supplemented with the 'CONTROLLED-NOT' (also called [4] the 'exclusive-OR' (XOR)), Feynman explains how any logic operation can be constructed. (See figures on our pages 503, 525.) We return to the physical realization of such operations in a later section.

2) Interference: Each $|x\rangle$ in Ψ corresponds to a Feynman path from start to finish, and the superposition is dominated eventually by those Feynman paths which evolve to have coherent relative phases.

3) Entanglement: Pairs of q-bits can be correlated as in EPR, so the state of one can be certified in an incomplete measurement in a non-destructive way by comparison to the other. This EPR information – termed an e-bit – is critical for error control [9].

4) Quantum measurement and uncertainty: A q-computation ends with a *complete* measurement of Ψ, thereby destroying it. This characteristic phase destruction during a (traditional Copenhagen-type quantum) measurement precludes the copying of q-bits – a no-cloning theorem for q-information. This is significant in encryption for detection of attempted break-ins.

§XVIII-2. Physical Model of Shor's Lemma.

The archetype problem for q-computation is integer factorization of large integers using Shor's algorithm [5]. Multiple parallel q-processing using q-interference between many intermediate states can be shown to provide exponential speed-up over c-computation [speedup]. It is easily believed that a direct classical numerical search for integer factors of integers $\sim 10^{200}$ is hopeless. Shor reduced the factorization problem to that of finding the period of a periodic function.

An elementary intuitive explanation of the q-computation of the latter problem is provided by Bennett [4], whose result we summarize in the following steps: Step 1) we need a q-computer with two registers X and Y of sufficient capacity. Register X with w q-bits for values of $x \leq Q = 2^w \gg r \leq \sqrt{Q}$, with r the estimated period; and register Y to store the corresponding values of the periodic function $f(x)$ whose period is to be determined. Start the computation with a uniform superposition of X-states, and Y-states set to zero, so the initial q-computer state is

$$\Psi(1) = \sum_x |x; 0\rangle.$$

Step 2) Evolve the Y-register to $f(x)$ by a coherent q-process, so now

$$\Psi(2) = \sum_x |x; y = f(x)\rangle.$$

Step 3) Fourier transform [ft] the X-register to

$$\Psi(3) = \sum_k |k\rangle\langle k|x; y\rangle = \sum_k e^{2\pi i k x/Q}|k; y\rangle.$$

where the k summed over are integers, and $\langle k|x\rangle \equiv \exp 2\pi i k x/Q$.
Step 4) Measure k, which must be an integer. The function $y = f(x)$ is periodic in $x \to x + r$. To get coherent contributions to the right side from all y, we must have closely the same periodicity in the exponential. This requires $2\pi k x/Q \simeq 2\pi k(x+r)/Q \bmod 2\pi$; so $kr/Q \simeq m$ where m is some unknown integer. The result is that the period Q/k is a fraction

$$\frac{Q}{k} = \frac{r}{m}.$$

Q is known, $k >\sim \sqrt{Q}$ is a measured integer and m can be found after a few guesses as a small integer constrained to give an integer solution for the period $r <\sim \sqrt{Q}$.

The key feature of the q-computation is that all operations (in particular, the Fourier transformation) are of algebraic complexity.

The simple intuitive arguments used here have a formal mathematical expression in modular arithmetic [5,6].

§XVIII-3. Basic Operations of Quantum Computation.

In his inimitably refreshing and physical style, Feynman [1] describes the essential operation(s) of reversible quantum computation. Following Toffoli [10], he requires

1) the *fundamental primitive* NOT-operation (see figure on our page 523) taking $a = 1 \rightarrow a' = 0$ and $a = 0 \rightarrow a' = 1$, pictured as a 'one-line' operation:

$$a \longrightarrow X \longrightarrow a' \neq a.$$

This operation is clearly reversible since $(NOT)a' = (NOT)(NOT)a = a$. From this Feynman constructs

2) the CONTROLLED-NOT which does not change the control a, i.e. takes $a \rightarrow a' = a$; and if $a = 1$ changes b, i.e. $b \rightarrow b' = b$ if $a = 0$, but $b' \neq b$ if $a = 1$. This is pictured as 'two-line' operation:

$$a \longrightarrow O \longrightarrow a'$$
$$b \longrightarrow X \longrightarrow b' \neq b \iff a = 1.$$

An useful equivalent description of the CONTROLLED-NOT is that b is changed to $b' = b + a \bmod 2$. The CONTROLLED-NOT (XOR) can be used to copy a into an initially empty register b; and can be used three times (with alternating control-lines) to exchange a and b. A third variant is

3) the CONTROLLED-CONTROLLED-NOT operation which is a 'three-line' operation with a and b as controls, and c is changed only if $a = a' = b = b' = 1$:

$$a \longrightarrow O \longrightarrow a' = a$$

$$b \longrightarrow O \longrightarrow b' = b$$
$$c \longrightarrow X \longrightarrow c' \neq c \iff a = a' = b = b' = 1.$$

The use of CONTROLLED-CONTROLLED-NOT operation is central to error elimination strategies [4,9] which filter erroneous q-bits from what is intended to be a perfectly identical array. One keeps checking control q-bit pairs $ab, bc, cd, \cdots$ against a test q-bit t, eliminating d if it changes t, etc, eventually distilling an error free array using fundamental QND-measurements. This test also serves to independently distill perfect e-bit arrays at two separated sites provided one assumes the separated e-bits are identically corrupted.

The NOT and the XOR operations – combined with simple spinor-rotations – can be combined to accomplish any logic operation [11]. Feynman exhibits a 'four-line' Full-Adder (see our page 525) consisting of two XOR's and three NOT's which we leave as an exercise while we return to the experimental physics of these operations.

§XVIII-4. Physical Realization of Basic Operations.

Wineland *et al* [12] first demonstrated a physical realization of the fundamental CONTROLLED-NOT (XOR) quantum logic operation. The two test q-bits $b, b' = |\uparrow\rangle, |\downarrow\rangle$ are two (internal) $2S_{1/2}$ hyperfine states of a $2S_{1/2}$ $^9\text{Be}^+$ atom. The control q-bits $a, a' = |1\rangle, |0\rangle$ are CM harmonic oscillator states of the atom when confined in an rf-trap. They were able to drive selectively all transitions among these states using an elegant array of off-resonant laser stimulated Raman transitions through the virtual atomic $2P_{1/2}$ state. The XOR operation

$$|0\rangle| \uparrow / \downarrow\rangle \rightarrow |0\rangle| \uparrow / \downarrow\rangle$$
$$|1\rangle| \uparrow / \downarrow\rangle \rightarrow |1\rangle| \downarrow / \uparrow\rangle$$

and its verification were attained with high efficiency ($\sim 80\%$).

In a further experimental realization of essential elements of q-computation, Wineland *et al* [13] have extended the above experiment to two trapped ions which they are able to prepare in specified trap-states and specified *entangled* internal

EPR-Bell spin-states. Again, the efficiencies of preparation, specification, manipulation, and verification are $\sim 80\%$.

No précis of ours can do justice to the *tour de force* accomplished by these authors. But if one wants to see the physics of the new millennium, there is no better place to look.

§XVIII-5. Concluding Remarks.

Optimism about the long term prospects of quantum mechanics waxes and wanes in spite of the truly remarkable achievements in theory and especially in experiment. The ability to manage quantum coherence in individual experiments is truly breath taking and leads to an ever-deepening understanding of the marvels of the EPR phenomenon and the subtlety of quantum mechanics.

Some reservations have been expressed recently [14] about the actual potential of liquid NMR q-computation. It seems – owing to the very small polarizations attainable in the liquid state (ppm) – that the liquid NMR systems so far investigated do not really exhibit entanglement and will not scale up to the exponential advantages of quantum computers. The suspicion is arising that the quantum description using unitary transformations is only a convenient framework for analyzing what is really very close to a classical probabilistic Boltzmann ensemble. Arguments claim that even though the unpolarized majority of molecules can be ignored and the few polarized ones can be considered as a pseudo-pure quantum state for the RF computation manipulations, there is no entanglement. And it has recently been proven [15] that there is no exponential speedup without entanglement. It may be that liquid NMR computation is not full q-computation, but still extremely promising in its own right as a basis for advancing c-computation.

Since the crucial idea of quantum computing is the superposition of linear complex amplitudes, any system with these properties can be the basis of a q-computer. In particular, wave optics is a natural choice. In a sense we have already carried out simple "q-computations" with light: examples are X-ray diffraction crystallography, Michelson interferometry, the Zernike phase contrast microscope, and optical holography. With two polarization states, a thin pencil of light is a q-bit.

In propagation through an aperture of area A, approximately A/λ^2 independent thin pencils of monochromatic light can be accommodated. In trying to locate and study a thin transparent microbe using phase contrast microscopy we make use only of the independent pencils and superposition of the signal beam with a phase shifted reference beam. This can be thought of as a search of a large data base with the field of illumination, for a specific small set, being the microbe.

In a separate evaluation of the future prospects of computing, Birnbaum and Williams [16] see a great and relatively [!] accessible revolutionary improvement in c-computation through nanotechnology. They use Feynman's estimate for the thermodynamic limit of c-computation and project that the limit will be set by a Teramac with quantum switches and (HTC bucky-?) wires. It will be a new technological revolution with c-computation reaching a speed one billion times the saturation speed of the present silicon integrated-circuit technology, projected by Moore's Law (doubling every $1\frac{1}{2}$ years) to occur in 2010. They see no practical use of q-computation before 2025.

Footnotes and References:

1) R.P. Feynman, *Int. J. Th. Phys.* **21**, 467 (1982); also *Foundations of Physics*, **16**(6), 507 (1986), with acknowledgement to C. Bennett, IBM J. Res. Dev. **6**, 525 (1979).

2) D. Deutsch, *Proc. Roy. Soc. Lond.* A **400**, 91 (1085).

3) D. Deutsch and R. Jozsa, *Proc. Roy. Soc. Lond.* A **439**, 553 (1992).

4) C. Bennett, Physics Today **48**(10), 24 (1995). This reference should be #1 on the reading list of every neophyte like ourselves trying to understand this subject; see also Ref(7,9) below, and references which can be traced from these.

5) P. Shor, *SIAM* J. Comput. **26**, 1484 (1997). For an excellent, succinct, and sophisticated review, see A. Steane, arXiv:quant-ph/9708022 v2 24 Sep 1997.

6) S. Fulling, quant-ph/9911051; also (http://www.math.tamu.edu/ fulling/chinese.ps).

7) S. Haroche and J.-M. Raimond, Physics Today **49**(8), 51 (1996) (see Ref(4) above: this reference should be #2); also W. Unruh, Phys. Rev. A**51**, 992 (1995).

8) D. DiVicenzo, Phys. Rev. A**51**, 1015 (1995); D. Cory, M. Price, W. Maas, E. Knill, R. Laflamme, W. Zurek, T. Havel, and S. Samaroo, Phys. Rev. Lett. **81**, 2152 (1998).

9) J. Preskill, Physics Today **52**(6), 24 (1999) (see Ref(4) above: this reference should be #3).

10) E. Fredkin and T. Toffoli, *Int. J. Theor. Phys.* **21**, 219 (1982); T. Toffoli, *Math. Syst.*

Theory **14**, 13 (1981).

11) D.P. DiVicenzo, Phys. Rev. A**51**, 1015 (1995); A. Barenco, C. Bennett, A. Ekert, R. Jozsa, Phys. Rev. Lett. **74**, 4083 (1995); A. Barenco, C. Bennett, R. Cleve, D. DiVicenzo, N. Margolus, P. Shor, T. Sleator, J. Smolin, and H. Weinfurter, Phys. Rev. A**52**, 3457 (1995).

12) C. Monroe, D. Meekhof, B. King, W. Itano, and D. Wineland, Phys. Rev. Lett. **75**, 4714 (1995).

13) B. King, C. Wood, C. Myatt, Q. Turchette, D. Leibfried, W. Itano, C. Monroe, and D. Wineland, Phys. Rev. Lett. **81**, 1525 (1998); Q. Turchette, C. Wood, B. King, C. Myatt, D. Leibfried, W. Itano, C. Monroe, and D. Wineland, Phys. Rev. Lett. **81**, 3631 (1998).

14) R. Fitzgerald, Physics Today **53**(1), 20 (2000).

15) N. Linden and S. Popescu, http:/xxx.lanl.gov/abs/quant-phys/9903057, also Phys. Rev. A (to be published).

16) J. Birnbaum and R. Williams, Physics Today **53**(1), 38 (2000).

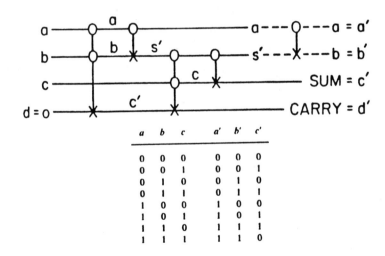

a	b	c	a'	b'	c'
0	0	0	0	0	0
0	0	1	0	0	1
0	1	0	0	1	0
0	1	1	0	1	1
1	0	0	1	0	0
1	0	1	1	0	1
1	1	0	1	1	1
1	1	1	1	1	0

Full adder with result table. See our ChXVIII, p. 522. [R.P. Feynman, *Foundations of Physics* **16**, 507 (1986).]

Chapter XIX

PLANCK'S QUANTUM:

THE NEXT 100 YEARS

§ XIX-1. Introduction.

What remains? where is it all going? and who can tell? Kleppner [1] has recently enumerated some of the many historic failed attempts to project the future of physics. The lesson to be learned is that such projections are impossible – even over very short time intervals. Kleppner too was trying to foresee the possibilities of the next fifty years and simply concluded with unbridled optimism that physics would continue its proliferating successes. His reasoning is that physics is an experimental science in which there are "ever more perfect eyes within a cosmos in which there is always more to be seen." Physics creates these more perfect eyes in new and more powerful experimental methods which will continue to drive our subject with ever more discoveries.

We share Kleppner's enthusiasm and optimism for physics in general; and also his wariness of predicting the future. The task we set ourselves is perhaps no more possible, but it is much more constrained. We restrict our speculations to the realm of quantum mechanics for its own sake, which we will try – for simplicity – to limit to the non-relativistic regime. Certainly the impact of the relativistic – even the extreme relativistic Planck scale – may well intrude and may even determine everything in a top-down flip-flop.

Is formal quantum mechanics *per se* – in the form left to us by the original creators Heisenberg, Born, Jordan, Dirac, Schrödinger, Bohr, Feynman, $\cdots$ – as far as it goes? Is there something deeper which underlies the original canonical structure inferred from the classical Action Principle? The answer is surely 'Yes!' and the progression through quantum field theory and the Standard Model to some form of string theory in an extended space-time is well underway. There no doubt will be surprises – both triumphs and disasters – for our contemporary ideas. An ongoing and persistent failure to find any explicit spectrum of Higgs or super-symmetric particles would be a political and financial disaster for the field of high-energy experimental physics but no doubt there would be a new Higgs with a new way of breaking symmetries not apparent to us now. In a sense, failure could be more fruitful than success – forcing physicists to think of more subtle and abstract ideas not so simply related to the structures employed in the traditional Standard Model. The ultimate boundary condition on any such construction is – thanks to Linde's proliferation of universes – no longer the mother of all correspondence principles. The top-down theory need no longer produce *uniquely* the world we see, but must only *include* it as a possibility. This seems to be the case for present string-theories, where there is no clear preference for one among many groundstates.

What *could* be the conceivable impact on the structure and results of non-relativistic quantum mechanics of the *full* realization of the *fondest* dreams of the advocates of string-theory? It is tempting to say none at all but such negativism precludes thought, so lets try a little harder. Wilson's renormalization group analysis shows how ostensibly fundamental theories (e.g., quantum electrodynamics) are in fact 'effective' phenomenological theories valid as low energy approximations in the Gell-Mann–Low running coupling analysis of more fundamental theories, which are similarly defined at a succession of higher energy regimes. The Feynman Path Integral formulation is the ideal way to 'integrate out' degrees of freedom as the energy is decreased through various thresholds and the more massive fields are 'frozen out' in a sequence of phase transitions. The sole impact of the degrees of freedom of each frozen out higher energy regime appears in the parameters of the lower energy theory (e.g., mass spectrum, running coupling strengths of induced interactions). The present situation is that there appears to be a small number of viable candidate theories at the Planck scale, but no way to choose one among many (billions?) of possible low energy limit theories. The criteria for a successful string theory (T.O.E. – Theory of Everything) include its ability to satisfy the correspondence principle requirement of producing the Glashow-Salam-Weinberg standard model. Presumably the success of the standard model at the scale of ~ 1 Tev or $\sim 10^{-4}$ fm insulates all of non-relativistic quantum mechanics and related phenomena from details of higher energy models. We are still left groping for any phenomenological or theoretical impact of the eventual T.O.E. especially on non-relativistic quantum mechanics but actually including even applications of relativistic quantum field theory in the laboratory domain.

One idea – suggested by the top-down derivation of the non-relativistic quantum mechanical world from the Planck T.O.E. – is the possibility to derive the *interpretation* of the theory also in a top-down way. In this way, the *meaning* or *interpretation* of quantum mechanics would be contained within the primary theory at each stage of the renormalization group analysis. Information from the higher regime (massive particle spectrum, interactions) is manifested in the succeeding lower regime in a 'jet experiment' where the massive particles appear as annihilation jets of the approximately massless particles of the lower energy regimes. This suggests that the fundamental measurement process is a jet experiment with the consequent loss of resolution and coherence at the lower level. The quantum coherence of the jet structure cannot get past a digital→analog conversion (in the optic nerve of the observer, if not sooner) so the ultimate thought processes involve neither the digital nor the quantum nature of the input information. The decoherence of the measurement event resides in an infra-red catastrophe in the jet event constituting the fundamental detection.

In summary: it seems inescapable that the top-down T.O.E. will have no impact on the results of quantum theory. We are as passionate as anyone about the pursuit of such knowledge and we are fully cognizant of the fact that no one can anticipate the unknown. None the less, our assessment of the situation is that this difficult goal must be pursued only for its own intrinsic interest. Looking back to Vilenkin's quantum theory of the origin of the universe as a tunneling event from nothing, we find the T.O.E. epoch hidden behind the potential barrier at $a = 0^+$ preceding the slow roll-over of Guth's inflation, and all at times much smaller than the Planck time. Linde's chaotic inflation makes our connection to the T.O.E. vastly more remote, since it requires only that the T.O.E. contain our realization of physics as one random chance in an incredible lottery.

§ XIX-2. Interpretation of Quantum Mechanics.

We find the present interpretation schemes – even with recent contributions –

to be lacking in any kind of beauty or elegance or imagination, and in that sense to fail Dirac's criterion for Truth. We look forward to a top-down interpretation of quantum mechanics as the ultimate limiting 'Quantum Information Theory' imposed by the quantum of action h. In this way, we would abandon the crutch by which the wave function is thought of as a quantum precursor to a physical field attached to a quantum system like the Coulomb potential is attached to a classical charge. Quantum mechanics would be more truly associated with our 'right to think' and with the limit imposed on valid logic by the existence of the quantum; and only secondarily with the physical realization of any logical thought progression. Quantum theory would be more about thought and less about matter. Not thoughts of a mundane sort about life, death, hunger, love, beauty, etc., which would be recognized as grossly macroscopic analog processes. But thoughts at the most elemental level – than which nothing can be simpler – which we presume to be governed by the most primitive digital (i.e., quantum) information theory contained in or containing quantum mechanics. Such an idea may be wrong, or vacuously tautological, or even already existing (like the proverbial pony buried in the dung-heap) but we find it a more significant role for quantum mechanics to be the fundamental limiting theory of how we *must* think, given the quantum.

Finally the ideal quantum information content is degraded by our very thought processes. In this view – admittedly not so different from all the others because, after all, they do 'work' – the information acquisition involves an unavoidably imperfect digital to analog conversion (in the optic nerve, in our example) prior to the analysis by the macro-molecular – classical, analog and non-quantum – thought processes. A possible merit of this interpretation would be to reduce every observation to a jet experiment – like a photo-multiplier cascade – in our own eyeballs. Whether or not this can be formalized in any way preferable to the existing interpretations, or whether it is worthwhile or even true, remains to be decided. The goal of making the wave function the unique optimal quantum information theoretic device governing the observer's possible thought processes, rather than into a left-over relic of 19th century classical field theory is as close to a top-down interpretation of quantum mechanics as we have been able to imagine so far.

§ XIX-3. Unresolved Problems.

There are a number of very long-lived puzzles in the class of applications of quantum mechanics which have resisted a generation of concerted effort toward their full resolution. There is no indication that these unresolved problems are fundamental matters of principle in the basic quantum theory, nor that these are problems which will occupy the main stage for more than another one or two decades; but each is a major goal of widespread interest in its own right. Among these goals, we would include a fundamental understanding of:

1) Color confinement in quark-gluon chromodynamics. This is perhaps not a problem of non-relativistic quantum mechanics. And indeed there is no shortage of suggestive models to mimic the result. But there is still no fundamental insight to connect this essential requirement of the quark model of hadron structure in any rigorous formal way to the universally accepted (?) underlying fundamental (?) gauge theory of quantum chromodynamics.

2) The spin structure of the nucleon. This problem is presumably not as fundamental as that of confinement but it is a bothersome experimental result that has stood without full understanding for a generation. Experiments indicate that some 50% or more of the nucleon's spin $\hbar/2$ actually resides in the gluons rather in the quarks. Whatever the precise experimental result turns out to be, there is certainly a serious inability to apply quantum mechanics to understand the detailed structure of the hadronic particles even in a qualitative way.

3) High temperature superconductivity. The impasse here has been comparable in difficulty and longevity with that in low temperature superconductivity. That problem was resolved eventually by the Bardeen-Cooper-Schrieffer theory based on the mechanism of an energy-gap due to the formation of Cooper pairs. The energy-gap dynamics of the mechanism is rather universal and might be applicable to high energy superconductivity and to confinement, or to explain any robust quantum state.

All these problems and many more are surely not any failure of quantum mechanics *per se*. They are more likely the subtle results of phase transitions in which the degrees of freedom are most effectively described in some collective form, witness the Cooper pairs resulting from density correlations induced by lattice vibrations.

These problems of application are all conjectured to be resolvable on the time scale of one or at most two decades. None could conceivably require fundamental changes in the structure of the underlying theory of quantum mechanics, but they all have resisted sustained efforts of great ingenuity, intuition and imagination. Their resolution is eagerly anticipated.

One common feature in *all* such problems – from the proton spin deficit through the BCS-theory of superconductivity to the fractional quantum Hall effect – is the lack of an intuitive explanation based on a model coordinate space wave function that is comprehensible to anyone except a physicist expert in the specialty. Talented students and other physicists are reduced to believing but not really understanding.

In their recent popular discussion *Doubt and Certainty*, Rothman and Sudarshan [2] play the devil's advocate in a refreshing and stimulating review of the endless controversies and debates inherent in the ongoing understanding of quantum theory *per se*. They make clear that each new generation must debate these problems anew, and in the process incrementally increase our collective understanding. This debate is the *sine qua non* for *any* revolutionary progress. But convergence is slow and – as in any civil war – closure on many emotional issues is never achieved. Dissatisfaction still exists on such problems as:

1) the origin of the direction of time in quantum mechanics. We would hold with Veneziano [3] that the origin of the direction of time is in the expansion of our universe, and must be imposed on quantum mechanics through the density matrix and measurement process by the 'collapse of the wavefunction'.

2) the interface between quantum mechanics and statistical mechanics involving the classical Boltzmann factor $e^{-E/kT}$ is regarded as *ad hoc*.

3) even the recent decoherence interpretation of quantum mechanics generates stubborn dissatisfaction, typified on one extreme by adherents of Bohmian trajectories and on the other by proponents of the Everett many universes interpretation. We view these (and others) as metaphysical constructs designed to have no consequences, and therefore not worthy of further attention but they show no signs of going away.

The above questions relate to inanimate systems. What about the role of quantum mechanics in the life sciences? The density functional description – developed for atomic-, molecular-, surface-, and solid-state effects – shows great promise [4] for fundamental model building of life mechanisms like cell transport which are on the cusp of quantum processes. There is the suggestion of new applications of quantum mechanics to understand in a more fundamental way the heuristic mechanisms constructed already in chemistry, biology, neural sciences, medicine, and even psychology. All – it seems to us – are reactive responses to 'explain' classical intuitive results obtained in traditional ways, rather than any pro-active creative strategy leading the way to new mechanisms. This is a near infinite realm of application of quantum model building but we foresee no reverse impact *on* quantum mechanics *per se*, but also no intrinsically quantum dependence of life processes on $\hbar$ itself, beyond the current understanding of quantum chemistry.

There still remains the possibility that the logic of quantum mechanics might have a parallel in thought processes. This would seem to be limited to the result of a coherence of reinforcing thought patterns analogous to the coherence of classical diffraction patterns in the scattering of sound or water waves, or of holography, and not of any intrinsic dependence on Planck's quantum *per se*.

§ XIX-4. Quantum Sociology.

In this section we confess in advance what will soon become obvious to anyone who reads on: we succumb to the inevitable pessimism of the old predicting a future they have failed to create and will not be here to witness. We focus in particular on the role of non-relativistic quantum mechanics in the education and intellectual life of – say – 2050 or 2100. What are the prospects? We assume that non-relativistic quantum mechanics is already a mature science encapsulated from any essential impact of anticipated developments such as an eventual Theory of Everything including super-symmetry, super-strings, space-time compactification, and so on; or further refinements of interpretation.

Quantum mechanics will live on with Newton's mechanics and Maxwell's equations as indispensable foundation subjects of engineering, technology and science. However it will no longer be a subject of primary interest to research physicists and most of those who teach it and use it will be concerned – as now, actually –

with acquiring a heuristic and intuitive understanding good enough f.a.p.p. (Bell's mantra: for all practical purposes). Quantum mechanics differs from these classical subjects, though, in being the perpetual target of critics unwilling to accept the abstractions of the subject. One fears for the erosions and degradations of the subject by unrefuted critics and philosophers and by unprincipled (literally) revisionists, free to run amok without dispute from anyone qualified by the singular experience of first-hand fundamental scientific contributions in the subject.

The educational system is inevitably forced to teach less and less about more and more until everyone knows something about everything but never enough to *really* understand anything beyond the level of rote response. We don't condemn! we sympathize. In the face of this pressure, generation after generation of text books and courses are forced to pander more and more. The material will be elided to the level good enough f.a.p.p.

On the brighter side, computerized multi-color perspective graphics of quantum systems evolving in time and space as full-fledged faithful solutions of Schrödinger's equation will (have) become wide-spread and powerful as an instruction aid and interest grabber. Translating this wonderful tool into a deep understanding for the self-selected few interested in the underlying quantum theory will remain a tremendous challenge. Again on the bright side, the computerization of even the algebra promises (as now, in a variety of symbolic manipulation programs) to relieve the theoreticians of overwhelming tedium and leave time, energy, and enthusiasm for more global thinking. The price paid is the intimacy of our involvement with the subject, attenuated as it must be by the 'black-box' of the computer manipulations.

With a world population of 10-15-20-billion people, surely there will be a small elite – bigger in actual numbers than now – capable of pursuing quantum mechanics to the same level of understanding that is presently obtained by the current small but (more than?) sufficient elite. So what is the problem? Maybe none. There is the story of the mayor of Calcutta welcoming an international symposium of physicists visiting his city and pointing out that there were 1000 beggars in the streets of Calcutta as smart as they. Will it be 10,000? 100,000? in 2050. So what is the problem?

One can't help but feel disheartened by the projected absence of any cumulative or even progressive impact on the collective human psyche of all the marvelous acquired knowledge and understanding of the first one hundred years of quantum mechanics. We must acknowledge a collective failure to interest people at large and especially educators and humanists in even the rudiments of our subject – physics as a whole – let alone in the more abstract and abstruse topic of quantum mechanics. Perhaps interest is not the exact word. People *are* interested but the physics community has generally failed to make its ideas attractive, palatable and digestible without removing its intellectual content [5]. Can anything be done? Hundreds of popularizations in book and video form have been put forth, but the key people – the K→12 teachers – have not been reached. Great efforts have been made in this direction – witness the Feynman Lectures on physics. But the Feynman Lectures – as even Feynman acknowledged – ended up preaching to the choir (i.e., the faculty) and not the congregation (in this case, of eager and gifted Caltech undergraduates), to say nothing of the unconverted. The conclusion has become ever more clear – that the primary act of creation of knowledge must be followed by an unending devotion to the process of – put bluntly – salesmanship. During the cold war era, when physics and physics funding were intimately connected to military power, there was no such need. But already – in spite of the intimate link between physics and the technology of the internet, which is the primary engine driving the economic boom at the turn of the 21st-century – physics as it exists is being declared sufficient f.a.p.p. At present, there is not enough respect and prestige given within the physics profession to the difficult, challenging, and most essential task of bringing an understanding of quantum mechanics to a wide audience *including* educators and children, and college undergraduates of all interests. Computer graphics, videos and lectures can be only part of the effort. The process has to involve interaction, visualization, verbalization, discussion, debate. An incredible commitment by all scientifically knowledgeable people will be required for this vast and ever increasing task [6]. This is the only conceivable way to teach enough quantum mechanics to enough people to insure the continuing viability of our subject, even enough for all practical purposes.

What about the future generations' Heisenbergs and Diracs? Schwarzes and Wittens? As now, we imagine the identification, education and nurturing of genius will continue to be a haphazard inefficient enterprise, as wasteful in the economically advanced countries as in the less privileged, but for complementary reasons: people on both sides of the economic divide are driven by the completely natural materialistic instinct to struggle from survival to security and equally franticly from security to luxury. Combined with a further limiting motivation to be somehow 'useful' (e.g., cure death), and the inevitable truth that 'a fat dog won't hunt', we are left with barriers to young people ever even aspiring to the ultimate heights of our quantum profession.

The problem at the research university level is a lack of will rather than any lack of resources. It can be – and in rare instances, even is – addressed in simple terms with the most informal of structures. Our best suggestion involves discussion groups of perhaps five or six people including able, interested and self-selected students of all levels – first year through graduate student – and junior and senior faculty. A brown-bag lunch or afternoon tea is a friendly environment, or a journal-club. It is not necessary or even desirable to have a fixed agenda and any lecture format is surely fatal. Simply talk, question, argue, explain, explore. The simplest questions can lead to the most profound concepts – which must then be related back to fundamental concepts accessible to everyone. It is no mean feat to avoid dogmatism and formalistic gibberish. The goal is to get an open discussion with lots of "I don't understand $\cdots$", "I don't know $\cdots$", "That can't be right because $\cdots$", "But what about $\cdots$", "Why $\cdots$", "How $\cdots$".

In our experience the over-riding threats to the success of this enterprise are two: the students are worried about a grade because they have standard class demands; and the faculty too is basically worried about a grade. They have to work on publishable research to enhance their status with peers and supervisors in the university, research group and funding agencies. For both, sitting around and talking is viewed with deep suspicion. Unarguably it is *not efficient*, but we argue that it is *desirable* and maybe even *necessary* in the higher goals of education and development, in learning by apprenticeship, and perhaps most important of all, in giving young people at every stage a *voice* in the process. If we are ever to recreate the environment in which Bohr and Born nurtured the prodigious development of Heisenberg, Jordan, and many others, there has to be more direct and informal communication across all levels of the hierarchy and at every stage of the process.

We need a new revolution in the presentation of our subject with an emphasis on economy, essential ideas, and direct and rapid progress to modern topics. We need a bridge between intuition and theory that is possible only for a select few authors. We must reduce the technicalities of the formalisms to reveal clearly their essential content; clearly illustrated with a few key examples. We need Feynman brilliance and charisma, but made accessible where – for whatever reason – Feynman was not. Perhaps Feynman with a readers guide, or a translation with illustrations. Nothing will be lost in the transition because our present approach is confounding practically everyone anyway, with its barrier of dull and repetitive problems; of narrow details as the foundation of great ideas. The reality perceived by the students is narrow details *instead of* great ideas.

Pedagogical reform – informed by active research participation – is absolutely essential to the continued life of our subject. There is also the question: from *whom* should we learn? If the answer is – as we believe – the creators of knowledge, then most of the creators need a lot of help to make their work widely understood. If magicians never explain their tricks, surely the creative magicians of our subject are not very revealing of their innermost creative impulses. We are confronted with the accomplished fact, with the mountaineer at the top of the peak and little indication of how to get there.

§ XIX-5. Concluding Remarks.

To quote a popular philosopher "It ain't over 'til its over." So when will the story of Planck's quantum be over?

Certainly the great progress was made in a few giant strides, although at first with great temerity and hesitation: witness the five year lapse between Planck's discovery of h and Einstein's light quantum, then eight years before Bohr's atom, and eight more before de Broglie. Then the deluge of Heisenberg, Born, Jordan, Dirac, Schrödinger, and Bohr, all within three years or so. One might argue that the sole new idea since that time has been the Dirac-Feynman Path Integral formulation of quantum mechanics which had a time lag of more than fifteen years.

The actual applications of quantum theory have come thick and fast, but *none* have required any fundamental change either in the theory or in the Copenhagen interpretation. In our view, we disdain *all* subsequent revisionist equivocation and fine tuning as purposely inconsequential and therefore metaphysical.

In all the above assessment we have accepted that non-relativistic quantum mechanics is the fundamental component of relativistic theories of everything which are inferred by a bottom-up deduction of a more fundamental structure. So far, attempts to proceed in a top-down way from a formulation of super-strings in a higher-dimensional space-time have not been uniquely constrained by any correspondence principle restriction to the world we see. Chaotic inflation provides a reason why this should be the case. This is the ultimate 'children's crusade' in physics, and an ever increasing army has been struggling across their own plains of Anatolia for more than twenty years, encouraged by occasional sightings of the Holy Sepulcher. But we – old, fat, listless, indolent and content friars that we are, sitting at home preaching our gentle gospel – wish them well in their pursuit of the greater glory of That which inspires us all.

When (if?) the T.O.E. crusade conquers Jerusalem, what then? We can look forward to mountains of coffee-table books and educational videos; all with multi-colored computer generated graphics of Planck size membranes spontaneously dividing and recombining and evolving and changing color and topology; looking for all the world like the amoeba we could never see through our eighth grade microscopes (which is why we became physicists in the first place), but were instructed to believe in by those who could. In spite of the fact that they defied any deeper understanding and were connected to the world we could see in ways unexplained if not inexplicable. Now the amoeba have become strings; we are firmly and in deepest tones instructed that this at last is all there is; and that it explains everything in ways that we could understand if only $\cdots$ we were smart enough, energetic enough, determined enough, patient enough, content to study our scriptures faithfully enough $\cdots$. *And* if only we would take on faith that $\cdots\cdots\cdots$. What? That the world consists of nine, ten, eleven dimensional amoeba? We don't argue that this is not a necessary end-game for theoretical physics; all we are saying is that it is not yet a pretty sight.

Let us return to the fundamental questions that might yet remain in non-relativistic quantum mechanics. We have referred to quantum mechanics in a heuristic way as an information theoretic device. Planck was forced to invent h in order to ascribe a countable number of possible configurations to the ideal gas, and thereby a probability and finally an entropy. The great result was that the Action integral must be quantized, and from this follows everything.

We leave you here while we try to understand Entropy as Action, to more deeply understand "Why h?" We are not the first to ask this question and it may already have an answer in the quantum generalization from classical Shannon information theory based on Boltzmann entropy, to quantum information theory based on von Neumann's quantum entropy [7] using the density matrix in place of the classical probability, and the concept of quantum bits of information with positive and negative handles. We might get lost, but it will be on our own plains of Anatolia.

The End.

Footnotes and References:

1) D. Kleppner, Physics Today **51**(11), 11 (1998); see also *Critical Problems in Physics* (Princeton, Princeton, NJ, 1997) V.L. Fitch, D.R. Marlow, M.A.E. Dementi, Eds., for the proceedings of a 1996 Princeton conference with similar goals.

2) T. Rothman and E.C.G. Sudarshan, *Doubt and Certainty* (Helix Books, PERSEUS, Reading MA, 1998).

3) G. Veneziano, see Ref. (7) in our ChXVI.

4) J. Bernholc, Physics Today **52**(9), 30 (1999). The promise has become a reality described briefly here in *Computational Materials Science: The era of Applied Quantum Mechanics.* – "The properties of new and artificially structured materials can be predicted and explained entirely by computations, using atomic numbers as the only input."

5) E.F. Redish and R.N. Steinberg, Physics Today **52**(1), 24 (1999), for current thinking on physics teaching.

6) H.J. Frisch, Physics Today **52**(10), 71 (1999), for a personal account of one physicist's missionary experience.

7) N.J. Cerf and C. Adami, Phys. Rev. Lett. **79**, 5194 (1999).

Index

FIGURES:

SUBJECT:

ERWIN SCHRODINGER

NIELS BOHR